Forests and Men

Forests and Men

WILLIAM B. GREELEY

DOUBLEDAY & COMPANY, INC.
Garden City, N.Y., 1951

Printed in the United States
at
The Country Life Press, Garden City, N.Y.
First Edition

To the men of the woods—

THE CRUISERS, LOGGERS,

RANGERS, AND FORESTERS

Foreword

REFORESTATION is a rising giant on the American land, and two of the country's 3,050 counties may be cited as examples. The people of Lee County, Alabama, are planting forty-five million tree seedlings on forty-five thousand acres of idle land as a co-operative effort of taxpayers and government. Far away in the Pacific Northwest, Mason County, Washington, has insured its future logging, sawmill, plywood, pulp, and fiberboard pay rolls by a hundred-year partnership of industry and government. In this plan the planting of trees by hand has only a small part in supplementing the natural reseeding of logged lands. Intensive protection from fire guarantees the success of the local reforestation program.

There were no stories of the kind to tell in 1905, when young U. S. Forest Ranger Bill Greeley hit the big-timber country of the West as one of Gifford Pinchot's young men. The stories of reforestation were few even in 1920, when he was appointed U. S. chief forester. Soon after his return from command of the twenty thousand forestry troops of the A.E.F. in France, Greeley had begun to advocate education and co-operation in forest conservation among *all* the people of the woods—government agents, lumbermen, state, county, and municipal authorities, lumberjacks, farmers, tourists, sportsmen, teachers—all who manage and use the nation's forests.

He is hard at it today, thirty years later. The tall ranger still rides like an old-time evangelist of the circuits, seeking converts to conservation. And his simple principle remains: local education and co-operative action must come first in conservation of the forests.

And that is how it is working. When the Soil Conservation Service was established in 1935 it was patterned on the principle that William B. Greeley had made his rule as chief forester in 1920. A soil conservation district is formed by majority vote of the farmers within the proposed boundaries, and twelve farmers are the district's governing board. The brand marks of Colonel Greeley's practical philosophy and years of labors for conservation are apparent in other phases of the movement. His leadership in forestry has brought forth the co-operative program of forest fire prevention called "Keep America Green," with state organizations and county committees in twenty-eight states. He is also to no small degree responsible for "More Trees for America," an educational program for farm-forest owners. In both Lee County, Alabama, and Mason County, Washington, "Keep Green" and "More Trees" are in force, as in hundreds of other forest counties of the nation.

Bill Greeley would be the first to stand up in meeting and deny that he is the Paul Bunyan of forest conservation. No such suggestion is intended here. But all through his forestry years his guiding faith has been more in people and the land than in laws, more in education and persuasion than in authority, more in local co-operation than in the powers of government. There he stands out. There he has led.

Today the woods are full of foresters, north, south, east, west. And their works are everywhere. Shelter-belt trees stand tall on prairies where the Indians of the early days saw only buffalo and grass. Thanks to pulpwood products and the forestry men of the laboratories, the timber pay rolls of the lake states are bigger than they were at the peak of the white pine lumber harvest. Stony hills of New England farms that were abandoned many years ago are now yielding tree wealth, and there foresters are also at work. Lumbermen hire foresters by the hundreds, because timberland management for good utilization in logging, for tree growing on cutovers, and for protection of tree crops from fire and disease has become part of the American lumber business.

Veteran Forester Greeley has not only seen it all come about, he has lived it, and in much of the great change on the green acres of the country he has led. In this narrative the story always comes back, from the great capitals and the big doings, to the land; to county, camp, town, and farm; back to forestry as a craft on the land. From first to last the Greeley forestry faith has been in the land and in the people who make their living from the yield of the soil—farmers, lumbermen, loggers, foresters. His book begins with an epic forest fire. It ends with an account of the people at work locally, in "the ways of democracy," to keep fire out of the cutovers and reforest the burns.

In the pages of this book a powerful story is narrated. The land is bright with promise at the end of the way. Yet a great challenge looms here too. Evil and menace still abide in the forest-land problem; and for the future Colonel Greeley relies on the people of the woods. The people are growing trees. The people are preventing fires. It is the Greeley way of forestry.

JAMES STEVENS

November 1950

Contents

Forests and Men

CHAPTER I

Smoke in the Woods

TRAVELERS today on the transcontinental trains of the Milwaukee Road sometimes wonder at the sea of young forest of pine and fir and larch which they traverse in crossing the Bitterroots east of Avery, Idaho. To me it is always a somber journey, crowded with memories of human effort and suffering and of loyal men who died on the job.

The spring of 1910 had been dry and parching in the rugged canyons of the Bitterroot and Cabinet ranges. Through the white pine forests of the Idaho Panhandle, little streams were drying up and the greening of early summer was half withered by hot winds from the dry plains to the south and west. Forest fires began two months ahead of the usual summer hazard. Several of the national forest supervisors had their hands full with fire calls in May. From the district headquarters of the Forest Service, in Missoula, we rushed guards over high snow fields to mountain lookouts 'way ahead of time, broke up road crews to spread men out where the line of fire patrols was far too thin, and alerted all hands to get their trails open and their tool caches in order. Trouble was in the wind.

In late May scattered showers brought a few days' respite, but by early June the Forest Service and all other protection forces in western Montana and northern Idaho settled down to a grim all-summer battle. The Chief of the Forest Service backed us up in emergency spending. Seasoned rangers from other national forest regions reported for duty. Upwards of ten thousand fire fighters were enrolled from logging camps, ranches, mines, and employment offices in Butte and Spokane.

Several companies of United States troops from Fort Snelling moved in to hold new fire fronts in the Bitterroots. We no sooner got one runaway blaze under fair control than several new ones broke out. Worst of all were the tantalizing thunderstorms. Thunderheads would pile up along the high ridges on a hot afternoon. Hundreds of weary fire fighters watched them hopefully. "Rain at last!" Yet the clouds gave us only a few drops, while thunderclaps rolled down the gorges and lightning flashed against the horizon. Next morning came the unfailing reports from the lookouts—new smokes in the high country, often forty or fifty from a single afternoon's lightning play.

I spent many days on the fire lines that rugged summer, trying to bolster up weak spots with stronger crews and work out reliefs for rangers who were kept on their feet only by sheer grit. We had a bad one up in the Blackfoot-Glacier Park country, where an east wind was driving half a mile of fire front through the lodgepole pine. Scarcely a man was to be had. By good luck I found a construction crew of the Great Northern Railroad laying new tie plates and respiking rails eastward from the summit of the Rockies. I got wires through to the division superintendent at Whitefish and finally to the chief engineer at St. Paul. They responded nobly, and as soon as the wind died down that night, the ranger and I trailed the forty-man crew single file through the woods to a fairly quiet fire front. We had it well trenched by morning.

Back at district headquarters in mid-August, the smoke-eating command was grimly exultant. We had lost a large acreage, but for two dry months we "had held her." Over three thousand small fires had been stopped and nearly a hundred big ones trenched in. But we exulted too soon. A forest fire "controlled" unfortunately is not a forest fire *out*. Fires have been known to smolder in punky logs all winter and to flare up, ready for business, on the first stiff breeze of a new spring. On August 19 the western forests in District One were spotted with columns of lazy smoke, "old smokes" in lookout language. Then disaster struck.

Air humidity dropped to the level of the Mojave Desert. The southwest winds blew up into gales. Hell broke loose in the Bitterroots and Coeur d'Alenes. The whole western sky seemed afire with a smoky, yellow glare. Then it turned dark with billowing waves and great mushroom upthrusts of smoke. A roar drowned out the crackling of the nearest flames. To me it sounded like a hundred freight trains rolling over high trestles.

Many miles of fire trench, built with such toil and sweat, had to be abandoned. The safety of fire crews and settlements became the rangers' first duty. A fire leader in the Thompson River country got his men out on the face of a bald granite knob where they hugged the rock on their bellies while the scorching draft of smoke and gases rolled over them.

Ranger Pulaski was caught with a large crew between two fires on the rugged St. Joe watershed. But he was an old miner on his home ground. He crowded his men into an abandoned mining tunnel and held them there by sheer physical strength when the roar and smoke of the fire threw panic into the crowd. There was a trickle of water in the old tunnel, enough to soak a blanket. Pulaski stood at the mouth and knocked down a couple of men who tried to bolt. With bare hands he held wet blankets over the opening to keep out smoke and fumes. When a blanket caught fire, he grabbed another one. When the fury swept past them, Pulaski was badly burned around his arms and head and most of his men were "out," but only one life was lost in that crew of fifty men.

Down on the Clearwater two fire-fighting gangs got to the river in the nick of time. The boss of one of them made each man duck into the water under a blanket or coat, coming up under this dripping hood to breathe when he had to. Rising for a breath himself, the foreman brushed against a sturdy body, took a peek under his blanket and discovered that his neighbor was a black bear.

Other crews were less fortunate. Several were struck down in their tracks as they fled in blind terror when the flames rolled upon them. Seventy-eight fire fighters were killed. Seven others, homesteaders or prospectors, lost their lives in that fearful day. Many little towns along the Coeur d'Alene branch of the Northern Pacific Railroad were wiped out, and a considerable part of Wallace, the North Idaho mining capital.

But most of the people were evacuated from this danger zone on special trains which the railroad had held, ready and waiting.

The holocaust ended as suddenly as it started. The fires raged beyond control for one full day and then were drowned in drenching rains and snows. Over three million acres had been burned, mostly in the national forests, and probably eight billion feet of standing timber. This included some of the finest stands of Idaho white pine in the Panhandle.

To a young forester, thrown by chance into a critically responsible spot on a hot front, that summer of 1910 brought home the hard realities of our job. Ideals, glorified by "the golden haze of student days" and the enthusiasm of inspired leaders, suddenly came down to earth. I had to count the cost in hardship and sweat, in danger and human lives. And I had to face the bitter lessons of defeat. For the first time I understood in cold terms the size of the job cut out for us. We had to overcome the habits and practices of a people who had taken their forests for granted—for two and a half centuries. And we had to engineer and organize the best resistance man can devise against terrific natural forces which at times are overpowering. From that time forward, "smoke in the woods" has been my yardstick of progress in American forestry.

The tragedy of the Coeur d'Alenes was but one in a long series of national disasters from burning forests. The worst in loss of life and in wiping out communities were the terrible fires in the pineries around the Great Lakes. One started near Green Bay in 1871, on the very night that Mrs. O'Leary's cow kicked over a stable lantern and set Chicago ablaze. This fire burned out Peshtigo and other Wisconsin lumber towns, swept over 1,280,000 acres and took a toll of six hundred lives. On the same day waves of fire rolled over two million acres in the Manistee and Au Sable valleys of Michigan and northward, destroying some fifty towns and settlements and taking many lives. Much of this country was reburned in 1881 when many lightning and other fires merged into a great front that tore through Tuscola, Huron, Sanilac and Lapeer counties. There were 169 deaths in Michigan fires that year.

The Hinckley fire in Minnesota, in 1894, burned out homesteads and villages for twenty miles and accounted for 418 lives. The city of Cloquet, Minnesota, was practically wiped out by forest fires in the Moose Lake region, in 1918, which ran over 250,000 acres and killed 453 people. There were many lesser fires in the lake states during the heyday of their logging and settlement. It all adds up to a grim and fearful record.

The forests of the West have also suffered from many fire catastrophes. They have been less costly in human life than the big burns of the North, but have destroyed even more of the virgin timber wealth. Fire played an important role in the life story of Douglas fir, long before the first sawmill was built on the Columbia River by the Hudson's Bay Company. Left to fight it out among themselves, with no break in the forest canopy from cutting or burning, the western hemlocks, cedars, and balsams tend to run Douglas fir out of the woods. They can endure more shade in the struggle of life. Many of the finest stands of Douglas fir are traceable to great fires of a couple of centuries ago. Doubtless started by lightning or by Indian hunters, they gave this sunlight-demanding aristocrat a break in its battle against hardier neighbors.

The virgin forests of the Northwest are hard to set afire, even in the usually dry weeks of midsummer. Early settlers felt comparatively safe from sweeping fires like those that wiped out many communities in Michigan and Wisconsin. But the idea that "green timber will not burn" soon proved a will-o'-the-wisp. The awakening came in 1902. It was a hot, dry summer when parching winds from east of the Cascades frequently drove the air humidity down 'way below its normal safety. Clearing fires and slashing fires had been burning up and down the Northwest coast in the usual casual fashion. In September came the spark that set off an explosion. For two or three days atmospheric moisture was down so low that the ground cover even in deep woods was sucked tinder-dry. Then rising winds; and men woke up to the bitter fact that *green timber was burning* in a dozen places from northern Washington to the Willamette Valley of Oregon.

Several fires joined in a sweep through the Lewis River Valley just north of the Columbia. This big, blackened area became known as the Yacolt Burn. Many other virgin timber areas also were turned into snag patches. Stump homesteads, backwoods communities, and saw or shingle mills were burned out. Probably seven hundred thousand acres on the west slope of the Cascades were swept by the fires of 1902 with a known toll of thirty-five lives.

The awakening of the people of Oregon and Washington to the hazard that hung over their greatest resource was rude but effective. The fires of 1902 started protection organizations, compulsory patrol laws, and the expanding forest fire code of the Pacific Northwest. The greatest hazard lay in slash burning and land clearing and in the almost universal use of fire by loggers, settlers, timber lookers, trappers, and hunters. Many timber owners did not wait for state laws but themselves shouldered the task of organized patrols and fire fighting. Often a group of them in the same county or watershed pooled their lands for protection and prorated its cost on the acreage of each owner. These private protection associations spread over Washington, Oregon, California, and the Idaho Panhandle between 1905 and 1912. They took the lead in combating forest fires and wrote a deal of forest history.

The fire associations had a large part in the development of western fire codes and gave them a practical cast of direct action. Time and again the associations persuaded their state legislatures to pass laws enforcing new techniques or methods of fire control which their own experience had proved efficacious. Thus the western states adopted such requirements as compulsory patrols during dry weather, obligatory fire-fighting equipment at all logging camps, compulsory slash disposal, closed burning seasons, snag falling and enforced shutdowns of logging operations in low-humidity weather. The fire associations also gave hundreds of forest owners firsthand experience in co-operating on the common problems of their woods. It was here that industrial forestry began in the West.

I saw much of the fire associations and their work in my years as district forester in Montana and northern Idaho. We had a good chance for practical co-operation in the Idaho watersheds where national forest lands were mixed with railroad grants and other private holdings. We worked out practical boundaries within which all protection work was carried out by either association or government rangers and shared the cost pro rata on the acreage of the private owners and Uncle Sam. It was a common-sense solution for the crazy quilt of land titles in that country.

One of the leaders in organizing fire associations was George S. Long, western manager of the Weyerhaeuser timber interests. This lanky Hoosier, Lincoln-like in figure and manner of speech, was endowed with rare foresight and a practical gift of getting things done. He grasped the full proportions of the protection job which confronted the forest regions of the West and of its many ramifications in state laws and co-operation between public and private agencies. In 1909 Long was prime mover in creating the Western Forestry and Conservation Association. It soon became the grand lodge of all protection agencies in the western states and British Columbia—of the state and federal services as well as the private associations. The genius of its manager, Edward T. Allen, gave "Western Forestry" far-reaching leadership in national as well as western forest affairs. For many years this unique organization was a regional training school in forest protection. Through its activities experiences were compared, post-mortems held on badly handled fires, fire weather interpreted and forecasted, new equipment designed and tested. It wrote the *Western Fire Fighter's Manual* and carried it through several revisions. In 1911 another handbook was published on *Practical Forestry in the Pacific Northwest*. For many years the association led the expansion and unification of state forest codes. It was a good example of American democracy and self-reliance in working out new problems. The success of the Western Forestry and Conservation Association and its constituent patrol associations in organized co-operation inspired the Clarke-McNary Act of 1924, which carried the same ideas into federal policy.

There has been plenty of hot action on the northwestern front to point up the necessity for its fire codes and to strengthen them with fresh experience in meeting emergencies. The worst attack came in 1933, when the Tillamook Burn gave us the last word in blazing destruction.

There was irony in the choice of a stage for this infernal carnival by the fire demons. The mountains bordering the Oregon coast have a yearly rainfall of 75 inches or more. They are often swept by summer fogs. The area was regarded as one of the "safest" in the state. Its fine stands of Douglas fir had been picked out by several experienced timber men for investments in future reserves. But in two blistering August days over three hundred thousand acres in the heart of the Tillamook went up in smoke. A logger was about to quit in early afternoon, following warnings of dangerously low air humidity. The ground cable pulling in his last log fell across a dry, rotten cedar windfall. Friction ignited the punky wood, and the handful of men could not hold the fire out of a heap of slashings nearby. In half an hour the blaze was up the canyon and over the hill. The flames killed twelve and a half billion feet of timber, or about as much lumber as the entire United States was then consuming in a year. And then, as if to mock the futile efforts of an army of fire fighters, rain and fog blew in from the Pacific and stopped the fire in its tracks.

East and West, the conflagrations that roar through miles of forest in a few hours have followed the same meteorological pattern. They have come in the fall or late summer after woods and ground litter are parched by months of abnormally dry weather. Then for a period, usually two or three days, the relative humidity of the air drops from its normal 50 or 60 per cent to perhaps 10 per cent. The thirsty air is sucking moisture out of humus, leaves, twigs, and shrubs. The great sponge of vegetable matter over the earth, as well as the soil itself, is relentlessly robbed of its hoarded moisture. The dryness of the woods reaches an explosive quality when touched by a flicker of flame. Fires that normally burn slowly through the ground litter race faster than men can run, "crown" up into the treetops and become sheets of flame that in places scarcely touch the ground at all. In rangers' language, "she blew up and ran away." All that is needed for a holocaust when air humidity drops so low is a falling barometer and rising wind. A fire burning in logging slash, a railroad crew cleaning up rights-of-way, a burning cigarette flung from an automobile, the friction of a steel cable drawn over a punky log—any of these may be the spark that ignites disaster.

In most forested regions, stations of the U. S. Weather Bureau now forecast "fire weather" like coming frosts that threaten crops or storms along the coast. State fire codes and commercial insurance policies on logging camps and equipment require shutdowns in the woods during these periods of acute hazard. Slowly and at bitter cost we are learning the technique of forest protection.

First and last, fire has been the greatest destroyer of American forests. But its greatest wreakage has not come from the spectacular blazes that write headlines in the newspapers. It is the cumulative destruction, year after year, of thousands of little fires set by the woods-burning habits and carelessness of our people.

Most of the pioneers in every westward trek across this country were woods burners. Settlers fired the forests to clear land for tillage or pasture; to make their homesteads safer from Indian forays; to facilitate hunting; to protect buildings and crops from fires running loose in the neighborhood. Every new community has had its own unwritten code, based on good frontier logic, for burning the woods. It went with the tradition of free land and the self-reliance of folk who had to look out for themselves. The legacy of these pioneer customs to later generations, down to the present time, has been the greatest of all hazards to American forests. It is the "man-caused" fire.

Throughout the South, man-set forest fires were well nigh universal. They were part of the accepted order of things. Farmers and loggers and turpentine orchardists all fired the woods. They set fires to "green up" the forage, to uncover oak and beech mast for their razorback hogs, to clean out litter before boxing pines for resin, to open up the brush for better hunting, to get rid of cattle ticks or chiggers or snakes, or just because the woods had always been burned now and then.

All the inventories of timber supply, which began soon after the turn of the century, estimated the known forest fires every year at from 150,000 to 200,000, burning twenty-five or fifty million acres. Other great losses could not be measured in acres. Among them was the effect upon the lumber industry of unrestrained woods burning and the fear of disastrous conflagrations. It was a risky business whose raw material might be wiped out overnight. Fire danger accentuated the migratory cast of the sawmills and the lumberman's philosophy of "cut out and get out." Forest land was an asset for quick liquidation, not a resource for long-range use and planning.

In my early days as a forest missionary I was often told by mill men in the piney woods: "Nothing can change the woods-burning habits of these people. Perhaps the government can stand the cost and risk of attempting to grow trees. But no lumberman can do it." The conviction was burned into me that fire prevention is the No. 1 job of American foresters—both to keep trees on the land and to put faith in the men who own it.

II

The first national attack on fire in state or private forests came just a year after the holocaust in the Bitterroots. The Weeks Law of 1911 carried an appropriation of $200,000 for federal co-operation with the states in protecting the forested watersheds of navigable rivers. At that time only eleven states were doing enough protecting of any forest land to qualify for the help offered by Uncle Sam. But the new proposal did not lack in sales effort. I was back in Washington as assistant forester, and my job included the direction of state co-operation. I had an able associate in Girvin Peters, diplomat-extraordinary and master strategist in unlocking the right door to a complicated state situation. I was spurred on by vivid memories of blazing canyons and smoking ruins of little settlements and rows of canvas-wrapped bodies out in the Northwest.

Aggressive service emissaries set forth to create a greater market for co-operative protection by inducing more states to set up enough of a fire organization to qualify for federal aid on at least a few thousand acres. We were not too particular about details as long as a state was willing to make a start.

We were evangelists out to get converts. The number of co-operating states and the acreage under organized protection gradually increased. Best of all was the educational impact of the new policy. Conferences and meetings on forest protection multiplied. Bulletins on fire equipment and methods appeared. Experiences in different states and regions were compared. The important role in forest fires of relative air humidity was broadcast. Slowly the technique of planned protection with specialized equipment began to take over from the rough and ready efforts of the early fire wardens.

Forest protection got more attention in the press and in commercial circles. The first Southern Forestry Congress was held in 1916. It was organized permanently under the leadership of Joseph Hyde Pratt, Director of the Geological Survey of North Carolina, and made the extension of state forest departments and fire prevention on forest land its No. 1 plank. Southern pine lumbermen joined the campaign and railroads whose previous interest in cutover lands had been limited to farming and colonization. Only five southern states had made even a passing bow to forestry by 1916, but the leaven was at work. Many southerners were beginning to realize the potentialities of their stump lands. Here and there lumbermen were employing wardens and seeking the co-operation of farm neighbors in keeping fire out of their woods.

More and more forest protection laws were spread upon the statute books. They showed great diversity between states and bore the democratic stamp of local experience. The states which had federal land grants or other woodlands frequently organized fire protection under their land commissioners. In other states the political influence of fish-and-game agencies made forestry the handmaiden of that department. In many instances the prevention and suppression of forest fires stood on its own feet as a new public responsibility, and "chief fire warden" appeared on the roster of state officials. As time went on and more forest activities were added, the trend became general to merge them all under a state forester.

Equally divergent was the practice of different states in paying the bills. The richer states, like New York and Pennsylvania, accepted the duty of forest protection as a public charge, like the maintenance of city fire departments. Other states, particularly in the South, followed the principle of county option and cost sharing. The state puts forest fire patrols in any county which pays half the cost and enacts the necessary police ordinances. Often private forest owners contribute to the county's quota. In the voluntary patrol associations of the Pacific Northwest, each forest owner paid for his own acreage. It was not long before these protection-minded timbermen were demanding that free-riding neighbors pay their share too, and compulsory protection was written into the fire codes of Oregon and Washington. If any owner now fails to protect his forest acres to the satisfaction of the state forester, the state takes over the task and its cost is added to the property owner's taxes as a special fire assessment.

Through the great body of forest law written by our states when we entered World War I, the red pages of fire prevention and organized suppression became more frequent. Gradually they followed a common pattern. Through all the diversities of differing state situations and backgrounds, the national objective of the Weeks Law gained recognition. "Federal co-operation" was strong medicine.

Fire was plainly recognized as forest enemy No. 1 in the next great stride of national conservation. The Select Committee of the United States Senate which wrote the Clarke-McNary Act of 1924 was told a hundred times, at public hearings across the length and breadth of the country, that the first and greatest commandment of American forestry is to keep fire out of the woods. Substantial gains had been made under the Weeks Law, but thirteen states with large areas of forest had yet made no start. Nearly half of all state and private woodlands were still without organized protection, and a fourth of our annual timber crop was going up in smoke.

The new law made the co-operative attack on forest fires a three-way alliance. The private forest owner was brought into partnership with his state and national government. The old constitutional swaddling clothes, which limited federal interest in state and private forests to their effect upon navigation, were cast aside. Forests became a national concern in their own right. The Clarke-McNary Act launched an all-out offensive to protect every acre of forest land in the United States. Since 1924, state after state has legislated and organized the protection of its forest lands and then constantly expanded the scope and intensity of protection. The state and private outlays for keeping fire out of the woods, which totaled $173,000 in 1911, now exceed $17,000,000. In mid-1949 over 80 per cent of the 439 million acres of state and private woodlands were under organized guard. On these protected acres the average yearly burn was under six tenths of one per cent. On unprotected forest lands, the loss was twenty-four times greater. One acre out of every seven was burned.

The magnitude of the co-operative undertaking in forest safety has exceeded all forecasts. In 1924 we estimated that "adequate protection" of state and private forests could be attained for $10,000,000 a year. The Clarke-McNary authorization was set at $2,500,000, on the theory that the landowners should pay half the cost of protection and that the state and government should divide the other half between them. The authorization is now $9,000,000 and legislation passed by the 81st Congress will advance it by successive stages, to $20,000,000 in 1955. The program now looks to the federal coffers to carry half the total cost of forest protection. The state and private owners jointly are expected to provide the other half. And adequate protection requires an estimated $40,000,000 annually. This fourfold increase in the cost of fire-free forests, since 1924, not only tells the story of a much larger field army of rangers, patrol men, and fire fighters than we dreamed of in the early days, but it sums up the manifold advances in the techniques of forest protection—the mechanization of the job with motor truck-tanks and electrically driven pumps, bulldozers for cutting fire lines, radio walkie-talkies, and paratroopers for the back country. Forest protection has become highly specialized engineering.

In large measure, twenty-five years of Clarke-McNary co-operation has accomplished its purpose. Most of America's woodlands have been made reasonably safe for timber cropping. In many parts of the United States forests have become, in fact, insurable fire risks. The best evidence of this fact is the establishment of commercial tree farms in twenty-nine states and the constantly expanding industrial investment of $5.00 to $20 per acre in forest planting. Forestry will never be free of fire hazards as citrus growing doubtless will never be free from danger of destructive frosts. But to many thousands of forest owners fire has become a calculated risk. Large tree farms have been set up in the Pacific Northwest on an estimated fire loss, in acreage, of one fourth of one per cent annually. These forest enterprises take no chances. They often spend twenty-five or fifty cents per acre—every year—on continuing plans of fire defense, building protection roads, keeping up lookouts and telephone lines, felling dead trees, and bulldozing fire breaks across old slashings. In fire weather they rely first upon their own rangers and construction or logging crews. Their men are armed with motor-driven pumps and tanks and as much specialized equipment as shock troops landing on a Normandy beach. The network of roads on some of the most advanced tree farms brings practically every square rod of forest within range of the powerful water stream from a tank truck. Behind the home guard, in a second line of defense, is the state-wide organization of lookouts, patrol men, and flying squadrons of fire fighters, equipped and waiting in every county. There is, of course, no absolute security against fire in either forest or city. Under extreme climatic hazards we may get another Tillamook Burn; but government, state, and industry are organized to cope with it. And we are not afraid to get on with our business of growing trees.

One hot day last June I took a leisurely drive over the back roads of Kitsap County, Washington, in which I live. Its virgin timber is practically gone but it still has 80 per cent of forest land and twenty sawmills. I frequently passed trucks loaded with second-growth logs. I saw half a dozen "gyppo" loggers. Every one of them had heard that morning, by phone from the state warden, that humidity was close to 30 per cent and dropping. The state forester had warned all loggers to use extra care. If humidity dropped even a little, he would close down all woods operations in three counties. Most of the loggers I saw were quitting "today noon." They would take no chances on starting fires and having to put their crews to fighting them.

At the District Forestry Hall near Port Orchard three big red trucks stood ready and waiting. Each of them had a tank and pump and five hundred feet of hose. There were axes, shovels, canvas water bags, pumps for men to carry on their backs. Fourteen men were eating lunch. They are at the station day and night in summertime, always ready for a fire call. The dispatcher showed me his season's record with pride. It carried the "elapsed time" in hitting every fire after a report was received. There had been thirty fires this year—most of them caught when less than an acre. The largest was twenty-nine acres.

A few days before I had stopped at the State Fire Depot just east of Olympia. Half a dozen bulldozers were sitting on their big trucks, ready for a call to any emergency fire line in western Washington. They were the last word in rugged, powerful machines, driven by caterpillar traction. They can gouge out a broad fire trench on any ground where men can stand up. The mechanics to operate these monsters were ready and waiting too. How many miles of fire trench had we grubbed out by hand with mattocks and shovels and axes, in 1910, yard by sweating yard, hoping to head her off before she went over the ridge?

On the way home late that afternoon I stopped beside a California car by the roadside. The party had a brisk campfire going a hundred yards off in the woods. I put on my deputy warden's badge and walked in to investigate. A woman looked up from the group busily preparing supper around a perfectly safe fire and said wearily: "Fish out your permit, Tom, here's another one." I was the third warden who had checked up on them that afternoon.

Thinking over the day's experiences took me back forty years. Eternal vigilance still is the price of forestry. There still is too much smoke in the woods. But we have moved far in the know-how and organization of protection since the August skies of 1910 were blacked out over the Bitterroots.

CHAPTER II

The Saga of Free Land and Timber

As a young man my grandfather worked on a land-clearing crew in New Hampshire. He often told me of the prime white pine logs and whole trees which he helped roll into the Connecticut River. It was the easiest way to get rid of them. Other men of his generation, land clearing on the Sangamon or Wabash, burned white oak and black walnut timber or split it into fence rails. One of them became President.

The American colonies were planted in a primeval forest. They grew into a nation among trees and surrounded by a wilderness of unbroken woods. To them, the forest was something to be overcome. It stood in the way of homes and farms. It was a menace to the safety of isolated settlements. The conquest of the forest through many years of early American history, renewed in every wave of migration westward, bred physical vigor, hardihood, and self-reliance.

Our forebears also found in the forest a storehouse of materials—for building, furniture-making, for wagons and farm tools and a vast array of human needs. As a boy exploring the old homestead in New Hampshire, I marveled at the wide pine boards which sided the woodshed; at the roof timbers hewn from oak and mortised into the upright framing with pins of locust; and at the beautiful cornices over the front entrance with their hand-wrought scrolls and beaded moldings. I wondered at the skill and patience of men who built such homes. Universal use of the materials at hand in the woods created all manner of American arts and crafts.

The forest was our first industrial resource. The little mill, pushing a slash saw up and down by running water, appeared early in New England villages. Often it also ground corn or wheat. One of my ancestors was deeded a mill lot by the town of Salisbury, Massachusetts, for just such an enterprise, in 1650. Woodworking shops followed, with furniture makers, coopers, wagoners, and boat builders. The forest, always near at hand, started innumerable village manufactures. It was the mother of free enterprise. Little woodworking plants, scattered over colonial America, were springs whose rivulets gradually formed industries of national proportions. From them flowed labor skills, new tools and equipment, machines that opened the way to quantity production, capital, managerial know-how and daring.

We were slow to realize other great gifts of our forests, gifts to American character. We appreciated them no more than children appreciate the motherhood which surrounds them. "Woods and templed hills" were part of the environment in which we were born. They were part of us. We had no conception of life without them or without the wild life they sheltered and the lakes and streams they fed. Forests have given us more than hunting and fishing; more than the recreation of woodland vacations and hobbies; more even than the inspiration of their beauty and serenity. During most of our national life a great hinterland of unbroken forest has challenged us as the New World challenged the gentlemen adventurers of Queen Elizabeth's day. It has quickened the zest and imagination of young men. It has kept alive the spirit of daring and overcoming, of meeting unknown obstacles with bare courage and resourcefulness.

On a trail far back in the Sierras I once came upon a penciled marker pointing to *"Weak Head Smith—His Sawmill."* Soon we rode by placer sluices boxed with freshly sawn pine boards. The "sawmill" proved to be a huge standing pine, topped off sixty feet or so from the ground. The miner had erected pole scaffolds around the sturdy trunk, and working on this footing with a long rip saw was sawing his log down foot by foot into inch boards.

A few days later we visited one of the great lumber mills of the San Joaquin Valley. It had flashing band saws, electrically driven motors, dry kilns, and resaws shooting out a stream of box shooks. The ranger with me said: "Weak Head is a bit out classed here, but for ingenuity and guts I still give him the prize."

At the start of things the continental United States contained some eight hundred thousand square miles of forest. The woods which hedged in the early Americans were as limitless as the air and water and game. Until reduced to private property, they were possessions in common. All the community shared in their use. New Englanders objected to the broad arrows of the British monarch on their tall white pines, not in revolt against the crown but because they resented any restriction upon their common rights in the forest.

As property lines were extended by settlement and by charters and grants of many kinds, the free land was pushed back; but there was still no end to it. And it was easily had. Any squatter could stake out his claim. Thousands of land grants were made to colonial and revolutionary soldiers. Land was given away under charters to build roads and canals, to found new settlements and endow colleges. Land was the universal commodity.

The magnet of this great free resource steadily drew the American people westward. Every wave of exploration and migration brought more vast stretches of land and forest within the grasp of men who would take them. However crowded the old neighborhood might become, there was always new free land out beyond the ranges or across the plains.

The people of America have never been forced to fit their ways of life into rigid barriers of geography like the nations of western Europe. There has always been new land somewhere. We are prone today to overlook realities and think of Alaska as another frontier waiting for the settler's ax and plow. It is proposed to reward G.I.s of World War II with homesteads in Alaska even as the veterans of the French and Indian War were granted lands in the Allegheny valleys. It is part of the American tradition.

Often in the young settlements of the West I have heard such questions as "Have you taken up *your* land yet?" or "Have you used *your* homestead right?" From the beginnings of our public domain, it was national policy to put public lands into private ownership as fast as possible. The land philosophy of England is based upon primogeniture, the inheritance of an ancestral estate by the oldest son. The land philosophy of the United States has been created around the right of every citizen to his share in the new free land. This American inheritance was confirmed by the long political battle for free homesteads, the battle led for years by Andrew Jackson against the joined opposition of the conservative Northeast and the slave-holding South. The Homestead Law, vetoed by James Buchanan but re-enacted and signed by Abraham Lincoln, is the cornerstone of our public land policies.

The national public domain began with the cession to the federal government of the westward lands claimed by many of the first thirteen states. It was expanded by purchase, treaty, or conquest until it took in about three fourths of the continental area of the United States. The vastness of these public lands and the desire to put them into private ownership and use continued the unstinted grants begun by the colonial legislatures. Thirty-seven million acres were given to the new states of the West; ninety-one million acres to aid the construction of railroads. It is not surprising that Uncle Sam's acres were not looked upon as "property" in the ordinary sense.

Public land was also a ready source of federal income. The Ordinance of 1785 authorized land sales at $1.00 per acre in minimum lots of a square mile. The Cash Sales Act of 1820 raised the base price to $1.25 per acre and reduced the minimum to eighty acres. This law started the public land auctions, which were colorful episodes in the early life of the new states and launched many schemes of vast speculation. The phrase "doing a land office business" comes from these days of Uncle Sam's activity in real estate. Two hundred and twenty-five million acres of public land were sold under legislation of this type before its repeal in 1891. Such sales started many timber baronies in the lake states and the South.

The fight for the lone settler against the big land company led to the Pre-emption Act of 1830. It gave any squatter first right to buy 160 acres at $1.25 per acre for his own use and benefit. The conception of our public lands as a resource to be widely distributed and to encourage individual settlement and homemaking came to fruition in the national Homestead Law of 1862. It offered 160 acres of free land to any person who was the head of a family or twenty-one years old and a citizen of the United States if he would live on the land and farm it for five years. An early wave of forest enthusiasm over the prairies made tree planting another means of winning a homestead.

A work-hardened rancher in the Sierra foothills once explained the Homestead Law to me in cold turkey. "It's like this, sonny! The government will bet its 160 acres against your sixteen-dollar filing and prove-up fees that you can't live on the durned place the first eighteen months without starving to death."

In 1878, lumber and other free land pressures passed the Timber and Stone Act. It permitted any citizen to buy 160 acres of public land for $2.50 per acre, with no required residence or improvements. The theory behind these laws was *individual* homemaking or use of resources. Each entryman took oath that he would occupy or use the land himself and that he was not acting for anyone else. Another set of laws authorized the practically unlimited entry of public lands shown to bear minerals and their patenting after a minimum of prospecting or mining.

Free land drove the covered wagons westward. It built the transcontinental railroads and sent wave after wave of migration to the shores of the Pacific. An aggressive people wrote this saga, courageous men and women trained in the democratic pattern to take care of themselves, impatient of restraint upon any man's opportunity to take what he wanted for his own enterprise. Every part of the raw West wanted more people, more industries, and pay rolls. Lusty young communities, hewing their own way in the wilderness, were intolerant of laws and bureaucrats in distant Washington. So came about the free-and-easy "public land conscience." It was an American tradition that gained fresh vigor with every migration into new country. And there was never a lack of politicians to cater to it or of scheming men to take advantage of it.

The forest story of America begins with the tradition of free land. Free timber was part of the tradition. It seemed to be inexhaustible. Everyone in authority was anxious to get it out of the government's hands. It is not surprising that venturesome spirits on the frontier took the trees just as they took the beaver pelts or panned gold nuggets from the rivers.

As lumbering pushed out into the public domain, particularly in the expanding enterprises of the northern pineries, government timber was simply stolen without pretense of legal entry. A favorite method was to buy a section or so and then log the entire countryside indiscriminately. Reports of the Secretary of the Interior referred, with increasing frequency, to "trespass" or "depredations" on public timber lands. Special field agents were employed to locate and check these inroads, and now and then individual trespassers were brought to book. Bavarian-born Carl Schurz, Secretary of the Interior under President Hayes, had an exceptionally clear understanding of the whole timber supply situation. He did his utmost, with the inadequate resources provided by apathetic Congresses, to stop the looting of the public domain.

After the repeal of "Cash Sales," there was no lawful way for a lumberman to acquire directly enough public timber to carry a sawmill of even moderate capacity. Often several thousand acres of railroad or other land previously patented could be purchased. Otherwise a unit of economic size had to be put together by buying many individual homesteads, timber and stone claims, mineral patents, state school sections, and what-not. Usually it was a clean transaction; but it was easily perverted into forms of collusion that violated the principle of *individual benefit to the entryman* written into the later public land laws.

The increasing vigor of federal prosecutions of timber stealing drove the freebooters to take cover under farcical land procedures. The professional timber locator kept close to the progress of public land surveys. When a new township plat was approved, he knew where the good timber claims were to be found. He or his employer, who might be an eastern lumberman or a local speculator, would provide the necessary number of entrymen. In the heyday of western timber grabbing they were brought in by the trainload—often under contract to turn over their patents to the arch conspirator or his agent as soon as legal title could be transferred. Each land-hungry citizen would be guided to the four corners of a timber and stone claim, thence to the local land office for the filing, entry fees, and affidavit of "personal use and benefit." Then he would sign a transfer of title and be sent happily on his way, all expenses paid and $50 in pocket for the use of his right to public land.

The same entryman, if he desired, could also file a homestead claim. This simply took another set of papers and an eight by twelve cabin of poles and shakes, thoughtfully erected by the locator's crew in advance, to prove occupancy and use. The story of a homestead in the Idaho Panhandle thickly studded with towering white pines is a sample of what went on in the orgy of free timber. The claimant's application to the land office read well enough. He had located and staked the corners of his homestead. He had cleared a part of it and planted a patch of potatoes. He had cut timber and erected a cabin. He had harvested a crop of potatoes and consumed them for his own sustenance while engaged upon the arduous labor of clearing and improvement. Surely here was a hardy pioneer, overcoming the wilderness in the best American tradition.

Patient investigation by a Forest Service land examiner rewrote this story in more exact terms. The timber locator with his man in tow tramped back into the benches of the St. Joe country, with packs on their backs. One of the packs included a small sack of potatoes. They located a public survey corner and paced off a quarter section with a hand compass. They drove a cedar stake in the ground at each corner and blazed a tree or two. They spent parts of two days in "residence and cultivation." On the first day they scooped out a hole in the thick wood's humus, dropped in the sack of potatoes, and kicked the litter back over it. Then they knocked together a doll house, four by four or some such matter, of sticks and little strips split from a down cedar. On the second day they dug up the potatoes, fried them for lunch, shouldered their packs and went home. The claim cruised, incidentally, over three million feet of choice white pine.

Many were the wiles and stratagems of the "timber lawyers" in gilding their banditry with some color of legality. The Congressional grant of swamp and overflow lands to the states was a choice hunting ground. The theory of this law was to encourage the drainage and settlement of swamplands and lands inundated by spring freshets through free grants to the states. Any state could obtain patent to a section of public land by offering proof that it met the watery specifications of the grant.

This law was made to order for the shrewd timber "looker." Any reasonable proof of "swamp and overflow" would qualify an application to the State Land Office for purchase of the land at a legal minimum of $1.25 per acre. The locator's affidavit became the basis of the states' claim for a federal patent, and the transaction usually rolled through to private title with routine smoothness.

Agents of the General Land Office finally checked some "S and O" claims in California, whose swampy character seemed to coincide most strangely, forty by forty, with choice stands of redwood timber. The locator had attested to the marshy nature of the ground by a sworn statement that he had crossed it in a bateau. What further proof could any reasonable official ask? His affidavit neglected to include a minor detail, that the bateau was mounted on axles and wheels and drawn across the sections of dry land by a yoke of oxen.

Timberlands often acquired, in land office files, showings of mineral wholly unsuspected by the geologists who had examined them. In 1905 it was one of my jobs to check the estimate of timber standing on the right of way of the projected Western Pacific Railroad across national forests in northern California. The advent of a new transcontinental line had drawn the land-grabbing and land-speculating gentry to that country like flies to a honey pot. The mineral entries engineered or acquired by one H. H. Yard were reported to include 265,000 acres of timber. In a newspaper published at Quincy, California, I read several advertisements of blocks of placer claims offered for sale. One of them unblushingly guaranteed that "these mining claims will run twenty thousand feet of sugar pine timber to the acre."

Ethan Allen Hitchcock, Secretary of the Interior under

Theodore Roosevelt, made a determined and successful effort to break up the public land frauds in the Pacific Northwest. He secured the appointment of the able, fighting Assistant Attorney General Francis J. Heney to lead his campaign. One of several federal grand juries impaneled by Heney, in 1904 and 1905, returned twenty-six indictments against timber groups, land speculators, locators, and conniving federal officers, principally in Oregon. The indictments usually charged "conspiracy to defraud the United States of public lands." They exposed and largely blew up the system of organized, fraudulent land entries and brought about the repeal of the Timber and Stone Act. Public indignation was heightened by the conviction of United States Senator Mitchell and former Land Commissioner Binger Hermann for complicity in some of the shady transactions.

The timber frauds threw more fuel upon the fires of political reaction. The dramatic exposure of the "Oregon timber steal" came as Gifford Pinchot was forming his lines of battle to create enormous areas of national forests and transfer their administration to a new service in a different department. Aroused public opinion was ready for forest conservation.

In the meantime, the sawmills were not far behind the public land surveyors and timber lookers. As the settlers moved westward in search of new, free land, the sawmills moved with them in quest of new forests and free timber. It was all part of the great saga.

CHAPTER III

Cut Out and Move On

TWENTY years ago I stood on the log deck of a sawmill on the lower Columbia. An order of special importance was tacked up on the head sawyer's board. It was a list of long timbers destined for the Brooklyn Navy Yard to remast the United States frigate *Constitution*. The mainmast of "Old Ironsides," I recall, required a stick three feet square and one hundred and twelve feet long. Her main yard took a piece two feet square and ninety feet long. "The old girl sure carried some canvas," the foreman remarked. "No wonder she could sail rings around the Britishers."

The boss of the log boom had picked his logs with extra care. The timbers must not only be true as a plumb line, they must stand up to the exacting grade of select structural in straight grain and number of rings per inch. As I watched the superb trunks of Douglas fir come up the jack ladder and roll down on the saw carriage, I thought of the first masts in the old warrior, hewed with broad axes from the great white pines of New England, and of the little water-run sawmills which cut the planking for her decks; and here was the same craftsmanship with different tools, cutting another set of masts and spars on the shores of the Pacific. The life of one grand ship spanned the migration of the sawmills from coast to coast.

Lumber milling appeared with farming, fur trapping, and fishing among the earliest industries of the American colonies. For a long time it was a local industry. Little water-run sawmills supplied their own villages and shipped their lumber to nearby settlements by river and coastwise boats. As population and demands for wood increased, large dams and groups of sawmills were built on the lower reaches of most of the rivers along the Atlantic seaboard. Streams and canals brought the logs down to them from inland forests. Large centers of lumber manufacture grew up on the Penobscot, the Merrimac and the Susquehanna. A hundred years ago a stretch of twenty-one miles along the Machias River in eastern Maine was almost filled with sawmills, timber docks, and shipyards.

Access to tidewater markets through the Erie Canal, completed in 1825, pushed the sawmills into the rich pine forests of the lake states. The lumberman became a pioneer railroad builder and his tracks often blazed the way for the common carrier lines which followed.

The sawmill gave full play to the American genius for machinery and production. In the early eighteen hundreds large mills with many "up-and-down" water-run saws under one roof operated on most of the eastern waterways. Steam engines brought the faster and more powerful circular saws and later the still faster band saws, but it was the distribution facilities of the railroads that made lumber "big business."

First Maine, then New York, and then Pennsylvania led the states in lumber production. In the vast pineries surrounding the Great Lakes, following the War between the States, lumber manufacture took the form which molded its destinies. It became the great nomad among American industries, driving from one virgin forest to another like a threshing machine from one ripe wheat field to the next.

The public lands of the lake states, with their incomparable forests of white pine, were readily acquired under loosely administered land laws conceived for quick distribution of the national domain. These vast level stretches of forest invited daring men and venture capital to plunge into timber ownership and lumber manufacture on a scale beyond the dreams of their eastern fathers. The country, too, was entering a period of unprecedented industrial growth. Railroads were being built at a prodigious rate. The prairie states were rapidly being settled. Big business was in the full flush of its youth, and the lumber industry fell in step with the times. It devised new tools like the band saw. New devices of many kinds speeded up production. The brains of the industry competed in building bigger and better sawmills. Quantity production became the industry's god and, when markets were slack, its devil.

The lumber from the big sawmills was cheap and well made. Keen, competitive merchandising pushed its sale into every nook and cranny of the land. It built most of the new homes in the westward sweep of population and gave the country an unlimited supply of inexpensive industrial material. Sash and door plants, furniture factories, box factories, vehicle and car shops followed the sawmills. The per capita consumption of lumber doubled in the fifty years following the Civil War.

Lumbering in the region of the Great Lakes became an industry of tremendous driving power. Its captains and kings were hardy and resourceful, skilled in organizing men and machines and in overcoming physical obstacles. Many of them were highly successful. The capital and self-reliance they acquired and the organizations they built up sought new fields to conquer with an aggressiveness scarcely paralleled in industrial history. Powerful economic forces were set in motion. Among them were speculation in timber, the amassing of huge properties, and a rapid increase in the capitalization of the industry.

The drive of large capital investments for speed and profit brought about the rapid skinning of enormous areas of timberland. Twenty years, and even less, became the common lifetime of a sawmill. Then—dismantle, junk, and move on. Not only did lumbering perforce become a nomadic industry; it became an industry with no permanent interest in the land. A logged-off section was in the same category as a junked sawmill—to be sold for what it might bring, or abandoned and forgotten. Hence forest fires were of small moment until they threatened merchantable timber or a logging camp. They swept at will over slashings and young growth. This devil-take-it attitude toward the land was strengthened by the common belief that most of it would soon be in cultivation anyway. The smoke of burning slashings, often indeed of burning timber, hangs like a pall over the vivid retrospect of humming lumber camps on the Saginaw and Wisconsin and the upper Mississippi.

Another characteristic was stamped upon the lumber industry in those colorful days. Its large capitalization forced not only maximum production, but constant production. Big sawmills and overhead organizations and obligations to capital could ill endure idleness. The industry lost its old-time flexibility to its markets. Adventurous-minded, chance-taking men built up sawmill capacity far beyond the normal requirements of the country. Periods of overproduction and intense competition became frequent. The market was either a feast or a famine. Magnificent virgin timber was too often forced upon sluggish customers at less than the cost of production. It was less costly to manufacture at a loss than not to manufacture at all. One of the pine kings of Minnesota advised his brethren to "keep your saw in a log and sell at the market." In serious depressions only the best and clearest logs were taken out of the woods; the rest went up in the smoke of slash fires.

For many years men's minds refused to concede that the timber of the lake states was exhaustible. Farsighted lumbermen realized the truth sooner than anyone else. Early in the eighties, lumber capital from Michigan and Wisconsin began to flow into the nearest virgin fields—the southern pineries. The South entered its period of active timber buying, speculation, and the blocking up of large holdings. Many tracts of state timber, public lands, and private patents were brought up, much of it at $1.25 an acre. In the following twenty years, company after company cut out in the North, junked the old plant, and trekked with its group of skilled hands to a new location in Mississippi or Louisiana.

Leaving the Great Lakes in its full-blown vigor, the lumber industry struck the southern pineries with still bigger mills, larger and more heavily capitalized organizations, and greater faith than ever in the creed of quantity production. Its strong points and its weak ones, its housing of the nation and its waste of timber, its opening up of new states and its aftermath of wrecked land—all became more pronounced. The southern industry was more heavily financed than the northern mills. Timber values were rising and larger investments were necessary to consolidate operating units. The talk of a timber famine sent many men scurrying after stumpage. A new form of lumber finance, the timber bond with unescapable interest and liquidation charges, appeared and assumed large proportions.

Everything combined to drive for large continuous production.

Hardy men from Maine brought the first American sawmill around Cape Horn to the Pacific coast in 1852 and began logging its huge timber with ox teams. Lumbermaking developed slowly in pace with the needs of western settlement, coastwise trade with California, and offshore markets in the Orient. But keen lumbermen from the East early took the measure of the vast coniferous forests west of the Rockies as the seat of a great industry. While their brethren in the craft turned to the South, these men came West—not as manufacturers but as explorers, locators, and timber buyers.

It was in the nineties that the concerted rush to the West began. The end of lake states timber was plainly in sight. We had learned that no forests are inexhaustible. The fame of western timber spread far and wide. Its procurement under loosely administered public land laws was simple enough. "Timber" cast the spell of easily gotten wealth over the public mind.

Time would fail to tell of the doctors, lawyers, merchants, and thieves who joined in the scramble; of the professional locators who filed claims for Nebraska schoolteachers and Chicago clerks; of the cruisers for eastern lumber companies who located the best with a shrewd eye to "logging chances" and then piloted waitresses, barbers, and drummers around the four corners of "their" claims. The old American tradition of "free land" was riding high. An easy "public land" conscience applied a standard of honesty to the national domain quite different from that expected in ordinary affairs of life. The game went merrily on until it was slowed up by the creation of forest reserves.

Hard upon the heels of the locator came the timber speculator, the investment buyer, and somewhat later the real operator blocking up logging units. Millions of dollars of lumber-made capital, released from the lake states, flowed into western forests. Large public land grants facilitated consolidations of ownership. The abundance of western timber and rising values in depleted forest regions of the East spurred the energies of men accustomed to playing for large stakes. Timber baronies were built up in the West which made the large operators of the lake states look like peanut vendors. When the Bureau of Corporations investigated timber ownership in 1907, it found 436 billion board feet of western stumpage in the hands of twenty lumber and railroad companies, 664 billion feet in the control of 131 concerns. The latter group owned nearly a fourth of all the timber in the United States.

The manufacture of lumber on the Pacific coast is the crowning glory of the craft. It is a kaleidoscope of vivid pictures, filled with primal energy and human daring—a dozen bulls straining on a huge redwood log down a corduroy road, logs flashing like meteors down miles of water flume, the log train creeping and switch-backing up a Sierra canyon, the railway incline pitched two thousand feet against the sky, the "bull" donkey engine whose steel tentacles grip giant trees and yank them home like jackstraws, overhead cables carrying tons of logs through the air, caterpillars dragging in raw material for a couple of houses—there have been few equals in sheer power, resourcefulness, and courage in the industrial history of the world.

Volume production in the sawmills of the West kept pace with the power of their logging machines. In the earlier years of the industry east of the Allegheny Mountains, a yearly cut of more than ten million board feet from any plant was rare. The band mills of the lake states manufactured twenty or thirty million feet a year, and the frequent combination of two band "head rigs" with resaws and other machines produced a total of forty or fifty million feet. The monster plant of the Virginia and Rainey Lakes Lumber Company in northern Minnesota operated five band head saws. Many sawmills of the South and West manufacture from 100 to 200 million feet of lumber annually, and a few mills on the Pacific coast can, at full capacity, cut a million feet of lumber a day.

On the other side of the railroad tracks are innumerable smaller mills—down to the "peckerwoods," "coffeepots," and "fire splitters" indigenous to every forest region. The census of 1909 found nearly fifty thousand sawmills of great diversity in size and equipment, but three fourths of them cut less than a million feet of lumber in any year. The little sawmill has held its ground in every stage of American economy. It has pioneered the building of new communities. It has cleaned up odds and ends of virgin timber that the big mills passed by. It has remained, to operate on second and third growth forests, after the larger companies moved on. It has been the "in and out" fringe of the industry, flooding the production of lumber on good markets and dropping out when the going is tough. It has come to the rescue of the country in every wartime or housing emergency, and it has constantly put new blood into the larger organizations.

Migratory lumbering was an inescapable part of the political and economic philosophy of free land. The virgin forests of the Northwest, like the open ranges of Texas or the placer gold of California, were waiting for enterprising men to take them. Cheap virgin timber dictated every phase of the lumber business. It was a one-product industry, using but part of the superb trees it felled in the woods. Its leftovers in the slashings looked like an enormous wastage, but for the most part were species and grades of logs that the lumberman, with his own facilities for manufacture and marketing, could not convert into anything salable at a profit.

I recall a doleful meeting of lumbermen during the slack market following World War I. A man from Wisconsin lamented the timber he had to leave in the woods because the glut of low-grade lumber had knocked the bottom out of the market. The dean of the West Coast delegation remarked that an average acre of Douglas fir slashing contained more wood than any acre of eastern forest ever grew. He spoke the literal truth. A study of the U. S. Forest Service at about that time showed that Douglas fir loggers were leaving on the ground 20 per cent of the timber that would be merchantable and used in most eastern operations. This abandoned material often ran up to twelve or fifteen thousand board feet to the acre.

In the fervor of conservation it was easy to indict the lumbermen of the migratory age for wasting a natural resource and junking cutover lands. "Devastation," for the time being at least, was too often the name for it. Yet their works were out in the open for all to see, and they were playing the game according to the rules of their day. The rules of the game, written and unwritten, were drawn from the nineteenth-century American creed of free land and free enterprise.

Lumber manufacture reached its peak at the very time when the crusade for forest conservation was gaining power under Theodore Roosevelt and Gifford Pinchot. For a decade, beginning with 1904, the cut of lumber exceeded forty billion board feet a year. Per capita consumption passed five hundred board feet annually in 1906 and 1907, by far the most lavish use of wood ever attained by any people in the world. Lumber was one of our important exports. Uses of wood for paper and many other purposes were on the climb. The forecasts of timber famine were not without reason.

Questions were being asked also on the other side of the fence. The tremendous expansion of lumbering in the verdant forests of the South and West had made it top-heavy with timber lands and installed sawing capacity. New building materials were coming into use, and lumber encountered hard sledding as per capita consumption fell off. The Federal Trade Commission held a clinic on the ills of the industry in 1914. Two years later I took part in a region-by-region study of the lumber situation, conducted by the U. S. Forest Service.

We found that the installed sawing capacity of the mills exceeded by one half the greatest cut of lumber they had ever made. We found that the private timberlands of the West, so lately in public ownership, had built up an assessed tax value of $358,000,000 and probably represented over a billion dollars in the capital structure of the industry. Overproduction and cutthroat competition were common ailments. The stratospheric climb of the industry had created a Frankenstein of capital charges and the machine was in process of devouring the weakest of its creators. It was evident that a more stable and less chance-taking forest industry was needed before conservation could get very far in business circles.

It is worth a pause to ask why, after two and a half centuries, our main timber-using industry had such a migratory cast. Why was it not profitable to grow some trees, at least on the Atlantic seaboard where commercial lumbering was begun by Captain John Smith in 1610?

Many little wood-using plants did stay at home. Pulp and

paper mills largely remained rooted in their first locations and took the lead in forest management. But the main source of national lumber supply moved every twenty or thirty years. The primary reason was "competition," competition created by the sheer vigor of an expanding national economy.

First, of course, came the wealth of virgin forest, stretch after stretch of free timber inviting exploitation. Then came the rapid extension of transportation. Efficient, low-cost rail and water facilities pushed out into new regions and brought their woods back to compete with the products of old manufacturing centers. A third factor was the mechanical progress in mass production of lumber. Larger and more efficient sawmills constantly reduced the unit cost of production.

These developments were speeded up by the rapid increase in lumber consumption during the fifty years after Appomattox. Lumber manufacture became "big business" and sharply competitive. Volume production was its mainspring. The competitive quest for the cheapest lumber constantly drove manufacturers into new stands of timber with the latest and most efficient mills.

Lumber financing had no small part in the cut-out-and-move-on fever of the times. The industry borrowed freely and often on a fixed schedule of liquidation. Every thousand feet of timber felled usually had to pay its quota into a sinking fund, and minimum yearly cuts were prescribed by the bonding houses. Debt requirements often forced the uneconomic depletion of timber properties.

All of these elements merged in the liquidation philosophy of forest conversion. It was a colorful chapter in the westward expansion of America. However we may regret or condemn its wasteful features, we must recognize its vast contributions to home and farm building and to our whole industrial development. At their peak, the logging camps and sawmills were exceeded in the employment of labor only by the farms, the railroads, and the mines. And it is hard to conceive how, under a free economy, some such process of using our vast primeval stands of timber could have been avoided before we reached a sound footing for industrial forestry.

The rush of the sawmills to virgin fields, even the destruction of the forest as they went, had much to do with the astonishing speed of America's westward expansion. To hundreds of wilderness outposts, lumbermen brought the first railroads or steamships, the first pay rolls, the first permanent communities, the first markets for farm produce. They opened up the country and started its economy rolling. My oldest neighbors in Port Gamble, Washington, whose sawmill was built in 1852, like to spin yarns of the days when folks from the little settlement called Seattle came to Gamble to get their mail, their supplies, and their news of the outside world. They also tell about the New England boys who came around Cape Horn in Pope and Talbot lumber ships to find jobs in the new mills and logging camps. Many of them, after making a grubstake, drifted on into business ventures of their own.

The water-shipping lumber mills at Port Gamble, Port Blakely, Port Madison, and a dozen other inlets on Puget Sound and the southwestern harbors created the state of Washington. In like fashion, lumber pioneered the means of living in much of the undeveloped West. The fast, sharply competitive drive of sawdust capital pushed the young states forward at an astonishing pace. If the industry had tarried to reforest the lake states and had limited its cut to the sustained yield of their fifty-five million acres of woods, following the practice of Germany or Sweden, lumber production in the West could hardly have risen today beyond the consumption of its local population and possible markets in the Pacific area. The construction of the transcontinental railroads would have been put off; the Panama Canal might not have been dug until many years later, all of which is water over the dam. What has been done has been done—in the unplanned, impetuous, self-seeking American fashion. In its doing we have traded a large part of our virgin forest heritage for national expansion to the Pacific in two generations.

Whatever its causes and justifications, the aftermath of migratory lumbering was inevitable. Its great treks to the South and West were all too plainly blazoned by ghost towns and stretches of idle land. Roving lumbermen did not invent the "ghost town," as might be inferred from some of the more condemnatory conservation literature. Transient communities have marked the trail of pioneer industries and early, unstable uses of land all over the world. Ghostly towns stand as monuments to mining camps and placer diggings throughout our West. The blight of shifting industries, moving closer to sources of raw material, has indeed fallen upon many old, well-settled manufacturing towns in New England and elsewhere. But migrating sawmills have left their full share of jobless communities. Dead mill towns were implicit in the liquidation philosophy of timber mining. Again it has been the lumberman's curse that his works stand out in the open for all to see. Nothing looks more like the realm of the dead than yawning, half-collapsed mill buildings standing precariously on rotten piling, with heaps of sawdust and rusted smokestacks or fragments of machinery strewn about. The ghostly aspect is heightened if the abandoned mill has a backdrop of blackened skeletons of trees. President Franklin Roosevelt struck a popular note when, on viewing such a scene of desolation in the Far West, he fervently wished "that the lumberman who did that is now roasting in hell!"

The tradition was deeply rooted in pioneering Americans that "the plow will follow the ax." It was often cited as a justification for migratory lumbering and runaway forest fires. As late as 1910 a senator from Idaho opposed appropriations for fighting national forest fires on the ground that the conflagrations were a means employed by the Almighty to make the land ready for the settler.

Often, indeed, on the better soils of the lake states and the valley lands of the South and West farming moved in as lumbering moved out. Many towns which started around logging camps and sawmills settled down to peace and permanence as centers of agricultural communities. The larger lumber companies usually had cutover land departments, whose duty was to sell as much of the logged-overs as they could get rid of. Often they made real contributions to the progress of their states in agriculture; and often they degenerated into speculative colonization schemes. To most of the lumbermen, once timber was cut, the land was a depleted asset, in the same class with a worn-out sawmill or outmoded donkey engine. It might be held if it had mineral or town-site possibilities. Otherwise it was dropped. If there was no market for it, the simplest way out was to quit paying taxes. Many millions of acres of cutover forest passed into state or county ownership by the route of tax delinquency. Like so many other things in the exuberant exploitation of our natural wealth, this form of land abandonment pursued its ominous course for many years before the country woke up to find an orphan on its doorstep.

One summer some twenty-five years ago I traversed a sixty-mile railroad in the southern Appalachians along which thirty-five sawmills, large and small, had been dismantled within the previous dozen years. Most of its stations were sawdust piles with clusters of vacant, rotting buildings. The forest industries of this region had passed on. Its towns were "one with Nineveh and Tyre." Here and there one could see a little group of bottomland farms or a patch of hillside pasture, but the great forests of hemlock, spruce, and oak which had brought the railroad and built the towns were gone. The region was without an industry and almost without a people.

I saw another sample in the Au Sable Valley of Michigan. The forest supervisor and I hiked over many miles of burnt-out pineries, looking up remnants of public land that might fit into a national forest. Between repeated fires and drifting sand, even the old stumps had mostly disappeared. It was hard to find any vestige of "tall, whispering pines." Here and there we found little hard black cones of wood forming a symmetrical pattern in the sand. They were the dense cores of knots in what was once a red pine tree or log. The rest of it had burned up completely. Here and there were pathetic evidences of attempted homesteading—a cellar half filled with sand, part of a stick and mud chimney still standing, or scraps of an old stove. The "pine barrens" of Michigan were well named.

No general reckoning of the forest lands where the plow has not followed the ax was made until Gifford Pinchot turned the spotlight upon all our natural resources. One of the most painstaking inventories ever made was conducted by the Forest Service in 1920, in response to a resolution urged by Senator Capper of Kansas. The "Capper Report" estimated that out of every ten acres of primeval forest, something over four acres had been converted into farms and pastures. Four acres more had gone down before the lumberman's ax where the plow did not follow. The other two acres were still in uncut woods. In round numbers, about half of the timber land we had logged, across the length and breadth of the country, had passed into farms and about half was still classed as forest.

Three fourths of the cutover and unplowed forest land bore trees of varying size and degrees of stocking, but the last fourth had been logged and burned into barrenness. This eighty-one million acres had become idle land—without trees or farm crops or people. The aftermath of booming sawmill towns in the pineries of Michigan had been repeated in many other forested regions. The amount of idle land lying about in the United States, even in areas of dense population, was astonishing. There were twelve million acres of it in New England and the middle Atlantic states. Over twenty million acres were strewn through the lake states, much of it not far from Detroit, Minneapolis, or Milwaukee.

It seems probable that the eighty-one million acres in 1920 marked about the extreme high tide of denuded forest land. A similar survey by the Forest Service in 1945 found sixty million acres of unstocked woodlands. The advances in preventing forest fires, along with gains in public and industrial forestry, were beginning to tell. On many areas, too, Mother Nature has had the last word. Given time, many "scenes of devastation" have disappeared under waves of young, green forest. But there still remains a staggering job of reseeding or hand planting. A concerted attack on this land problem through federal, state, and industrial forest nurseries, educational drives and new techniques, like sowing tree seed from helicopters, is one of the most encouraging recent happenings in our forest story.

Idle land and ghost towns are the devil's pay-off on the American traditions of free land and free enterprise. They are the debit entries, the sacrifices to speed and momentum in national growth. They are part of the price of democracy. They also have contributed powerfully to the awakening. One of the surprising things about the crusade for forest conservation was its popular acceptance in the West—the West of

1905, still young, undeveloped, individualistic. These people were familiar with idle forest land and ghost towns. They also had seen silted reservoirs and overgrazed ranges, and they knew what Theodore Roosevelt and Gifford Pinchot were talking about.

II

Migratory lumbering has been the national scapegoat for land idleness; but the pattern of abandonment in every region is crossed by other currents and shifts in our happy-go-lucky use of land. Agriculture as well as lumbering has often been migratory. The cut-out-and-move-on fever of the woodsman has had its counterpart in the restless American farmer. Often indeed his has been a migration of necessity. Often it has been a grim story of "work out and move on" if not "starve out and move on."

One of my first lessons in forestry, as a greenhorn in the north woods of New Hampshire, was to identify on the stumps of freshly felled spruce the minute growth rings that recorded the four historic cold summers of New England. They came just after the War of 1812, years with killing frosts in every month. They started the first notable westward migration of New England farmers. It has been going on irregularly ever since. In preparing the Capper Report, the Forest Service found that New England had lost 32,000 farms and a million acres of tillage in the decade ending with 1920.

The ebb and flow of cultivation is one of the most absorbing stories of rural America. Millions of acres of farmland have been abandoned in older communities of the East because they could not compete with the richer soils or kindlier climates of the South and West. The lure of the city, the growth of manufacturing towns, the urge to seek adventure across the plains have all had their part in the abandonment of eastern farms where living was hard and close. The War between the States had a profound influence upon the shifting tides of American agriculture. Many farms left by men entering Union and Confederate armies were never plowed again. In the aftermath of every war restless men have not returned to the old furrows but have sought something new and venturesome, often a raw homestead in the West. The gold rush of '49 to California, the gold strikes on the Yukon in 1897, any event which offered escape from the humdrum of settled life—has led to the reversion of farmland somewhere to woods or scrub or barrens. Overshadowing all other forces in the migration of American farmers has been the lure of new land—free land under the federal Homestead Law, allotments in Indian reservations opened to settlement, cheap stump land offered by lumber companies.

Intermingled with the idle forest lands of today are twenty million acres of abandoned and impoverished farmlands that should be put to work again growing trees. Many more acres are hanging in doubt between farm crops and wood crops. Slowly and painfully we are working toward a more rational use of all our soil. I have just seen some of the counties in northern Wisconsin which had the courage to draw zone lines beyond which settlement is not permitted. You can still see the log cabins of the pitiful homesteads in some of these old rural slums, but they are disappearing in young forests. The land is getting back into its best employment. The old adage that it takes only three generations to pass from shirt sleeves back to shirt sleeves seems to hold true on a good bit of our land, in the cycle from the first stumps back to trees again.

CHAPTER IV

Ground Swells of Conservation

FEW Americans are aware that our first forest reserve was set up to foster the young U. S. Navy and particularly to keep its sturdy frigates and ships of the line on the seven seas.

Directly after the Revolutionary War the growing shipbuilding industry of New England was alarmed over the scarcity of local hull and mast timber. The General Court of Massachusetts prescribed a penalty of thirty pounds for cutting any white pine tree two feet or more in diameter, on lands of the state, without prior license. Soon after the War of 1812 Massachusetts authorized rewards for growing oaks and other trees necessary in shipbuilding. Doubtless the anxiety of Massachusetts for her shipping industry was the background of the concern of President John Quincy Adams over adequate supplies of live oak timber for the United States Navy. In 1828 he ordered a survey of the oak forests on the southeast coast, and later established a live oak reservation of thirty thousand acres on Santa Rosa Island in Pensacola Bay.

President Adams had definite and practical ideas on the administration of this forerunner of the national forests. His plans included the thinning of dense young stands of live oak and experimental plantings of acorns. Unfortunately politics stepped in. The project was attacked and the reservation abandoned soon after Andrew Jackson took over the White House. One can readily imagine the language of Old Hickory's followers in denouncing the "New England aristocrat" for locking up lands that belonged to the people. Doubtless it had the same vigorous ring that was often heard in the

West seventy-five years later when federal forest reserves were set aside by the millions of acres. However, the idea of timber reserves for the Navy persisted. For the next forty years live oak reservations were made, fought over, and, from time to time, abandoned. They stirred up frequent political turmoils and scandals over the appointment of "timber agents." The sound policy conceived by President Adams was frittered away in futile embroilments. In 1868 Secretary of the Navy Gideon Welles reported 264,000 acres of oak reservations on the Gulf coast. It was not until 1923 that the last of them was officially erased from the land office records.

However inexhaustible the great hinterland of forest might appear to the colonists in America, it was not long before many of them realized that timber near their settlements was becoming scarce. Too much timber stealing and woods burning were going on. Before the end of the seventeenth century, the records of Massachusetts towns carried complaints of the dearth of building timber. Plymouth Colony adopted several ordinances to prevent or control the firing of woodlands. In 1681 William Penn's well-known decree ordered that, in land clearing, at least one acre out of every five should be left in trees. Two and a quarter centuries later the American Forestry Association tried to write the same idea into the national Homestead Law.

By the time of Bunker Hill a substantial body of local ordinances and acts of colonial legislatures sought to restrain forest fires and timber trespass. At least two colonial statutes, in Massachusetts and North Carolina, not only stressed the protection of useful timber but referred to the damage caused by fire to soil and young trees. Apparently our national carelessness with "smoke in the woods" had early beginnings.

The little forest shortages and fears for the future, even in these early years, had their part in the slow schooling of experience. But the main currents of the day passed them by. Men were too engrossed in the opportunities offered by new land. There was too much far-reaching forest in every direction to worry much over local scarcities. Two more generations spread westward over the free land before there was a serious awakening to the onslaught of fire and ax upon America's forests. But not long after the War between the States a ground swell of education and public opinion set in. More and more segments of American life took part in it. Democracy went to work. It acted through many local ordinances and state laws. It created national leadership through education and association. It laid the cornerstone of a new forest order by setting up a system of federal forest reserves. And it made the people ready to follow brilliant leaders who came to the front at the turn of the century.

State commissions were set up to study the extent of forest destruction and recommend remedies, in Michigan and Wisconsin in 1867, in Maine 1869, New York 1872, Vermont and New Hampshire 1881, Pennsylvania 1887. Our forests made their first appearance in the United States census of 1870. In 1873 the American Association for the Advancement of Science petitioned the Congress and the state legislatures to enact laws for the protection of forests. In 1876 Congress authorized the employment of a forestry agent by the Department of Agriculture, and scholarly Franklin Hough became Uncle Sam's first forester, with an appropriation of two thousand dollars. By 1891 the department had a small Bureau of Forestry, with purely educational duties.

Effective education was started at the grass roots when Governor Sterling Morton founded Arbor Day in Nebraska in 1872. Many other states followed his example, and Arbor Day became an American institution. The ceremonies, speech-making, and publicity centering around thousands of tree plantings every spring, from the yard of the little red schoolhouse to stately capitol grounds, had a far-reaching influence upon public thinking about trees and forests.

By 1880 nineteen states had enacted special tax exemptions or other forms of legislation to encourage tree planting. Congress fell in step with the popular demand from the prairies by adopting the Timber Culture Act. It offered 160 acres of public land to any citizen who would plant trees on twenty acres of it. The Middle West was learning the value of farm wood lots and shelter belts. In the meantime, forestry associations sprang into life in several states, beginning with Minnesota in 1876.

On the call of Dr. John A. Warder, physician and horticulturist of Cincinnati, forest-minded men and women from many sections of the country met in Chicago in 1875. The American Forestry Association was formed. Other groups of like minds merged under this national green standard. In a few years the association became a potent agency, not only in education but in politics. It engaged Bernard E. Fernow as general secretary. Dr. Fernow was an able, strong-willed German forester, who organized his campaigns with the thoroughness and determination of a Bismarck. The ground swell of conservation began to show practical results.

The immediate returns were a series of state laws, setting up forestry boards or commissions, and taking first steps toward a forest policy. New York, Ohio, Colorado, and California enacted such laws in 1885. They are of interest mainly in reflecting the popular demand that something be done. The Adirondack Forest Preserve of New York, however, was created by her first forestry legislation and set before all the states an example of direct and forceful action in meeting a local situation.

Often spurred by local disaster, state legislatures strove to prevent and control forest fires. Pennsylvania led off with a forest fire law in 1870; New York, Colorado, and Alabama followed in 1885; Connecticut and Massachusetts in 1886; Idaho in 1887; Maine, New Hampshire, and Washington in 1891. The Hinckley and other great fires in the lake states prompted the first forest protection laws of Minnesota and Wisconsin in 1895.

On the stage of national politics, the forces of the American Forestry Association centered on a campaign to set aside forest reserves from the public domain. They had a stout ally in Carl Schurz, refugee from the German revolution of 1848 and staunch friend of Abraham Lincoln. Schurz was forest-wise from his upbringing in Bavaria. From firsthand experience in the Department of the Interior, he knew just what was going on in the great pineries of the North and South. In official reports he challenged the tradition of "inexhaustible timber" and repeatedly warned of forest depletion from cutting and burning on the public lands.

Other immigrants brought to reckless America old-world respect for land and the instinct for conserving its resources. Among them was a young German named Frederick Weyerhaeuser. From piler and roustabout in a St. Paul lumberyard, he became one of the great forest owners and industry organizers of the Northwest. Mr. Weyerhaeuser was seldom seen at conservation meetings. But his influence in organizing fire control was one of the quiet forces helping to shape the new order. He was one of the first to put patrol men and post "No Fire" signs in his own woods. In 1903 Mr. Weyerhaeuser requested the federal Bureau of Forestry to examine all of his company's lands in the Pacific Northwest and advise him on their future management.

Other allies of the American Forestry Association appeared in the young irrigation industry of the West. The campaign won its objective in 1891, when Congress, in the fewest possible words, authorized the President, by executive order, to withdraw public lands as forest reserves. It is perhaps typical of the ways of democracy, at least of American politics, that this far-reaching bit of law, which threw into reverse one hundred years of free land legislation, was passed without benefit of oratory or even a clear-cut vote. It was slipped through as a rider on a current appropriation bill.

President Harrison signed the first forest reserve order in 1892, creating the Yellowstone National Park Timberland Reserve. During his administration thirteen million acres were set out from the public domain in this new classification of federal lands. The forestry advocates, however, were unable to induce Congress to adopt administrative policies and procedures for running the forest reserves. They remained *reserves* in fact, and the whole program was imperiled by the mounting opposition in the West against "locking up" its natural resources. Failing in Congress, the indefatigable Forestry Association prevailed upon the Secretary of the Interior to invite the National Academy of Sciences to study the whole subject of forest reserves. And thus, in 1896, a committee was named to offer recommendations on the extension of the forest reserves and their administration. Its chairman was the eminent dendrologist of Harvard, Dr. Charles Sprague Sargent.

Of greater potency was its secretary, a young forester recently home from studies in France. His name was Gifford Pinchot.

At this point the young "Lion of Judah" enters the story of America's forests. He brought into it a fervor of religious intensity and a magnetic personal leadership that have rarely been equaled in the American drama. For the next fourteen years the astonishing vigor in the planning and execution of successive moves for national conservation largely expressed the zeal and energy of Gifford Pinchot.

The Committee of the National Academy made a thorough job of it. Its members examined many parts of the public domain. It recommended the addition of twenty-one million acres to the forest reserves, which was done promptly by executive orders of President Cleveland. The academy also proposed a common-sense, practical plan of administration, designed to make the timber, forage, and other resources of the reserves available for use with reasonable safeguards. This plan was adopted by Congress in 1897, as an Act for the Administration of the Forest Reserves. It was a well-drawn law which has remained the basis of operating the national forests to the present time. One of its significant phrases was the declaration that the reserves were created "for the purpose of securing favorable conditions of water flows and to furnish a continuous supply of timber for the use and necessities of citizens of the United States."

The first clear, realistic step in a public policy of conserving, rather than distributing, forest resources had been taken. Perhaps of equal significance was the fact that many thousands of Americans, from the intelligentsia of our learned societies to dirt farmers on irrigated land, had been sold on the proposition that forestry is a proper subject for federal action.

Two of the earliest reserves endeared themselves particularly to the enthusiastic young foresters in Gifford Pinchot's bureau. They gave us an immediate chance to do real business in selective logging. One was the Black Hills, the green island of pine rising from a sea of Dakota prairie, the storied land of Wild Bill Hickok, Calamity Jane, and Crazy Horse. Here the needs of Homestake and other mining companies required large sales of ponderosa pine, carefully marked and remarked to cut only mature trees and leave all the promising striplings. Here too we had many a battle with "The bug Dendroctonus, who lives in the heart of the pine."

Another proving ground for American forestry was the great stretch of lodgepole pine on Wyoming's Medicine Bow. The Union Pacific Railroad drew heavily upon "The Bow" for crossties, hewed by hand in those days from the best and straightest lodgepoles. This was another fascinating job in selective marking. It was just as fascinating to watch a Finn "tie-hack" snap a chalk string on a prostrate trunk and hew to his line with huge broad ax, down the length of the tree, five or six inches to a stroke.

II

Other forerunners of conservation were quietly appearing here and there in the woods themselves. While the larger sawmills were mostly on the march, seeking new worlds to conquer, some forest industries were getting rooted in the land. The earliest and most general was the village woodworking plant. It was usually a settled institution, continuing a modest volume of production under generation after generation of family craftsmanship.

The retention of wood-using industries in the northeastern states has been helped by the low forest fire losses of the region. Its fire record is a good one, because of climate and also because of the conservative temper of the people. Furthermore, their first great timber tree, white pine, has fairly held its place in second growth and the reseeding of abandoned fields. Through generation after generation it has supported many small sawmills, box factories, and woodworking shops. White pine was widely planted in early New England forestry.

Thousands of stable little industries in the eastern half of the country have been supplied with timber from their own land or their own neighborhood, often by a third or fourth cutting in the same woods. They and the woodcutters they maintain have practiced a deal of homespun, common-sense forestry. It has usually been ignorant of the canons of silviculture and has often lowered the quality and growth of the stands of trees, but it has at least kept the land in a fair degree of production and provided woods and mill employment through many years.

At the opposite end of the industrial scale are wood-converting plants that cannot pull up and move on because of large investments in site, power, water sources, buildings, and specialized equipment. Wood pulp factories, whose investment in a single plant runs well into the millions, are outstanding examples of this type of forest industry. The requirements of an economic unit often range upward from two hundred cords of pulpwood daily. Wood supplies must be assured for a long operating life. A forest becomes essential to the capital structure of the enterprise, a forest at hand and under company control. Industrial forestry on large holdings made its first appearance in America, at the end of the last century, on the woodlands of pulp and paper companies. They gave American foresters the first chance to test their principles of continuous growth and long-range production against the cold profit standards of Wall Street.

Back in 1902 I had an exceptional chance to see long-range forest planning in the making and also to form a lifelong friendship. Two forestry students were trudging up a snowy tote road through spruce and maple woods. We were in the Androscoggin country of northern New Hampshire, headed for a logging camp on the Little Diamond. Behind us we could hear the peculiar whine of dry snow crunching under heavy sled runners on a cold day. The four-horse tote teams were enveloped in clouds of frosty vapor. Down the road ahead came the jingling bells of a one-horse pung. From behind its shaggy buffalo robe appeared a head and beard of shaggy red hair. A hearty voice with the twang of down-East Maine hailed us. "Where do you boys think you're goin'?" Our friend was Austin Cary, forester-extraordinary of the north woods.

My classmate, Ralph Hawley, and I spent two rugged and exciting weeks in the spruce forests, trying to keep up with Cary on bear-paw snowshoes. In camp we tried to hold our own with the French-Canadian lumberjacks in assaulting giant pots of baked beans and dishpan pyramids of doughnuts. Cary was cruising and mapping woodlands for the Brown Paper Company, studying growth rates and experimenting with selective cutting. He soon had us enthusiastic over his job. He was shaping up a pattern of forest ownership to balance the production planned for the pulp mills.

Austin Cary was about the first American forester who did not get inspiration or credo from the classical schools of Europe. He was self-trained in his native New England woods, a nonconformist on "timber famines" and much of the current conservation propaganda, and a lone wolf always. He lectured occasionally at Yale and could talk book forestry with the best of them, but he endeared himself to the students by his rugged individualism and his down-to-earth Yankee mind. Outdoors the "man from Maine" was a bull moose, always at his best when charging through the woods with a bunch of boys, ax in hand.

Cary started several New England pulp companies on forest management by working out land policies adjusted to long-range plant requirements. In later years he was a roving missionary of the Forest Service in the pine forests of the Southeast. That country is still full of stories of the Maine Yankee who stomped through the woods and left behind him a trail of forgotten mail, clothing, baggage, and what-not. Nobody could keep up with him, but he had a marvelous way of getting hold of men under their own trees and selling them on forestry. Cary did yeoman service in helping the turpentine orchardists change their woodland practices to a system of permanent tree growing and cropping for resin. His many friends among them, and among all ranks of American timbermen and foresters, erected a monument to his memory, not in Machias, Maine, where his work began but in Waycross, Georgia, where it ended.

Not many large industries were thinking seriously about permanent wood supply for their mills in the days when I saw its beginnings in the north country. But the idea was working. Gifford Pinchot himself had put the Vanderbilt estate in North Carolina under a sustained yield cutting plan in 1898. Mr. Pinchot and Henry S. Graves made working plans for the spruce properties of paper companies in the Adirondacks and started two or three of them on a policy of cutting no more than the growth of their forests. Probably twenty-five substantial industrial owners throughout the country were getting advice from the federal Bureau of Forestry. Out on the Columbia River, the Crown Zellerbach Corporation was acquiring more and more forest to assure continuous supplies of wood for its expanding production of pulp and paper. This company began test plantings of cottonwoods, spruce, and other pulp-making species in 1901.

All industrial forestry, at the start of the new century, made but a small spot of green on a national map of migrating mills and blackened cutovers. The first ground swell of conservation was a wave of public sentiment demanding public action. But little currents were starting to form, deep within industry itself, that in another fifty years would roll up a second great wave of their own.

CHAPTER V

The Great Crusade

ANCIENT astrologers foretold world-shaking events from the proximity of heavenly bodies. But who could predict the consequences of the simultaneous presence in Washington of two very dynamic and forest-minded men—Theodore Roosevelt in the White House and Gifford Pinchot in the Bureau of Forestry?

Both were men of great idealism and men of action. They lived the "strenuous life" in body and mind. Both saw, in laws and situations, not the limitations but the opportunities. Theodore Roosevelt said on many occasions that laws are made to get results, and to be interpreted and enforced for the public good. When confronted with alternative courses, Roosevelt and Pinchot invariably chose the most aggressive and daring. Each of these men had a rare gift of personal leadership and a dramatic instinct for using situations and setting the stage. They made a great team as crusaders of conservation, and they put over one of the most effective selling jobs in our history.

My first glimpse of "T.R." was at the meeting of the American Forestry Association in Washington, in 1905. This meeting sounded the bugles for legislation which transformed the reserves into national forests and transferred them from the Department of the Interior to the Department of Agriculture. As a young recruit in the Bureau of Forestry, I was thrilled when the President threw down his manuscript and strode across the stage. With shaking fists and flashing teeth he thundered, "I am against the man who skins the land."

"G.P." had been Chief of the Bureau of Forestry since

1898. His scouts explored every gulch in the public domain, and as their reports streamed into Washington, executive proclamations of new forest reserves rolled out of the White House. By the end of T.R.'s administration, President Harrison's thirteen million acres had become 120 million. There is no better illustration of the teamwork of the two crusaders than the way they met the "crisis" of 1907.

Messengers raced through the Forest Service's Atlantic Building summoning G.P.'s staff. His orders were clear and crisp. "Just back from the White House. They've loaded a rider on the department's appropriation bill. It would stop all additions to national forests in the six most important states without approval of Congress. The President can't veto, but he will hold the bill to the last minute. Now we get busy."

Each of us was assigned a state or part of a state. We studied all the reports on its public lands and pored over every old map. Wherever we found reasonable evidence of forest cover, we redrew national forest boundaries. Midnight oil was burned in most of the Forest Service offices that fortnight. Wires flashed back and forth to supervisors and rangers in the West. Stenographers dropped everything and typed proclamations. And G.P. soon began daily trips to the White House.

Several of the gang had experiences like the one over which Billy Cox, later State Forester of Minnesota, chuckles to this day. He was summoned to go with G.P. to the White House, his arms filled with maps. Soon the maps were spread over the floor of the President's study. The thickset President and his lanky chief forester were both on their knees, working over the state of Montana, watershed by watershed. T.R. fired many questions, showed an astonishing knowledge of western topography, and drifted into stories of early experiences on the range or in hunting big game. "Have you put in the north fork of the Flathead? Bully! Up there one winter I saw the biggest yard of black-tailed deer! . . ." Finally he dusted off his knees, swept papers from his desk, dipped his stub pen in the inkwell, and grinned broadly. "Bring 'em on." Thus Theodore Roosevelt added eighteen million acres to the national forests and on the tenth day signed the bill which ended the President's power to create national forests in six western states.

G.P. had the happy faculty of inspiring his young Forest Service with the same spirit of team play in everything we undertook. After the Transfer Act of 1905, we all pitched in and rewrote the regulations for the administration of the national forests. We started with Secretary James Wilson's first commandment, "to serve the greatest good of the greatest number in the long run." In contrast to the former "reserves" the national forests were *for use*. And we surely put them to use—timber, forage, water, agricultural lands, minerals, recreation facilities, and wild life. We coined the phrase *"multiple use"* to express our zeal for the utmost public service from a section of land. We had the thrill of building Utopia and were a bit starry-eyed over it.

One of G.P.'s finest qualities was his capacity to understand and work with men whose background was totally different from his own. To the young knights of his round table he added a wholesome mixture of practical, seasoned men, like Albert Potter, experienced in western livestock, and Eugene Bruce, a tough logger from the Adirondack camps.

Around the solid, realistic job of protecting and administering a hundred million acres of federal forests and ranges, Gifford Pinchot built an organization of three thousand people and inspired it with genuine zeal for public service. He made crusaders of us all. He threw together college-trained foresters, cattlemen off the ranges, lumberjacks fresh from the woods and fused them into a fighting team. He made the merit system of Civil Service a vital, living thing. The absence of personal or political influence in the growing organization gave the bond of public service between its members reality and a cutting edge.

The service had a strong family atmosphere. The Washington cabinet met every week; the whole crowd, down to stenographers and messengers, every month. Everybody knew what was going on and the latest drives that G.P. was planning. Every fortnight all the service men in town were invited to the Chief's spacious home on Rhode Island Avenue. There we met and heard President Roosevelt, Seth Bulloch from South Dakota, members of Congress, visiting foresters from overseas. There we told of our own doings—a supervisor in town from Arizona or a green lad with new ideas on germinating tree seeds. And there the gingerbread and baked apples made the evening a lifesaver to the junior assistants who were putting in their first stretch with Uncle Sam at $25 per month.

II

Gifford Pinchot was the last man in the United States to settle down to the administration of the national forests, important as that job was. His eager mind never quit searching for new worlds to conquer. When he became Chief of the Bureau of Forestry, the campaign to sell forestry to the American people rolled on with redoubled power and an astonishing range of activities. Help to farmers in wood-lot forestry was launched in many places. Surveys and management plans were offered to industrial forest owners. Timber cutting recommendations were made to southern Appalachian coal companies. An enormous project of timber cruising and growth estimates was carried out on lands of the Kirby Lumber Company in Texas. Every opportunity was seized to make some start in commercial forestry.

In research and fact gathering, Mr. Pinchot left no stone unturned. The important forest trees and the forest needs of many regions were subjects of study. Sample working plans for forest properties, providing for a sustained timber yield, were published, with demonstrations of good cutting methods and the advantages of selective logging. The first estimate of standing timber in the country was made by the Forest Service in 1907. It showed a stand of 2,500 billion feet, a yearly lumber cut of forty billion, a yearly growth of only one fourth as much.

The enthusiasm and showmanship of the Roosevelt-Pinchot team kept the political crusade rolling. New ideas and points of attack were brought into play. A general conference of state governors, called in 1908, headlined the conservation of all natural resources and pointed up the need for aggressive national leadership. It was on this occasion that a southern governor declared "the doctrine of states' rights is dead beneath the feet of a million marching men." Among many reactions which strengthened the hands of the crusading team was a widespread demand for the extension of national forests beyond the public domain, by the purchase of key areas in the eastern states. A potent and indefatigable group of New Englanders, banded together as the Society for Protection of New Hampshire Forests, insistently urged a national forest in the White Mountains. A federal forest of some kind had been advocated for the Pisgah area of North Carolina since 1885; and other sections of the southern Appalachian Mountains joined in the demand for protection of their scenic watersheds and control of floods. The issue became national under leadership of the American Forestry Association.

Congressional lawyers debated learnedly over the constitutional grounds on which the federal government might embark upon the business of buying forest lands not needed for national defense or public improvements. It was agreed there was only one: "to protect the navigability of navigable streams." Since many navigable rivers ignore state lines, obviously it was a federal duty to protect the downstream states from the recklessness or indifference of their upstream neighbors. And since the courts had construed "navigation" as embracing the movement of a skiff, a bateau, or a raft of logs, the limitation not only satisfied constitutional lawyers but was almost perfect for practical purposes.

On this principle the Weeks Law of 1911 was added to the conservation statutes. The Pisgah and White Mountains National Forests were created. By 1941, purchases were under way in eighty-one new national forest units. Tucked away in the Weeks Law was another bit of legislation which has proved even more important than the forest purchases in steering the course of federal policy. It authorized the Secretary of Agriculture to co-operate with the states in protecting the forested watersheds of navigable rivers from fire. This was the first step in a great co-operative expansion of national forestry.

The Weeks Law added significantly to the scope and prestige of the Forest Service and its ability to influence local developments. It took the service into a dozen new states, as active land buyer and administrator. The negotiation of state-enabling acts, which the law required, sharpened public discussion and legislative action on their own forest programs. Every new purchase unit became a demonstration of forest protection and cutting practice. The policy exemplified national leadership in conservation, which was a cardinal feature of the Roosevelt-Pinchot crusade.

Other moves of Pinchot strategy had a grimmer aspect. One of the high priesthood of conservation in Washington was a fellow member of T.R.'s "Tennis Cabinet," James R. Garfield, son of a President and Commissioner of the Bureau of Corporations. It was not by accident that Congress directed the Bureau of Corporations to investigate timber supply and concentrations of timber ownership, at the very time when the rebellious West was fighting extensions of the national forests. The bureau substantially confirmed the Forest Service estimates of standing timber and rate of depletion and added explicit facts on the heavy concentration of western and southern stumpage in the hands of relatively few lumber and railroad companies. Its findings were soon known as the "Timber Baron Report."

Early in his crusade Mr. Pinchot coined the phrase "timber famine." He used it often in speaking and writing of the catastrophe toward which we were rushing headlong. Several times he predicted the virtual end of our timber supply in thirty or thirty-five years. These forbidding forecasts were made at a time when the lumber cut of the United States was running at an all-time high level of forty-four or forty-five billion board feet a year. The end of the great virgin pineries in both the lake and southern states was in sight. Some of the most astute lumbermen of the country made their own calculations and came to about the same answer. A dozen years later there was a good deal of "timber famine" in the Capper Report, in which I took a part. None of us, foresters or lumbermen, had as yet any conception of the reproductive power of the logged-over forests, especially in the South, or how the growth rate was increasing as young trees replaced old timber. The "coming timber famine" was a telling argument in support of any forestry measure, and G.P. used it again and again in his forceful, dramatic way. With it he always coupled the dependence of the country upon an abundant supply of cheap and available wood. The lumbermen found themselves in the pillory. They had a monopoly of a vital, fast-diminishing natural resource. They were wasting it and leaving idle land and ghost towns behind them. The stage was being set for more drastic things to come.

To an astonishing degree the crusade for conservation, led by Roosevelt and Pinchot and joined by many other able men, changed American public opinion and American politics. It all but ended the tradition of free land, free timber, and inexhaustible forests. The tradition lingered on, of course, and many political skirmishes and rear-guard actions were still to be fought under its banner. But a basic change in national thinking was brought about, and we have never turned back.

Not only did the crusade give the United States new laws and forceful agencies, with their direct means for influencing public opinion; it also created a powerful conservation cult and a conservation press. A new school of writers from intelligentsia to muckrakers explored and worked a great new field of public interest. Every kind and shade of conservation sentiment got aboard the bandwagon. T.R. assured the "baked apple club" one night that there was "a lunatic fringe" on every great movement. But his crusade drew solid and continuing support from the irrigation and flood control interests; from the awakened national concern over soil conservation; and from the many groups throughout the country devoted to wild life and the preservation of "forests primeval." "Conservation" meant many different things to different people, but the Pinchot campaign gave them all a common rallying point. And something new was doing in the big tent every night. All of scientific Washington flocked to the Cosmos Club to hear the debates between Gifford Pinchot and George Willis Moore, Chief of the Weather Bureau, on the relation between forests and climate.

The support of the conservation movement by educational institutions soon took a practical turn of great importance to the undertaking in the long run. This was the training of men for technical forestry, for a profession which, in the words of Dean Henry S. Graves at Yale, "did not exist in the United

States." The first schools of forestry were established in 1898, at Cornell University and on the Vanderbilt estate in North Carolina, by its German forester, Dr. C. A. Schenck. The Yale School was endowed in 1900 by the Pinchot family. In 1905 eleven colleges and universities were teaching forestry; in 1910, seventeen, and in 1915, twenty-four, covering every forest region in the United States. There is no better evidence of the virility of the forestry movement or of its acceptance by the country than this extraordinary growth of educational facilities within a couple of decades. Young men began to train themselves by the hundreds for the new profession, and another potent source of education and leadership was brought into play.

The solid grip of the crusade of Roosevelt and Pinchot upon the country was soon tested. Under President Taft, T.R.'s vigorous administration of public lands perceptibly cooled off in the White House and Department of the Interior. Gifford Pinchot's love of battle and high sense of chivalry involved him in controversy with Secretary of the Interior Richard Ballinger and took him out of official life. The Forest Service carried on under the capable but less dramatic leadership of Henry S. Graves, and forestry as a national movement lost none of its momentum. The prevailing public opinion was solidly behind it. Congress remained conservation-minded. Contrary pressures might slow its pace or bend its course, but our general line of march was set.

Lumbermen and other industrial forest owners were put on the defensive. The left wing of the conservation school made them the scapegoat, not only for their own sins but for the crimes committed by the nation in the name of free land and westward expansion. "Devastation," "ghost towns," "cut out and quit" became familiar terms in the magazines. One of our real trials, in early service days, was the demand of free lance writers for stories about devastation. Our patience was sorely tried by pen wielders whose ignorance was appalling and who often had no interest in forestry beyond whipping up a lurid tale with some greedy lumberman as arch criminal.

I tried several times to satisfy a determined woman who insisted upon something "graphic," something that would catch the "public imagination." Finally an imp of Celtic ancestry prompted me to find out how far her gullibility and thirst for drama would take her. "Probably the worst example of forest devastation in the world," I told her, "was the clean cutting of North Dakota by the most notorious of our early loggers, one Paul Bunyan." Her pencil was scribbling busily in the fat notebook as I added details. There was some kind of deal between Bunyan and the King of Sweden, I told her, although the State Department has never permitted that phase to be fully aired. It seems that the King wanted to get some thousands of stubborn Swedes out of his hair. At any rate, Bunyan moved in his huge camps from the Red River of the North. The job was in charge of his logging boss, Ole Olson, the hardest man driver and most ruthless timber butcher the world has known. "Why, he even made his axmen chase granite boulders rolling down hillsides and sharpen their axes as they ran." The desolation of North Dakota went on apace.

The literary lady snapped shut her notebook and thanked me with shining eyes. I never saw her story in print. But old Paul had done me a good turn. She never darkened my door again.

Many writers contrasted the forest controls of Europe with the reckless exploitation of American woodlands. New legal and philosophical ideas about the status of forests gained circulation. "Are not all forests invested with a special public interest which transcends the rights of private property?" Some writers held that forest lands have the nature of public utilities and that no use of them which does violence to the public welfare should be permitted. The Supreme Court of Maine, under its authority to advise on general questions of law, held that the state has constitutional power to control the use of its forests in order to perpetuate them for the benefit of its citizens.

When Theodore Roosevelt declared war in 1905 upon "the man who skins the land," he started one of the main offensives of the crusade. By the time Gifford Pinchot followed his chief into strenuous retirement, the lines of battle were clear. The property rights asserted by forest owners to cut their woodlands as they pleased were under attack. Many groups of people were demanding that the public have something to say about it.

III

> Now the Four-way Lodge is opened,
> Now the Hunting Winds are loose— . . .
> Now the Red Gods make their medicine again!

To many gray-haired range men, logging engineers, and foresters scattered about the United States, the Golden Age came to pass in 1905. We were privileged to become G.P.'s rangers, inspectors, and supervisors in the thrilling job of setting up national forests on a hundred million acres of the raw West. It was a job packed with strange experiences and high adventure. Not since the days of Dan'l Boone had more alluring trails been opened to the feet of young men than when Pinchot's task force took over the western forests and ranges.

The chief expected us to be supermen. He wanted everything done at once. Protection of the national forests from fire had to be airtight. Trails and fire lookouts and ranger stations must be located. The sheep and cattle grazing on these mountain pastures must be brought under federal permits and their numbers reduced to the estimated carrying capacity of the range. Pending land claims were to be investigated. Mature timber must be cruised and its sale and use under controlled cutting begun wherever possible. We must show the people of the West that the national forests were for use; and we must show Congress their possibilities as producers of federal income.

It was a rare opportunity to bring down to earth the beautiful ideal of national forests run "for the greatest good of the greatest number." We had to hammer out a realistic administration of these federal properties, geared to the rough-and-ready ways of life of many different western communities. The greatest joy of it was that we were on our own. Detailed regulations had not yet appeared, nor had the minutiae of legal and fiscal procedures. The men on the ground were running the show, and they stood or fell on over-all results.

Once an inspector from Washington searched the office of an ex-cowhand forest supervisor for a record of his year's authorization and expenditures, and searched in vain. On the trail next day the supervisor fished something out of his pocket and handed it back over the cantle of his saddle with the remark, "I believe this is what you were looking for." His accounts for the year were all there, in pencil, on the back of a blank grazing permit; and they were right to the last penny.

Of course we made mistakes. The wonder is that in the multitude of things we tackled and our zest for getting them done, the Forest Service did not have a multitude of blunders to live down. One spring a supervisor in the Southwest asked approval of Washington to make new range available for lambing some thousands of ewes. The spring was late and cold and the regular lambing grounds unfit for use. The action of one of G.P.'s youthful aides on this request was shrouded in mystery; but suspicion stoutly maintains that the *first* telegram instructed the supervisor not to shift the range but to defer lambing for two weeks.

In 1905 I was dispatched to my home state, California, as timber inspector. Mr. Pinchot instructed me to check all the going national forest sales of timber for fair appraisals of value, methods of cutting and disposal of slashings. I was also to do a selling job of my own among the lumber companies who were fighting shy of business with the government under the new rules. They must be shown that it was practicable to cut and log only the larger and older trees, and to pile and burn the slashings, without going broke.

In the first summer of that assignment I rode my big sorrel mare from Ventura through the Mount Pinos country where the borax mines were requiring large quantities of timber; thence across the San Joaquin Valley and up the length of the Sierra from Walker Pass to Mount Lassen. Along the route I went over the current timber sales with the local rangers and their supervisors. We pored over the service manual, marked sample areas for cutting, threw up sample piles of slash, measured the height of stumps, scaled and rescaled logs of doubtful merchantability, worked out problems in timber appraisal like schoolboys at their arithmetic.

One of the supervisors had made a "trial" sale of forty acres to a lumberman who was prominent in the pine industry of California. His protests over selective marking, "pulling up your steel and leaving good timber behind," had reached an explosive stage. I finally induced our irate buyer to go out into the woods and mark timber with us for a day. It was a stand of beautiful ponderosa pine, well mixed in ages. Our plan was to cut two thirds of the footage, by marking the larger and older trees, but to leave a thrifty forest of younger trees that would give us another cutting in twenty or thirty years.

At first I thought my little stratagem had failed. Twice during the morning a new Stetson sombrero was dashed to the ground with profane outbursts. "No timberman in his senses would try to log on such a long-haired, pink-tea proposition." But I kept a running tally of the merchantable trees we had "cut" and "left." We went over it that evening. I was able to convince the practical mind of our purchaser that, after all, he was getting the cream of the forest in volume and especially in lumber quality. The quality argument went home. He exclaimed suddenly: "By George! The way you cut will give me a bigger percentage of shop and selects. My boy, you've got something there." Before the evening was out we were going over a map and listing the government forties and eighties that would fit into the company's logging operation during the next year and the dates when they should be advertised for sale.

Whenever possible, after that experience, I took the rebellious lumbermen out on the job. We got much farther in talking about *this tree* and *that tree* where they stood than in discussing diameter rules and selective logging in an office. The trees themselves seemed to lend understanding. And the patches of young "bull" pine gave me a chance to point up what the service contracts were trying to accomplish in controlling the prevalent ground skidding with donkey engines and steel cables. We must save the young forest at all costs.

The foresters learned a deal on their side from these arguments out in the woods. I became convinced that we were trying to accomplish too many changes all at once. It seemed most important to get the lumbermen to go along with a reasonable degree of good forest practice and thereby put the vast federal timberlands into use and production. Otherwise they would still be "reserves" in fact and what we should be harvesting as a crop would rot on the stumps. For one thing, the service contracts were crowding the operators too hard in the use of inferior timber species. Mingled with the prized sugar and ponderosa pines on the western slope of the Sierra were many old, snaggy-topped white firs and rot-pitted incense cedars. We were requiring our purchasers to take them all, at a price. I found many logs of these species paid for and lying in the woods because they would not return the cost of the haul to the sawmill. It was not hard to sell the higher-ups in Washington that it was poor business to fight the industry over details of perfection when we had much bigger fish to fry.

After a few rounds in the timber in a spirit of give and take, most of the lumbermen grudgingly admitted that our plan of cutting was probably right on government lands, where everything was done "for posterity." But whenever I had the nerve to ask why not do the same in their own timber, I was all but thrown out of camp. What a thunder of expletives about taxes and carrying charges, mill investments to be liquidated, stockholders demanding dividends! It might be all right for the government to grow trees, but it was no game for a businessman. My preaching was only fifteen years too soon. By 1920 or soon after, some of the same men I argued with were ordering their logging foremen to leave the bull pines under eighteen or twenty inches uncut and to anchor their ground lines with corner pulleys instead of letting them swing free, so as to knock over the fewest possible young trees.

I formed a liking and respect for these plain-speaking, direct-acting lumbermen with whom disagreements could usually be settled once you met them face to face. As I came to know something of their problems and harassments, I realized there was another side to the forest "devastation" of which we heard so much in conservation circles. It was not all beer and skittles. The mills seemed to be shouldering far too heavy a load of timber investment and carrying costs. Both the forest and the industry appeared to be caught in the gears of highly competitive, ruthless economics.

Many of these knotty questions I could not answer. But it was clear that, with its far-flung acres throughout the western timberlands, the service had a marvelous opportunity for practical education and leadership in forestry. It could at least show the way in some of the first things, like co-operative fire patrols and lookout towers, and make every timber sale an example of the ABC's of good cutting practice. We got on pretty well with the men of the industry in those early days. We fought and argued, won a point here, gave a point there, and usually kept our mutual good will and sense of humor. Once a tight-lipped sawmill manager came out on one of our marking parties. We were cutting a mixed stand of sugar pine and white fir and wanted to keep enough of the pine to hold the ground for this species in the next crop of seedlings. After watching us blaze and stamp the trees to be cut for a couple of hours, our silent observer asked suddenly for one of the marking axes. He examined it with a critical frown. "I would like to know," he said, "whether it's the way these axes are ground or the temper of the steel that makes it impossible for them to cut through the bark of a sugar pine."

In 1906 I was made supervisor of the Sierra South, a national forest of some two and a half million acres between Kings River and Tehachapi Pass. It covers the watershed of the Kern River and a vast stretch of westerly Sierra slopes from the crags and mountain meadows down through heavy timber into the oak and digger pine foothills. On the east side of the Kern the country piles up into the towering rampart of the Whitney Range, fronting steeply the valley of Owens River. That alpine region offers one of the rarest and most beautiful sights out of doors. Myriads of tiny golden trout flash all summer in the crystal pools of little streams flowing over white granite basins. The national forest surrounds the Sequoia National Park and includes several other groves of giant redwoods. On the trail one day Mr. Pinchot renamed the forest "The Sequoia."

The human activities spread thinly over this great stretch of mountains and canyons were typical of the back country of California. Water was its greatest resource. Los Angeles was harnessing the Owens River with dams and aqueducts for municipal supply. Irrigation and power engineers were probing back into the feeder canyons and reservoir sites of Kings River. The Kern waters were fully used to irrigate ranches in the San Joaquin Valley, but there were still many applications for ditch rights of way and little upstream dam sites. The prevailing industry was livestock raising. Many little foothill ranches grazed a hundred or two head of cattle on the national forest, from the winter range under the oaks and digger pines up to the high mountain meadows and lodgepole pine parks in late summer. The upper San Joaquin Valley had many thousands of sheep, and large flocks of them had, from the days of the Spanish ranchos, moved up into summer pastures in the southern Sierra. Some years before the water users had prevailed upon the Secretary of the Interior to prohibit sheep grazing. But many Basque herders, whose reply to all questions was "No comprend Angleesh," observed the fiat of the honorable Secretary chiefly by slipping over the forest boundary whenever the ranger's back was turned. Their underground intelligence was perfect.

It was very much a "horse and saddle" national forest. The highways were few and short. Four fifths of our work was out in the open and reached by the most direct trails over the ridges and up the canyons. Headquarters was at a ranger station on a canyon "flat" well up within the forest, to which we brought our water by the bucket from a spring across the foot bridge. There I lived with my two saddle horses, my pack mule, and big sheep dog. The world was mine!

The foremost job was to tighten up action on forest and brush fires—all along the line. We had a good bunch of cooperating stockmen. They wanted no fires in their ranges and put out many small blazes themselves. But that great stretch of wilderness was sadly lacking in lookouts and communication. Often a cowhand would ride in to a telephone somewhere and report a fire that had been burning for two days way back on the Kern Flats. I drove the rangers hard to build lookout cabins and telephone lines and made life miserable for the guardians of the exchequer in Washington. We had to build some key trails and pack-mule bridges, too, where bad canyons had left serious gaps in the general network of stockways.

In the summer months there was lots of hard riding day and night to get to fires and make sure they were out before all hands rode happily away. We organized minutemen among the grazing permittees in each district. We persuaded the scattered mines and sawmills and the summer cow camps to provide caches of fire-fighting tools. By hook and by crook, we stretched the miles of telephone line back from ranger stations to the summer guard cabins and lookout peaks. I had a fine crew of rangers, real mountain men who could do anything from shoeing a horse to surveying a township line. They quickly caught the spirit of a hard-hitting fire organization.

Between fire calls I was mostly on the trail, with Sancho, the pack mule, paddling behind under the observant eye of the big dog. There were always stock ranges to examine and little timber sales where I could give the ranger a hand in selective marking. And there were always new trail locations to be checked and supplies to be packed in to construction crews. Sancho taught me much more about packing than I ever learned in throwing the diamond hitch over the dummy mule at the forest school—and some of his lessons were fast and sudden. We carried some respectable packs at that. One load for the bridge-building crew on the upper Kern was a puzzler. It included fifty pounds of dynamite and a nine-foot crosscut saw. But it got through.

Something had to be done about the trespassing sheep. They were eating out ranges allotted to permittees and putting the whole system of controlled grazing in disrepute. The rangers came in for a council of war and the campaign was planned. The migrating flocks left their winter pastures in the San Joaquin Valley in early spring, moved slowly through Walker Pass, and turned northward along the east slope of the Sierra. The herders grazed them in the national forest for days at a time, then brought them out again. They played hide-and-seek with the forest rangers all summer.

During the following winter, one of the young foresters on the staff organized a surveying crew and reran a hundred miles of the east boundary. "No Trespass" signs in English and Spanish were put up every half mile. In the spring secret lookouts were posted at points commanding the woolly line of march. There were enthusiastic volunteers for this duty from young cow punchers. Soon dust clouds carried the news that sheep were moving up Walker Pass, and headquarters had daily reports on the location of the lead flock. The district ranger ostentatiously left his station with his pack string, for a long trip over the high country.

We waited for days while the sheep grazed along but never across the boundary. Then suddenly they moved in. We held back until the flocks had made a day's drive and a night's bedding within the national forest. Then the deputy supervisor rode up with three or four deputized marshals and a Basque interpreter. There were nine thousand sheep in trespass under care of a dozen herders and camp tenders. Through the bedlam of South European expostulations, the deputy made the boss herder pick three men to take care of the sheep. He arrested the others and marshaled them before the United States commissioner at Bakersfield.

Suspicions of long standing were confirmed when top-flight attorneys from San Francisco magically appeared as counsel for the sheepherders. They represented some of the largest land companies in the state. They challenged the power of the Secretary of Agriculture and all his minions to control grazing on public lands of the United States. However, the commissioner bound the Basques over to the federal court and unwonted peace descended upon the south Sierra ranges. Many months later the Supreme Court received the case of United States *vs.* Grimaud, Cazazous, and Inda and settled for all time the authority of the Secretary to regulate grazing on national forests.

All of our grazing troubles, however, were not over. Some of the cattle ranges were overstocked, especially an area west of the Kern River which was used in common by herds from a dozen ranches. We became convinced that many more cattle were running in this country than the permits called for, although all the brands we could find belonged there. The stockmen had turned in the "best count they had" and kept stalling on a general roundup. After they had run us around the barberry bush several turns, one of the rangers—an old cowhand himself—organized a roundup of our own. Most of the force had a saddle in it; and there was some wild riding and rough tumbles that week. But we got enough of a count of the principal brands to show at least 30 per cent of unpermitted stock. Armed with these figures, we called the permittees together for a showdown.

The cowmen accepted our tally with surprisingly little argument and offered to pay the grazing fees on their surplus. However, they stood tough on the real issue: that the range was overgrazed and a lot of cattle had to be taken off. Finally a couple of the oldest ranchers came to my support and told the bunch it was time to quit cutting their own throats. They could remember the days when the grass brushed the stirrups of their saddles. We reached a compromise and settled on the number of cattle that should be permitted the following year, but in the course of it I got the master squelch of a lifetime.

The debate over the number of cattle which a section of that type of range would carry had been heated. One of the cattlemen said that the supervisor of the next national forest to the north was allowing a larger number than we demanded. My Irish blood retorted that it made no difference what other supervisors were doing, "on *my national forest* the range is in bad shape and the numbers of stock have got to be cut." When it was all over, the oldest of the stockmen put his feet upon the table and leaned back comfortably to roll a cigarette. As he sifted the tobacco from his sack of Bull Durham, he said in the most casual way:

"When the young supervisor just now talked about *his* national forest, it sort of reminded me of the time when the old Devil took Jesus Christ to the top of a high mountain. He offered Christ all the kingdoms of the earth if he would fall down and worship Satan. All of 'em, mind you. The old s.o.b. didn't own a damn acre!"

IV

One of the highlights of my days as forest supervisor was an inspection trip with Gifford Pinchot. I got to know G.P. as men can know each other only in camp and on the trail.

We marked timber together on one of the sale areas; we scaled a couple of peaks where fire lookouts were planned; we met rangers and cowmen and prospectors. And my admiration for the boss grew with every mile. Pinchot was very much a man's man. He could outride and outshoot any ranger on the force. If camp was within a mile of a stream of any size, he invariably had his morning plunge; and if the stream came from a snowbank a few miles up-canyon, all the better.

The enthusiasm which fired us all in Washington was doubly infectious out in the open. G.P. thrilled over a stately sugar pine or a glimpse of high, snowy peaks or a well-built piece of trail or a well-thinned stand of young timber like a boy just out of school. He carried a new Luger pistol with which he practiced on every possible target. Whenever we rode over one of the paths made by a snake wriggling across a dusty trail, G.P. would finger the Luger and look about eagerly for a rattler. Once we sighted a big rattlesnake which had crossed the trail into the chaparral above us. I offered to herd the snake back into the open, on a bet with G.P. that he could not nail the reptile with one shot from horseback. I was counting on the skittish roan mare he was riding (which had thrown me a couple of times) to win the bet. I crawled through the chaparral with a long pole in hand, circled the rattler, and headed him back toward the trail where G.P. and the mare were doing a war dance, Luger at the ready. Sure enough, he drilled the snake's head on his first shot and held his horse steady in the bargain.

Around the campfire G.P. talked to us intimately and with undimmed enthusiasm. He told us of his plans for the service and the next moves in the national march of conservation. He made us all—rangers and fire guards and Mexican boys building trail—feel like soldiers in a patriotic cause. Pinchot was an outstanding evangelist and a great leader of men.

The Forest Service performed a remarkable job in translating the idealism and generalities of the Roosevelt-Pinchot program into a working system of land management. It was localized and closely geared to the industries and livelihood of many thousands of western Americans. By personal education and example it carried the meaning and sense of conservation into everyday land uses and occupations. And it was largely a democratic process. The mailed fist of authority was kept in the background. The approach to the timbermen, stockmen, water users, and thousands of other people who wanted some part of these national resources was self-interest in the long run. That and Secretary Wilson's "greatest good of the greatest number." Nine times out of ten, this approach was effective.

The human relations of service men with the people of the country where we worked lent infinite variety and spice to the job. Whenever a venturesome mountaineer got himself lost or a child strayed from a summer fishing party in a canyon, the nearest ranger was called to the rescue. I was an enthusiastic amateur in first aid and usually carried a leather case of essentials in a saddlebag. Unfortunately I made too sudden a reputation by successfully sewing up a bad gash, cut by barbed wire, on the chest of the deputy's favorite mule. I took calls to doctor ax cuts on settlers' feet and to put splints and tape on broken arms—as part of the job. But I quailed one day when a cowman rode a lathered horse into camp and begged for help. His "missus was expecting" momentarily. The office telephoned the nearest upcountry doctor, while I rode back with the rancher, my jaw set, racking my head for everything I had heard or read about childbirth. But by good luck the doc arrived in time.

One of the rangers in the northern Sierra came to the aid of a prospector who had had something of a tussle with a grizzly bear. The story got into the newspapers with more bloodshed and heroics than had happened in real life. Soon after the ranger had a letter. As closely as I can recall, it read this wise:

"I seen in the paper that you-all reskered a man being et up by a bar. I want to know whare this man is at. I think he is a distant husbind of mine. You can tell by an anker tattered on his chist if so that part of him was not et up by the bar."

The success of the Forest Service came from its sincerity and its realism. As I knew it intimately for twenty-four years, it was a field organization. Its approach to any problem was a demand for the facts. Its thinking and spirit were scornful of armchair routine in running affairs. The most characteristic attitude of government organizations, "It has never been done," did not get far in the service. Rather—"What are the facts and what is the common sense thing to do?"

This direct, unequivocal facing of situations enabled the Forest Service to master the tough problem of agricultural land in the national forests. Most of the timbered valleys within the reservations had occasional flats or benches of tillable soil. Some of the upland meadows would make good farms. "Locking up farms from land-hungry people" was one of the most common attacks upon the Pinchot program. In parts of the West, it rolled up a political crescendo which threatened the breakup of some of the national forests. The old "free timber" buccaneers climbed aboard the caravel of the pioneer farmer, deprived of his American birthright by a heartless bureaucracy.

Much of the land hunger was genuine. During my days in Montana a veritable wave of homeseekers flowed over the Northwest. We heard much about "dry farming." Homesteaders were filtering into the upper Missouri basin and throwing up log cabins on land that former settlers had passed by. I met up with families "squatting" in isolated little mountain valleys, self-reliant men and women, ready to carve out their homes with their own hands. It was the real thing, and it made me proud of my race. But we had a hard political situation on our hands. Half a dozen bills were in Congress like the one introduced by Senator Carter of Montana. As the homesteader's champion, he demanded the elimination from national forests of all land within three miles of the Kootenai River, along its entire length in Idaho and Montana. There were some nice flats along the Kootenai, but 90 per cent of the senator's proposed elimination was timbered canyons and mountain slopes.

The answer of the Forest Service was: "Get down to cases. The highest use of any acre of land will govern. We will open to settlement whatever land, in actual fact, is valuable chiefly for agriculture." A special Forest Homestead Law authorized the Secretary of Agriculture to open to homestead entry any national forest land which he found to be more valuable for agriculture than for other purposes. A forest homestead might range from one acre to 160 acres. It could be fitted to the topography of mountain country. If a public survey was made, we shaped a homestead to an alluvial river bottom by legal ten-acre squares. If there was no public survey, we ran out the prospective farm by metes and bounds, following the actual lay of the terrain.

Patiently the land examiners worked over watershed after watershed, setting out the land of real value for farming, holding back the many acres applied for whose chief worth was in standing timber. It was a fine example of realistic administration on the ground, acre by acre; and it met the situation.

The clamor for farms locked up from deserving settlers brought Secretary of Agriculture James Wilson to Montana in 1910. I took the gray-haired old Scotch-Iowa farmer directly to the settlements in the Kootenai and Priest River valleys that were hotbeds of unrest. It was a rare show to watch the keen old veteran of many political battles meet the country folk on their own level and put over homely truths with shrewd wit. One vociferous spokesman demanded for settlers in Montana the same opportunity accorded to settlers in Iowa: that of taking up whatever land they pleased, not just what government men said they could have. If the land carried good timber, the government owed it to the homesteader to help him make a start.

The old Secretary shot right back: "In Tama County, Ioway, we were glad to homestead the bare prairie. What you want, my friend, is not to farm but to skin a grubstake off the land and move on somewhere else. That kind of farming will do this western country no good." Then "Tama Jim" likened the contentious gentleman to the Scotch lad who bargained too hard over the dowry of his bride. The girl's father finally demanded whether his prospective son-in-law was "after the siller or the lass." After long reflection, the boy replied: "Sin I can no get the siller wi-out the lass, I take the lass."

Secretary Wilson was anxious to see some of the fine timber that was involved in the "land hunger." I drove him in a buckboard through miles of unbroken virgin pine forest in the level Priest River Valley. After an hour's silence the old gentleman turned to me, slowly shaking his head: "There are no woods like these in Ioway." During the day I asked his opinion of the craze for dry farming in eastern Montana. He peered at me over the rims of his steel spectacles. "It works fine, my boy, as long as God sends down his rain."

Secretary Wilson gave Chief Forester Graves fine support in keeping the hands of Congress off the agricultural land question and letting the Forest Service work the problem out in its own painstaking, factual way. His support was needed in the letdown and wavering of the Taft administration.

Constantly facing new problems and settling them, not in armchairs but on the land itself, the administration of the national forests grew with experience and was changed to meet situations. Official regulations were broad and flexible and were often revised. To a surprising degree the Forest Service kept free from the strangulation of individual initiative so common to "bureaus" in either government or business. It remained a field organization which looked facts in the face and wanted new ideas. And so it retained vitality and leadership.

CHAPTER VI

World War Touches Off New Forest Explosions

ACCORDING to the unimpeachable authority of Stewart Holbrook, "It took an uncommon amount of rum to begin a war in 1775." However that may have been, it certainly took an uncommon amount of lumber—not to begin two world wars but to finish them.

While the United States was still on the side lines of World War I, ordnance officers brought to the Forest Service the problem of wooden stocks for army rifles. To make thoroughly seasoned gun stocks that would hold rifle barrels true, it had long been the practice to keep black walnut flitches in dry storage for a matter of two years. And all the flitches then in the country, processed beyond the standing tree, would make less than two hundred thousand army rifles. "You must tell us how to season walnut gun stocks in sixty days; and you must find enough walnut timber to arm three million men." The wood seasoning experts at the Forest Products Laboratory took up the first challenge; and the Walnut Manufacturers' Association, the second. It was not long before dry kilns with exactly controlled humidity, temperature, and air currents were delivering perfectly seasoned gun flitches in forty-five days, and the timber for millions of walnut stocks had been located—down to an individual tree in Farmer Jones's back forty.

When it became clear that the United States could not keep out of the war much longer, the War Department named a committee of lumbermen on national defense. I was drafted as secretary, and we went to work day and night on the problems which the Army brought to us. One of the first was the drawing up of standard lumber specifications for troop cantonments. I worked many hours on a set of interchangeable lumber grades that would permit the filling of cantonment orders from any construction wood, cut at any of the twenty thousand sawmills in the country. In the thick of this job, the Balfour Commission landed from England to advise on American participation in the war. One of its urgent requests was a regiment of forest engineers to join the British colors in France and speed up the cutting of trench lumber, wire entanglement stakes, road plank, and the many other kinds of timber needed in operations at the front. I dropped the mobilization of lumber for American cantonments and began the mobilization of lumberjacks for the 10th Engineers, A.E.F.

The twelve hundred men first planned to relieve the timber shortage of the British Tommies grew into twenty thousand engineer troops and supporting labor battalions, constituting the Forestry Division of the American Service of Supply. When the Armistice was signed in November of 1918, we were operating ninety sawmills scattered over France behind the trenches. Many other detachments were cutting piles, telephone poles, entanglement stakes, and fuelwood. We had delivered to the engineer dumps and construction jobs in the American sector something over six hundred million board feet of sawn lumber for hospitals, depot warehouses, duckboards, and planks for trench walls and bomb-proof shelters. There was a constant demand too for railroad sleepers and road plank to carry the trucks over marshy or shell-torn ground. The Forestry Division never caught up with its orders. The ex-lumberman in charge of shipping priorities had to stand off furious complaints, often from brass of high rank. He said wearily one night, "I wish we could have a market like this in the states, sometime."

The 20th Engineers, to which all forest troops were assigned for military organization, became the largest regiment in the history of the American Army. It acquired a reputation for delivering the goods. At one of its mills in the Vosges Mountains an impressive American officer arrived one morning, with staff car and orderlies. The sawmill was humming like a beehive, but the captain of the company was on his back, working with a monkey wrench on one of the machines. A rattled sergeant shouted in his ear: "There's a big guy out here with a flock of stars on his shoulders." The captain came out hastily, wiping his greasy hands on a piece of waste. He got a look at his visitor as he saluted and stammered out: "Aren't you General Pershing?" The big general laughed: "If you see anyone else around here wearing four stars, let me know." The C.O. inspected the whole layout and asked many questions. He wrote a nice letter to Regimental Headquarters, complimenting the outfit for the way it was hitting the ball, with everyone at work on the job, *including the commanding officer.*

When the Meuse-Argonne offensive was planned, the Forestry Division put all its wits to work to devise quickly movable sawmill and logging units that could follow closely behind an advancing front and cut whatever timber items were needed at the moment. Our supply officers and master engineers worked out the most portable sawmill the world had seen. A big truck carried a little sawing unit, which dogged short logs at the ends and "cut them up alive," without turning the log or squaring the edges of the planks. Behind the truck, piled high with tools and camp equipment, trailed a couple of eight-wheel Kentucky log wagons that could plow through any mud in Europe. And behind the wagons trailed a dozen big logging horses. The thirty or forty men rode wherever they could find a seat.

These little mills went to work wherever the Division Engineers wanted them; cut what timber they could find into road plank or bombproofing; then picked up and moved on. Several of them moved out in record time, some minutes under two hours, when German shells got their range.

Every month of the war brought new demands upon the Forestry Division. One day a general inspector from G.H.Q. came to my office at Tours. He could have been an old cavalry officer right of a Remington painting, mustaches, bow legs and all. With genuine boot-and-saddle profanity the inspector wanted to know why the Forestry Division was forever demanding more men and more equipment. "Damned outlandish stuff at that! Thirty years in the Army and never heard of such tomfoolery!"

I got out our file of orders and explained them to the colonel, one by one. There was lumber for the great engineer depot at Dijon; lumber for half a dozen new hospitals; railroad sleepers crowding first priority with planks for the truck roads; piling for new docks at half a dozen embarkation points. The old cavalryman followed me patiently until we came to the last order. The Quartermaster Corps could not find enough hay or straw to stuff the doughboys' mattresses. The Forestry Division, S.O.S. with due official endorsements, was instructed to procure, install, and operate enough machines to produce fifty tons of excelsior per month. At this point the inspector from G.H.Q. exploded. With a burst of oaths he jammed his campaign hat on his head, mouthed his black stogie into a belligerent angle, and strode out of the room. At the doorway he gave me a parting shot: "All I've got to say to you, Colonel, is that waging war has come to be a hell of a complicated proposition."

The longer we operated in France the more we learned to respect the French forests. Not even our boys from the far Northwest could sniff at the firs of the Jura, where the French service allotted some compartments of mature sapin, ready for a harvest cutting. Beech and white oak on old seigneurial estates of central France rivaled the best hardwoods of our southern Appalachians. And the pineries of the southeastern Landes, started by Napoleon's engineers to stop the invasion of sand dunes rolling inland from the Bay of Biscay, were the most impressive achievement in man-made conservation the American foresters had ever seen. Their coastal strips of maritime pine, planted for protection from wind and sand, had spread into a productive forest of two and a half million acres. It had made the province a healthful place in which to live and had given it great, permanent industries producing lumber and naval stores.

We found somewhere, in the forests of France, literally every size and kind of timber that any technical branch of the Expeditionary Force demanded. My heart dropped to my knees one day when Major General Mason M. Patrick handed me the list of piles required for the huge docks at Saint-Nazaire. Eighteen thousand piles, *90 to 120 feet in length.* I said to myself: "They are not to be found in La Belle France." But the American Command had ordered the docks to be built in preparation for three more years of war, and had further ordered that none of the precious cargo space be used to bring timber from America. I took the list and saluted in silence. One seldom discussed and never questioned an order from General Patrick.

We combed the battalions for junior officers and noncoms who could speak French and "knew their way around." They set forth to locate piling trees, down to two or three on a farm or small wood lot. Soon Battalion Headquarters was dispatching ten-man logging crews over the back roads and hedge lanes. Every 20th Engineer logging boss was selecting and marking the piling trees in his operation. Long piles rolled into Saint-Nazaire until we got orders to hold off. We found our eighteen thousand; I believe we could have found a second eighteen thousand long piles. Shortly before the Armistice our scouts located fine stands of old-growth fir on the northerly slopes of the Pyrenees. They were far away from the theater of operations, but the forest engineers would have been in there had the war continued much longer.

The strongest impression brought home from my two years with the A.E.F. was the hold of forest conservation upon the French people. It was part of their instinctive thrift. The magic words *"C'est la guerre!"* never availed to shake the established rules of forest cutting and sustained production. We had many arguments with the French foresters over cutting requirements and I found myself on the other side of the table from similar controversies with loggers back home. The Frenchmen were understanding and realistic—and mighty good woodsmen. But in issues between their established regime of timber culture and the exigencies of Allied manpower or speed in getting wood to the front, the forest always won out. In the last cutting of a growing cycle of white fir, after the ground is covered with saplings, the old seed trees come out. In these stands we had to shinny up the veteran trees and limb them before felling, so as to break down as few as possible of the youngsters on the ground. And in no compartment could American troops cut more than five times the annual *possibilité,* that is, not over the estimated growth of five years. A grizzled *conservateur* said with a fatherly smile, to a bunch of impatient Americans: "Our forests have fought several wars before this one."

I was often struck by the degree to which an understanding of forestry reached down through the French people. It is a national creed, with its roots in the soil. Walking on a country road in the Landes, I fell in with a bright-eyed French lad of twelve or fourteen years clopping along in his wooden sabots. I tried out my stumbling French, and with his quick intelligence we got along quite famously. He lived on a farm and his father was a small vintner. But this lad knew all about the pine forests which were around us in every direction. Responding to my interest, he told me eagerly how the "resiniers" cropped the trees for pitch; how they thinned the young forests for pit props, shipped to the coal mines in England; when the time came for *gemmage au mort* (bleeding to death) and logging and sawing lumber. It was a tale made vivid by flashes of boyish enthusiasm, and it was about the best résumé of maritime pine forestry I had ever heard.

While twenty thousand troops were doing their best to keep the Expeditionary Force supplied with wood, the lumber industry in the United States was more and more closely centered upon the sinews of war. The Spruce Production Corporation was set up by the War Department in the Pacific Northwest. It bought timberlands, built railroads and sawmills, and organized the industry for maximum production of Sitka spruce lumber. The needs of the United States and her Allies for fleets of airplanes suddenly made the light, resilient spruce of the West Coast the most sought-after wood in the world.

The Lumbermen's Defense Committee at Washington was duplicated at several milling centers. They were all linked up in an improvised network for passing out war orders, holding down prices, and rawhiding the timbermen into cutting Uncle Sam's lumber ahead of more lucrative civilian trade. It was all done through rough and ready co-operation, without benefit of law or regulation. Most of the lumbermen cheerfully took orders from their committee. The holdouts could not be brought to hook by official process, but they might find their string of empty railroad cars mysteriously diverted to other mills or a log supplier unexpectedly short on deliveries. Immense quantities of lumber moved from the sawmills to cantonments and shipyards and to arsenals and factories for packaging munitions. The toll upon the forests of the country was heavy, possibly as much as a full year of peacetime consumption. There was no great bulge in the over-all cutting of timber, but civilian uses of wood were sharply curtailed. The national economy evened up the score in the building boom of the early twenties, and in the final accounting our forests paid their quota to the war.

The war also blew the lid off several explosive situations in and around the woods. It fired up new movements and forces in American forestry. Among them was the labor movement in logging camps and sawmills.

Labor organizations made little headway in forest operations before World War I. Lumberjacks have generally been too independent a breed of men to lend themselves to union control. When tough organizers of the International Workers of the World were invading western mines and railroad construction crews in the early nineteen hundreds, the "Wobblies" tried hard to break into the logging camps. Missoula, Montana, was a hot center of propaganda. I listened one night while young and beautiful Elizabeth Gurley Flynn, on her soapbox, flung impassioned indictments at the slave drivers in the woods and called upon the "timber beasts" to cast off their chains. There was considerable fisticuffing and street rioting. The police finally decided to break up all Wobbly demonstrations. Speakers were arrested as fast as they mounted the soapbox. It became a contest in endurance.

I was called by the police chief one morning to bail out a young Forest Service timber cruiser who had got caught in the melee the night before. He had come to town after months in the woods and was drawn to the excited crowd without knowing what was going on. He saw man after man and a couple of women yanked off the improvised platform as soon as they opened their mouths. His New England blood could stand no more. He jumped up on the box himself. "I don't know what these people are trying to say," he shouted, "but the right of free speech——" Bang! A big Irish policeman had him by the coat collar, down on the cobblestones.

In a long running fight for shorter working hours and better food and quarters, the Wobblies got a foothold in northwestern logging camps. On the rising demands of world war for timber, they struck hard, with many forms of violence. They all but paralyzed the war effort in a lot of Douglas fir logging operations.

The I.W.W. menace was finally blocked by one of the most unique organizations in industrial history. It was the Loyal Legion of Loggers and Lumbermen, created by General Bruce E. Disque, wartime head of the Spruce Production Corporation. The 4L was built on the principle of equal representation of workers and employers at every level of action, from a camp committee to the top Board of Governors. A tie vote, which occurred but rarely, was settled by the president, Dr. Norman F. Coleman, universally respected head of Reed College in Portland.

The Loyal Legion gave West Coast sawmills and logging camps the eight-hour workday, reduced from the long-prevailing ten hours. It made several advances in wages and vastly improved living and working conditions in the logging camps. From the simple starting point of equal representation of men and management in everything undertaken, a spirit of cooperation developed that made the 4L one of the most successful of our many experiments in industrial relations. It gave the workers wide experience in the affairs of their industry. I have never had more thoughtful discussions of Northwest forestry and its possibilities than those with 4L locals in Oregon and Washington.

The Loyal Legion was sacrificed to New Deal conceptions of industrial relations. It fell under the ban imposed by the Wagner Act upon "company unions." But the labor movement has carried on with increasing power in northwestern forests under rival unions, the A.F. of L. and the C.I.O. Organized labor has taken a strong hand in many phases of regional forestry. It is a powerful molder of public opinion and local political action. The labor unions are represented on almost every state forestry committee and official agency. They appear at every public hearing on any forest question. They pulled a strong oar in the enactment of the Washington Conservation Act of 1945.

The industry's treatment of its forest has no more caustic critic than the foresters on the staff of the C.I.O. C.I.O. indeed has entered the arena of national forest policy. On every forum it can reach, it stands for drastic federal regulation of all private forest practices. It sponsored the Hook Bill of 1946, the most forceful scheme yet proposed for federal regimentation of American forests.

While the war brought organized labor into the field, demanding more forest conservation, a political aftermath threatened to tear down the structure already built.

II

President Harding hailed the end of the war as a "return to normalcy." Some Americans of the old free-land persuasion saw an opportunity to capitalize upon the political reaction and public revulsion to the disciplines of war. The days of free range and unreserved timber might be brought back. The setting was a bit like a western movie where "Gentry rides again." I succeeded Colonel Graves as Chief of the Forest Service in 1920. Soon after the inauguration of Harding I sensed that his administration would give the Pinchot-Roosevelt conservation program some rough going.

Seldom has a more incongruous team been assembled in Washington than the first Harding cabinet. The leonine Charles E. Hughes, outstanding in moral and intellectual stature, rubbed elbows with foxlike Harry Daugherty, the mastermind of political intrigue. The Secretary of the Interior, Albert B. Fall, was an unreconstructed apostle of the free-land, come-and-get-it traditions of the young West. But his opposite number, in the chair of Agriculture at the cabinet table, was an able and forthright exponent of all that the conservation of natural resources stood for. Henry C. Wallace[1]

[1]Father of Henry A. Wallace, Secretary of Agriculture in the first cabinet of Franklin Roosevelt.

was not a crusader of the Pinchot type. He was a conservative who kept his feet always on solid ground. He had a farm-bred knowledge of land and people, a marvelous endowment of political sagacity, and a vast fund of humor. He was Scotch to his bristling red eyebrows, and he was the finest man I ever worked for.

At the first conference with his bureau chiefs, Secretary Wallace told us about the traveler in the British Isles who wanted more sugar in his tea. If his hostess was Irish, she would pass the whole sugar bowl. An English hostess would ask: "One lump or two?" But a Scotch hostess would look him straight in the eye. "Hae ye stir-r-red up the bottom of yer cup?" The new Secretary made it quite clear that he expected the bottoms of all the cups to be stir-red up when we submitted our estimates for the next year's budget.

The new administration was not three months old when Secretary Wallace told me gravely that we must be on our guard. Some of his many newspaper friends had tipped him off. A scheme was brewing to turn the national forests back to the Department of the Interior. A buildup in the press was coming, in the hope of rolling up political strength for repeal of the Act of 1905, which placed the forest reserves under the jurisdiction of the Department of Agriculture and set up the Forest Service for their administration. The main line of attack would be that "conservation" as then applied meant only locking up from use. The economic development of the West, her timber and power and livestock industries, were blocked by the theorists in the Forest Service. The prize exhibit of a shackled empire was Alaska. And the scheming centered in the Secretary of the Interior.

A number of attacks on the way the national forests were being administered appeared in the newspapers. They could all be threaded on a common string of impractical management. Portions of the western livestock industry were in revolt against reductions in the sheep and cattle admitted to national forest ranges. In overgrazed areas the Forest Service was holding generally to a gradual curtailment in the number of permitted animals, in the hope of restoring the carrying capacity of the forage. Many stockmen accepted and co-operated in this practical conservation of their own future, but often the grazing cuts were violently opposed. They brought forth unlimited argument over the unwritten "right" of the stock ranch to the use of the adjacent public ranges which formed a vital part of the pioneer's enterprise.

Some power companies, too, were restive over the discovery that Gifford Pinchot had beaten them to the draw by plastering all known dam and reservoir sites in the national forests with administrative withdrawals. This left them no way to acquire title to the highly important lands on which their millions must be invested. The only recourse was to lease and operate under federal permit. Government ownership and control of the sources of hydroelectric power was a high plank in the Pinchot platform of conservation.

There were always a few complaints over the handling of national forest timber. Stumpage appraisals might be too high. The scale of logs might be too close. The restrictions on operating logging machines might be unreasonable or of excessive cost. These situations occasionally drifted into Washington on appeal from a district forester's decision, and a sharp newspaper man could make a story of bureaucratic mismanagement out of any one of them.

All of these criticisms and attacks were part of the daily meat of the Forest Service. There seemed to be many of them in the first year of the Harding administration, but not enough to sound a general alarm. After all, they appeared to be just the usual rustling forays and not a range war. But our doubts were soon removed. Secretary Fall himself rode blustering into the roundup, *chaparajos* sweeping the grass and a gun on each hip.

The Secretary began to make speeches and to give press interviews in which he directly attacked the going conception of natural resource conservation and the administration of the national forests. "Conservation," he told us, should mean "use," "development," "the satisfaction of human needs." The practical stockmen and lumbermen of the West were the true conservationists, but theorists of the Pinchot school were driving them out of business and locking up the West for future generations. He wanted the national forests back in the Department of the Interior, where they belonged with the rest of the public domain. When that had been done, and not before, the due and needed development of the West would be resumed.

In most of his pronouncements Secretary Fall cited Alaska as the outstanding example of great natural riches and opportunities, locked up by impractical conservation theorists. His speeches carried echoes of the Pinchot-Ballinger controversy of ten years before. It became apparent that the vast national forests of southern Alaska were the first objective of his attack.

The Forest Service itself had done a good bit of worrying over Alaska. The Territory was clamoring for people and industries. The momentum of gold mining and other mineral extractions was running down. The fishing industry had apparently reached its peak. The opportunities for agriculture were pitifully small. Many eyes turned longingly to the great stands of timber in the accessible coastal forests. Their sixty billion feet of Sitka spruce and West Coast hemlock, with water power in many streams cascading through them into the sea, should someday give Alaska a wood pulp industry like that of Sweden or Norway. But Alaskans wanted their industries and jobs right now.

One of my first undertakings as Chief of the Forest Service was a firsthand study of how we might bring wood pulp mills to Alaska. I took to the woods with our able district forester, Frank Heintzleman, and top experts from the Forest Products Laboratory. The Tongass National Forest, with a timbered coast line of some ten thousand miles around the islands and inlets of the Alaskan Panhandle, was supplying half a dozen sawmills with logs for the lumber needs of the Territory, the boxing of its salmon pack, and an occasional export cargo of high-grade spruce. It was furnishing the piling for hundreds of docks and fish traps. But the establishment of a pulp industry with an investment of five millions upward was an entirely different order.

We mapped and cruised the most promising areas of pulp timber. We drafted pay-as-you-go contracts for a fifty years' supply. We drew up prospectuses, advertised in the paper trade journals, and interviewed pulp and newsprint manufacturers on the West Coast. The conviction was painfully forced upon me that we were twenty years ahead of the economic pendulum. Like most Alaskans and Secretary Fall himself, we wanted the future to join us at tomorrow's breakfast table. This effort was renewed by the Forest Service after World War II. The first large sale of Alaskan pulp timber appeared to be assured in 1948.

In the spring of 1922, the cold war became hot. With belligerent confidence, Secretary Fall declared to the press that he proposed to take over the national forests and the Forest Service without further shilly-shally and that his first official act would be to fire the impractical theorist who was then mismanaging the whole show. This pronouncement brought a tart rejoinder from redheaded Victor Murdock, congressman from Kansas and a potent leader of the conservation bloc. In a press release Mr. Murdock advised the Secretary of the Interior to ponder well the familiar recipe for hunter's stew which begins: *"First catch your rabbit."*

The declaration of personal war put me in a quandary. I was not afraid of the outcome. The conservation policy and the Forest Service had many friends in Congress and throughout the country. And I would match the political generalship of Henry Wallace and Gifford Pinchot against the field. But the fighting traditions of G.P. were stirring within me. I did not want to take any more of it lying down. All of us in the Forest Service had been restrained by the fresh example of the Pinchot-Ballinger controversy, which came very near forcing Secretary James Wilson into a cabinet fight which he did not want. I would have cut off a hand rather than drag my stanch friend Henry Wallace into a feud that might take him out of the government. I told him I would resign and then take on without gloves the old Texas cowman who was running Interior.

Henry Wallace heard me through. The friendly wrinkles around his eyes were smiling with inward chuckles when he said: "My boy, don't ever get yourself into a pissin' contest with a skunk." In a few moments he added: "This is my fight as much as yours. We'll take it on; but let's pick our own ground. Get the sun in the other fellow's eyes." When I left the Secretary's office that afternoon, he put a hand on my shoulder. "It may take something sharp to settle this thing," he said. "If that time comes and you feel you must quit, you and I will step out of the government together."

We both feared that the charming, friendly man in the White House, who stood so loyally by his cronies, might be committed to Fall's scheming before there was a chance to fight it in the open. Secretary Wallace got the assurance of the President that he would make no decision in the matter without a full cabinet discussion; but the Secretary watched all the currents of White House gossip and influence like a hound dog at a coon tree. Meantime the conservation press took up the Fall challenge, and the most powerful farm organizations in the country made themselves heard with remarkable unanimity. Secretary Wallace was picking his own ground, and the winds soon blowing around the White House would have given pause to a politician much less astute than Warren G. Harding.

Plans were shortly announced for the President's visit to Alaska. Secretary Wallace saw to it that I was included in the official entourage. This, the Secretary told me, "will likely be the showdown on Fall's grab for Alaska. Bring plenty of powder and keep it dry." But just as we were briefing facts and arguments in the Forest Service, like a clap of thunder Albert Fall and all his schemes blew up. It was some time before we got the full story of the Teapot Dome oil scandal, but Fall dropped out of the cabinet overnight; and his name became anathema to the administration. It was a subdued and saddened President with whom we journeyed to Sitka on the transport *Henderson*. Our conferences on the federal forest policy in Alaska were pushovers. Secretary Wallace whispered, at one of them, "There is not an ounce of fight in this whole ship."

On a blue, sunshiny afternoon the *Henderson* steamed slowly back down Puget Sound and past the Pacific fleet in its brightest array. The crew on every battleship and cruiser "manned the rail" in their smartest uniforms as the President's ship moved majestically down the line; and each vessel in turn fired the Presidential salute of twenty-one guns. Then came the colorful ceremony of "piping aboard" all the commanding officers of the fleet, in order of seniority, with each receiving his due and sacred number of ruffles and flourishes as he came up the *Henderson's* ladder. On the next day, in the Seattle stadium, President Harding delivered his long-awaited address which laid down the administration's policy for Alaska. Gifford Pinchot himself never packed more conservation into a single speech. It was almost the last official act of Warren G. Harding.

Henry C. Wallace followed his chief into the beyond a year later. The day of his funeral in the White House was one of the darkest in the lifetime of the Department of Agriculture. Every one of us not only mourned the leader who had won our admiration but felt the sharp personal grief of losing a loyal friend. Many grown men wept unashamedly in the White House that morning.

The Albert Fall fiasco left the national forests stronger in public standing than before. The foundation work of Pinchot and Roosevelt was secure. "Conservation" again became strong political medicine. The Department of the Interior tried hard to live down its lapse from grace. Secretary Wilbur publicized the claims of Interior as the first and major conservation agency of the government. Secretary Ickes, a few years later, tried to change the name of his department to the Department of Conservation.

Meantime old controversies flared up, like a half-dead fire that has smoldered for weeks in a punky log, and led the country to a momentous decision on the direction its forest policy should take.

III

The most dynamic chain of forest actions and reactions touched off by the war was the reawakening of acute public interest in timber supply. Many groups of people asked: "Where do we stand in this unfinished business of forest conservation and where do we go from here?" The chain of events brought sharply into the open two opposing political philosophies—the growing of trees by regimentation and by free enterprise. It started a second forest crusade.

We are a demanding people and have never accepted the necessity of going without butter when we must have guns.[2] Even while our boys were in the trenches of World War I we did not like to have home building shoved aside by military priorities for cantonments and overseas freighting. We looked around for a scapegoat. It was the lumber barons who had wasted our great forest heritage and left us short of wood. Newspapers and magazines revived the forecasts of "timber famine." Forestry was listed in most of the proposals for postwar reconstruction. Discussions of forest policy became more tense and insistent. Controversy was in the air.

Colonel Henry S. Graves, then head of the Forest Service, told a meeting of lumbermen in 1920 that regulation was necessary to stop destructive logging and asked their co-operation in working it out. I had scarcely warmed the colonel's chair when requests came from several directions for a "clarification" of the government's intended program. The new and untried bureaucrat sorely needed the wisdom of adders if not the harmlessness of doves.

Gifford Pinchot, no longer in the public service, set up a high-powered bureau of his own. Old crusaders, lawyers, and publicists, flocked about him. He had many friends and stanch supporters in Congress. Mr. Pinchot now demanded federal regulation of all private timber cutting in the United States. In earlier years he had often referred to commercial forestry as a business "that must pay its own way." But he was impatient over the delay and inertia of the forest industries and their habitual defense that "we cannot grow trees until they are worth the cost of growing them." He was also irritated by the alibi of hard times and tough competition in the lumber business. His crusading zeal demanded *action now*. He held that the public interest in all forest lands was paramount and must be protected whether the owner made money or went broke. He proposed to meet the competitive angle by the uniform imposition of cutting restrictions under federal law. They would apply to all commercial forest operators and put all

[2]Franklin Roosevelt promised the country both guns and butter in 1941.

competitors on the same footing. It would be like subjecting them all to a uniform federal tax, but the money would go into the future productiveness of their forests rather than the national treasury.

Mr. Pinchot's lawyers searched the courts for constitutional grounds on which the federal government might exercise this police power. They wrote several bills which were introduced as trial balloons by Senator Capper. All of them gave the force of law to regulations of the Secretary of Agriculture concerning forest cutting practices in any state or region. One bill denied shipment in interstate commerce to any forest products not cut in conformity with the Secretary's rules. Another bill proposed a federal excise tax of $5.00 per thousand board feet on all timber cut on private lands in the United States and the remission of $4.95 of this tax in the case of timber which was felled in compliance with the requirements of the Secretary.

Mr. Pinchot's challenge reverberated through the forests like the silver trumpeting of a bull elk from a Teton meadow on a September morning. The fight was on. The stand-pat wing wanted simply to "stop Pinchot." There were many groups, however, who sincerely desired a more positive national policy toward private forest lands but sought a co-operative approach rather than Pinchot's big stick. Most of the opposing factions joined in setting up a remarkable institution called the National Forestry Program Committee. It was an excellent example of the ability of widely scattered people and organizations to get together overnight on a new idea or problem. This "committee of committees" represented the American Forestry Association and most of the organizations which had taken part in the earlier campaigns for conservation. Its most indefatigable worker was Royal S. Kellogg of the National Lumber Manufacturers Association, and its master strategist was E. T. Allen of Western Forestry and Conservation.

After much medicine making, the committee drafted a comprehensive charter of federal forest policy, which put Uncle Sam in the role of educator and co-operator rather than that of police officer. Most of its features were incorporated in the bill soon introduced by Congressman Bertrand E. Snell of New York. This bill authorized the federal government to lay down essential requirements for the control of forest fires and the cutting and removal of timber by states or regions. It proposed that the government match dollars with any state in enforcing its code of fire control and forest cutting, whenever the state system met the standards of the Secretary of Agriculture.

The Snell Bill would also have strengthened the position of the federal service by inaugurating much larger purchases of forest land under the Weeks Law and removing their limitation to watersheds of navigable streams.

The most significant testimony at the hearings on the Snell Bill was offered by men appearing for forest industries. They did not accept Gifford Pinchot's "paramount public interest in all forest lands," but several of them recognized a degree of public interest which the private owner must take into account in what he did on the land. Their main plea was for *local interpretation and administration* of public safeguards. They feared the distance and bureaucracy and domination of the federal government. Even so, the industry's position was a far cry from the free-land and property-right traditions of not many years before.

In the meantime I put several Forest Service men to work with state foresters, lumbermen, and other local folk, jotting down in black and white just what the "minimum requirements" were in regional cutting practices. Just what was necessary to keep the forests growing and productive? Since the stage was apparently being set for the control of destructive logging by some form of public action, it seemed only common sense to put what it was all about into practical, understandable terms.

I was hoping also that some common ground might be found on which the opposing camps could join forces. Surely we should take advantage of the strong public interest to make some solid advance in forestry, even if it did not write the last word. Why should we let our forests burn up while arguing over who should control the manner of their use? My hardest work, in those days, was trying to keep the lumbermen in the co-operative tent. They might easily have stamped off the reservation completely and taken the warpath, as some of them did. I thought the industry had gone pretty far, in the Snell Bill, to meet the public halfway. There was danger of their throwing that overboard and telling the government to leave them alone.

For months a wordy battle raged between proponents of the Snell and Capper bills. The first Capper Bill, which rested upon federal control over interstate commerce, was given a serious setback by the decision of the Supreme Court, which denied use of the interstate commerce clause to enforce prohibitions against child labor.[3] But G.P.'s resourceful lawyers soon offered another constitutional peg for policing forest practices. The second Capper Bill invoked the taxing power of the national government.

Those months of hot debate brought to a head a long-smoldering division in the thinking of foresters. A split started within the ranks of the service itself. Many of our most sincere men stood fast with Gifford Pinchot on his ivory battlements. They were the ardent crusaders, the left wingers impatient for fast action. Some of us, on the other side, could not thrill to the call of the trumpets. Perhaps we had done more grubbing in the dirt, trying to make the beautiful ideal work. Perhaps we had labored more closely with the lumbermen in the rough and tumble of fighting fires and cutting timber. Perhaps we had been too close to the economic troubles of forest industry. At any rate, we doubted whether worthwhile forestry could be brought to pass on the free soil of the United States by federal police methods.

It was a hard wrench to break with our inspired leader to whom we owed so much and felt such strong personal allegiance. But the march of events forced some of G.P.'s boys to face up to it. To my regret, I found myself crossing swords with Gifford Pinchot on the second Capper Bill before the House Committee on Agriculture. I could not accept his starting point, that direct police action by Uncle Sam is necessary to bring about decent treatment of our forests. Mr. Pinchot told the committee that the lumbermen had pulled wool over

[3]This decision was reversed by the Supreme Court in 1940.

my eyes. Whether it was wool or sawdust, I could not visualize practical results in growing trees on the 75-odd per cent of forest land in private ownership by abandoning co-operative methods and trying to force the change down the throats of the landowners. I did not think it would work. In much of the United States, furthermore, commercial forestry was dependent upon the prevention of forest fires and the nature and weight of forest taxes. It seemed to me that the same public agencies which directed how a forest should be cut must also be responsible for its protection and its taxation. The government could go far through friendly leadership and co-operation, but it rested with each state, I thought, to determine the place for police regulation in its own integrated program of forestry. Besides, I was convinced by that time that the Pinchot plan was heading Congress into no action at all; and twenty or thirty million acres of forest were burning up every year while we debated.

Opposing arguments and lack of aggressive political leadership brought the situation in Congress to a stalemate. Then a new chieftain quietly entered the field. He was Charles L. McNary, senator from Oregon. "Charlie Mac" had spent his life among the Douglas fir forests of the Willamette Valley. He had planted many forest trees on his own farm and had a keen interest in forestry as part of the agriculture of the West. Withal this sandy-haired Scot was endowed with an uncanny political sagacity. Of all the men I have known in public life, Senator McNary had the surest intuition of when to do something and just how to go about it.

The senator told me one day that the Capper Bill was dead and the Snell Bill bogged down. "It's too big a dose for one bill," said he. "What's the first thing to be done?" I replied: "Stop the forest fires." "All right," said McNary, "we'll write a bill around that. But first we've got to build a fire under Congress. We must get more public interest behind this thing." It was not long before the Senate adopted a resolution directing the appointment of a Select Committee on Reforestation. It was instructed "to investigate problems relating to reforestation, with a view to establishing a comprehensive national policy for lands chiefly suited for timber production in order to insure a perpetual supply of timber for the use and necessities of citizens of the United States."

Under Chairman McNary, the Select Committee made a thorough job of it. Twenty-four hearings covered all the important forest regions of the country; and they were down-to-earth hearings. Down South, old-school lumbermen told the committee that no businessman would risk the hazards of holding land and regrowing timber, but Henry Hardtner spoke enthusiastically of the young pine forests on his lands in northern Louisiana. "If the government will help us stop the fires," he said, "the South will bring back its piney woods." Out in Minnesota, Congressman Bede and others told the committee the northern pineries would come back if "we can keep the fires out and the taxes down."

I confess to packing the stand at the committee hearings with fire witnesses. Whatever else the honorable senators might learn or ignore, I was determined they should get first-hand, over and over again, the urgency of forest protection as the place to start. The National Forestry Program Committee had active scouts in the field, and not many men with a real forest-fire story to tell escaped the witness chair.

I also had the naïve idea that a committee instructed to study forest problems should see a lot of the woods. Between hearings we made strenuous detours through forests, logging camps, and old burns. Chairman McNary accused me of trying to show them every tree in the United States. At an early morning start in Montana he demanded why we were taking a roundabout road to Spokane. Senator Moses of New Hampshire was always ready with a quip. "The top sergeant says there are a million trees on this road, Charlie. We simply must see 'em."

In the Pacific Northwest the protection associations of forest owners told their story of successful co-operation. George Long urged the senators to have confidence in the industry. "Help the lumbermen with forest fires and forest taxes," he said, "and they will find a way to regrow the timber." Through the smoke of after-dinner cigars he reminded the distinguished legislators of the Missouri proverb, "There is honey in a dung heap, but it takes a bee to get it out."

Listening at most of the hearings, I was impressed by the number of lumbermen who were growing trees or holding their cutover lands in the hope of growing trees. The greatest handicap was the universal hazard of fire. Usually cited as the second barrier to commercial timber growing was the risk of unpredictable taxes over the fifty or sixty years necessary to produce a crop. The present tax burden was not holding up forest replacement so much as the uncertainty of future taxes and the prevailing county philosophy of getting all the traffic would bear. One wise old logger said that plenty of trees would be grown when they were an insurable fire risk and you paid only a fair tax on what you got for the crop.

In the older forest regions of the East we also heard a great deal about rising timber values, about western lumber coming through the Panama Canal, and the cutting of pine which had grown of itself on fields abandoned during the Civil War. The conviction became strong in my mind that industry itself was the greatest latent force for reforestation if the government could give it a lift, shake off some of its shackles and get it going.

The Select Committee gave Congress a bird's-eye picture of America's forests in 1923. The primeval forest had shrunk to 138 million acres of virgin timber and 331 million of culled, second-growth and burnt-over woodlands. The standing timber had dwindled to 2.2 trillion board feet. It was badly distributed in relation to population, and eastern lumber prices had advanced far beyond the all-commodity index. The growth in the forests was believed to replace only 25 per cent of the drain upon them by ax, fire, insects, and disease. The progress in commercial forestry thus far was little in actual extent but encouraging in its possibilities. It was mostly in the southern states and in New England, where the depletion of local supplies had brought sharp advances in timber values.

All forms of public ownership made up 21 per cent of the forests in the country. A part of this area, in the unreserved public domain and state land grants, was running wild as far as forest management or protection was concerned. The Select Committee declared that the creation of public forests, including two million acres purchased under the Weeks Law, "represents the most effective step yet taken in the United States to assure a stable supply of timber." But public ownership was lagging far behind the rate at which the forests were being cut.

Senator McNary's committee offered Congress two lines of attack upon the forest problem:

(1) Extend public forest ownership in areas where special public interests or responsibilities are involved, like the protection of navigable rivers; and also where the natural difficulties, costs, and hazards attending reforestation render it impracticable or remote as a private undertaking.

(2) Remove the risks and handicaps from private timber growing as far as practicable, in order to give the greatest possible incentive to commercial reforestation.

A bill was ready when Congress convened in January 1924. The heart of it was protection from forest fires, with the framework of co-operation which had been presented so forcibly by the testimony and experience of the Pacific Northwest. The three-way co-operation of government, state, and forest owner became the law of the land.

The McNary Bill freed forest purchases under the Weeks Law from the limitation to watersheds of navigable streams. It authorized the President to establish national forests on military and other federal reservations where the production of timber might be combined with other public uses and to add unreserved public lands to the national forests upon recommendation of the commission which approved Weeks Law purchases.

Under the astute guidance of the senator from Oregon, the bill had remarkably easy sailing through Congress. It was championed in the lower House by an enthusiastic farm-forester from New York State, John W. Clarke, and so became the Clarke-McNary Bill. The diversion of Gifford Pinchot's interest to a new arena in Pennsylvania politics left the supporters of forest regulation without a leader. They made no fight against the new and popular proposal from the West. It was ten years before federal control of logging practices was heard from again. The old western rebellion against T.R.'s wholesale creation of national forests from public lands, however, again raised its head. It struck out the section of the bill that restored this power to the President. Otherwise the measure went on its way triumphant, with scarcely a ripple of opposition.

On the day of its passage I invaded the sacred precincts of the House of Representatives. Congressman Clarke smuggled me into the Republican cloak room, where I could look directly into the chamber and hear the debate from the side lines. One of the pages passed on to Mr. Clarke and other members in the thick of the fray my penciled replies to questions from the floor. After the four years of controversy, it was a great thrill to be in at the kill—even if the victory was bloodless.

In addition to its drive against forest fires, the act opened up new fields of co-operation between federal and local agencies in forestry. One was in growing and distributing forest seedlings. The Select Committee had found the scale of forest planting lagging far behind its commercial opportunities and the public need for reforestation of unproductive land. Out of more than 100 million acres of denuded woodland and abandoned farms, only 36,000 acres were being planted annually with forest trees. The Department of Agriculture became the ally of the states in providing trees and seed, on the familiar pattern of matching dollars. This bit of legislation itself proved a "tree planted by the rivers of water." The distribution of forest seedlings from state nurseries in the year ending June 30, 1949, exceeded 200 million trees.

The act also made farm forestry a part of our national program for agriculture. Forestry was fitted into the system of rural education already established and under which some of the states had started work on their wood lots. The Select Committee was impressed by the opportunity offered by three million farm woodlands. Here again the act set in motion a co-operative endeavor that grew to large proportions.

For twenty-five years the Clarke-McNary Act has led a federal policy of co-operation in dealing with state and private forest lands. Its passage cleared the air of controversy and launched an era of good will and joint effort. Federal and state foresters and private owners of woodland were given a national pattern, and co-operation is infectious. Federal men had opportunity to aid and advise state foresters on local problems of all kinds and found themselves co-operating with lumber companies on almost every phase of forest management and utilization.

One of the fruitful forms of co-operation centered around the U. S. Forest Products Laboratory. Much of its work dealt with fundamental wood science, like the basic mechanical properties of American timbers. But much of it had direct and practical application to industrial processes and profitable forest use. The laboratory, for example, worked out kiln-drying techniques for many tough, resistant woods and greatly expanded the field of raw material for furniture and woodwork. It determined the pulping qualities and yields of a large number of species which the pulp mills had never used and opened up new horizons for this industry in many directions.

Hundreds of wood-using concerns took their technical problems to the laboratory, served on its committees and contributed to co-operative research products. Many reactions were carried back to the logging and growing of trees. The profitable marketing of timber is the obvious key to industrial forestry and the crusading founders showed rare foresight in the emphasis placed upon wood utilization in the earliest days of conservation.

It may be drawing a long bow to assume any connection between developments of this kind and the governmental policy expressed in the Clarke-McNary Act. Much of the technological progress was under way years before the bill was written, and most of it would have gone on regardless of how Congress dealt with fires or with forest owners. Its driving impulse was economic—the growing necessity, as our virgin forests contracted, to find new sources of raw material and new uses of capital. At the same time, the settlement of the controversy over regulation had an appreciable effect.

The industry men lost their resentment over the threatened interference from Washington and their fear of domination by distant bureaucracy. They welcomed men from the Forest Service, talked and worked with them freely and, incidentally, took many foresters and wood technicians of government training into their own employ. The spirit of teamwork took hold and spread.

Senator McNary put more power into the co-operative solution of forest problems by joining with Congressman John McSweeney of Ohio in another basic piece of legislation. The McNary-McSweeney Act of 1928 was dubbed by some foresters "The Hoot-Mon Law." It set up a permanent plan of forest research with authorization for continuing funds far beyond what the Forest Service had been able to wrangle in its current supply bills. The resources of the Forest Products Laboratory were greatly increased. A series of regional Forest Experiment Stations was provided, where the tree-growing techniques and forest economics of the local woodlands could be worked out for the benefit of all owners. "Co-operation" was written in every paragraph of this constructive law. In almost every instance, the tree-growing stations located study tracts on private forests. Industry men were put on advisory committees, proposed study projects, and helped in the preparation of reports. Soon most of the experiment stations were demonstration areas of local cutting and planting methods. They became meeting grounds for farmers and lumbermen. Here was more education and co-operation of a practical kind.

The McNary-McSweeney Act also set in motion the gathering of vital statistics on national timber supply. It ordered an inventory of standing trees, their growth and the drain upon the resource from cutting, fire, and natural pests. It was to be kept current like the national census of people and manufactures. The "Timber Survey" was of obvious necessity in guiding state and federal policies, but to a surprising degree it began almost at once to serve industrial purposes. The large areas of regrowing pine lands found in the South opened the eyes of pulp and paper men to a source of pulpwood which previously had been largely "written off." In many parts of the country the survey gave us the first accurate measure of the growing power of our cutover forests.

Another job blueprinted by the McNary Committee to put American business into timber growing was the adjustment of forest taxes. This has been no simple or easy task. An exhaustive study of the problem was made by Fred Rogers Fairchild, tax economist of Yale University. Dr. Fairchild's investigation carried him into most of the heavily forested states. He made many spot checks of the tax burden borne by forest lands and its effect upon timber cutting and replacement. He worked with state tax commissions, devised model tax laws, and conducted a veritable educational campaign of his own against killing the forest goose that lays golden eggs for the community.

A yearly property tax on woodlands, related to their commercial value, has the effect of taxing the same growing crop many times. Unless very moderate, it may build up a tax load that forces an owner to cut his timber prematurely and discourages him from holding the land to grow any more. The countries with long experience in forestry and stable land economies have generally solved this problem by taxing the yield or income from a forest property. But it has proven well-nigh impossible to adopt such forest taxes out of hand in the United States. The very counties where the problem is most acute depend largely upon forest taxes for their livelihood. They could not change over to an equitable yield tax without going bankrupt. On the other hand, a yield or income tax fair enough in itself would usually mean confiscation if placed on the same stumpage which has already paid forty or fifty annual property taxes.

Dr. Fairchild and his collaborators worked out several plans for gradually shifting the principal forest tax to the value of the crop at time of cutting. One of these plans was adopted by Oregon and Washington. It makes the change with the start of a new crop of trees. In each of these states cutover areas may be registered as "reforestation lands." Thereupon the owner pays a low, fixed tax on the land itself. In Oregon the tax is set by law at five cents per acre. In Washington the assessed value of the land is fixed at $1.00 per acre, which means a tax of four or five cents. In both states any timber subsequently cut from "reforestation lands" pays a yield tax of 12½ per cent of the stumpage value at time of felling.

The enactment of these laws about 1930 checked the high tide of forest tax delinquency that had been rolling up for many years. Many lumber companies paid up back taxes on cutovers that were still redeemable. In many counties the owners did not have to exercise their option of classifying logged areas for reforestation. The assessors adjusted cutover valuations so that such lands would stay on the regular tax rolls. The net result of this competitive interplay of tax systems has been to stabilize the tax cost of growing timber and put confidence into long-range forest enterprises. It was right here that the foundation was laid for the West Coast tree farms which came into the story a few years later.

Several other states have given owners of cutover or regrowing lands the option of putting their holdings under some form of harvest tax. The greatest value of Dr. Fairchild's far-ranging investigations was to give the taxing authorities an awareness of the problem. Forest-minded tax commissions and assessors have often found practical solutions under the laws they now have. And in many cases the forest owners, once they quit opposition in generalities and got down to cases on what their land could actually produce, found that taxes were not the bogey they had previously pictured. This has been true particularly under the levels of stumpage values prevailing since World War II. The problem of forest taxes, like the hazard of forest fires, will always be with us. But, like fire, taxes have become a relatively stable and calculable factor in the cost of growing trees. In most of the United States today neither fire nor taxes are the "major risks and handicaps on commercial reforestation," which the McNary Committee regarded them in 1923.

CHAPTER VII

Forestry at the Grass Roots

THE young communities of the Chehalis Valley were deeply concerned over their spreading acres of logged-off land. The smoke of burning slashings clouded their skies every spring and fall, and too often in midsummer the sun was obscured by the smoke of runaway fires. The residents of the valley called a meeting in 1911 to see what could be done with their growing empire of stumps. Many wanted to raise money, advertise stump farms in Chicago and Minneapolis, and start colonies of new settlers westward.

"Reforest them," said George S. Long of the Weyerhaeuser Timber Company. "The Pacific coast is the nation's great woodlot. Someday timber will sell for what it costs to grow it, as other crops do. If lumbering can be made perpetual in western Washington, and I think it can, it will afford the greatest basis for continual prosperity of any asset we have."

Here and there in the South and in the Far West, even in the heyday of cheap virgin stumpage, a few timbermen were feeling their way toward something permanent for themselves and their communities. These men had a blind faith in the land and they hung onto it. Mark Reed of Shelton laid down the rule for all the logging companies with which he was associated in the South Olympics: no land would be dropped for taxes. The timbermen owed it to their communities to carry through.

The Weyerhaeuser Timber Company held its cutover lands, selling only areas of agricultural value near established communities. The company started its own studies of soil quality and potential forest growth, which led to a classification of all cutovers from the standpoint of best permanent use. Thirty years later the lands of these and a score of other pioneer timbermen became West Coast tree farms, committed to the "continuous production of forest crops."

Years ago I had a memorable tramp with Major Griggs of the St. Paul and Tacoma Lumber Company, through some of his cutovers on the westerly slopes of "Mount Tacoma," as he persisted in naming Puget Sound's towering peak. Signs along the road told the world: "New forests are growing here." We found a respectable logging show of young timber where the company did its first cutting in 1891. The major radiated enthusiasm; *his* company would never quit and move out of the country. One night I joined him in talking about Douglas fir forestry with his mill crew and the permanency of jobs which had growing forests behind them.

A cardinal point in the Pinchot crusade was to reach and convert the forest-using industries. Its immediate weapons were education and the pressure of public opinion. Many industrialists listened and were interested. Some of them, reading about the impending timber famine, rubbed their chins and bought more forests. More companies began serious study of their woodlands. A few pulp mills, railroads, and water companies in the eastern states started to plant trees.

But on the surface of things there was little change in the ways of forest industry at the time World War I thrust its urgent problems upon us. Mr. Pinchot told his friends that as far as the lumbermen were concerned he felt like John the Baptist preaching in the wilderness. There were measurable gains in controlling forest fires in the North and West and a start in the South, but woods burning still took a fearful toll. Migratory and destructive lumbering was still in full cry. The sawmills went right on cutting out, junking their land and moving to new pastures.

At Mr. Pinchot's request several of us undertook a survey, by questionnaire, of the current practices in timber cutting. Our form queried each logger on his timber holdings, operating life, and so on. Its final question read: *"What provision do you make for reproduction?"* Most of those questioned left this line blank. One operator in Minnesota wrote in: "Nothing of the kind allowed in my camps." As far as renewing trees was concerned, this reply fairly summarized the situation.

Even in those days the industry was still dominated by "free timber." Its shadow hung over the lumber markets from coast to coast. Any sawmill in the West could pick up incomparable old-growth stumpage in the next valley or the next county for four bits or a dollar per thousand feet. Men conserve things that are scarce and valuable; and notwithstanding the prophecies of shortage, timber remained a cheap commodity by the cold standards of commerce.

Furthermore, at the height of the conservation crusade, lumber encountered the first great invasion of its markets by "substitutes." Steel, concrete, brick, and paperboard ware threatened the reign of King Timber. The per capita consumption of lumber fell off from the peak years of 1907 and 1909. Timber values wavered and dropped in most of the forest regions. The opening up of the vast new resources of the West threatened all the softwood markets of the country with oversupply and instability. In 1912 the lumber industry entered a five-year slump. Hard times shut down many mills and forced most of the others into a struggle for survival. With an overburden of milling capacity, uncertain markets, risks of fire loss and unpredictable tax burdens, the lumber industry was unready—indeed, largely unable, to settle down and grow trees.

I sat on a log one day during these troublous times, talking with one of the most successful and farsighted timbermen in the Pacific Northwest. He had built up a barony for his eastern employers, much of it bought in the early days at ten or fifteen cents per thousand feet. But he feared that even those shrewd purchases would not prove him ahead in the end. Costs were piling up at a rate that threatened to wipe out any profit. And now lumber was losing many old customers; its days of widespread use might be ending. My friend felt that his duty to his backers required abandonment of his cherished program of gradual cutting over a long period. He must build another sawmill, cut off the timber as fast as possible, and bail out the enterprise before it was too late. A hard man to sell on forest conservation!

Gifford Pinchot and I looked at the economic side of the forest picture through different glasses. He saw an industry so blindly wedded to fast and destructive exploitation that it would not change. I saw a forest economy overburdened with cheap raw material. Mr. Pinchot saw a willful industry. I saw a sick industry. G.P. disagreed sharply with the 1916 Forest Service report on lumber, in which I tried to give a factual picture of the underlying economic troubles. He called it a "whitewash of destructive lumbering." Chief Forester Henry S. Graves thought the report was sound and published it, but I lost caste in the temple of conservation on Rhode Island Avenue.

Mr. Pinchot was keenly disappointed by the apparent indifference of the lumber industry to his appeal for conservation, but his crusade accomplished far more than could be seen on the surface. It forced the lumbermen to defend conceptions and practices of fifty years' standing, like the abandonment of cutovers through tax delinquency. It rubbed their noses into the stark realities of idle land and ghost towns. They had to justify these things not only to the public generally but to their own local communities and to themselves. They were on the spot. And a slow change in the thinking of the industry, in its attitude toward its own land and raw material, was set in motion. In nearly every forest region leaders appeared among the timbermen. Practical minds brought the high-sounding gospel of conservation down to earth. With an instinct for doing first things first, they organized fire patrols, set up state protection codes, worked for better ways of taxing cutover lands.

Lumber-borne communities which had no relish for a ghost-town future added to the force of public opinion. Several western towns, viewing their horizons of stumps with foreboding, urged the government to take over the old slashings and reforest them. This concern led to special acts of Congress which authorized the Forest Service to exchange government timberland for private holdings and to trade public stumpage for the cutover land of lumber companies. The buying of forest land under the Weeks Law started a wave of community petitions for national forests that would give their industries more assurance of a future. I went to Florida on such a request from the Jacksonville Chamber of Commerce. A group of us explored a stretch of sand hills and scrubby pines, looking for national forest possibilities. I recall stopping at one hilltop where our courteous guide was anxious that I should be duly impressed. "Here, suh, you are standing on the highest mountain in the state of Florida. Three hundred and twenty-five feet, suh!"

These currents of public concern had their effect upon the timber people, however indifferent they might seem. Perhaps the most telling force of all was the example of the men who kept their forest land after logging. With a permanent stake in the region, they led its progress toward a new forest economy. It was a slow and halting evolution—not the quick reform demanded by revivalists and crusaders. It was to take another quarter century and a second world war to make the change decisive.

II

A year or two after the Panama Canal was open for commerce, I visited a huge lumberyard on Chesapeake Bay. It had been set up by a Northwest milling company to distribute Douglas fir and West Coast hemlock lumber up and down the Atlantic seaboard. While I watched, an intercoastal freighter was warped into one of the docks. Her decks were gleaming yellow with Douglas fir freshly sawn at mills on Puget Sound. It was a fascinating sight to see her derricks swing bundles of lumber over shipside and down onto the pier. There straddle-bug carriers pounced on the parcels and scurried away with them to storage sheds or loading platforms along a maze of railroad tracks. The boat disgorged three and a half million feet of lumber, more than enough to build three hundred houses. Most of this lumber, brought 6,500 miles from North Pacific harbors, was not the choice high-grade items like ship decking and wide finish, for which we must turn to virgin forests. It was the studs and joists and sheathing of everyday home building.

The sturdy intercoastal freighter steaming up Chesapeake Bay told a lot of forest economics. Some of the strings of boxcars rolling out on the dock were to take her lumber as far back toward the forests which grew it as Buffalo and Pittsburgh. The region she served, which was roughly the states east of the Mississippi and north of the Ohio and Potomac, consumed nearly half of all the lumber put under hammer and saw in the United States. This region contained within itself three great softwood forests. The timber of New England, the Alleghenies, and the Great Lakes had more than supplied its needs for two hundred and fifty years; but now less than a fifth of the lumber used in these sixteen states was of home production. Even the vast pineries of the South were no longer able to fill the breach. The intercoastal fleet of a hundred and fifty ships had become a vital source of supply, and thousands of freight cars, loaded with far western lumber, were rolling over the transcontinental railways every year into Minneapolis, Chicago, Omaha, and St. Louis. British Columbia was shipping lumber cargoes to New York and Boston, and a few shiploads of spruce were landed from Russia.

Our use of timber species has changed from generation to generation—to fit the woods actually available. And there has been no lack of Paul Bunyan guile in dressing up woods new to the market with familiar trade names and sales appeal. Douglas fir broke into world markets as "Oregon pine" because *pine* was the softwood universally known and accepted. An enterprising manufacturer in British Columbia sells West Coast hemlock around the world as "Alaska pine." And the ponderosa pine of the West sought for years to link itself with the white pine of the lake states. It paraded through the market place as "western *white* pine" until the Federal Trade Commission ordered its manufacturers to stop using the word "white" in any way, shape, or form.

About the shrewdest trick of all was pulled by a Yankee whose mill was located in Norway, Maine. Running out of white pine timber, he turned to the harder and less popular red pine of the northern states. His lumber was conspicuously labeled *Norway pine;* and who could deny that it was made in Norway? Thus he quietly appropriated the high standing, in the trade, of Norway iron and other Norwegian products. The man from Maine put over his stage name so successfully that all the lumber made in half a dozen states from this native American tree became known universally in the markets as "Norway pine."

But no substitutions of new woods or new products could conceal the cold reality that our main supply of timber was far removed from the people who use it. The McNary Committee was staggered by the national freight bill on lumber, some $250,000,000 a year. It was costing $16 a thousand feet for the steamers to bring Pacific coast lumber through the Canal to Baltimore or New York and probably $3.00 more for distribution by trains and motor trucks to the points of actual use.

The steady retreat of the timber front, to the North, the South, and the West, has been well screened in the United States by our marvelous transportation system and by the driving power and competitive structure of the lumber industry. But heavy lumber hauls of 6,500 miles by sea and 3,000 miles by land made plain the cold fact of forest depletion. The center of American lumber production has shifted every twenty or thirty years; and every shift has pushed it farther from the regions of largest consumption. Transportation has taken a larger and larger bite out of the consumer's dollar. And transportation costs from afar have set the value on little trees grown at home. Commercial forestry often has had its start in the competition between a second-growth woodlot and a boxcar.

The McNary Committee heard many lumbermen tell about the rise in eastern stumpage values during the preceding twenty years. Almost every commercial timber species had at least trebled in value. In Virginia and North Carolina second-growth pine accessible to Washington and Baltimore was selling on the stump for $7.00 to $10 per thousand board feet.

White pine, forty or fifty years old and suitable mainly for box lumber, was generally valued at $10 in Maine and $16 in New Hampshire and Massachusetts. "The 600 billion feet of culled and second-growth timber in the eastern states, although of inferior quality, is the most valuable forest growth we now have," said the McNary Report, *"solely because of its location."*

In the nineteen twenties I had a chance to revisit the old family homestead in New Hampshire. Only one small piece of bottom land was still in cultivation. All the upland, running high on the slopes of "Peaked Hill," had become a white pine forest. Roaming over the old pasture, I recalled how the village boys used to gather there for Indian fights. The main attraction then was the hundreds of slim little pines. Trimmed and pointed, they made wonderful javelins. I still carry the mark of one which pierced my cheek. Armed with these near-lethal weapons, the tribes would dodge and ambush each other behind the granite boulders. Now the old pasture was a pine forest of ten- and twelve-inch trees. It had become the most valuable part of the farm. There were many other pine lots like it in the county and half a dozen little mills cutting boxboards and sheathing.

One of the amazing characteristics of most types of forest in the United States is their tenacious hold on the soil. On enormous areas, after cutting and burning to the point of seeming destruction, the forest has repossessed the land. The author of *Gone with the Wind* wrote vividly of the disheartenment of Confederate soldiers who returned to homes in Georgia only to find that the pines had taken over most of their farms. Loblolly pine became known as "*old field* pine" because it so often seeded up abandoned farmland where old rows of corn or cotton hills could still be traced. I once saw a pathetic old apple orchard in Vermont, in the first wave of invasion. The ruthless little white pines and spruces had moved in. They were crowding up through the branches of the gnarled old apple trees, dying and dead, like hunting dogs pulling a valiant old stag to earth.

West of Puget Sound, around my home, farming on a substantial scale began in the seventies and eighties when logging companies put cutover land on the market. Three tides of settlement have flowed and ebbed in the little valleys leading back from the bays and inlets of the sound. After each ebb of the tide, the waiting Douglas firs have thrown a scouting force over any untenanted homestead or abandoned pasture and then, in seven or eight years, closed in with solid ranks. Today over 80 per cent of the county is again in forest.

The new forest that has come back on old burns and cuttings and abandoned fields is not always the same forest that was there before. In some places hardwoods have taken over old coniferous woodlands. Logging in New England and New York, since colonial times, has cut down the proportion of pine and spruce and turned the north woods into more of a hardwood forest. On old burns and reburns of the lake states, jack pine and aspen have often pushed out the valuable white and red pines. But on three acres out of four the persistent forest has come back in some form and started of itself to replenish the depleted timber supply of cutout regions.

With the self-renewal of the forests has come also a growing and changing technology in their use. Many of my forest-school conceptions of "weed trees" and "inferior woods" have been knocked into limbo. My early ideas of how wood can be used have been constantly revised. This does not apply merely to the progress in chemical pulping and plasticizing. It goes as well for the use of sawn lumber, for adapting construction practices to new species and grades of woods, the jointing and gluing of little pieces into big pieces, laminating heavy beams out of inch or two-inch boards, and all the rest of it. Of course we still have plenty of uses for the big, clear logs grown only in old forests, and we miss them when they become scarce. But an astonishing part of the forest story of America is the fitting of wood-using practices in every region and industry to the forests actually within reach. The useless trees of one generation become useful material to the next. The timber cruiser on the Columbia River who told me in 1905 that hemlock trees "aren't worth counting" could not foresee the use of hemlock in making ladders and airplanes or rayon yarn. Technology is the philosopher's stone that turns all our trees into gold, *when we need them.*

Technology joined forces with Mother Nature in bringing about the outstanding forest recovery of the world, the resurgence of 150-odd million acres of southern pine. In spite of clean cutting and wholesale burning, the returns under the timber survey of 1930 showed a great tide of young trees sweeping back through the old piny woods. The restoration of southern pine as a great source of raw material, backed by the expanding production and techniques of the pulp and paper industry, is the most striking event in the story of American forests.

We have, of course, always drawn upon the growth in our woodlands as well as their stock of old timber. From the days of William Penn, timber has been *a crop* to farmers, village sawmills, and innumerable small woodworking industries. When big business became migratory and moved on to virgin fields, little business stayed at home with the regrowing woods. The significance of the comeback of southern pine was its magnitude in acreage and footage. Second-growth pine became "big business" in its own right. It hit the milling industry on the virgin timber fronts of the West in the spot most certain to command attention—by competition in their own markets. It knocked into a cocked hat all the painstaking estimates of the "end of southern pine." By 1930 second-growth logs furnished nearly one half of the pine lumber manufactured in the South. These regrown forests were then yielding over five billion feet a year, or 17 per cent of all the softwood lumber cut in the United States. Probably another 10 per cent came from the second-growth stands of New England, the Allegheníes, and the Great Lakes.

It has always seemed to me that the turn in our forest economy from the timber *mine* to the timber *crop* came in the decade ending with 1930. Cost of transportation had clearly become the true measure of forest values. The old migrant, "King Timber," and his slogan, "The cheapest source of raw material," were challenged on their own ground. The little trees were taking over.

III

A number of things combined to make the "climate" favorable for private tree growing in the ten years or so before the depression of the thirties. Forestry seemed to generate itself at the grass roots in many different places, like young pines popping up here and there in a New England pasture.

The stumpage value of young timber and many examples of annual growth at the rate of two or three hundred board feet per acre made landowners wonder if growing trees might not be a paying business. The housing boom and strong lumber market of the middle twenties gave weight to the idea of money in trees. Wood pulp production was increasing and ranging afield for new supplies.

Protection from forest fires gained ground steadily under the co-operative inducements offered by the Clarke-McNary Act. Often forest owners found their old woods-burning neighbors surprisingly responsive to stopping fire once they were convinced "the company" was really going to grow more timber and *keep the mill running*. It was discovered that one of the best ways to cure incendiary fire setting was to employ local farmers and their boys and girls in planting trees or to give them a few thousand to plant at home. The South began to shake off the old bugbear that "nothing can stop the burning of the woods."

One of the old-school lumbermen who became converted to the possibility of growing young pines was Colonel Sullivan, who managed the huge mills and 250,000 acres of the Great Southern Lumber Company at Bogalusa, Louisiana. He had once told me with explosive Irish that only "Uncle Sam could regrow southern timber and Uncle would probably go broke at it." I was genuinely surprised to learn that the colonel had hired a forester and started a forest nursery. But Great Southern was one of the first of the pine operators to salvage small logs and woods slash for the manufacture of kraft pulp. They knew the value of small wood. I used to marvel at the trainloads of little rough logs and branches coming in from their woods, following the cars of magnificent old-growth longleaf.

On my next visit to Bogalusa, Colonel Sullivan took as much pride in showing me his nursery and young pine plantations as he had ever displayed over his export timbers and rift flooring. When the last virgin log had rolled in and the Great Southern sawmill had been dismantled, the Gaylord Container Corporation took over. I had the good fortune to visit the property again twenty-five years after Colonel Sullivan and I were down on our knees examining the shaggy mane of needles and marvelously protected buds on the yearling longleaf seedlings. The company had planted, all told, 54,000 acres out of the 250,000 of the Great Southern cutovers. They found that natural reseeding and their demonstrated success in fire prevention would restock the rest of it.

I drove for most of an afternoon through an unbroken forest of young pine. The annual growth on the property was estimated at upward of 300,000 cords of pulpwood. The first cutting in plantations started by Colonel Sullivan had begun. Foresters were taking out about half the trees, ten cords or so to the acre, leaving the other half to grow and reseed the ground. Some of the little farms we passed were also growing pulpwood, with the help of company foresters. The fire patrols and lookout watchers were local farm boys. A large part of the success at Bogalusa has been in good public relations.

Back in the twenties many other southern mill men were finding new supplies of logs in their culled woodlands and young forests on old clearings. One of the most effective of the early missionary campaigns of the Forest Service was its study of logging and milling costs on logs of varying size and grade. Compare the cost of hauling and manufacturing each log with its actual yield in lumber footage and dollars! Why is it not good business to leave standing and growing in the woods the trees that will not pay their own way through the sawmill? Many mill men applied these tests to their own stands of timber and drew a practical line on the trees which were money losers if converted into logs. In course of time, the trees left standing often paid unforeseen dividends in the added footage, quality, and value of their wood. Many southern forestry enterprises started from just this firsthand experience. "Selective logging" became the art, not only of picking the trees which would pay out best by standing and growing a while, but also of leaving a growing stock on each acre that would put on all the wood the soil could produce.

In 1902 the Crossett Lumber Company began cutting its Arkansas shortleaf pine, down to trees about fourteen inches on the stump. The growth of the small trees interested these farsighted men. Summer camps of forestry students from Yale made maps and estimates and determined rates of growth on trees of different size and spacing. Recommendations of Yale's H. H. Chapman in 1922 laid the basis of the Crossett plan. It moved progressively from fire protection to selective logging and tree planting where necessary to fill gaps in the growing stock. The Crossett forest has now attained a yearly net growth of three hundred board feet of pine to the acre. The raw material is converted into lumber, kraft paper, charcoal, and chemicals from the distillation of hardwoods. In twenty-five years Crossett has become one of the most impressive industrial forest enterprises of the world.

Many lumbermen and foresters made pilgrimages to Urania, Louisiana. It was a mecca of piney woods forestry. There we shared southern hospitality and the enthusiasm of lumberman Henry Hardtner for growing trees. Roaming over his "longleaf pastures" with "Marse Henry," I learned more dirt forestry of the South than I had ever gleaned from textbooks or lectures. He had an indefatigable zest for finding things out himself. He had learned when the longleaf seedlings should be lightly burned to free their needles of the brown-spot fungus. He had areas of saplings, side by side, one regularly burned, the other kept meticulously free of fire, to show the actual effect of the prevalent woods burning upon mortality and growth rate in young pine forests. He had seeding plots and thinning plots, running the whole gamut of southern pine silviculture. And he told his wealth of experience, the things he had learned by trial and error, in a homey, down-to-earth way that took hold. My days with Henry Hardtner were inspiring. He had the unconscious power of a man whose taproots grip Mother Earth. He was a great leader of southern lumbermen toward a new order of land use, and I felt signally honored when he put my name on one of his "pastures" in commemoration of our visit.

The new forest of saw logs caught some of the southern timber barons unawares. A good story is told on one of them, highly renowned for business astuteness. By counting the remaining holdings of virgin timber he had forecast the end of southern pine lumber production to the last year. This gentleman asked the manager of one of his sawmills where he found the small clean logs that were being hauled in by the truckload. They were bought, he learned, from old holdings of his own company, which he had junked as worthless land.

But the leaven was at work. The growing army of conservation was moving out in widening circles. There was an increasing flow of new blood from the forest schools into sawmills and logging camps. In the twenties, three or four hundred trained foresters were graduated every year. Most of them found jobs in the public services, but an increasing number seeped into the ranks of industry. They worked in logging camps, and many of them became top logging engineers and timber cruisers. They worked in sawmills and pulp mills. Many became lumber salesmen. They recruited the technical processing of wood, the fast-developing fields of wood preservation, wood lamination, wood pulping, and so on. They largely took over the protection organizations, private as well as public. And often they made their way into the land and timber departments of forest industries and the ranks of management itself. In the long run the educational impact of this invading army was enormous. In 1949, thirty-four schools were training foresters and 2,700 of their graduates were in private employment. As the young trees were taking over the woodlands, so the young foresters were moving in on the wood-using industries.

In 1920, 40 or 50 per cent of the western timberlands were under federal administration. The timber sales on national forests and Indian reservations, the arguments and squabbles, the constant give and take of ideas between government and industry men gave the lumbermen a practical schooling in forestry. Some of them began to copy Uncle Sam's methods in their own logging. Concerns with timber ahead for many years saw possibilities in leaving a growing stock on their land as the government was doing. We drew the Secretary of Agriculture himself into a memorable quick deal on the Lassen National Forest. A citrus growers' co-operative had large timber holdings there, a sawmill and factory for making their own orange boxes. Their aggressive manager wanted a ten- or fifteen-year contract to cut the government's sections adjacent to the co-operative's own logging operation. Striking while the iron was hot, I propositioned this Californian of large ideas on selective logging and slash disposal on company lands exactly as the government required on the national forest. He finally agreed. Seeing a chance to cut through departmental red tape in Washington, we persuaded Secretary Houston to give the Forest Service instructions, in longhand, that very night. A scrap of paper authorized a long-term sale of government timber with a stipulation that the same cutting and logging practices be applied on company lands. That bit of negotiation was the germ of the co-operative sustained yield policy which Congress enacted into law thirty years later.

The pioneering of the federal foresters in cutting methods and tree-growing research had much to do with the turn of western lumbering from timber mining to timber cropping. The decade beginning in 1920 saw the start of a dozen industrial undertakings in western pine and Douglas fir forestry. The first step always was tighter control of forest fires. The Diamond Match Company had a forestry plan prepared for its California pine lands in 1915, but for several years attempted only to lick a bad fire hazard. Fires were burning an average of two and a half per cent of Diamond Match lands every year. Then the company revised its logging methods and began leaving a growing stock of small pines and sapling thickets. After working out effective fire protection with its neighbors, the Michigan-California Lumber Company, also on the western slope of the Sierra, started selective cutting and the saving of advanced young growth, in 1924. A company study of the value of growing trees led to a policy which leaves uncut the young, "tapering" sugar and ponderosa pines up to thirty-four inches in diameter.

In the white pine of the Idaho Panhandle, Potlatch Forests, Inc., began its long-range program of continuous production. Consolidation of land holdings was the first step. This was followed by intensive protection from fire lookouts and patrols and slash piling after logging. Many tests of selective cutting developed a plan of tree-by-tree marking, to guide the fallers, under which a third of the merchantable white pine timber was left for growing stock.

Our pine forests generally lend themselves to selective harvesting. Groups of vigorous, spire-topped youngsters crowd in among veterans of three hundred years, whose flat crowns bear the stamp of maturity and slow growth. A fourth of the trees may contain three fourths of the merchantable volume and nearly all of the high-quality grades. It is seldom difficult, in the American pineries, for a forester to select the prime old veterans for cutting and leave a fair stand of respectable trees. If the loggers handle their machines well and the slashings are burned with care, the result is a picture-book forest. There is no evident devastation. Travelers on the highway see young timber, green and vigorous among the stumps, and rejoice. To most Americans, and to all magazine writers, this picture is *forestry*. This and the planting of trees outright.

But in the tall Douglas firs of the Pacific Northwest forestry has no parlor tricks. From these woods nature has largely excluded the family type of forest, where sons and grandsons cluster around the aging parents. She has given over the land to a mass stand of trees of nearly uniform age, in dense, unbroken ranks. The size and height and density of old-growth timber give the logger an engineering problem of removing two or three hundred tons of wood from an acre. Harvesting such a crop is not possible without wrecking the forest. And after the heaps of slashings have been burned, the landscape becomes a fitting approach to Dante's *Inferno*. No passer-by, unfamiliar with the ways of Douglas fir, can believe that this sort of thing is *forestry*. To him, it is devastation, pure and simple. The forest seemingly has been destroyed.

Although I should have known better, once in my crusading days, I used an old photograph of just this kind of logging and slash burning to illustrate a magazine article on destructive lumbering. Not long after its publication an old friend and fellow forester on the West Coast brought in a picture of a young forest, twelve or fifteen feet high, whose density and vigorous growth would delight any man of the woods. "That's the real stuff," said I. "Where are you growing trees like that?" "You've seen that forest before," said my friend, David Mason. After peering at the picture several minutes I recognized a conspicuous black snag. It was my horrible example of destructive logging, fifteen years later.

Along many highways through the Pacific Northwest, old scenes of logging desolation have disappeared under waves of young, green Douglas fir and West Coast hemlock. Foresters have had to learn the patience of Mother Nature. Douglas fir produces a heavy seed crop only once in six or seven years. If the dry fall winds are on time and a heavy mast on the trees, seedlings shoot up far and wide, in city lawns, gardens, and plowed fields as well as old burns and cuttings. Land long denuded suddenly greens up with young forest. Seventy-five or eighty per cent of Douglas fir forestry, we have learned, is keeping fire out of the woods. Enormous areas have become restocked with junior forests through no skill or planning of the owner beyond good protection. And where blank spots occur in this young timber, they usually bear the scars of repeated burning. The cutting practices on national forests have worked with the natural habit of Douglas fir to reproduce en masse in large, clean openings. "Clear cutting by patches," we call it, leaving wind-firm seed trees or blocks—within easy range of the fall winds.

In 1920 thoughtful men were discovering that the greatest forest asset of the Douglas fir region was not its virgin timber but the growing power of its soil. The Western Forestry and Conservation Association offered its services to any forest owner who wanted to explore possibilities for long-range tree growing. Several companies, up and down the Coast, employed the association to study their properties. One of its most interesting undertakings was an investigation of the regrowth and industrial future of the Grays Harbor region. Grays Harbor was then booming like a new placer camp, with eighteen sawmills and lumber shipments close to a billion feet a year. Western Forestry's survey and recommendations started future planning on four or five of the larger holdings in southwestern Washington. They became West Coast tree farms twenty years later.

I came to know many of these western timbermen in 1928,

when I left the federal service to work for the West Coast Lumbermen's Association. I was impressed by the number of them who knew that the day of commercial timber growing in the Pacific Northwest was not far away. Mark Reed, head of the Simpson Logging Company, told me of his dream to create, from virgin timber and cutovers and young growth of the South Olympic country, a sustained yield forest that would insure the perpetuity of the community which he had built up. He was looking for a forester who could translate the dream into reality. Mr. Reed found his practical builder. He was George Drake, an experienced logging engineer of the Forest Service. The Simpson enterprise moved on in planned timber growing and new ways of using wood.

Much was doing at the ground level around the states in those fruitful years before economic depression struck the country. New York and Michigan were replanting state lands on a large scale. The manufacture of box lumber from young white pine was a growing industry in New England, and many farmers, small estate owners, and box companies were busily planting pine. Pulp and paper companies in the Wisconsin River Valley began tree planting in a large way. The interest of the Goodman Lumber Company at Marinette, Wisconsin, in long-range production of hemlock and northern hardwoods, led to the adoption of a forest working plan in 1927. It was based upon selective logging of about half of the standing timber. Trees under twenty inches in diameter were mostly left as growing stock. I had a glimpse of the Goodman Forest in 1950 after the first cutting was completed. It is growing two hundred feet of timber to the acre every year. Under the fine maples, birches, and basswoods, a young forest is pushing up into all the openings. The company logs lightly every ten years and has a continuous cut of high-grade timber of near-virgin quality.

This "working plan" idea seemed to be catching. In 1930 the Society of American Foresters undertook an enumeration of the "starts" in industrial timber growing. In the entire United States the society found 288 forest owners in all industrial categories, including railroads, mines, water, and power, employing some degree of forest management. Their holdings ran up to about thirty million acres. This was perhaps a fifth of the total industrial forest acreage.

One of the interesting signs of the times was the co-operative forest nursery installed and operated by redwood lumbermen in California. The undertaking was planned and worked out by David Mason, a West Coast consulting forester. Here several companies grew seedlings of Coast redwood and Port Orford cedar, which were planted on the barest of their cut-over lands. The leader was C. R. Johnson of Union Lumber Company, for years the dean and seer of the redwood fraternity. The potent influence of conservation propaganda had some part in this enterprise. When Mr. Johnson told me about it, he said, "I had to do something to live down that terrible word devastation."

And then, as the sun's rays are gradually shut off by total eclipse, the shadow of depression slowly blacked out many constructive undertakings in the lumber woods. We never dreamed, in those dark days on the West Coast, that out of the industry's despair would come a strong forward march in growing trees.

IV

Early in 1933 a delegation of West Coast lumbermen were bound for Chicago in a Great Northern Pullman. We were to meet other delegations of harassed mill men, from the northern and southern timber districts. Should the industry apply for a Blue Eagle code of self-government under the National Industrial Recovery Act? It was a grim company, for the times were hard and necessity presided at our councils. Fully a third of the Douglas fir sawmills were shut down, and the rest were running under a slow bell. The lumber industry had degenerated into a struggle for survival.

At the meeting, in Chicago, a few days later, of lumbermen from all parts of the country, appalling stories were told of how some sawmills kept going. An earnest plea for a *wage floor* of some kind was made by a hardwood manufacturer from Mississippi. He told us of a little outfit up the creek from his plant. "All the hands there get for their work," he said, "is ten cents an hour. But the boss starts the mill off every morning with prayer."

In our Pullman sessions we wrote and tore up and wrote again drafts of many sections that might find a place in an N.R.A. Lumber Code. Control of production came first; the authority would assign so many running hours to each sawmill every quarter. A schedule of minimum wages would surely be required by the New Deal administration, and everyone was for it. We had ready a list of minimum prices, item by item, which we hoped, with many misgivings, might get past the gantlet of General Hugh Johnson and his Blue Eagle economists. When the many ingredients of a lumber code had gotten fairly boiled down, I had my turn. "President Roosevelt," I said, "is almost certain to want something in this code on *forestry*. Let's beat him to the draw. It will help us get the rest." It was agreed. I was commissioned to draft a clause which would commit the industry to a reasonable program of forest conservation.

So it came to pass, after all the discussions and compromises. Article X of the Lumber Code committed the industry to leaving "its cutover lands in good condition for reforestation." The Forest Service concurred. In a final session at the White House the President beamed his approval and urged the lumber delegates to adopt the "top lopping" of slashings which had greatly aided fire control in the Adirondacks. We left the Capitol with shiny new halos on our heads. The proceeding reminded me a bit of the baptism of armies of Franks and Gauls, in the Middle Ages, by the simple rite of wading through a river. But, by and large, the lumber industry took Article X seriously. We made an honest effort to live up to it through all the strains and frustrations of our none-too-successful plunge into industrial self-government. For the timber-using industries of the Northwest at least, progress in tree-growing practices and know-how was the only enduring legacy of the Blue Eagle.

On the West Coast the Pacific Northwest Loggers and West Coast Lumbermen set up a Joint Committee on Conservation with the duty of making Article X the actual law of the woods. Many of the ablest men in the industry have served on the Joint Committee. In the past fifteen years it has been one of the most forceful agencies of the Northwest in all phases of forestry, going far beyond its original purpose as an instrument of the short-lived Lumber Code. It has set in motion many industry-born undertakings which are moving in on the forestry problem from one practical approach or another. From an agency of government, enforcing its regulations under sanctions of federal law, it became a striking sort of internal combustion engine, creating its own power from forces within the industry itself.

As field director, the Joint Committee engaged Russell Mills, an experienced "calked boots" logger and head of logging engineering at the University of Washington. A *Handbook of Forest Practice,* in loggers' language, was written by experienced industry and government foresters. Director Mills and his foresters held meetings with the loggers up and down the Coast, discussing and explaining the manual in everyday terms of the woods. Then the code inspectors went from logging show to logging show. To woods goers in the Northwest, a "show" is any distinct logging unit, like the area of timber tapped by a truck road or railway spur. The foresters got down to cases with the camp foremen on their own jobs. They advised each woods boss how his particular timber and ground could be "worked" most practicably to meet the requirements of code forestry. It was education of the most direct kind, and it was carried up the forks of every creek. The basic requirement was to leave at least five per cent of the timber for reseeding and no cutover land over a quarter of a mile from a source of seed. Otherwise good forest practice was protection from fire and preparedness in fire-fighting equipment, snag felling and controlled slash burning—as laid down in the fire codes of Oregon and Washington.

The hardest part of the new schooling in the woods was selecting the sources of seed for the next crop. It was as much engineering as forestry, for the blocks of seed timber must be picked with an eye to topography and to the roadways and machine settings required in logging the stumpage to be cut. On a few areas of gravel soil, or with many defective trees of poor lumber quality, scattered seed trees could be left. But in the great bulk of our wet and rugged North Pacific country such trees would be blown over before the end of the first winter. Seeding timber must be left in blocks or patches of windfirm size.

The most workmanlike application of this hop-skip-and-jump style of logging and the one most widely used on the larger holdings is the *staggered setting*. A complete road system is projected through an area of timber. The forest is broken up into logging blocks around machine settings or tractor landings. As logging progresses, enough complete blocks are left uncut to reseed the ground stripped clean around them. This checkerboard method of logging, as far as experience has shown, is about tops in Douglas fir forestry and fire control. It enables a logger to get first into his overripe timber or bad windfalls. The access roads and reserved blocks of unbroken forest make it easier to stop fires and to burn slashings safely. And the whole scheme fits the renewal of the Douglas fir forest. To the leeward of any bank of green timber, the seedlings move in—thousands to an acre, their very density producing the greatest volume of wood and the tallest and clearest trunks.

It requires no apprenticeship in either forestry or engineering to perceive that what the Joint Committee on Conservation undertook in 1934 was no child's play. The committee had to translate the beautifully simple dictum of Article X into a scheme of practical logging, workable in the heaviest stands of timber on the most rugged terrain in the United States; workable too in the hands of a tough breed of loggers whose gods were volume of production and machine power in smashing through the wilderness.

The enforcement of code forestry was not altogether seasoned with brotherly love. The committee's foresters were thrown bodily out of a couple of logging camps. But most of the loggers took a degree of pride in making the d— thing work. It was a challenge to their skill and resourcefulness, just like opening up a hard show. Moreover, "the industry" had taken on this job. Nobody else was telling us what to do. We were telling ourselves. He was a poor sort of s.o.b. who would let the industry down. In the spring of 1935 the director reported that 65 per cent of the current logging, in acres, was being done in substantial compliance with the rules of Article X.

Then the blow fell. The notorious "case of the sick chicken," under the poultry-marketing code of New York, reached the Supreme Court. At one stroke the court killed the Blue Eagle and all its fledglings. The Joint Committee was left on the end of a limb and the tree was swinging through the air to the familiar call of "Timber!" But the committee had done a better job of selling than it realized. The directors of each of its parent associations voted to adopt Article X as an association policy. All members of the two organizations were urged to conform voluntarily with the code rules of forest practice. The forestry assessments and the Joint Committee were continued. The *Manual of West Coast Forest Practice* was revised and republished. The committee foresters renewed their rounds of the back roads and logging shows. Shorn of legal authority, they preached forestry as doing right by the land, as good public relations, as paying its way with a new crop of timber.

Most of the other large groups of lumber manufacturers had a similar experience in the short reign of N.R.A. and its aftermath. About 200 million acres of woodland were covered in this first general effort of lumbermen to put some minimum of forestry into their treatment of the land. The general reaction to the summary execution of N.R.A. by the high court was to adopt forest conservation as part of the industry's own business creed. Most of the associations set up permanent forestry committees, with field staffs, recommended cutting practices and advisory services for their members. The death of the Blue Eagle started a spontaneous sweep of industrial forestry which has carried on to the present time.

"The Lord moves in mysterious ways, his wonders to perform." An industry was sick from cutthroat competition. It sought public sanction for controls of production and prices denied by the Anti-Trust laws. It swallowed Article X of its recovery code in order to get Articles I to IX. Out of depression, almost as a last appeal for public help, came the first organized attempt of American industries to apply the rudiments of timber cropping to privately owned forests. And the attempt rested solely upon the powers and injunctions of federal law.

Then, the miracle! The law folded up. Uncle Sam threw up his hands and said: "As you were, men!" But the loggers seized the idea and turned it into a voluntary plan of their own. What started as a federal mandate ended in an upsurge of private enterprise.

Many influences were at work in the late thirties to make timbermen forestry-minded. One was a revival of faith in their own industry as the depression clouds began to break up. Another was the renewal of the threat of federal forest regulation. Most potent of all was the growing realization that the end of the virgin forests was in sight. In the Far West, cheap stumpage no longer could be picked up in the next valley or county. Timber was more and more strongly held and more closely tied to substantial sawmills, pulp plants, or plywood factories. Speculative holdings and open-market loggers were disappearing from the stage. Second-growth logs were crowding into the mill ponds, and the market for cutover lands was strong. The hoary deities of the market place, Supply and Demand, were taking sides with the growing of trees.

CHAPTER VIII

City Boys Raking Leaves

THE summer of 1949 took me again to the Black Hills of South Dakota. Here one may check every stage of selective cutting, tree growth, and reseeding through fifty years of technical management. I saw how the "crop" trees had grown on an area where I marked virgin timber in 1909. But the most impressive forestry of all was on the 300,000 acres, bordering the highways for mile after mile, where dense stands of young pine had been thinned by the Civilian Conservation Corps. Overcrowded thickets, stagnant in growth, had been transformed into an even young forest of four- and six-inch trees standing about eight feet apart. The most exacting German *Forstmeister* would commend the pains and craftsmanship of these young woodsmen. The Black Hills area is a sample of the nearly four million acres of forest which were thinned and put in better growth by the C.C.C.

Many automobile travelers through the northern reaches of the lake states marvel at mile after mile of highway from which they look out over a young Norway and jack pine or spruce forest. Some of them recall the desolate stretches of burnt-out and barren land which the young trees have now reclothed. Many agencies have been busily replanting the vanished forests of the North, but no autoist will traverse many miles without seeing the handiwork of the Three-C boys. Nearly two million acres were planted or seeded by them and a third of their new forests cover the old pineries of the Great Lakes.

Like the Blue Eagle codes for distressed industries, a vast scheme of conservation by hand labor stemmed from the depression of the early thirties. Franklin Roosevelt came to the White House with a definite plan for putting idle men to work on the land. When accepting the nomination for the Presidency, he proposed to relieve the distress created by the shutdown of factories by employing a million men on the restoration of forests and other natural resources. Within a month after Roosevelt's inauguration, the Emergency Conservation Act authorized the establishment of a chain of forest camps where idle men between eighteen and twenty-five could be employed in forestry and other land betterments. The President insisted on immediate action. Robert Fechner was made director. The Departments of Agriculture, Interior, War, and Labor furnished advice and personnel. The Labor Department started the mobilization of 250,000 unemployed young men, and in April 1933 the first camp of the Civilian Conservation Corps was set up on the George Washington National Forest in Virginia.

Like so many "noble experiments" on the American scene, the shining aims of this double conservation of men and natural resources were tarnished by a lot of weird politics. Some of us old foresters who had sweated in building roads to fires already burning shuddered to see man power wasted on fancy buildings and elaborate stonework at public camp grounds. Nevertheless, in its ten years of life, the corps shouldered through a volume of work that stands above anything in the annals of conservation the world over. Its strength ranged from 300,000 to 500,000 men. At its peak in 1935 the corps operated 2,650 camps, and of these, 1,300 were working on forestry projects. These lads formed a standing army of forest fire fighters, with camps in every state and on almost every national forest and national park. They repaired over 500,000 miles of protection roads and trails and built 118,000 new miles of truck trails and fire control roadways. In many sections of the Northwest their shovels and bulldozers cut through hazardous areas of logged-over land and second growth and gave all protective organizations better access to their back country. Many miles of abandoned railroad logging grades were converted into permanent truck roads. Fifty-six thousand miles of fire breaks were cleared, many of them along the chaparral ridges of southern California watersheds.

Aside from their extensive tree planting and forest thinnings, the C.C.C. workers fought white pine blister rust, gypsy moths, and bark beetles. They felled snags on a million and a half acres. They built telephone lines and ranger stations and improved many public camp grounds. They carried through a prodigious task of soil conservation by building thousands of check dams to control erosion and by seeding or sodding gullied lands. The safety and productiveness of American forests were raised to a much higher level by this farsighted diversion of idle men back to the land.

The C.C.C. program was of direct help to many industrial forest owners in extending their facilities for fire control. The new roads, telephone lines, and safety strips, cleared of standing snags, gave both private and public holdings a better grip on the No. 1 enemy of all forest management. More than that, the program, with its tremendous scope and diversity throughout the land, carried a powerful educational punch. Dirt forestry was demonstrated throughout the American countryside on a scale never seen before. Pulp manufacturers in the lake states and in the South, concerned over future wood supplies, were shown immediate examples of mass production of pulp stock by tree planting. Even hard-boiled lumbermen who scoffed at the "city boys raking leaves" were getting more conscious of the new order of things in the woods.

Another dramatic combination of forestry and relief employment grew from the lively interest and venturesome spirit of Franklin Roosevelt. This was the planting of shelter belts on the Great Plains. The imagination of pioneer homesteaders in the Missouri Valley had clothed the prairies with the abundant farm woods and forest hinterlands of the homes they had left behind. A nostalgia for trees and woodlands was very marked in the first generation of Midwestern farmers.

The early agriculture of the plains stressed the importance of shelter woods around farmsteads and belts of trees to break the force of winds and check evaporation of moisture from the soil. There were many tree plantings on individual farms, as well as co-operative and community plantings and experiments by public agencies. The federal Bureau of Forestry established a nursery at Halsey, Nebraska, in 1903, and by dogged persistence through many trials and errors created some sixteen thousand acres of pine forest in the Sand Hills. Most of the railroads crossing the plains planted wind and snow breaks of trees along exposed stretches of their tracks. Many foresters and agriculturists cherished the dream of great north-and-south forest belts through the treeless West to alleviate the rigors of wind and drought. But tree planting in the prairies was costly and precarious. There were many failures, and many doubts whether the results would warrant the cost of a major regional scheme of windbreaks. It was characteristic of F.D.R. to cut this Gordian knot and put the dream into action.

The Shelter Belt Program, or Prairie States Forestry Project, was launched by the Forest Service in 1934 with emergency funds allotted by the President. It was carried on largely with labor and resources made available by W.P.A. under the direction of federal foresters and soil conservation experts. A strip of land one hundred miles wide was marked out from Texas into the Dakotas. Within this strip, wherever the cooperation of landowners could be obtained, shelter belts of five to ten rows of trees were planted across the prevailing winds. The ranchers within the project were asked to do most of the work on the land, with the government furnishing planting stock and supervision and replacing trees that died.

This bold attempt to push a belt of forest across the heart of the country was continued for seven years and attained gigantic proportions. Two hundred and seventeen million trees were planted in shelter belts on over thirty thousand farms. Eighteen thousand six hundred miles of forest windbreaks were created. All the lore of seventy-five years' experience in trying to grow trees on the plains was drawn upon. The world was searched for the hardiest trees and shrubs, and planting layouts were carefully designed for each locality. The cottonwood, state tree of Kansas, was extensively used. Ponderosa pine, jack pine and eastern red cedar were selected from our native conifers, together with drought-resistant Austrian pine from eastern Europe. Chinese and American elms were used extensively with catalpa, Osage orange, green ash, and both black and honey locust. Russian olive and mulberry were drawn upon for "fill-in" rows between taller native trees.

It was a critical job to select the trees or tree mixtures best able to survive under these exacting conditions. And it was successful. A field check in 1944 showed a survival of 90 per cent in the shelter-belt plantings.

Nearly half a billion dollars of federal and state funds were expended by W.P.A. between 1933 and 1941, for relief employment on forestry and soil conservation. Aside from the shelter belts, some five million acres of farm wood lots were planted. Extensive areas were also planted on public forests in the lake states and elsewhere, and many other protection and improvement jobs were carried out in the woods. There is nothing in world forestry to compare with the C.C.C. and W.P.A. mass attacks in the United States. The nation seemed determined to make up at one stroke, in sheer physical labor, for the centuries of carefree exploitation during which its forests had been taken for granted.

CHAPTER IX

The Universal Raw Material

WITH a wealth of trees for the asking, the early Americans probably used wood for a greater variety of things than any other people in the world. Their little water-powered mills turned the trees into lumber, boxes, bobbins, spools, furniture parts, shoe pegs, tool handles, oars, barrels, wooden bowls, and spoons. Anyone familiar with New England has seen many of these little mills or shops at the ancient, moss-grown dams on the village streams. My Vermont cousins had a little sawmill well up on the Onion River that through several generations cut "spruce frames" for local housing or shipment to Boston. A few miles below them the same stream powered a mill that turned ash and maple billets into bobbins and shuttles for the textile industry. I recall a little factory in western Maine that for three generations in the same family has turned spools out of yellow and paper birch. During most of the time it has sold spools to the same firm of thread spinners in Scotland. When examining lands for the projected White Mountain National Forest, I visited a neat little shop in northern New Hampshire. It was cutting paper birch into thin blocks, scoring them with the precision of a watchmaker and splitting out shoe pegs by the million. While the American market had mostly dried up, this little New England mill seemed to be pegging shoes all over the world. I saw a dozen sacks of heavy pegs consigned to Hamburg, Germany.

The mass productions, inventions, and competitions of our expanding economy have constantly increased the diversity of wood uses as well as the over-all drain upon our timber supply. Forest conservation through better use and preservation of wood was headlined in the Pinchot program. It was kept in the front through expanding research and co-operative experiments in many directions. Once I spent several hours on my knees examining the metal tags fastened to hundreds of railroad ties. The crossties were samples of several different preservative treatments, laid under the rails for the acid test of durability in service. One of the most constructive moves in the whole gamut of conservation was the Forest Products Laboratory, projected by Gifford Pinchot and opened for work in 1910 by Chief Forester Henry S. Graves.

Every decade has seen wood used in the United States for more different things and by new and more intricate processes. But it took the scarcities and discoveries of World War II to drive home the truth of the term often applied to wood by German science, *der Allgemeinerohstoff*, the universal raw material.

During the rush of lumber to the front, I went through one of the timber-fabricating plants in Portland. That day it shipped out twelve carloads of beams, joists, and rafters, cut to exact dimensions, shaped, bored, and slotted to fit together with bolts and ring connectors. The steel connectors, or collars, were pressed into the faces of the beams or planks at every joint, around the belts, and gave the structure about three times the strength of old-fashioned joinery. American lumbermen had brought the device from Sweden a dozen years before, but the war gave them their first real chance to put it to work.

At the glue house of this big plant, curved trusses were being shaped and laminated from one- and two-inch boards, and shipped out by the carload, ready to carry the roof of a new shell factory. One huge press was rolling out a continuous beam, 24 inches by 18. It was also built of two-inch boards glued together, and it was stronger than any solid timber that earlier generations of lumbermen cut from the heart of a giant fir. This monster stick rolled out from its press like a hemp cable at a ropewalk, and the cutoff saw snipped off any length desired like the scissors of a ribbon clerk at her counter.

A few days later I saw the miracle of converting a hemlock log into silk thread. I stood beside a vat of western hemlock pulp, which had been alkalized to the stage of viscose. It had about the consistency and the allure of axle grease. Pressure was forcing this stuff through minute apertures into a vat of fixing liquor. The tiny filaments of spiderweb were caught up by little whirling spindles, and there was spun, before my eyes, the most lustrous, shimmering silk I had ever seen. It was rayon yarn, now second only to cotton among our textiles. The silk thread and the giant beam came from the same forest. They might have come from the same tree.

The great bulk of our use of timber has been as sawn lumber and the myriad kinds of woodwork fashioned from lumber for finishing and furnishing homes and for farm and industrial needs. Other marvels of wood technology have come to pass in expanding ancient crafts into assembly-line production and fitting their wares to the innumerable requirements of modern life. No industrial saga is more fascinating than the story of paper. It did not start with wood, but wood has largely taken it over.

Over eighteen centuries ago the Chinese invented the art of macerating vegetable fibers and then matting them, on woven wire screens under water, into thin, flat sheets. The materials first used were mulberry bark, old fish nets, hemp and rags. The papyrus of Egypt was probably the most widely used writing material of ancient peoples and gave its name to paper, but it did not meet the specifications of paper as the Chinese made it. Inner filaments from the stalk of the papyrus plant were flattened out, wet, laminated in several thicknesses, pressed together, and dried in the sun. Papyrus was really a light and leafy form of plywood.

Paper is still made by hand in China and elsewhere, and very much the same tools that Ts'ai Lun devised in A.D. 105 are used. Experts tell us that no more durable paper has yet been made than the hand-wrought sheets of linen and cotton fiber on which the Gutenberg Bible was printed in 1450. With a thousand technical improvements and the substitution of giant, complicated machines for human labor, modern paper mills turning out five hundred tons a day still use the same basic process.

Papermakers have searched the vegetable kingdom for

cheap and plentiful raw materials. The extraction of cellulose fiber from wood chips by cooking with various chemicals began in 1854 and expanded rapidly. "Chemical" pulps have long and strong fibers. English experimenters made book paper by grinding wood into "mechanical" pulp as far back as 1800, and the Germans made it a commercial process. The first groundwood pulp mill in the United States was built at Curtisville, Massachusetts, in 1867. Bolts of soft, fine-textured wood, like spruce, are ground into pulp by pressure against whirling stones under water. This is short-fibered pulp, used as a filler in many grades of paper. Mixed with enough long-fibered, chemical pulp to give the tensile strength needed in running over high-speed presses, it makes the great bulk of modern newsprint.

The finest grades of paper are still made from cotton and linen rags. Bible pages and cigarette papers are made from flax. Paper is made from straw, from cornstalks, from sugar cane refuse. But the heavier yields of fiber obtainable from wood have turned the scales in mass production. Consequently, paper experts concentrated on the pulping of woods, and today 95 per cent of all the paper made in the United States starts with wood as its principal material.

Papermaking is a dynamic industry, unceasing in employing new technologies, using new woods, and bringing out new products. In 1909 the manufacture of the heavy brown Swedish kraft paper from sulphate-cooked pulps began in the United States. It has expanded into wrapping papers, paper bags, laminated paper sacks which carry much of the heavy commerce of the world, and thick papers used in construction or built-up panels. Paperboard, fabricated into containers of every conceivable size and shape, now makes up nearly half of the national production of paper. Research at the U. S. Forest Products Laboratory laid the groundwork for converting all of the southern pines into white as well as brown papers and opened the way for the extraordinary industrialization of southern woodlands by pulp and papermakers in the last twenty years. In like fashion, federal and industrial laboratories aided the reconstruction of the pulp industry in the Lake States, after their prized coniferous woods had all but disappeared. Through the magic of the chemist's retort, the lowly aspen, jack pine, and other Cinderellas of the north woods have become the source of a revived industry and a permanent forest economy.

Paper is used today in over fourteen thousand commercial products. Their diversity is bewildering. Horseshoes are made of paper, in layers impregnated with water-proofing oil and laminated with powerful cement. Gas pipes and electrical conduits are made of heavy paper laps, dipped in melted asphaltum and wound and laminated over wooden cores. The last use of paper I have heard of was reported in the farm radio broadcast this morning. The up-to-date dairy cow is now cleaned with a tissue "udder towel" before milking—a highly sanitary process with a separate towel for each bossie.

Verily the "age of paper" is here. Since 1880, paper production has grown from 400,000 tons to twenty-two million tons a year and its consumption per capita from eighteen to 358 pounds. Our enormous use of paper is almost without parallel, even among the industrial marvels of the United States.

Another product of ancient craftsmanship has likewise grown into a great forest industry. The Egyptians apparently created the art of veneering furniture with thin overlays of beautiful wood. A bedstead belonging to the grandparents of Mrs. Tutankhamen has been retrieved from the tomb. Its headpiece is veneered with laburnum wood, inlaid with gold and jewels. It is not known how the laburnum was cut or glued to the solid core, but the basic modern skills of veneering and lamination were obviously employed. While the glue doubtless was not applied with the electrical hotplate of today, it has held grandmother's bedstead intact for thirty-five centuries.

Some ancient Egyptian coffins contain six layers of thin wood, securely glued together. One of them is overlaid with beautiful cedar veneer, pegged to the wooden base. In the days of Roman luxury, tables were veneered with the rarest and most costly woods of the Roman world. Chippendale, Sheraton, and other English and French furniture craftsmen of the eighteenth century reveled in beautiful veneers. They mastered the art of building up any part of a desk or chair to its desired shape and strength by laminating pieces of thin wood, one upon the other. And they learned the most important secret of modern plywood—to prevent warping, shrinking, and splitting lay each veneer with its grain at right angles to the grain of its neighbors.

For many years veneers and laminated woods remained a part of the cabinetmaker's art. The thin strips and panels were cut by planing, by sawing, or by slicing with powerful knives. Then, in 1890, it was discovered that sections of a tree trunk, after steaming or soaking, could be mounted and turned in a great lathe and that veneer could be peeled by a stout knife blade, set at the right angle against the rotating cylinder. The thin sheets were called "rotary cut veneer." They were made at first from the softer woods and used for basketry and berry boxes. But the skill of wood technicians and machinery makers rapidly extended rotary cutting to timber trees all over the world. Fast progress in glues and presses followed. The strength and stability gained by laminating the thin sheets of wood at right angles to one another were retained. *Plywood* in wide panels and long lengths, laminated to any thickness and strength desired, became a commodity of general use in the world's construction. Ninety-five per cent of American plywood is rotary cut.

Like paper, plywood manufacture has been a dynamic industry. It is constantly pushing out into new markets with new technologies. The last twenty years have produced "exterior plywood." The plies are bonded (i.e., glued) with waterproof phenolic resins in hotplate presses. The strength of some of these laminations is past belief. I have witnessed laboratory tests where plywood was torn apart in the tentacles of giant machines. The wood structure itself, in one of the plies, gave way before a resin-bonded joint would yield.

Plywood in panels or built-up boards and beams has come into common use in outdoor construction. There are even plywood panels faced with aluminum or copper. Plywood has a thousand uses, from cement forms to the most costly furniture or library wall; and, like paper, its uses are constantly multiplying.

Many other technical wonders bear witness to the versatility of wood as a raw material. Wood flour has been known to American industry for thirty-five years. It is a filler or extender in all manner of compositions, like linoleum. Wood flour is combined with synthetic resins in making many kinds of plastics from which are molded our fountain pens, telephone receivers, and no end of things in everyday use. After the disastrous fires of 1918, Rudolph Weyerhaeuser and other resolute lumbermen rebuilt the wood-using industries of Cloquet, Minnesota, from the leavings of raw material they still had. They turned to the forest weeds of the old days—aspen and jack pine. Years of research at the U. S. Forest Products Laboratory were put to work in developing new products from mechanically ground wood fiber. It was obtained from mill waste, logging slash, or cordwood from farm wood lots. A synthetic board appeared in the trade, called "Nu-Wood." It was used for sheathing, corestock under veneers, and acoustic tile for interiors. Another new product under the tradename of "Balsam Wool" entered the growing market for insulation. The galaxy of wood-using plants that now cluster around Cloquet, including pulp and paper mills, uses the timber crops from the industrial holdings of the region and also from thousands of farm woods. It is a striking instance of regional forestry created by wood technology.

The technique of separating wood fibers by mechanical devices and remolding them into panels or boards had a much later start than papermaking or veneering. It is now building up another great forest industry with its own marvelous skills. Molded insulating boards were made from flax straw in 1906 and from wood-pulp screenings in 1914. This was the beginning of "Insulite" and the commercial drive for insulation in the properly built home. The Wood Conversion Company of Cloquet started the defibering of wood in the United States by grinding down little chips with machines patterned after an upper and nether millstone. William Mason invented an ingenious "gun," in which wood particles are blown apart, into their constituent fibers, by sudden explosion from high pressure to a vacuum. American inventive resourcefulness has run riot during the last twenty years in devices for defibering wood, cornstalks, refuse sugar cane, and what-not, and for rolling out commercial wares from loose, fluffy insulation fillers to stiff boards for sheathing houses. Some grades, indeed, compete with sawn lumber for flooring and hard finish.

About half of the wood formed by a growing tree consists of *cellulose* fibers. The other half is *lignin,* nature's great adhesive. Lignin is the filler and binder, surrounding the fibers of cellulose and giving strength to the whole structure. Chemical pulps are made by separating the cellulose fibers from the lignin and matting them into sheets. *Pulp* mills have generally washed most of the lignin downstream and thereby created problems of river pollution. *Fiber* mills do not separate cellulose from lignin. Both wood elements are retained in rolling out and molding their panels of "insulation" or "hard board." Some part of the natural gluing qualities of the lignin may thus be saved and put to work again in the new composition.

There is no telling how far the laboratories will carry the techniques of separating and then reassembling the natural constituents of wood or what strength and hardness can be wrought into the new substance by heat, pressure, and impregnation with resins. Some of the hardest and densest types of fiberboard approach the weight, strength, and durability of metals. Someday we will bring to pass the technologists' dream of a perfect structural material—a plastic that contains all of nature's qualities of strength and resiliency but none of the defects of knots or cross grain—a plastic, furthermore, which can be molded into any size and shape desired by man, without the limitations of tree trunks, and which can then be pressed and fused to the density required for a particular use. This is one of the things that the Nazis were after in Hermann Göring's grandiose wood conversion factories, but today American engineers are not far from it.

No small part of the impact of World War II upon the forest economy of America was the lift it gave to our wood-using technologies. Most of the new ideas had long been known to the laboratories, but the war put them into factories and everyday use. Others were perfected under the pressure of urgency. The armies would not wait. The search of the U. S. Forest Products Laboratory for the ideal material for airplane propellers produced one of the technical marvels of the war. It was a wood alloy, wood laminated of many plies, filled and fused with resins, under heat and pressure. "Compreg" does not warp or rust or shrink or swell, and it has the hardness of mild steel.

Still other technologies were learned from the chemists and engineers of Europe. The war put a stronger and broader economic base under American forestry, not only by the volume of wood it consumed but by the manifold ways in which wood was made to serve the necessitous peoples of the world.

I have frequently been amazed by the resourcefulness and vigor of the industrial technologists at work on wood. Early in the war our imports of cork from the Mediterranean were cut off. Foresters at the University of Washington pointed out that the layers of cork flakes in Douglas fir bark might serve in the pinch. We laid upon the shoulders of West Coast lumbermen the patriotic duty of picking out their corkiest bark and shipping it to a shredding factory where the 15 per cent or so of cork flakes could be salvaged. One of the large companies put its research men on the recovery of Douglas fir cork and ended by processing all the bark into half a dozen new products, including plastic powders and fertilizer. The quest for cork to meet an emergency of war led to a new industry.

The take of World War I from the forests of America was a love pat compared with the shock of World War II. There was no better illustration of an all-out war of production than the tremendous demands for wood in a thousand different forms and the regimentation of the industry and domestic trade to make it available. In the six years of war and preparation for war the United States consumed 215 billion feet of lumber. Half of this vast footage, enough to build ten million American homes, was taken for direct and indirect military requirements. And besides the lumber, great quantities of paper, wood pulp for explosives, plywood, poles and piling, naval stores and other forest products disappeared in the world holocaust. We can picture the war as a grim reaper, consuming the harvest on at least ten million acres of American forests.

Forty-eight billion feet of lumber went into war construction—cantonments, munition factories, docks and shipyards. Forty-three billion feet were demanded for boxes, crates, and dunnage in shipping military supplies around the country and across two oceans. Most interesting to woodsmen and technologists were the ten billion feet of lumber used directly in fabricating weapons and other military equipment. The forests and farms of the central states again were searched for black-walnut gunstocks. The north woods were combed for the choicest birch and maple trunks for peeling into airplane veneers. Pacific northwestern loggers strained every resource to get at and bring out 100 million feet of airplane logs, the finest Sitka spruce, Noble fir, and Douglas fir we had. It was the cream of the West Coast forest. Every stick of "aero" had to be fine-grained (with at least twelve annual rings to the inch) and straight-grained its entire length. That meant a sixteen- or thirty-two-foot piece of lumber whose grain lengthwise did not vary from absolute straightness more than one inch in twenty.

To build the Allied air fleets, the Forest Service rafted forty million feet of Sitka spruce logs from Alaska to Puget Sound at the cost of a king's ransom. A typical log in these rafts measured thirty-six feet in length and three feet across the small end. Several West Coast mills installed special "pony bands" for resawing clear sections from their logs. They inspected and resawed every likely-looking clear cant; turned and resawed it and inspected again, to recover even a single piece of precious "aero." I'll never forget the pride of a little mill operator in the Willamette Valley when, almost reverently, he pulled back a tarpaulin to show me the ten or twelve pieces of airplane grade he had accumulated in a month's run.

In like fashion hardwood manufacturers in the South were picking out "aero" lumber from their yellow poplar and the exacting grade of "bending oak" from their finest white oak logs. It made the ribs for a dozen types of naval craft. Foresters scouted Latin America for balsa, the lightest wood known, to build floats and life rafts. The choicest grades of Douglas fir, next to "aero," went into the decking of battleships and airplane carriers, along with all the teak that could be imported from southern Asia. One hundred and fifty million feet of mahogany were brought from Central and South America for sheathing and finishing boats of many kinds.

World War II also took American industrialists into unknown worlds of wood chemistry. Dusty formulas from the shelves of our own laboratories and processes, heard of vaguely in Germany or Sweden, became realities in our use of wood.

The Nazis lost no time in drafting the wood science of Germany for their war preparations. Hitler had not long been in power before a suave Prussian forester, Von Munroy, toured the wood-using centers of the United States. Von Munroy tried to interest me in the possibility of organizing immense shipments of low-grade wood from the Pacific coast to Germany. This wood would provide raw material for the great new industries Germany was developing in cellulose, textiles, and foods. Forestry would usher in another great advance in civilization, with German science and technical skill in the lead. We had no inkling then that the Nazi high command was searching for every means of providing itself with the war potentials the lack of which crippled the fighting machine of Kaiser Wilhelm. Or that Von Munroy himself had been commissioned by Göring to plan factories in five major lines of wood conversion. They were to manufacture food, motor fuel, textiles, plastics, and structural parts.

Four years later I had an inside glimpse of Nazi ruthlessness in seizing the fruits of German wood chemistry. In my Seattle home, a German refugee, Erwin Shaefer, told us the story of his factory at Tornesh. It was here that the first full-scale demonstration of the Scholler process of producing wood sugar by the hydrolysis of sawdust or other waste wood took place. Wood sugar is something like a thin, crude molasses. It is a mother liquor from which industrial alcohol is derived, at half the cost in Germany of wartime alcohol from grain or potatoes. It is also fermented into protein-rich yeasts, used in Germany for feeding livestock and experimentally as ersatz meat for human consumption.

"I was just a businessman," said Herr Shaefer. "I had no interest in German politics. I paid no heed to warnings from my friends that I should warm up to the local Nazi bosses. You see, I had one grandfather too many. And then one night, with no warning, I was in a concentration camp."

After twelve weeks of bitter, degrading imprisonment, Shaefer bought his liberty by deeding to Nazi chieftains his wood-sugar plant and everything else he owned. But he got out of Germany in time, fled to England, and finally reached the United States. Here his first act was to offer his services to the War Production Board.

Dr. Glesinger, of the United Nations Staff on Forestry and Agriculture, tells us that the recovery of wood sugars in the form of ethyl alcohol, from waste liquor used to cook sulphite pulp, was begun in Sweden in 1909. The product has gone largely into the highly potent beverage of Scandinavia, *aquavit*. When a wood-sugar plant was planned for the Willamette Valley, some lumber wag proposed a Paul Bunyan Brand of Oregon gin, *"aged two hundred years in the wood."* The Swedish process was installed by the sulphite pulp mill of the Chicago *Tribune*, at Thorold, Ontario, in 1941, and at the Bellingham, Washington, plant of the Puget Sound Pulp and Timber Company in 1944. These installations were to help supply the tremendous war consumption of ethyl alcohol in the manufacture of synthetic rubber.

In the war years Swedish pulp mills also made large quantities of "cellulose fodder" for livestock. High protein feeds were produced by fermenting sulphite sugars with yeast bacteria. Cellulose "hamburgers" appeared on Swedish tables but soon were classed with the horrors of war and retired into limbo. Meantime, the pressure for ethyl alcohol led our War Production Board to authorize the first commercial wood-sugar plant in the United States at Springfield, Oregon. It was to be run by a co-operative company of lumbermen and supplied by the waste wood from a dozen Willamette Valley sawmills. The plant was built after many delays and made enough ethyl alcohol to prove the process and an attainable yield of fifty-five gallons from a ton of fir sawdust or shavings. After a period of idleness the enterprise is again converting sawmill wastes into wood sugar and ethyl alcohol.

In the days when the "war of substitutes" was relentlessly driving down the consumption of lumber, I often heard wood relegated to the "horse and buggy" stage of America. Or it was "passing out with the dodo bird." The most discouraging roadblock to forestry in the Pacific Northwest, in those days, was the immense quantities of useless wood which clogged our operations at every turn. It piled our cutover lands with great heaps of slashings—the most serious hazard of the region to its forests and its forestry. And it dotted the milling communities with tall, black waste burners whose columns of smoke by day and glowing sparks by night ascended like perpetual incense at the shrine of a god of savage exploitation. It was largely unavoidable wastage. It was necessary under the limitations of the time if our virgin forest were to be converted into building and industrial materials. And it is doubtful if the wastage at logging camps and sawmills ever caught up with the loss of overripe timber rotting to the ground in the woods.

But the utilization of raw material is the key to a successful and particularly a stable forest industry. Even more is utilization the key to successful forestry. The war market for low-grade lumber taken out of West Coast logging shows 15 or 20 per cent of the poor logs previously left in the slashings. Where pulpwood could also be marketed, another 20 per cent of the waste wood was harvested. Each stage or degree of utilization turns a larger part of the growing capacity of the land to actual account. Foresters for pulp and fiberboard companies, on the strength of postwar markets, are beginning to thin their young stands for pulp and fiber stock. They are recovering more of the potential growth of their woodlands, which otherwise would die out in nature's overcrowding, and at the same time increasing the growth rate of the trees they leave. And so we go, step by step, with utilization calling the tune for the forester.

The steady expansion in wood uses and technologies is the strongest ally of forestry in the United States. Not only are we consuming many more forest products, but we are making many different things from the same tree. Some of the great tree farms of the West are parts of industrial enterprises which integrate the manufacture of everything the forest grows. Some logs are turned for plywood; some are sawn into lumber; some are chipped for pulp. From some logs, the better textured, clear wood is sawn into lumber while the poorer grades are chipped into pulp stock. Logging slash and mill wastes feed a pulp mill or a fiberboard mill. Left-overs from one product become raw material for another. Most everything is used except the whisper of the pines.

The greatest long-range effect of World War II upon the forests of America will be their extension and perpetuation through better management. Every advance in pulping, spinning, laminating, or plasticizing wood is a gain in forestry. Every new industry for converting wood into textiles, new kinds of paper, alcohol, yeast, or lamb chops strengthens the business of growing trees. War-born uses and markets for wood have given American forestry an economic base it never had before.

CHAPTER X

Forest Industry Is Settling Down

SURROUNDING my home at the head of Gamble Bay are portions of the Hood Canal Tree Farm. This forest of some 75,000 acres has one small watershed of virgin trees, but the great bulk of it is logged-over land in every stage of regrowth from saplings to eighty-year timber. It borders the bays and inlets of the North Sound country around Port Gamble, where Captain W. C. Talbot anchored the good ship *Pringle* in 1852. Captain Talbot belonged to the firm of Pope and Talbot. These shrewd and venturesome Yankee traders had sawn lumber and built ships on the Machias River of eastern Maine for many years. Their masters sailed ships on the seven seas. Pope and Talbot founded a trading post on San Francisco Bay in 1849, but they did not rush to the gold fields—not these Yankees! They sold clothing, tools, and provisions to the miners and emptied ship after ship as fast as it made the Golden Gate. They soon foresaw a new city in the making and, being sawmill men, looked around for a source of timber.

One story goes that a Pope and Talbot sailing master had tacked up the Strait of Juan de Fuca after a stiff gale, looking for new spars. He told the folks in Machias that if they thought there was timber in the state of Maine, they had better take a look around Puget Sound. Being seafaring men, they may have read the log of Captain George Vancouver, who explored northwestern waters for the British Admiralty in 1792. The captain's journal contains many notes on the "thick woods" and "stately forest trees" around the shores of the sound which he named for Lieutenant Peter Puget. At Port Discovery he records luxuriant "Canadian and Norwegian hemlock and silver pine."

At any rate, the ship *Pringle* brought a muley sawmill from Machias to Port Gamble in 1852. Pope and Talbot were the first mill men to saw Douglas fir timber into "Oregon pine" lumber and ship it down the coast to build docks and houses and plank sidewalks in San Francisco. Now, after running nearly a century and being rebuilt and remachined half a dozen times, the old mill is assured another century of buzzing life by the Hood Canal Tree Farm.

It was about 1939 that reviving building and lumber use quickened the pulse of industrial forestry. Out in the far Northwest, the field men of the timbermen's Joint Committee on Forest Conservation were called in by woodland owners who wanted faster action in growing trees. Large pulp and lumber companies employed more foresters of their own and tightened up their protection from fire. All sorts of new ideas were simmering. Down on Grays Harbor the 120,000-acre Clemons tract of superb virgin fir was about logged out. Its owner, the Weyerhaeuser Timber Company, wanted to make Clemons a testing ground for intensive methods of protection and tree growing. The company was out to learn Douglas fir forestry and its costs on a scale sufficient to guide future planning on all of its properties. How could they gain public understanding and support and stop the careless woods burning by hunters and berry pickers? How could they put over the idea that this was not just a run-of-the-hills piece of cutover land but an area under intensive timber culture? The keen young editor of the village newspaper said, "Call it a 'tree farm.' Put tree farm signs all over the place. People know what a 'farm' is. They'll understand what you have set out to do."

The Clemons foresters found out what the land would grow in trees, what their fire losses had been under former systems of protection, and what it cost to make good these losses by planting. How much should be spent for better protection, as a matter of cold business, to save the losses and replacement costs? The foresters came up with a recommended investment of $1.00 per acre in roads, fire lookouts, telephones, tank trucks, and power pumps. The former eight cents per acre for yearly protection became twenty-five cents, for an enlarged force of lookout guards and patrols and for keeping roads and machines ready for action. This was simply sound fire insurance on the values at stake. It was the first law of American forestry brought down to practical business.

A new term appeared in American land usage. The Joint Committee liked the tree farm idea and made it part of the West Coast industry program. The committee defined a tree farm as an area committed "to the continuous production of forest crops" and "protected from fire, insects, tree disease, and excessive grazing." It might be a forty-acre farm wood lot or the four-hundred-thousand-acre holding of a lumber or pulp company. It might be an operating forest in process of cutting or a stretch of cutovers and second growth held for future supply. The essential point was "continuous production." It must be *land in the business of growing trees*. A tree farm certificate, issued by the Joint Committee, meant that the land had been examined by one of its foresters and the sincerity of the undertaking vouched for. If the proposed "farm" was of any size, the owner himself had to appear and satisfy the committee on his plans for growing trees.

The campaign for tree farms, vigorously pushed by the Joint Committee, had a double edge. It offered a *second step* to the forest owner who wanted to move beyond the minimum requirements of the industry code and the state conservation laws. It gave him definite things to do and obligations to meet. At the same time, it was a way to show the people of our two states that the industry was serious about growing trees. It was good publicity and the best possible reply to the clamor for forest regulation. "Ye shall know them by their fruits. Do men gather grapes of thorns, or figs of thistles?"

The idea seemed to fit the temper of the industry. It expressed what many operators were working toward in their own way. Large tree farms were soon set up by the concerns which had led the way in West Coast forestry. Soon interest was shown by some men in whom I had least suspected it. It was a memorable morning when Peter Schafer came to the association office. I had long admired Peter as a straight-grained logger of the old school. He had come up the hard way. In his office at Montesano I had read a framed piece of writing in the fine script of Peter's father. It was an agreement to lease to one of the logging companies on Grays Harbor, for the season of 1887, seven yoke of bulls with "one boy to work same." The boy was Peter Schafer. From bull skinner, he became hand logger, jacking and winching big Douglas firs into the Satsop River and rafting them to sawmills at Aberdeen. Now, near the close of a life of unremitting labor and well-earned success, Peter wanted to talk about tree farms.

I spent the morning trying to answer the old timberman's shrewd questions. How fast would young timber grow on the different types of ground—the bottoms, benches, and upper slopes? What would it cost to carry a stand of ten-year trees to merchantable size? How long should he wait for old burns to reseed before planting them by hand? Would one of the Joint Committee's foresters examine the Schafer cutovers and tell them what to do? He certainly would.

It then came out that what should have been the core of the Schafer Tree Farm, some twenty thousand acres of the finest tree-growing land in the Olympic country, had gone tax delinquent a bit too long and reverted to the ownership of the county. But Peter said: "Perhaps I buy it back." The land was put up for bids on the steps of the county courthouse. Another lumber company from a couple of valleys westward was also interested in tree farms. A dark horse appeared, reputed to represent some big pulp concern. The bidding was spirited for a day and a half, but Peter Schafer raised every bid and recovered his cutover land, at slightly over four dollars an acre. Soon one of the junior Schafers was mapping and cruising the property and organizing the planting of trees on burned-out spots. I never saw more quiet satisfaction in a man's face than when Peter Schafer offered his tree farm for certification by the Joint Committee.

And then some Washington farmers wanted to put in their wood lots. The Soil Conservation Service had started real farm forestry in several counties. One of these conservation groups in the Snohomish Valley sold their timber through a farm co-operative and under the guidance of S.C.S. foresters were carrying out about the best Douglas fir forestry in the state. One of these tree-growing farmers was Jasper Storm, a keen woodsman of Swiss blood. Jasper had worked many years in northwestern logging camps. He knew and loved the woods. When he retired he picked a farm with two hundred acres of second-growth trees. Here he tried out many tricks of forest craft and showed their results to his neighbors. For fuel wood, he did not cut the nearest and easiest trees, but picked the corky-limbed, slow-growing, spreading-crowned wolf trees that were crowding out half a dozen others. He carefully thinned thickets of straight saplings to grow the best poles and piling. He measured enough trees to satisfy himself how long it would take a forty-foot piling to grow to sixty feet and thereby command a much higher rate on *every foot of length.* Jasper Storm became the untitled *Forstmeister* of his community, and when the demands of national defense brought an unprecedented market for long piling, Jasper and the other forest farmers of Snohomish went to town.

The Joint Committee welcomed every wood lot, however small, that met the tree farm standard of "continuous production." It was a happy evening, up in Snohomish, when after clearing the boards of a country dinner, we handed seventeen tree farm certificates to Jasper Storm and his neighbors.

Meanwhile the tree farm idea was spreading far afield, in farm and industrial forests alike. First the National Lumber Manufacturers Association and then American Forest Products Industries promoted it in country-wide campaigns. The conservation committees of the two great associations of pine lumbermen, in the West and South, got heartily into the game. "American Tree Farm" became a familiar sign on many country roads. Tree-farm dedications vied with county fairs in popular interest, at times even in the glamour of brass bands and youthful queens. In every state, tree farms are sponsored by some well-known forest agency. Often it is the State Department of Forestry or Conservation, or a state forestry association. The conservation committees and foresters of lumber associations carry the torch in several states. The sponsor in every case is responsible for holding its tree farms to a fair level of "continuous production" and for withdrawing a certificate from any tree farmer who falls by the wayside. In nine years, ten West Coast tree farms have been canceled, but those in good standing now cover 30 per cent of the Douglas fir woods in private ownership. By November of 1950, the tree farms had spread to twenty-nine states and something over twenty-two million acres.

There is bound to be a certain amount of "window dressing" or bunkum in a movement of this kind which seeks popular support. Its integrity rests upon how closely many different local agencies hew to the line of the founding fathers. My own experience in a dozen states has convinced me that not all of the listed tree farms merit the name, but the great bulk of them are serious undertakings in timber cropping. The movement well expresses the change now under way on the private forest lands of the United States and in the outlook of the men who own them.

The diversity of the tree farms tells its own story of the many ways in which forests bear upon American life. A goodly number, in the hands of substantial lumber, pulp, or plywood companies, are first steps in long-range, sustained yield management. They mark the industrial transition from timber mine to timber crop. One of the newest American tree farms is the eighty-six-thousand-acre Pioneer Forest of Missouri. It is owned by the National Distillers Products Corporation and its purpose is to assure a perpetual supply of oak whisky barrels. The historic land grant to the Northern Pacific Railway, signed by Abraham Lincoln, was the start of five Northern Pacific Tree Farms. They are strung along the N.P. rails from western Montana to the Green River Valley in the Washington Cascades. The railway's Upper Yakima Tree Farm of 136,000 acres is set up to do three things. First, it will furnish water for the irrigation ditches in the rich Yakima Valley; second, it will grow a perpetual supply of timber for the company's coal mines near Roslyn; and lastly, it will help maintain a dozen near-at-hand pine sawmills and box factories. From its coal properties, the Northern Pacific mines around five hundred thousand tons of coal yearly, and an average ton requires eleven and a half feet of shoring timbers.

Several tree farms are in the business of growing Christmas trees. The Mart Weaver Western Pine Farm in northern Montana grows Christmas trees by "stump culture." When a tree is cut, at least one branch in the first whorl is left on the stump.

This soon straightens up and starts a new tree. Often Mr. Weaver grows two new trees from limbs on the same stump. He puts a spreader between the branches left so that each will have plenty of room and "bush out" to desired Christmas-tree proportions.

One of the most enthusiastic western tree farmers is Emmit Aston, a logger of Omak, Washington. Emmit runs his own tree farm of 2,200 acres besides the Biles-Coleman forest of 23,000. Anyone who rides out into the woods with him must expect to find an ax and "Swede" bucksaw in the car; and to stop at some group of promising young pines while Emmit prunes his daily quota of ten trees. He maintains there is no need to wait one hundred years to grow clear pine logs if you help the saplings shed their branches in early life. I was explaining the same idea one day to a West Coast logger, on my little Douglas fir tree farm at the head of Gamble Bay. I showed him fast-growing forty-year trees on which the pruning scars had disappeared under four years' growth and which every season are now adding a layer of clear wood. "In another ten years," I said, "I will be selling you 24-foot and 32-foot clear logs from this young timber." My logger friend snorted, "You and your clear logs! You'll make every log scaler carry around an X-ray machine, to find out what's inside your timber."

The Governor of Michigan launched the Upper Peninsula Tree Farms in June 1949, presenting eighteen certificates to the owners of 620,000 acres of woodland. One hundred and six tree farms were enrolled also in the Lower Peninsula. Wisconsin lumbermen and paper men, working with the State Conservation Commission, put the same idea to work on their own initiative in 1944. The "Industrial Forests" of Wisconsin, covering 544,000 acres, have planted over forty million trees.

Over half of the tree farms, in number and area, are in the South. They represent every type of southern forest industry from the great holding of a pulp or lumber company to the little farm wood lot. There are tree farms in the rich hardwood forests of the river deltas, which produce bolts for peeling or slicing into oak and gum veneer. There are tree farms over the uplands of Kentucky and Tennessee, which produce railroad ties and hickory handle stock and logs for making furniture or oak cooperage. And there are hundreds of tree farms throughout the piney woods from Virginia to the Gulf, growing saw logs, pulpwood, piling and naval stores. The tree farms add up to but a fraction of the well-managed wood-growing land in the South, and the perverse little trees seem to grow just as well on many "forties" unblessed by this formal baptism. But the tree farms link the South with the North and West in a common crusade for good forestry on private lands.

In May 1950 tree farms were formally launched in the redwood region of northwestern California. I roamed over the first redwood tree farm to receive an official certificate. On the Van Duzen River, the Hammond Lumber Company has been logging virgin redwood selectively since 1935. In the recent cutting which we saw, all the trees four feet across the stump and less were left standing. These are the "crop trees" and they contain about a third of the volume of merchantable timber in the original forest. Hammond's foresters estimate that their crop trees are adding 5 per cent in volume with every year's growth; that they will find double the present footage on the ground when the loggers return with their long felling saws twenty years hence. Logging Boss Blackie Freeman was on his knees, showing us the redwood seedlings that were coming up in the openings to carry on the cycle of growth and cutting for more hundreds of years.

What impressed me most was the skill and pains of the loggers in dropping their six- and eight-foot giants, two hundred feet tall, and dragging them out of the woods—with so little smashing of the crop trees that were left. It is no boy's task to take a hundred tons of tree trunks off an acre of redwoods and still leave a vigorous and impressive forest on the ground. But that is just what is happening on the Van Duzen River.

The tree farms of greatest interest to me are the wood lots of land-wise farmers, which show the care and skill of many patient years. Some of them have been in the hands of the same family for two or three generations and, like an old house, reflect the frugality and craftsmanship of the men who made their woods a part of their lives. Many tree farms carry the imprint of old-world forestry, brought to America by German, Swedish, or Lithuanian immigrants. "Yah," said my Swedish neighbor on Hood Canal, when I complimented the careful thinnings in his wood lot. "I learn that in the old country." And very often, in the case history of tree farms, we find the hand of state and federal foresters, farm extension workers or district leaders in soil conservation. The work and planning of these laborers in the vineyard were, many times, the real start of today's outstanding examples of successful forest management. There is a deal of soil-bred forestry in the land know-how of America.

This was brought home to me afresh on a snappy fall day when we dedicated the first tree farm in Massachusetts. The State Commissioner of Conservation selected for this distinction the Lawton farm woods up in the Connecticut River Valley. This 620-acre forest has been in family ownership for two hundred years, and the plantings of white pines have been going on for fifty years. It was a wonderful example of the land instinct of America. The Lawtons, past and present, did not worry about forest depletion or timber famine. They simply kept steadily at work, making the best use they knew of their New England hills.

The three and a half million acres of tree farms now scattered through the Douglas fir valleys and ranges are giving the regional forest industries real leadership. The tree farmers make up a grand lodge of forestry-minded men. They interchange ideas, try out new methods, and organize joint undertakings in timber growing. Some of them became convinced that it would pay to plant seedlings in the burnt-out spots on their tree farms where natural reseeding had failed or was long delayed. The average large Douglas fir tree farm, with the usual blocks of seed trees and the usual experience with woods fires, will reseed 85 or 90 per cent of the ground, by itself. Once these men had gone into the business of growing trees, however, they wanted a full crop. Planting stock could not be obtained from state or commercial nurseries, so they made long-term contracts with the West Coast Lumbermen's Association for five or six million trees a year—at the cost of growing them. The association provided the nursery and the Joint Committee took on its management. The Lumbermen's Nursery at Nisqually, near the capital of Washington, has brought its production up to nine million seedlings a year. These are grown on individual order—Douglas fir for the lumber companies, Noble fir and Port Orford cedar for the tree farms experimenting with these valuable species. The average cost of the seedlings, packed for shipment, is close to half a cent per tree. Every spring the company trucks roll out from Nisqually with their loads of tightly packaged seedlings, bound for tree farms up and down the Coast. The forest marches on.

Once, as a youthful forester with Uncle Sam, it befell me to explain the planting program of the Forest Service to Secretary of Agriculture James Wilson. I showed the wise old farmer the maps of our nursery locations and planting projects, the estimated costs, yearly budgets, etc. Suddenly Tama Jim interrupted, "How many acres have you got to plant?"

"Something over three million, Mr. Secretary."

"Young man, you'll never do it with a spade!"

For some weeks thereafter my progress through service headquarters in the Atlantic Building of Washington was marked by heads emerging from doorways and gleeful shouts down the hall: "Young man, you'll never do it with a spade!"

From the planting of the first fig tree, foresters have dreamed of reforestation in a big way by the direct sowing of tree seed. They have sown it broadcast like wheat. They have stuck it into the ground with Iowa corn planters, as Secretary Wilson wanted us to do. They have dropped half a dozen seeds into a little spot where the leaf litter has been scratched up with a hoe. But the vast majority of these attempts to reforest have failed—and often not from drought or the competition of other plants, but from industrious rodents who eat up the seed before it has a chance to germinate.

Every forest ranger has a tall story of the exploits of squirrels. A ranger in the Bitterroot laboriously gathered bushels of ponderosa pine cones for spring seeding in his nursery. He robbed a few squirrel caches to make up his quota. On returning from a trip to town, he was aghast to find his cone bin all but empty. A little sleuthing showed that a line of gray squirrels were busily moving his hoard back to their own hollow logs in the woods.

The tribute levied upon the seed traffic by the little four-footed corsairs of the woods is terrific. Biologists estimate that on the Oregon burns, where only planting or seeding will restore the forest, one acre's population of white-footed mice consumes thirty-two pounds of Douglas fir seed, when they can get it, in a single winter. Paul Bunyan would doubtless have solved the problem by mobilizing regiments of ravenous wild cats, but only recently have foresters found the answer. Poisoning must prepare the way for seed sowing like an artillery barrage opening up an advance of infantry.

Rodents have forced tree planters to turn to the forest nursery and the "spade." Such reforestation in the Pacific Northwest costs $15 or $20 to the acre, and the planting job still runs into many thousands of acres. Keen minds in the state and federal services and among industry foresters are working on combinations of rodent poisoning and the scattering of tree seed that will bring the forest back at a much lower cost. The latest exploration by Oregon tree farms and the State Department of Forestry is in the use of helicopters; first to scatter poisoned bait over a planting area and then, before the mouse population can recover, to sow a third or half a pound of Douglas fir seed per acre. The helicopter bids fair to become a potent ally of reforestation.

First and last, a tree farm stands for awakened interest in the land that grew the timber. It has started a new kind of timber cruising. Many foresters have been employed to inventory cutover and second-growth lands. They are mapping the growing quality of the soil, its present stocking with young trees, the areas in need of planting. They are locating snag patches and other fire hazards. They are tallying growth rates. The industry is appraising its long-neglected asset—the growing power of the soil. It is setting up the records and employing the techniques of long-range forest management. It is learning the first precept of forestry: *Know your own land and what it will grow.* Study it as a good farmer studies his fields. That is where forest management begins.

It is interesting to observe how often this process of fact

gathering, of studying land for growing trees as a geologist might study its possibilities of yielding coal or petroleum, has sold an owner on forestry. One result of the new style of timber cruising is very evident. Cutover land no longer is a drug on the market or a large item on delinquent tax rolls. Everywhere, in the Douglas fir country, it is in demand.

II

World War II was a lumberman's carnival. There was a strong market for every species and grade of wood in our forests. Military demands for steel threw a vast amount of heavy construction back on timber. The sawmills were flooded with "heavy cutting"—the beams and columns and thick plank that give a sawyer fast production and effective use of logs. The "auctions" of the procurement officers constantly brought to the lumbering centers around the map tantalizing spreads of new orders. The "competitive bidding" turned usually into appeals, to sawmills already gorged with business, to fill this or that urgently needed requisition at the highest price sanctioned by O.P.A.

But the war years were far from being years of happy hunting. There was an unending battle with the Selective Service to hold enough skilled men in mills and logging camps. There were battles to get enough gasoline and motor trucks and especially truck tires that would stand up under heavy loads on rough mountain roads. It was a continual fight to keep a logging camp in saws and wire cable and tractors, even in heavy boots and work gloves.

As the food supplies at logging camps were exhausted and the iron fist of O.P.A. rationing was more severely felt, the West Coast logging industry was threatened at its most sensitive point—the stomachs of its brawny woodsmen. The loggers' table proverbially has groaned with heavy foods, especially with meat of many kinds. Nothing will start dissatisfaction or send a crew "down the hill" so quickly as "lousy chuck." But the wisdom of O.P.A. experts persisted for months in bracketing logging camps with city mills and factories in their meat rations. The situation was saved by one of the unsung heroes of the war. A physician, nutrition expert with O.P.A., came to a Puget Sound camp to see for himself.

With the true spirit of scientific research, Dr. Alpert donned "tin pants" and calked boots. He worked his own shifts with double-bitted ax and crosscut saw. He dragged steel cables over windfalls and through thickets of devil's club. After ten days, the doctor was holding his own with the lumberjacks in the woods and more than holding his own with them in the cookhouse. He argued with "Frenchy," the cook, about food conservation and menus, and after a few rounds the two were fast friends. In the end O.P.A. decreed that it takes six thousand calories a day to log timber, more than to dig ore in the mines or longshore the ships. That is ten pounds of solid food that a faller or rigging slinger burns up in a day's work.

The war years were no time for lumbermen to go to sleep on a fat order file. Breakdowns in production were threatened constantly by one shortage or another. They could be forestalled only by the most alert and resourceful management. All the machinery of production and marketing labored under a welter of official controls and limitations. I served on the Lumber Committee of the War Production Board and sat in on many hearings and investigations of the Office of Price Administration. It often seemed that more energy was being consumed in regimenting the industry than in felling timber and sawing logs. And at times the ferocity of battle between the officers fighting for military needs and stout defenders of John Q. Citizen, like the Federal Housing Agency, would have done credit to shock troops at the front.

Every sawmill cutting over one hundred thousand board feet annually was under government control. No farmer could use more than five thousand feet a year from his own wood lot. Anything more was subject to Uncle Sam's beck and call. Many types of logs, lumber, and plywood were "frozen" for specific war uses—it was unlawful to sell them for any other purpose. The lumbermen struggled through a maze of control orders and priorities governing the disposition of their product and another maze of edicts and rules which determined what they might charge for it. To keep their mills running, they had to fight a third maze of allocations and procedures covering everything from beans to monkey wrenches. It was, of course, just part of the country's drastic schooling in regimentation for defense. *"C'est la guerre,"* we said in World War I. But most of my lumber acquaintances would have seconded the retort of the mill man whom I was congratulating on his war record of earnings. He cut me short. "For God's sake, give me a chance to go broke on a free market!"

The second war did not crash into the forests of the United States with a sudden sweep of destructive cutting. The production of lumber was strong during the war years but not as high as in our active periods of peaceful construction, like the first decade of this century or the housing boom which followed on the heels of World War I. The war machine was fed with lumber chiefly by denying it to civilians. The country was on half rations for six years. A tremendous want for the goods of the forest grew during the war. It grew again with the lack of homes for returning veterans and all the other lacks in postwar construction. The domestic shortage of wood was made greater by the immediate needs of European restoration, like the emergency housing in Britain, the rebuilding of bombed-out docks in France and the restoration of Holland's dikes. It was swollen by the cutting off of former imports of wood pulp from Scandinavian countries. The net impact of all these forces was ten years of unparalleled demand for every kind of wood that American trees produce. The war reversed the chronic oversupply of forest products. It doubled or trebled the value of every kind of stumpage in the United States. And it gave timber growing an economic footing which it had never had before. It took a world conflict and vast destruction to usher in the golden age of American forestry.

We can, with fair historical accuracy, set out three events which mark our steps in forest progress during the past half century. The first was the crusade of Roosevelt and Pinchot. It generated a national consciousness and policy, expressed in great public forests and a Forest Service. The second was World War I. It expanded the national policy into a new field of co-operation and education. And the third was World War II. It gave forestry a conspicuous place in the industrial structure of the country and ended the era of cheap timber and migratory lumbering. Forest industry settled down.

III

A couple of months ago I visited the tree-seed extraction plant at Camp No. 1 of the South Olympic Tree Farm. Trucks were rolling in piled high with sacks of tree cones. Some brought Douglas fir cones from the Weyerhaeuser operation on Toutle River and the Clemons Tree Farm near Grays Harbor. Two trucks came in with Sitka spruce cones from the Crown Zellerbach camps in Pacific County. The sacks were emptied into bins in the storage sheds, and from the bins a steady stream of cones flowed through the dry kilns and whirling barrel shakers and the battery of seed separators. In the shipping room, clean tree seed was being weighed and sacked and labeled with the exact locality where the cones were grown. The brown, triangular-shaped seeds of Douglas fir ran forty thousand to the pound, but they seemed huge beside the minute, poppylike seeds of Sitka spruce. It took three or four hundred thousand of them to make a pound.

This is all part of the lumbermen's co-operative nursery at Nisqually, Washington. One trick of tree planting which foresters have learned is that seedlings thrive best when they are native to the locality where they are planted. So the Nursery Committee instructed each company planting trees on any sizable scale to gather seed from its own home grounds. The collecting of cones and planting of trees where nature's seeding has been burned out are now as familiar industrial operations as sorting lumber grades on a set of running chains. Only one thought was uppermost in my mind when I left the extraction plant: "And this is the *migratory industry* over which we wept and prayed forty years ago!"

Eight years of forest planting and seeding on West Coast tree farms exceeding 61,000 acres is but one of countless examples of how forest industry is settling down to continuous timber cropping on the same land. Trees have been planted on a far greater scale on the industry's woodlands in the Lake

States and the South. Over 190 million trees were planted on state and private lands in the southern states in the fall of 1948 and spring of 1949. Fifteen million came from industrial nurseries. A large part of the rest, grown by state forestry departments, were purchased by pulp and lumber companies, either to plant on their own land or to distribute among their neighboring wood lots. Several types of mechanical tree planters are used in the South and in the Lake region. Two men and a tractor-drawn machine can set out ten or twelve thousand young pines in a day. And for every acre reforested by tree planting, a score of acres is kept in continuous production of timber by forest-wise cutting practices.

The check on harvesting methods made by the Forest Service in 1945 and 1946 gives us a tangible measure of how far forest industry is outgrowing its roving youth and settling down to the business of growing trees. Of the large forest ownerships in the country, the service found 68 per cent whose cutting practices are "Fair and Better" and 39 per cent which are managed for a sustained yield of wood. Eighty-six large southern forest operations and sixty-one operations in the lake and northeastern states are keeping their lands in continuous production. In the Pacific Northwest probably a third of the private forests are on a permanent tree-growing basis or headed toward it.

Economic interest in forestry increases in proportion to the plant investment per unit of raw material. Many concerns, starting as one-product sawmills, have gradually added factory departments, timber preservation plants, or fiber by-products. At some point in their expanding investment and range of markets, they take root. They have too much at stake to pick up and move on. Tree growing at home becomes economic and essential to their business future. Often this conclusion has been aided by the discovery of how fast young trees are growing on their old cuttings, when given half a chance. Such plants are a far cry from the sawmills of two generations ago which moved on every twenty or thirty years.

Fifty years ago forest industry was busily building bigger and better sawmills. Today the same profit motive is at work for bigger and better crops of timber. Fifty years ago venture capital accumulated in old lumbering centers usually followed the lure of new virgin forests somewhere out beyond the ranges. Today much of this sawdust capital is staying at home and finding its opportunity in new technologies for the profitable use of second-growth trees.

In 1949 Ernest Kurth of Lufkin, Texas, was proclaimed first citizen of the South. Mr. Kurth was thus honored not because of his success as a manufacturer of machinery and of southern pine lumber, but because he took the risks, pioneered the techniques, and gave the South its first newsprint paper mill. While many of his old lumber confreres took their earnings West, some to prosper and some to go broke, Mr. Kurth stayed at home and created a new forest industry. One phase of his far-seeing enterprise was the growing of trees. Company lands were put under forest management. Millions of pine seedlings were distributed among East Texas farmers. A sawmill might pick up and move to Oregon but not a pulp and paper plant costing well up in the millions.

The invention of "Masonite" created another new forest industry at an old "cut-out" lumber center of the South. Masonite consumes the waste from many sawmills and provides a market for small wood from the farms of the countryside. Here again *technology and tree growing* replace the virgin forest as a source of raw material and as a field of investment. There is an expanding partnership between young trees and the latest "know-how" in wood use. The pulp and fiber industries of the South consume eleven million cords of wood annually, and not a stick of it comes from virgin timber. A great market for little trees has brought profitable forestry to the doorstep of every wood-lot owner in the South, down to a farmer with forty acres in trees and a team of mules. Everywhere you drive in Dixie, pulpwood is coming out of the back roads and piling up at the railroad sidings.

The old, durable profit incentive is bringing some wonders to pass in American woodlands. One of the problems of southern pine forestry is to keep the pines in top place on types of soil where scrub oaks, gums, and other low-grade hardwoods like to crowd in and take over. Sometimes the pines can be given the break they need by burning out small, weedy hardwoods, just before a good fall of pine seed. On thousands of acres every year crews of men work through older forests. They kill out the inferior hardwoods by girdling or by hacking gashes in the trunks and dropping poison crystals in these cuts. This little forestry job costs the owner four or five dollars an acre. Where a few pine seed trees are around and a good crop of cones is ripening, some operators bulldoze out most of the understory of unwanted hardwoods, at a cost of around seven or eight dollars an acre. They are usually repaid by a marvelous carpet of pine seedlings. Good forestry, indeed; but, more important, *good business!*

A year ago I rode through the great sweep of ponderosa pine forest in the Mount Lassen country of northern California. The region had special interest because it was the scene of one of my early Forest Service battles to induce the lumbermen of forty years ago to cut government timber and leave a part of the forest for growth and reseeding. We had also struggled, usually in vain, to get private owners to join Uncle Sam in emergency fellings to stop epidemic outbursts of pine-killing beetles. "The bug Dendroctonus" was pretty much the timber boss of the country in those days, and his handiwork could be seen in every direction. Some of the older cuttings we now saw had the familiar aspects of near forest wreckage; but soon we drove through a very recent private logging show. It was hard to believe there had been any cutting at all. You could see the tractor roads and occasional brush piles; but you had to ramble through the woods and fairly stumble over a low stump to realize that logs had been taken out. The owner, Collins Pine Company, was building roads and working rapidly through its virgin timber, cutting only the trees marked by its entomologists because of their *high risk* of bug attack. This meant taking only a quarter of the merchantable footage. The tractors crawled away from thousands of the highest grade lumber trees—to wait their turn in a later cut.

Industrial owners are waking up to the tribute exacted from their tree growth by forest insects. Soon after the war, pulp companies on the Oregon coast successfully combated the myriad swarms of leaf-eating loopers, moving through forests of West Coast hemlock. It was chemical warfare—spraying the infested woods with D.D.T. from airplanes. Forest owners in Idaho joined forces with the federal government in breaking up an invasion of the white tussock moth in a great mountainous stretch of lodgepole pine and Douglas fir by attack from the air. In 1948 and 1949, similar co-operative campaigns with air power cleaned up infestations of the spruce budworm which had struck westward into the domain of Douglas fir. The skill of the entomologists in timing those offensives almost on the exact day when the tender little green worms emerge from cocoons and start chewing on the foliage around them resulted in a "kill" of unbelievable thoroughness. Better control of the bad bugs of the forest is one of the constructive gains from our postwar reappraisal of woodland values, and the marvelous technology of insect attack from the air has become one of the great assets of American forestry.

World War II taught the West Coast logger half a dozen new techniques. In *pre-logging,* he takes out poles, "pee-wee" logs, and pulpwood before the main job. This saves the breakage of small trees when the heavy stuff comes down and out. In *salvage logging,* he moves in after the big timber has gone, to gather up the small logs, chunks, and broken trees that can be marketed for one purpose or another. He has indeed gone back into many slashings three or four years old, on former logging shows, and recovered fifteen or twenty cords per acre of low-grade saw logs, pulpwood, and short, clear chunks that can be turned for rotary-cut veneer. In a few operations the logger is learning an art that is new and untried in the Pacific Northwest. This is the *thinning* of the "junior forest," the dense stands of second growth, thirty years and up in age. His experience with ten- and fifteen-ton logs is of little use on this job. Along with his lightest tractors, the northwest logger is turning back to the sturdy old "Dobbin" of the north woods. We may yet see the bull teams of 1860 back on Douglas fir skid roads. Several companies are determined to find a practical way to put to use the immense quantities of wood now going to waste on our overcrowded acres of young forest.

A visit to the woodyard of a Puget Sound paper mill lately gave me a picture of the current progress in West Coast utilization. About half of its pulp stock was still chipped from old-growth logs. But the other half was what interested me. Truck after truck rolled in, loaded with green pulpwood from farm wood lots. This was in eight-foot billets, four inches and up in diameter. Other trucks were piled high with dry pulpwood, salvaged from three-year-old slashings. As we watched them unload, an electric dolly pushed in three huge freight cars filled with thick, barky slabs. They were gathered up from the refuse dumps of little crosstie mills back in the sticks. More cars were spotted in, bringing clean waste, trimmings and edgings from sawmills in nearby Tacoma. I thought the possible gleanings from the West Coast forest industry were about accounted for, but then I noticed a new bin under construction. "What is going in there?" I asked the yard foreman. "Pulp stock, chipped for us at two plywood plants, out of their waste peelings and rejects."

The utilization of West Coast forests is still very far from complete. The average logging job now takes from the woods perhaps a quarter of the material that was left to feed slash fires before the war. Even this is a great boon to regional forestry. So is the trend toward making plywood, lumber, pulp, or fiberboard under the same management—indeed, from the same tree. Great plants that turn out many different products, like those at Shelton and Longview, Washington, are not only the last word in using the entire tree, even to its bark. Forestry is, of necessity, a part of the enterprise. Such industrial structures are not conceivable without a backlog of forest land and management that assures wood supplies for years ahead.

There is a pine mill I love to visit up in the Okanogan Valley of northern Washington. It manufactures fifty or sixty million feet of logs every year from its own tree farm or from the Colville Indian Reservation. But except for the needs of the local people, this sawmill sells no lumber. Every board goes from the dry kilns into its own factory. It is cut or laminated or glued up into a knocked-down casket, a window frame or sash, a table, a chair, furniture for a breakfast nook, a kitchen cabinet, an ironing board, or any one of twenty other things of everyday use. Not a scrap of wood is wasted.

The latest device of these resourceful millwrights is a special sawing rig for utilizing the great forest weed of the West, the lodgepole pine whose dense stands of six-, eight- and ten-inch trees clothe many million acres of our high country. Little lodgepole logs, eight feet long, and four and one-half inches or more at the top end, tumble down onto a moving chain cradle. They are split in half by a thin, fast horizontal band saw. The half sections spin around and are split again. A stream of little lodgepole boards finds its way through the dry kiln and factory. They are edge-glued into panels, shaped into caskets, or laminated into square pieces which may be turned on a lathe into piano legs.

The most impressive thing about this establishment is not the mechanical marvel of its machines but its army of skilled workers. From every thousand feet of logs cut in its woods, this plant creates more than double the man hours of labor that one finds at the usual western sawmill of the old style. It is in employment that forest industry is making one of its great contributions to the economy of the United States. The trend of today does not stop with more trees to the acre and more wood out of every tree. It carries on into the employment of more workers to the cubic foot or ton of raw material.

The average West Coast logging camp and green lumber mill employ seven year-round workers for each million feet of timber felled in the woods. The number of employees maintained by each unit of raw material steps up in the wood pulp and plywood industries. And now integrated operations producing pulp or fiberboard and factory items, along with lumber and plywood, are employing twenty year-long workers on each million feet of timber taken from the forest. Better markets and advancing technologies are giving us three times the pay roll from the same tree.

The increasing yield of employment from our woods is restoring the economy of forest regions whose virgin timber is gone. The pulp and fiber industries developed from young forests have given the South an investment of more than a billion dollars and an annual pay roll in excess of $200,000,000. Washington passed her peak of log production in 1929. Employment in logging camps is now less than half their number of workers twenty years ago. But the employment in processing and fabricating a log brought in from the camps is steadily gaining. Our state now maintains one third more people on each billion feet of timber taken from its woods than when the logging camps were running full.

These characteristic developments of American business could not be foreseen when Gifford Pinchot read the portents of timber famine in the skies. They are not reflected in any official estimates of tree growth and drain. The American genius for technology and industrial organization is taking over the woods. Through the union of many techniques in manufacture with intensive timber culture, this country is taking the lead in industrial forestry.

The progressive advances of some forest owners and manufacturers in every part of the United States have set the trend of forest industry. But it is still a *trend,* not a completed change. The situation is very mixed. As texts may be lifted from the Bible to support any system of theology, examples may be picked from the vast expanse of the U.S.A. to support any theory of forest progression or retrogression. Many sizable woodland holdings are still drifting in the old currents of a temporary industry. There are still fifteen thousand or more little sawmills, many of which move about from wood lot to wood lot, crowd lumber production on good markets, and go out of business in hard times. Behind the little sawmills are over four million farm woods and other little forests. The practical task of swinging this host of potential tree growers into step with the present-day economics and know-how of land use is being undertaken with fresh vigor. But the condition of most of the little woodlands, and of many large woodlands as well, simply points up our whole forest picture. It is a picture of change. Irresistible economic forces are moving us out of timber mining and into timber cropping. The change is slow and ragged. Some men lead and others hold back. The old and the new go on side by side. It is like our evolution from primitive farming to scientific agriculture. And it is just as American as a Presidential campaign.

Nothing has ever fascinated me more than to watch this change all about me in the Pacific Northwest. Some of the captains of industry just cannot take the mental hurdle from big timber to little trees. And some of the quickest to grasp the economics of forestry are hardheaded loggers who have lived in the woods all their lives and "seen 'em grow." Today a screening process is under way between concerns that are winding up and pulling out and those determined to stay in business. Every now and then, some well-known lumber company carries its old policy of liquidation to the bitter end. The last log is hauled up the jack ladder and the last whistle is blown. When that happens, the old sawmill goes to a junk yard, but there are two or three bidders for the cutover land.

Doubtless the lumber industry will always have a marginal fringe of migrating mills. They will pick up and move about from place to place, just as they move into production and out again with the rise and fall of lumber markets. But the great treks of big sawmills are ended. The speculation in free public timber is over. The lumber market will not again be knocked topsy-turvy whenever a new railroad is built into the West. The industry has sown its wild oats and is settling down.

Forest industry is becoming more technical. The same plant or company is making more products because it cannot afford the old wastes. Competition compels it to keep up with the new technologies. Plant requirements are increasing the capital investment, and rising capitalization makes forestry more and more necessary to the whole industrial structure. In 1949 the investment in our factories for manufacturing and processing wood exceeded $7,000,000,000—$15 for every acre of commercial forest land. It was twelve times the investment in 1899 when much more of our forest was virgin. These invested billions exert the greatest force in the country for the continuous growing of trees. For one thing, outside of Alaska and Latin America, no new and available timber resources are left. Industrial forestry may take the form of direct land ownership and management or it may organize the growing of wood supplies in the neighborhood. Often it will combine both methods. In either event, the industrial demand for enough wood *in place* to carry present-day manufacturing and investment requirements is strong assurance that our available land will be kept in trees.

CHAPTER XI

The Role of the States in American Forestry

ONE of the most imperative missions laid upon the shoulders of Gifford Pinchot's young scouts was to get forestry recognized and undertaken in some form by every state. I remember urging a group of interested people in Mississippi to get behind a bill setting up a forestry commission and starting protection from woods burning. With missionary zeal I stressed the importance of a *nonpolitical commission.* "Keep politics out of your forestry," I insisted.

A charming old southerner answered me. "I think we should caution our enthusiastic young friend from Washington," he said, "that politics is in everything in the state of Mississippi unless it be a freshly laid egg; and I suspect the political implications of each new egg."

Two of us were called to G.P.'s office one morning and instructed to draft a letter "full of punch" to the governor of every state in the Union. It must detail the forest conditions and needs of his state and the dependency of its people upon continuous supplies of wood. It must recommend the major points in a state forest policy. Soon an Office of State Cooperation appeared in the Forest Service. The widespread enactment of state laws making some start in forestry became an educational force. Twenty-five states had something going when the Weeks Law was before Congress in 1911. Most of these statutes dealt with forest fires or set up conservation departments or launched surveys of local forest situations. But several states struck out aggressively on original lines, such as creating state forests, operating forest nurseries, and offering tax exemptions on land planted with trees.

Of possibly greater influence than legislation in creating a forest consciousness in many localities was the power of a remarkable group of early state leaders. The conservation cult drew and inspired men of crusading temper. Dr. J. T. Rothrock, forest missionary and first Commissioner of Pennsylvania, tramped his state from end to end, selling the people on conserving Penn's "woods." Lumberman Henry Hardtner put the seed tree and reforestation contract laws through the Louisiana legislature almost singlehanded. There was Governor Sterling Morton of Nebraska, with his zest for tree planting on the prairies; and tough Governor George C. Pardee of California, who made forestry a successful political issue. Joseph Hyde Pratt of North Carolina was an evangelist in spreading organized fire control throughout the South. And up in Massachusetts, Forest Commissioner Raines at some point in every address displayed a little bottle of white pine seeds, trickled them into his hand, and told his listeners that in these little seeds lay the restoration of the forests of the old Bay State.

Federal co-operation in protection from forest fires, offered first by the Weeks Law and on a constantly expanding scale by the Clarke-McNary Act, gave all state undertakings in forestry a common line of march. The great expansion of state services in the last twenty-five years has centered on forest safety. Meantime another problem, common to all states, has moved up into second place. All states ask: How can we make woodland-owning citizens grow more trees. The universal answer, of course, is "Show them how." From the earliest days of the federal Bureau of Forestry, its men were out around the country, examining wood lots or industrial holdings and advising owners how to take better care of them.

I was assigned to one of these forestry-selling missions in the summer of 1904. It was on a coal property in the Cumberland Mountains of Kentucky. It was my first cruise in the oak and beech and poplar forests of the southern Appalachians, and I enjoyed it hugely. But the incident which stands out most sharply in memory had nothing to do with timber. It occurred one evening as I was trudging homeward. I was following a well-used bridle path up a long slope toward a ridge where the skyline was still bright. A horseman appeared in the patch of light ahead of me, with a long rifle across his saddle. The moment he saw me on the shadowy trail below, up came the rifle to his shoulder and I was covered.

All I could do was to stand out in the brightest spot I could find and spread out my hands to show that I was without weapons and of peaceful intent. The man on horseback rode toward me with that confounded rifle barrel aimed at my chest and growing bigger by the second. At my side, he stopped his horse, leaned over, peered intently in my face, raised his gun and rode on without a word.

Uncle Sam's co-operation in teaching agriculture through the land-grant colleges opened the door for the states to take an active hand in promoting farm forestry. Federal co-operation in educating wood-lot owners and growing trees for them to plant was vastly expanded by the Clarke-McNary Act and later by the Norris-Doxey Act of 1937. It now reaches into the management of farm woods on the ground, into timber estimating, marking trees for cutting, even finding purchasers for them.

The states have been pulled from above by these federal leads and the loosening purse strings of the national treasury in dollar-for-dollar grants. At the same time they have been pushed from below by the swelling tide of local farm and community interest in cutting and growing trees. The result of the two forces is taking forestry to the small woodlands by all the devices of rural education and technical service. This is the second phase in the expanding role of the states.

The Farm Forestry Committee of Kitsap County has just had a meeting with the first "farm forester" to land in our part of Washington. We are all giving him lists of farmers who are interested in doing something worth while in their wood lots. We are planning a tour of the county this fall, on which interested folk can see half a dozen examples of thinning and careful logging in our stands of second-growth Douglas fir. Our forester will mark a couple of demonstration acres and explain just why each tree should be cut or left to grow. We are studying ways and means of assembling and marketing the vast quantities of pulpwood that should be cut from our overcrowded young forests and fed to the huge mills around

Puget Sound. This sort of local teaching and planning is going on today in hundreds of counties around the United States.

The Chief of the Forest Service reported for the fiscal year 1949 the distribution of seventy-seven million trees from nurseries in the co-operating states for farm plantings. "Extension agents have enlisted approximately 23,000 local leaders to assist in promoting forestry in their respective communities. . . . Many thousands of farmers last year conducted such forestry activities as planting, thinning, weeding, pruning, selective cutting, and improved practices in naval stores and maple syrup production." Under the Norris-Doxey Act, the chief reported: "Improved management practices were applied to 1,769,240 acres of small woodlands by 17,140 individual owners. . . . Requests for assistance from 3,121 woodland owners remained unfilled because of the heavy demands on the farm foresters' available time."

It is heartening to see how many states have adopted the idea of helping their woodland owners grow trees and have carried it beyond the co-operative programs of the government. The idea has tremendous vigor. Management service to forest owners has become the most actively expanding field of state forestry today, as protection from fire was twenty years ago. The states are co-operating with federal agencies in maintaining 325 extension foresters and farm foresters on the job and, in addition, are employing 280 foresters of their own for advice and direct technical help in the wood lots.

Extension Forester Barraclough of New Hampshire, with the co-operation of the State Commission, has put a forester in every county of his state. One of the results is the establishment of over eight hundred demonstrations of good wood-lot management, covering every section of New Hampshire. Vermont has a forester at work in the woodlands of every county. Michigan is creating great interest among her thousands of owners of little forests by a series of "model farm wood lots." The forest and extension services of Texas are conducting intensive forestry schools for the county farm agents, so that they can make tree growing an effective part of their local educational programs. North Carolina is one of the leading states in well-planned demonstrations of wood-lot forestry. Extension Forester Graeber began working on them in 1926, shortly after the Clarke-McNary Act launched the co-operative program, and he has established them by the hundreds in farming and small mill communities.

The Forestry Division of New Jersey started a solid, businesslike drive on the wood lot in 1938. After every request for help, a wood lot is examined by a state or federal forester. The trees that should be felled are marked. The owner is advised just what he has and what he should cut to leave a good growing stock. All this is free service, but the owner must agree to abide by the cutting plan and to recoup the state for its costs should he fail.

The Jersey plan puts the actual cutting in the hands of a "timber agent," an experienced woodsman of proven trustworthiness. For a commission of 10 per cent, the agent gets the highest bid for the timber to be cut, makes the sale, supervises logging and cleanup, and turns the proceeds over to the owner. About half of the state's cut of timber is now harvested under this voluntary acceptance of state control.

New Jersey has gone farther than any other state in carrying forestry on the wood lots of her citizens through to the dollars delivered on the stump. But many other states are moving in the same direction. Pennsylvania keeps seventy-five foresters busy in advising wood-lot owners, doing sample marking, making plans for tree planting and the like. Most of the states offer this service without charge; others give the owner free advice on his general scheme of management but charge him the cost of marking timber to be cut, tree by tree. The field foresters of the South Carolina Commission made 711 woodland examinations last year and completed marking jobs on about twelve thousand acres. These owners were charged fifty cents for each thousand feet of saw timber marked and thirteen cents for each cord of pulpwood. With the volume of wood marked for sale, each owner receives a list of prospective buyers, a sample sales contract and such other help as the badly rushed state men find time to give him. There are 102,000 small woodlands in South Carolina, and the commission cannot keep up with the demand for this practical help.

In the policies of several states, tree growing by education has moved on into tree growing by special inducement or by compulsion. This is the latest chapter in the story of state forestry. It is of special interest because it goes to the heart of the issue which has run through federal forest politics since Mr. Pinchot demanded regulatory laws in 1920. The controversies in Washington have been re-echoed in many state capitols. "States' rights" has raised its voice in opposition to federal encroachment. The solutions thus far undertaken by the states show great diversity in ideas and traditional differences between conservative and radical tempers. They range all the way from easier taxes on tree-growing land to laws which tell a logger just what he may cut in his own woods. It is the democratic process in action, following the traditions and squirming around the inhibitions of the local electorate or its dominant political forces. Of special interest are the resourceful devices to induce a forest owner to grow trees without actually compelling him to do so. Our states are showing great ingenuity in getting the horse up to the watering trough, but only a few are forcing him to drink if he doesn't want to.

In the early days of the conservation movement, there were numerous attempts to curb destructive logging by legislation. Controls of cutting were sometimes forced onto statute books by aggressive minorities or singlehanded champions of forestry in advance of public understanding or demand. They largely failed in practical results. Surprisingly enough, Nevada led off in 1903 with a prohibition against selling wood from any conifers of less than one foot in diameter at stump height. Few people ever heard of it. New Hampshire legislated, in 1921, the leaving of at least one pine seed tree per acre. Apparently the statute was never enforced. A more or less similar fate overtook the noble experiments in legislated forestry of Louisiana, New Mexico and Idaho.

Minnesota made a brave attempt to save the young trees on her private woodlands. Its legislature specified a minimum cutting diameter of six inches for spruce, balsam, jack pine, and tamarack, and of ten inches for white or Norway pine, birch, maple, and oak. These restrictions did not apply to cutting fuelwood or Christmas trees or when land was cleared for farming. The law proved so difficult to enforce that the State Commission has more or less given up the attempt.

There was, of course, a considerable by-product of education and forest awareness from these early undertakings in forest conservation. But it is a rather dismal record of good intentions still-born. It gave the possibility of state control of bad woods practices a black name, and this has not been overlooked by proponents of regulation under the strong right arm of the federal government. There is a different story, however, in the reforestation of the earth banks left after strip coal mining in Indiana. That began with several years of voluntary co-operation between coal producers and the state forester in planting spoil banks with trees furnished by the state. In 1940 the Coal Association and Conservation Department worked out a compulsory law with a legislative committee. It required an application for each new stripping, a bond of $25 per acre to assure its forestation to the satisfaction of the state, and the progressive planting of old strippings of the same operator. Twenty-three million trees have been planted under this statute of Indiana.

One of the most courageous undertakings of any state to get to the heart of its land-use problems was the rural zoning law of Wisconsin. It was brought about by the aftermath of wholesale logging and burning—in tax-delinquent lands, attempted settlement of poor soils, and near-bankrupt counties. Under state-enabling legislation, the Wisconsin counties have zoned five and one-half million acres out of agriculture and settlement just as a municipality zones factories out of its residential districts. The official labeling of this acreage aided the remarkable reconstruction of Wisconsin forests. Other steps were airtight protection from forest fires; the distribution of trees from state nurseries, now running eighteen million seedlings a year; and the launching of many counties in the forest-owning and tree-growing business.

In the early sweep of conservation ideas, many states offered tax inducements for tree planting. Oregon and Washington took a long stride in forestry by offering a low, fixed land tax on cutovers, together with a harvest tax on the next crop of timber. Several states have offered tax concessions in return for commitments by owners to keep their land in trees. Most of them start with a low annual tax and then add a harvest or yield tax whenever timber is cut. Louisiana, Minnesota, Wisconsin, and Missouri have substantial areas of "reforestation contracts," "auxiliary forests," or "forest crop lands," on which the main tax load is deferred until timber is harvested. Meantime the owners must carry out the plans of tree growing and protection prescribed by the foresters of the state.

New Hampshire was the last state to grapple the tough old bull of forest taxes by the horns. After a hard-fought legislative battle in 1949, Governor Sherman Adams, ex-lumberman and fighting conservationist, won the day. All forest lands in the state are placed on a fixed assessment of $2.00 per acre, plus a harvest tax of 10 per cent of the value of any forest products cut from now on. However, if cutting conforms with the rules of a committee of local woodland owners and state foresters, the severance tax is reduced to 7 per cent. Most of the rules require an owner to keep a minimum growing stock on his land of ten per cent of its merchantable volume of wood.

New Hampshire has made the most forceful stroke of any state to promote tree growing by use of taxation power. There are no "ifs, ands, or buts" in this law, no speculative choice between different methods of figuring taxes. You save 30 per cent of your timber tax if you follow the forest prescription of your own county. No shrewd Yankee need bite this dollar or turn it over to see what is on the other side. The New Hampshire plan also incorporates the grass-roots type of cutting control which is gaining favor in many states. A *local committee* of woods owners working with neighborhood foresters who represent the state is the prime mover in several recent laws which are closing in upon woodland management from one direction or another. The same kind of local agency plays an important role in the forest controls of Finland and Sweden. It will be used widely in carrying out the forestry programs of this country. It is the way we like to do things.

A New York statute of 1948 couples the local committee plan with a different incentive to the owner. The inducement here is all-out help from the state in managing a forest property. The law sets up fifteen District Forest Practice Boards, composed mostly of local forest land owners. A forester from the State Conservation Department works with each district board. Its rules of approved woodcraft must be sanctioned by a state board on which each district is represented. The State Conservation men examine woodlands upon application. If the owner undertakes to carry out the approved forest practices of his district, he receives from the state timber-marking service, help on any forestry problem, marketing assistance, and planting stock—all free. The state legislature has as yet been unwilling to throw any form of tax concession into the bargain, but New York leads all the states in the free technical aid given to forest owners who sign the code. Up to mid-year of 1949, the department had received applications for co-operation covering 730,000 acres of woodland.

New York, by the way, well illustrates the importance of local forestry to a thickly peopled and highly industrialized state. Wood-using industries rank high in her economy, carrying 137,000 employees and a pay roll of $422,000,000 in 1947. New York consumes a billion and a half feet of lumber annually for construction and fabrication, including her outstanding furniture industry, but imports 69 per cent of it from other states and from Canada. New York ranks first among the states in the manufacture of paper, but imports four fifths of the pulpwood or pulp which makes it.

Nearly half of the Empire State is forest land, but 30 per cent of that half is classed as "idle or unproductive." Throw into the scales the needs of her vast population for domestic and industrial water and for recreation, and it is clear that effective use of all the forest talents at her command warrants the care and interest they are receiving in New York.

California has elevated her Forest Practice Committees into legislative bodies. Their rules, if approved by two thirds of the acreage in their own constituency and by the State Board of Forestry, have the "force of law" over the woodlands of the district. In theory, California thus abandons the inducement approach to forestry on private lands and calls upon each committee to write the legal ticket for every woodland operator in its district. However, the law as yet carries no provision for police enforcement. The state forester must rely upon his powers of persuasion and the force of public opinion. The record, however, appears to sustain the reputation of the state for phenomenal things. On the third inspection, after previous explanations and warnings, three fourths of the cutting operations were found to be in compliance with the law. Every timber operator in California must register with the state. In lieu of the diameter limits and other cutting rules of the district, he may follow an alternate plan of his own if the state board is satisfied it will accomplish equivalent results.

To Maryland should be awarded the distinction of carrying the district-committee form of forest control to its highest point in the United States. Maryland has made of it true local self-government in forestry, with the unequivocal legal backing of the state. The Maryland law of 1941, foundation of the committee structure, was haled into court as a violation of the constitutional rights of property. In a sweeping decision, the Garrett County Court applied to forest controls the legal principle which underlies municipal zoning ordinances. The court said: "There is no such thing as absolute ownership of property in the sense that it is held free from any possible restrictions of any kind. For as property owes its value to the protection of the state, it is held subject to the right of the state to impose such restrictions upon its use as may be necessary to enable it to afford that protection."

Maryland's Commission of State Forests and Parks has authority to appoint District Forestry Boards, representing forest and woodworking interests, owners of farm woods, and the Farm Bureau or Grange. The board drafts its own local rules of forest practice. When approved by the commission they have the force of law. Every operator must apply to his district board for a cutting license thirty days before going to work. The board has the woodland examined and studies the wood cutter's plan for leaving a growing stock or, if clear cutting is intended, for restocking. It advises the operator how he may cut, and its "advice" is enforceable by fine or imprisonment.

Massachusetts has a single state-wide Committee on Forest Practices, also composed of woodland owners and public foresters. It has drawn up a standard code of acceptable forestry for the state. The code permits clear cutting only where the ground has a count of one thousand healthy seedlings to the acre. Otherwise seed trees picked from an approved list of twenty "desirable species" must be left. Massachusetts has also adopted a unique and original way to get the horse up to the water trough while leaving the drink itself to his own free will. The know-how of good forestry is compulsory; the swing of the ax is voluntary.

Any forest owner who intends to cut over forty thousand feet of wood in a calendar year *must* give advance notice to the state. A forester from the State Division examines the woodlands and draws up a cutting plan for the prospective logger. He tells the owner just how to apply the State Committee's Code to his particular acres, and then the wood cutter exercises his New England right of self-determination and cuts as he jolly well pleases. This system of compulsory advice, however, has a powerful educational punch. It usually brings the wood-lot owner, the wood cutter, and a state forester together under the trees. The state men report what actually is done, and their findings show that three fourths of the current logging meets standard requirements.

The last among the diverse patterns of our states to get good forest practice on the lands of their citizens is a revival of the early attempts to curb destructive logging by specific statutory requirements. Mississippi in 1944 specified by legislation the minimum number and sizes of trees to be left when cutting pine or hardwood saw timber and pine pulpwood, or when tapping trees for naval stores. In 1948 Virginia prescribed the leaving of pine seed trees in forests where loblolly or shortleaf pines make up 50 per cent of all the trees. A logger may cut his woodlands under any alternate plan of his own, if the state forester approves. The Virginia legislature recognized the strong pull for a local voice in forestry affairs by limiting the application of its seed tree law to the counties whose own supervisors concur. Within a year after passage of the law, forty-eight county boards out of fifty-five pine counties approved local enforcement of seed trees.

The conservatism of the South is one of our immutable traditions. As late as 1920, no part of the United States impressed me as more wedded to woods burning and all the old abuses of the land. However, not only have two southern states taken the lead in statutory controls of timber cutting, but the whole South has moved up in fire protection, wood-lot services, and industrial forestry faster than any other section of the country. With her natural advantages as a tree growing country and the backing of her wood-using industries, I am betting on Dixie to lead the nation in profitable forest management.

The Forest Conservation Acts of Oregon and Washington were direct legacies of the Blue Eagle Lumber Code. They became possible politically because the pine and fir lumbermen of the Northwest had made a serious effort to live up to their commitment under the Recovery Act and had carried on Article X as a voluntary undertaking after the death of its parent law. The industry had been well prepared. Still, it was a bitter pill for many of the rugged, old-school loggers to take. I put in a lot of hard work to get the industry to go along with these state laws and had some stiff rebuffs. A strong argument was that if we did not work out the forestry problem ourselves to the satisfaction of the public, it would only be a matter of time before Uncle Sam stepped in with a heavy hand. And the industry should hold its lead in the situation.

The key provisions of each state law were lifted from the forestry sections of the lumber code. The renewal of ponderosa pine forests was decreed by diameter limits in cutting, and of Douglas fir by reserving at least 5 per cent of the standing timber on every quarter section. Satisfactory forest plans of a woodland owner can be substituted for these legal specifications. Most of the tree farms, for example, leave from 25 to 60 per cent of an old-growth forest in the first cut. These seeding blocks are fitted to the topography and alternate with patches of clean logging.

The most distinctive feature of the forest mandates of the far Northwest is their excellent equipment of sharp teeth. As the laws stand today after considerable amending, no one can log commercially in either Oregon or Washington without a license from the state forester, renewable every year. If any cutting is found in violation of the Conservation Act, the operator may give bond to make his deficiency good by tree planting or he will be subject to a penalty of $8.00 per acre. Continued violation may forfeit a logger's license, i.e., his right to remain in business. This is forest control with a big stick. A county judge in northeastern Washington, on a plea from a small logger, enjoined the state forester from requiring a license to utilize a man's own timber, but the Supreme Court of Washington and the Supreme Court of the United States sustained the statute in 1949 as a proper exercise of state police powers. The counsel for the West Coast Lumbermen's Association intervened in this battle of briefs in support of the right of the state to forestall destructive logging.

For the first several years, each state centered the enforcement of its considerable power over private forests upon education. Inspection in the woods, show by show, brought the law home in specific terms to each operator. After this period of schooling Oregon led out with vigorous enforcement. At the end of 1947, 309 violations of the act were reported and $114,439 collected in penalties. The inspectors found reasonable compliance on over 95 per cent of the acreage logged. One loophole was plugged by the Oregon legislature in 1947. Lands cut clean on grounds of pending use for agriculture may now be reclassified by the state forester after five years have elapsed and the owner found in default of the Conservation Act if the plow has not, in fact, followed the ax. Washington's enforcement is still largely on the educational side. There have been five convictions in Washington and two delinquent operators have posted bonds to replant their cutovers.

At least half a dozen additional states are working out their own patterns of inducement or compulsion to grow more trees on private lands. We will hear much more from the states in the future story of American forests. They are stepping out aggressively in this big field of public direction or control of the use of land. Their aggregate power over the future of our forests is enormous, and in the feel and growth of it they are writing a new chapter on the ways of democracy.

When Oregon wrote compulsory cutting rules into her forest code, the Joint Committee on Conservation of the Douglas fir industries centered its efforts on making this trial of state control *work*. Our field men followed up cases of noncompliance and tried to set rebellious operators on the right track. I took a hand with a stubborn Swede logger. He was one of the most efficient of the craft and prided himself on using his timber to the last scrap of merchantable wood. He could not bring himself to "valk avay and leaf good stumpage."

Did he not know how the forests were cut in Sweden? "Yah, but that's a different country where everything has to be saved. In America a man has a right to do what he vants with his own land."

What would he do with his land after logging? The county would probably get it as it had the rest. Did he know what lumber and pulp companies were paying for cutover land if it had lots of little trees on it? Why should he drop his land for nothing? Did he know what pee-wee second-growth logs were selling for in Portland? He did not—but would look it up.

A few months later, at a lumbermen's meeting, my Scandinavian friend greeted me with a grip like the vise in his own machine shop. "Come and see my voods again," he said. And then, quite confidentially, "You know, the longer a man lifs, the more, by Yimminy, he finds oudt."

I have asked many a state forester about the effect of *his* program in getting at private woodland practices. Every answer lays stress on what "our plan" has accomplished in *education*. All of these state roads seem to be heading for the same terminus. Whether the starting point is forestry by tax concessions, by self-government through local committees, or by statutory rules, the pay-off is forestry by the free choice of informed men. Cutting controls of one sort or another serve as stop-losses until individual forest management takes over. A state may help or hinder by its tax system or its degree of protection. Whatever the route, the measure of progress is the intelligent use of land for growing trees from self-interest. *Education* is the marker on every milepost.

CHAPTER XII

Our Public Forests

PRESIDENT Benjamin Harrison doubtless did not realize that he was starting a chain reaction when he signed the bill which authorized the creation of federal forest reserves. In 1949 the national forests had grown to 180,373,000 acres. They took in 9.5 per cent of the country's land. If blocked up solidly, the national forests would cover New England four and a half times or all of California and Oregon with half of Washington thrown in. Eighteen and six tenths million acres in Uncle Sam's big wood lot have been purchased under the Weeks Law, for a total of $90,214,000. Five million, six hundred and ninety-seven thousand acres have been acquired by trading public land and timber for state or private woodlands. The rest of our national forests were carved chiefly from the public domain.

No colony of beavers was ever more persistent in wood cutting or stream damming than the Forest Service in its determined expansion of the national forests. Federal ownership of woodlands has spread into every type and condition of forest and near-forest, from the spruce woods of Vermont to the chaparral canyons of southern California. And it has brought examples of forest protection and good cutting practice home to an astonishing number of American communities.

The expansion of the national forests has not been without political skirmishes. After the first flush of enthusiasm for buying up forested watersheds of navigable rivers, the Congressional purse was tightened. John W. Weeks, able Massachusetts banker and author of the law, stoutly defended the policy as Secretary of War under President Harding. He offered to buy all the lands acquired in the White Mountains at the price paid by Uncle Sam and called them one of the best investments the government had ever made. Early in my days as chief of the service, I was put on the spot by the belligerent Society for the Protection of New Hampshire Forests for not pushing purchases in the White Mountains fast enough to stop devastation by Yankee lumbermen on the sacred slopes of the Presidential Range. The society threatened political action to turn the job over to, of all men, Secretary Albert Fall of the Department of the Interior. They turned hopefully to the Secretary for faster action and less insistence upon looking into the horse's mouth before buying.

Of all our Presidents, Franklin D. Roosevelt most vigorously supported the enlargement of the national forests by land buying. From 1934 to 1938, F.D.R. enriched the Weeks Law budget with fifty million dollars from emergency and relief appropriations placed in his hands by Congress. The depression was certainly a shrewd time to roll up forest acreage at low cost.

A few of our most self-sufficient states, particularly New York and Massachusetts, have maintained their independence of federal apron strings by refusing the legal concurrence in forest purchases required by the Weeks Law. Other states have limited the total acreage that may be acquired. By and large, however, the forest-buying dollars from the treasury have been welcome. And the presence of the National Service as local forest administrator and co-operator has been helpful to the progress of the state.

Exchanges of land or timber offered a practical means of consolidating and extending the national holdings. It was sanctioned by Congress in several special acts and in the general Exchange Law of 1923. The forest supervisors have had many chances to block up scattered lands by trading acre for acre and to add an acre of cutover land in trade for a thousand feet of standing timber. The latter transactions were soon dubbed "stumps for stumpage." They became popular and have substantially swelled the national forest acreage.

The Forest Service has created not only a vast national estate but a remarkable structure of administration. Here is the realization of our starry-eyed dream at Gifford Pinchot's round table. Here is multiple use of the land, a technique and guardian for every use! Everything is provided for—the harvesting of timber, the grazing of forage, the mining of ores, the homesteading of arable land, the conservation of water and wild life, recreation, replanting run-down forests, reseeding run-down ranges, and protection from fire, forest bugs, and tree diseases. Constant research bores into every resource and activity. The traffic officers must be puzzled at times in routing the diverse users and specialists. But it all adds up to an outstanding achievement in the co-ordinated use of natural resources carried out in practical and human terms.

The methods of protection and management which have proved efficacious on the national forests have filtered into other federal timberlands under the Department of the Interior. There are sixteen million acres in Indian reservations, administered in trust for the various tribes. Nine million acres of indifferent forest, largely without commercial utility, remain in the unreserved public domain, excluding Alaska. Seven million acres in the national parks and monuments are devoted to the preservation of natural beauty and reserved from any form of commercial use. One of the most interesting national holdings is two and one-half million acres of checkerboard woodlands running north and south through the heart of West Coast lumbering in Oregon. It consisted originally of the alternate sections of public land within a twenty-mile strip on each side of the Southern Pacific Railroad. The revested Oregon and California grant was first given by Congress to aid construction of the Southern Pacific and then taken back because the company did not live up to its terms. This busy stretch of forest is administered by the Bureau of Land Management, Department of the Interior. It is covered by the most clear-cut and mandatory law yet passed by Congress to limit the take of timber from federal lands to the volume of wood the forest is currently growing. The Act of 1937 fixed the maximum annual cut of timber from the revested grant at 500 million board feet until the estimated growth is revised by more exact surveys. It also prescribed that this total be broken down into allowable cuts from the several watersheds and other natural units supporting existing lumber towns.

The twenty million acres of rich forest on the coast of southern Alaska, equal in area to the state of Maine and destined to become one of the great paper factories of the world, are in national forests. To the Department of the Interior is entrusted the thankless watch and ward over vast stretches of low-grade spruce, birch, aspen, and willow spread over inland Alaska. Yet notwithstanding their slow growth, these too, in the main, are productive forests and have their place in Alaska's future.

All of the forests in federal ownership contain 39 per cent of the standing timber of commercial quality and accessibility in this country. About half of Uncle Sam's wooded domain covers mountains too rugged and hills or mesas too sparsely timbered to be included as yet in our inventoried woodpile. These great stretches of the out of doors have, of course, many other highly necessary social and economic services. They protect water sources, conserve topsoils, and provide range for both game animals and domestic livestock. They keep for the oncoming generations something of the great inspiring and re-creating hinterland that has contributed so much to American character and resourcefulness.

The extension of managed federal forests, replacing the unreserved public lands of earlier days, has had something of a parallel in the slower spread of state forests. By 1949 over one thousand state and territorial forests had been established on a fairly permanent basis of public ownership and administration. They totaled nearly 13.5 million acres. These are woodlands dedicated to all-around use, including the growing and harvesting of timber. They do not include state parks or the two-and-a-half-million-acre "preserves" in the Adirondacks of New York where no cutting is permitted. Nor do they take in the state timberlands still subject to sale or disposal without provision for future management. They are, as far as present policies go, part of the permanent wood-growing estate of the Union.

Many local influences have joined forces with the currents of national conservation in creating state forests. New York was one of the first in the field. Her purchases of forest land began in 1885 and now include 512,000 acres of state forests, managed for timber crops along with other public benefits. At about the same time, Dr. Rothrock started Pennsylvania on her state forests which now cover 1,736,000 acres, or 6 per cent of the state.

Michigan launched a farsighted program of forest restoration in 1900. It was aimed directly at the hardest-cut and hardest-burned land in the state, then largely on the delinquent tax rolls of the pine counties. Much of this land was taken over by the state, consolidated by exchanges, and converted into permanent state forests. Other woodlands were purchased with money derived from the sale of hunting licenses. In 1904, the planting of these lands began. A few years later I tramped over some of Michigan's plantings and marveled at the miles of furrows which big plows had broken through the pine barrens. Sturdy little red pines were growing six feet apart in the furrows. Michigan leads all the states in her 3.8 million acres of state forests, comprising a tenth of the state, and in the 225,000 acres of state lands which have been planted with trees.

Minnesota is second in the area of state forests, with 2,011,-270 acres. The seed was planted back in the days of "free land," when enormous grants from the public domain were still in process of reckless distribution. In 1899, the state legislature authorized the acceptance and management of forest reserves. Governor John S. Pillsbury gave the state its first reserve, 1,000 acres of cutover land in Cass County. In 1903 the federal government added the Burntside State Forest, 20,000 acres in the Iron Range country.

For many years the constitution of Minnesota practically compelled the sale of the original federal land grant of 8.5 million acres. But by 1914, the accumulation of burnt-out and tax-delinquent lands brought a change in public sentiment. The people amended their constitution, permitting the legislature to "set apart public lands of the state better adapted for the production of timber than for agriculture." The legislature promptly placed some 350,000 acres of northeastern Minnesota in state forests.

The legislature of 1931 also put into state forests a large

acreage of lands granted by the federal government in trust for public institutions and lands which had reverted to the state through uncompleted purchase contracts. Many more acres were acquired when the state assumed the drainage bonds of northern counties, and the counties in return relinquished tax-forfeited lands to the state.

The sale of timber from Minnesota's state lands has had its lurid chapters and scandals in high places, but through trial and error and frequent political upsets a stable policy has finally emerged. Regulatory controls over cutting state timber have been adopted. The state sells only the stumpage and keeps the land for continuous production. In this wise, a typical lumber state, indeed a cradle of the timber-grabbing, free-wheeling industry of the roaring nineties, has worked out its own evolution. State forests have become a factor of consequence in rebuilding the forest economy of Minnesota.

Federal land grants were the start of many western state forests, like those of South Dakota, Montana, Idaho, and Oregon. A common form of grant conveyed to the state two sections in every township. These were usually Sections 16 and 36, commonly known in the West as the "school sections." An early undertaking of the Forest Service in several states exchanged solid blocks of land for these scattered parcels, and the legislatures fell in step with the idea of permanent management by making their new and impressive properties "state forests."

One of the most constructive of these trades gave Washington some 200,000 acres in the southwestern Olympics in return for the school sections scattered through most of the national forests. The legislature promptly christened this splendid tract of thickly standing hemlock and silver fir "Sustained Yield Forest No. 1," and ordered that it be cut only in keeping with a sustained yield plan of management.

The medicine making which grew out of the N.R.A. Lumber Code brewed a proposal that the government move more aggressively in building up the state forests. This was written into the Fulmer Law of 1935, even to the point of drawing upon the federal treasury for half the cost of additional state forests set up under satisfactory standards. But this act was permissive, not mandatory. The idea of putting hard cash into the purchase of state forests did not set too well with Congress or the federal budget, although the Forest Service recommended it year after year. The proposed monetary aid was finally shelved, but in land consolidations, co-operative protection plans, and many other practical phases of management, the Forest Service has strongly backed the enlargement of state forests.

Meantime, several states, like Michigan and Wisconsin, are showing fine initiative in attacking local problems by direct state action. Outstanding is the undertaking of Oregon to rehabilitate the Tillamook Burn. Three great fires, the first in 1933 and the last in 1945, left a "sea of snags" in the heart of Oregon's rich coastal forest. Three hundred and eleven thousand acres, burned and reburned, were apparently without hope of natural reseeding. And no seeding or planting by the hand of man could be done with any assurance of survival until the unbelievable fire hazard created by millions of standing dead trees had been conquered. Indeed, some plantings bravely undertaken after the first Tillamook fire were wiped out by the second or third. The State Board of Forestry accepted this ugly smear as its own job. The state holdings in the area were extended by voluntary gifts from private owners and cessions of tax-delinquent land from the counties. By 1949, Oregon state forests covered 84 per cent of the burned acreage.

The ashes of the 1945 blaze were barely cold when Governor Snell appointed a committee to study the problem and offer a plan. The legislature of 1947 adopted a scheme of forest research, supported by a severance tax of five cents per thousand feet on all timber felled in Oregon. The state forester drew upon these funds to expand the studies in reducing fire hazards and in methods of felling snags more economically. Much research was also carried out in rodent poisoning and the sowing of tree seeds by hand tools and by airplanes.

In 1948, by public referendum, the voters approved the issuance of Oregon Forest Rehabilitation and Reforestation Bonds. The authorized issue will run to about $10,000,000. The plan now under way follows three steps in working over each unit of the burn:

First: Salvage logging to get all the wood possible off the ground.

Second: Over-all fire control through access roads and "snag-free corridors" from 200 to 2,000 feet wide.

Third: Rodent poisoning and sowing tree seeds from airplanes. On a few of the brushiest areas, seeds or seedlings will be hand planted.

Oregon made ten thousand acres ready for aerial seeding in 1949 and 1950. Four tons of Douglas fir seed were collected at a cost of $15,000 per ton. On the heels of thorough poisoning, a third of a pound of Douglas fir seed is sown to the acre. Above elevations of 2,500 feet, the planes will scatter a quarter pound each of Douglas fir and Noble fir seed to the acre. The tests to date indicate that good results in airplane poisoning and seeding can be obtained within $5.00 an acre. These Oregon folks mean business and they know their woods. The restoration of the Tillamook will rank with the irrigation of the Gila and Columbia River basins and the redemption of the French Landes from sand dunes as one of the great human achievements in engineered conservation.

Down still closer to the lives of everyday Americans are the county and community forests. Two thousand, nine hundred and fifty of them were listed by the American Tree Association in 1949, with a total of 4,244,000 acres. They carry us back to the communal forests of France and the old and famous town forests of Germany and Switzerland. The local public forests of the United States reflect many facets of civic and rural life, and their over-all influence of example and education is a strong force in making us a forest-minded people.

The oldest known community forest in America was established at Newington, New Hampshire, over three hundred years ago, and it is still maintained by the town. The great awakening of conservation in the early years of this century started a wave of community forests. The Massachusetts Town Forest Act of 1913 authorized its towns to own and manage forests for timber production. Many other states adopted similar laws and offered their towns free planting stock and technical service. Often the watersheds which supplied towns with water became community forests. Often the poor farms of New England were expanded to take in neighboring woodlands, and forestry became an activity of the establishment. The concern in the northeastern states over the abandonment of farms by their westward-migrating people led a good many communities to start forest projects in the hope of finding other work for the old "hill farms" to do. Many county forests were started with tax-delinquent lands to which the county had fallen heir. Then came school forests, memorial forests, and recreational forests to swell the numbers. Whole communities turned out for spring tree-planting bees; and through picnics, outdoor festivals, and wood gathering in the fall, the community forest made its own place in the life of the town.

Today New England has over four hundred community forests with 136,000 acres. Typical of Yankee thrift is the town forest of Russell on the lower slopes of the Berkshire Hills in Massachusetts. Russell is a hill town and its 1,200 people and their forebears have maintained themselves chiefly by making things out of wood. Its present thriving paper industry was started in 1872. Town-owned lands were incorporated in a community forest in 1923. The problem of serving run-down hill farms with roads and schools was gradually solved by acquiring them and putting the land into the town forest. Thirty thousand trees were planted annually over a stretch of ten years. The town forest has grown to three thousand acres and has built up stumpage values in saw timber and cordwood of $44,000.

The *Daily Express* of Newport, Vermont, comments editorially, in an issue of last November, on the $2,700 net return just realized from selective cutting on the nine-hundred-acre town forest of Vergennes. "Vergennes," says the *Express,* is "Vermont's oldest city," also "one of the smartest." This is proven by a "perpetual timber income" from the town's land, along with plenty of water.

New York's first community forest at Gloversville, established in 1909, has now grown to 697 town, county, and school woodlands, containing 83,000 acres. The State Conservation Department reported in 1941 that over seventy-seven million trees, furnished by state nurseries, have been planted on these public forests. Many of them are watersheds for municipal reservoirs, which also under state law afford sanctuaries for birds and mammals.

One of the best-managed forests in the Pacific Northwest is the Cedar River watershed of the city of Seattle. After prolonged study and consultation of experts, the City Council laid down a policy of dual use. The production of water for municipal needs and production of timber for municipal income are combined in carefully planned cuttings. The city forest of ninety thousand acres now keeps the reservoirs filled and provides a sustained yield cut of thirty-seven million board feet a year.

The Lake States are well sprinkled with community forests —1,400 of them, taking in nearly three million acres of woodland. Twenty-seven counties in Wisconsin own more than two million acres of "forest crop land." In many states, as today in the Far West, county tax lands have been the most temporary and shifting type of forest ownership. But Wisconsin has developed a strong program for converting the forgotten lands of migrating lumbermen from liabilities into county assets. Its own forest pays Marinette County $25,000 a year. Through good management and protection, tree planting and the purchase of submarginal farms, the counties are putting their shoulders to the task of forest reconstruction. The state helps its forward-looking counties by money grants and technical service.

Wholly apart from wood and water, the town and village forests offer great spiritual values to the people of America. They have many subtle ways of allaying the stresses of life and reawakening its aspirations. The most recently dedicated community forest, at Shannondale, Missouri, was acquired through the joint sponsorship of community churches. The St. Louis *Post Dispatch* explains the project as designed "to see what happens when the people of the community, working together, regard the hills, the trees, the valleys as the gifts of God."

II

For sixty years the onward march of the national forests has been one of the great forces shaping our progress in conservation. In every decade, consciously or not, the activities of the nation as a forest owner have influenced the course of private industry. The first reaction to the spectacular withdrawals of Pinchot and Roosevelt was violent rebellion among the land-looking, timber-grabbing fraternity then in its full flush of western conquest. There was a great deal of lumber wirepulling back of the Congressional revolt which sheared off Roosevelt's power to create national forests by Presidential order.

Then the woods settled down and more stable operations with large investments in logging and milling facilities gradually took over from the timber locators and speculators. The attitude of many lumbermen toward Uncle Sam's intrusion into the timber game changed. They realized that the national forests were holding back vast footages of stumpage that would otherwise have been sucked into the whirlpool of speculation and top-heavy investment. These reserves cut down the wild flurries of "shoe-string competition." Not a few western timber barons of the old hierarchy told me they wished Gifford Pinchot had been born twenty years sooner.

The favorable attitude of the more responsible men in the industry was generally strengthened as they learned to know the Forest Service as neighbors and its ideas of administration. Most lumbermen applauded the limitation in national forest sales to the sustained yield of the unit. While they groused over methods of marking and cutting restrictions in government contracts, they usually admitted that the "white hat" boys were doing about right by the timberlands which all the people owned. As the West drifted into one period after another of overcutting, when mills ran half time and marginal operations went to the wall, it was obvious that the national forests and the other federal timberlands were about the only points of stability in a very disturbed and often distressed industry.

President Hoover listened sympathetically to a lumber delegation from the West, which urged him to shut off sales of public timber while private owners were struggling to keep their heads above water. His orders to the executive departments substantially held federal timber off the market during the depression of the thirties. David T. Mason, forest engineer of Oregon, proposed to relieve the lumber headaches from glutted markets by the purchase of additional national forests in the best timber growing areas of the West. Each of these islands of federal ownership should become a core of co-operative sustained yield management, extending over the private lands around it and holding their cut of timber within reasonable limits. The national forests stood fairly high with the western men who went to Washington in 1933 to hammer out a Blue Eagle Lumber Code. At Secretary Wallace's conferences, they concurred in the Forest Service proposal to add another 150 billion feet of timber to Uncle Sam's woodlands. In lumber's distress, this appeared the most direct road to a stable and profitable industry.

But the same proposal became a sour note five years later. It was then coupled with Acting Forester Clapp's vigorous demand for federal regulation of private forest management. To many industrialists, the joint program of federal ownership and regulatory power spelled domination. They could visualize its leading on, step by step, to a nationalized industry, a police state or what have you. Then, too, private forestry was recovering its wind. Tree farms were in the making. Many companies were beginning to see a future for themselves in growing trees. The vision of industrial leadership in nation-wide reforestation was breaking through the clouds.

III

The official program of large additions to the national forests has had faint support from industry during the last decade. Federal policies of acquisition have been challenged at several points. Many ask why the government should be buying land in the southern pineries when pulp or lumber companies are ready to acquire it and put it under probably more intensive timber culture than federal foresters could finance. There is much questioning in the West over the extensive acquisition of cutover private lands in return for timber-cutting rights in national forests. In some cases high-powered government salesmanship appears to have extinguished promising industrial tree farms. The offer of standing virgin timber may be too alluring for a logged-out lumberman to resist. He surrenders his cutover acres, takes their worth in government stumpage, cuts it and is through. The government seems, in such situations, to encourage the very "cut out and quit" sort of forest ownership which it had so long and so roundly condemned.

The expansion of state as well as national forests in recent years has sometimes brought the public agencies into actual competition, as land buyers, with private owners who are expanding their tree farms. This sort of situation may well raise a question or two. Under the conditions of today and the spread of commercial tree growing, what types of land should be taken into national forests? And how should state forests fit into the future land pattern, between federal holdings on one side and forest-conscious private ownership on the other?

We would do well to read again the recommendations of the McNary Committee in 1924. The committee strongly commended the national forests. They stood out as our greatest accomplishment in forest conservation. They should be extended further *where special public interests are involved,* like the protection of watersheds; and *where difficulties, costs, and hazards make forestry impracticable or remote as a private undertaking*. Elsewhere, under the committee's philosophy, private tree growing should be encouraged to carry on.

We still need a strong core of national forests. They give security, stability, and leadership to the whole undertaking. It would be foolish to attempt arbitrary limitations upon their extent. But the time has come for careful, over-all planning on the future pattern of forest ownership. The McNary Committee offered a sound starting point. Public ownership, as a general rule, could well be limited to the areas where special public interests must be protected or where there is no reasonable assurance that productive private forestry can be maintained.

The rest of our forest land could well be left an open field for private ownership and initiative.

The state forests are bound to grow in acreage and to move ahead in technical administration. They have a natural place in the expanding forest interest and activity of many states. They are rallying points for local responsibility in working out tough problems like the Tillamook Burn in Oregon or the reclamation of spoil banks in the wake of strip mining. The more our states step into such situations and take care of their own needs in forest restoration, the better for over-all progress. In many states, also, forest lands are the source of school funds and of income for universities and other public institutions. There is every reason for intensifying sustained yield management on lands of this class. American forestry will grow in strength as it spreads outward into every level of public activity as well as of private land usage.

Where state forest ownership should stop and private ownership take over can be determined by no formula or rule of thumb. It is a problem for intelligent planning in each state, in the light of its own conditions and resources and the potentials of its land. Nor can any slide rule show us where federal ownership should stop and state ownership begin. Such lines can be drawn only by experience and blueprinting as we go. Charting the future pattern of forest ownership offers an exceptional opportunity for the three-way co-operation of government, state, and citizen, which was the inspiration of the Clarke-McNary Act.

CHAPTER XIII

Teachers or Policemen in the Woods

I WAS once persuaded to serve on a radio panel on forestry. My good friend, Lyle Watts, Chief of the Forest Service since 1943, sat across the table. It was a friendly discussion. We batted timber supply back and forth a time or two, skirted warily around industrial forestry, but fetched up hard against the American wood lot.

"Here," said Mr. Watts, "is the toughest part of the whole problem. How are you going to make these millions of little woods owners grow trees? Their score thus far is very poor."

"Send them good, practical teachers," I replied. "Men of the soil-conservation and farm-extension stripe."

"I agree on the teachers," said the chief forester, "but I want some policemen to back them up."

The question Lyle Watts asked me was the one which fired the controversies over forest policy immediately after World War I. The Clarke-McNary Act put it to sleep for a time. But during the last fifteen years it has flitted on and off the forest stage in a dozen guises. Several states now have policemen in their woods to enforce compulsory forest practices. Forest industries have often supported their own states in resorting to police power. The general attitude of private woodland owners encourages each state to work out its forest problems in its own way—by education or inducement or compulsion, as its own people may choose. It is their affair. But the hackles rise whenever it is proposed to pin the badge of forest police upon a federal officer. That carries wholly different implications. It stirs up something in the blood which goes back to the broad arrows of His Britannic Majesty blazed upon the white pines of New England.

Federal control of private timber cutting came back to life in the turmoils engendered by the Blue Eagle. Article X of the N.R.A. Lumber Code provided for a conference of government and industry men. It was to explore all public and cooperative actions still needed to back up the undertaking of the industry to reforest its lands. The conference was called by Secretary of Agriculture Henry A. Wallace in late 1933. It brought F. A. Silcox, new Chief of the Forest Service, into the leadership of the government forces. Mr. Silcox was a trained forester, seasoned in the administrative ranks of the Forest Service. My warm admiration for "Sil" dated back to our strenuous labors in organizing the Montana-North Idaho District and fighting the fires of 1910. He had also gained much experience in labor-management relations, back in the Northwest logging camps in the turbulent days when the International Workers of the World were gaining power. In temperament and outlook, Silcox was every inch a New Dealer.

Secretary Wallace was quite colorless at the Lumber Code conferences. He seemed to harbor a suspicion that the industry men were still timber barons of a medieval cast and that they professed religion only in dire extremity. The Secretary's attitude implied that "when the devil was sick, the devil a saint would be." He was frankly opposed to the whole N.R.A. conception of establishing minimum, stop-loss prices, even as temporary expedients in emergency. If the Secretary had any sympathy for what we were trying to start in forestry under Article X, he did little to show it.

Chief Forester Silcox was keen, quick-minded, lively, and forceful in discussion. He was suspicious of the industry's sincerity and constantly on his guard lest the lumbermen "put something over." Silcox and his associate, Robert Marshall, seemed more concerned over our conformance with the current precepts of collective bargaining than our success in growing trees. They were disposed to exact concessions, to trade government aids in forestry for New Deal social and industrial aims. There was a colder atmosphere in the Federal Service.

The industry was put on the defensive. The warmth of co-operation present in the early Clarke-McNary days was missing.

However, we plowed through our job. Half a dozen committees, each with industry and government members, brought in reports. Most of them were ultimately accepted by the Department of Agriculture and embodied in an over-all forestry program. This program stood for several years as the supposedly joint legislative platform of government and industry. Several Congressional bills were based upon it. The number and diversity of subjects dealt with gave them a common nickname—the "Omnibus Forestry Bill."

Prominent in the list was a larger authorization under the Clarke-McNary Act. Organized protection was spreading steadily and the requirements in men and machines for good fire control were mounting. Another plank was designed to build up co-operative defense against forest insects. Bark beetles, which can girdle veteran trees as effectively as an ax, were probably taking a greater toll from the forests of western pine than fire itself. Many other insect pests would suddenly break out like the hordes of Genghis Khan and sweep through great swaths of timber. It was high time that a forest-minded nation organized to fight these enemies on the instant of an attack.

Another "omnibus" proposal dealt with long-range forest financing. American finance had not yet learned to meet the needs of a forest owner who chose a sustained type of operation. Loans of the Reconstruction Finance Corporation, like old-fashioned timber bonds, were based upon the liquidation of assets and the retirement of capital on a rigid time schedule. The Forest Service agreed to work out a scheme of credits that would help more forest owners shake off the cut-out-and-quit financing of the migratory sawmill.

The Lumber Code conferences also brought to a head the idea of co-operative sustained yield units including government and private woodlands. This means of spreading good treatment of the forest could be put in effect by long-term contracts between an administrator of Uncle Sam's forests and the owner of nearby private lands. Sustained yield from the combined area would be assured, for the government would call the turn on the rate and methods of cutting. A continuous supply of wood would underwrite local mills, pay rolls and communities, since the contract would stipulate where the logs should be manufactured.

The omnibus platform represented a deal of constructive labor on the part of government and industry, and it had no small influence on the events of the next ten years. More sinews of war for co-operative protection were provided. Research was expanded at the forest experiment stations and Forest Products Laboratory. Beyond that point, however, the joint program bogged down. There seemed to be no steam behind it. Federal leadership was swinging to the left. Chief Forester Silcox took the line that our forest policy must include public control of private timber cutting. The federal regulation demanded by Gifford Pinchot in 1920 came again to the fore.

I was never sure how seriously Gus Silcox wanted the government to step into the role of forest policeman, patrolling a beat of four and a half million woodland owners. Silcox was a past master of verbal fencing. He loved to thrust and parry with the lumbermen, to show up weak spots in their position and to needle them into action. He carried it off with a personal charm and aplomb that blunted the edge. Silcox never offered specific legislation on forest regulation. His proposals were general and vague. But he obviously enjoyed playing up the idea and holding it over the heads of the industry.

Whatever doubt shrouded the intentions of Chief Forester Silcox, there was none whatsoever as to the purpose of his successor. Earle H. Clapp headed the service as acting forester for four years after the death of Silcox in 1939. He had come up through thirty years in the organization, and was an able, hard-hitting administrator. I worked with Earle Clapp for many years; he was one of the straightest-grained men I ever knew, wholly sincere and forthright in wanting forest regulation and single-minded in pursuit of it. Where Silcox parried with a rapier, Clapp slashed with a Highland claymore. He organized all the resources of the Forest Service behind an unequivocal program, attempting even to command the personal support of service men in their local public relations.

It now became the official dictum that there could be no "Omnibus Bill" nor any other major legislation on forest policy without public controls of destructive logging. It was, indeed, strongly implied that the service would not sanction larger authorizations for co-operative protection unless cutting controls were accepted as a *quid pro quo.* Why should the government, asked Clapp, continue to pour money into the protection of private woodlands with no certainty that the lumbermen would leave any trees to protect?

A second string to Acting Forester Clapp's longbow was a major enlargement of the national forests. To the current 177 million acres, he proposed to add 150 million. He would give Uncle Sam ownership of half of the commercial forest lands in the country. This, the aggressive head of the service declared, was necessary to protect the forest economy of the United States. It was not difficult for critics to read into the Clapp program a purpose to give the government economic controls of wood supply together with regulatory controls of private forest use.

In 1938 President Roosevelt sent a special message to Congress, asking for a "policy and plan of action" on the forest problem. He urged an inquiry, including the need for extending public forest ownership and "such public regulatory controls as will adequately protect private as well as the broad public interests in all forest lands." After the pattern of the Select Committee of 1923 under Senator McNary, a Joint Congressional Committee was named to follow the President's lead. It was headed by Senator Bankhead of Alabama. The committee held extended hearings throughout the land and brought in another weighty review of forest facts and needs. Most of the hearings were conducted by the vice chairman, Representative Hampton Fulmer of South Carolina. Mr. Fulmer had championed the little sawmills of his state in N.R.A. days, against the "lumber barons" who were seeking to enforce their will by the Blue Eagle Code. He led the committee westward in an avowed hunt for the lumber trust that was destroying the forests with one hand and fixing prices for the consumer with the other. In the hearing at Portland, Oregon, Chairman Fulmer questioned every witness on how lumber was sold, who wrote the price list that every mill man followed, and so on. One lumberman patiently explained the sharp competition in the lumber market, how buyers would dangle orders before a dozen mills to draw out the lowest bid, and how sales of the same item on the same day would often show a three- or four-dollar spread. Still unable to satisfy the Honorable inquisitor, the exasperated witness finally retorted: "Mr. Congressman, we sell our lumber exactly as you used to sell slaves in South Carolina, on the block to the highest bidder."

For the Joint Committee's use, the Forest Service put together an up-to-date estimate of standing timber, growth and drain. It showed a saw-timber supply of 1,763 billion board feet, about 6 per cent greater than the last preceding compilation. Yet the committee's report had a strong "timber famine" flavor. It laid stress upon the 96 per cent of the country's wood supply that came from forests in private ownership and on which destructive cutting widely prevailed. The report referred to "individual examples of good private forestry" but gave meager recognition to the over-all gains in industrial timber cropping. More than fifty-eight million acres of private forest land were found to be unproductive.

The Joint Committee's report charged that "destructive exploitation of the timber resources, together with unfavorable economic conditions in many forest regions, accentuated unemployment, decreased income, lowered purchasing power and living standards, and rendered idle land and ghost towns more frequent." Concern was also expressed over the woodland acreages that had become tax delinquent in recent years. The committee seemed to make private forest ownership the scapegoat for the rural ills of depression and the idleness of half the sawmills in the country.

The committee painted a black picture and prescribed strong medicine. It urged federal ownership of much more forest land and federal regulation of harvesting methods on what remains in private hands. A specific prescription for regulation was offered in a bill by Chairman Bankhead. The Secretary of Agriculture was to determine the cutting practices deemed necessary in each state. The state would be granted a period of five years to put them into effect to the Secretary's satisfaction. The Secretary might pressure a recalcitrant state into good behavior by withholding co-operative grants in aid of forestry. After the period of grace, the Secretary could move into a defiant state and directly enforce compliance with his rules by the strong arm of federal marshals and courts.

There were no hearings on the Bankhead Bill. The high tide of New Deal enthusiasm for regimentation was on the ebb, and the country was soon absorbed in preparations for national defense. But in high levels of the Forest Service, forest regulation after the Joint Committee's pattern has remained the gospel if not the law. True to his fixity of purpose, Acting Forester Clapp sought to have compliance with departmental regulations on timber cutting included in orders of the President under his emergency wartime powers. The Joint Committee's Report inspired a number of Congressional pens. The Wallgren Bill of 1942 and the Hook Bill of 1945 would have ignored the states completely and set up the Secretary of Agriculture as direct overlord of forest cutting.

To industrial forest owners the regulatory propaganda of the Forest Service has been worse than the regulatory bills in Congress. Cato was not more persistent in telling the Romans that Carthage must be destroyed than the chief of the service in telling our people that timber cutting must be regulated. It has been the official "line." No report or summary of the forest situation fails to carry it. Usually coupled with the insistence upon regulation are two other burnished lines of federal propaganda. One is the demand for a great increase in the area of national forests. The other is the durable bogy of timber famine. It stalks through the tables and text of every official pronouncement on timber supply. The country is told again and again that our forests are being overcut, that critical shortages of wood are in the offing.

It is the law of physics that actions and reactions are equal, but a surer formula in human relations is that he who sows the wind shall reap the whirlwind. Men in forest-using industries are as much of human flesh and blood as any other class of people. They have been under criticism and attack, and they have replied in kind. It has been distressing to witness the partial breakdown of co-operation between these men and their government. Distrust has replaced much of the old spirit of confidence; and, as always in the reactions of *homo sapiens,* distrust has gone to unwarranted extremes.

There is, of course, a natural resistance to rules from Washington on what may be done in their own woods, from men brought up in the strongest American traditions of individualism and self-reliance. But the reaction against Forest Service propaganda for federal ownership and regulation goes deeper. The lumbermen believe that a powerful government agency has veered away from its constructive functions in promoting forestry and set its sights on domination. To attain this goal, they believe that the service has not fairly presented the facts of timber supply and replacement; that it has played down the forest recovery of the country and the progress in private timber growing; that much of what it says is colored by a political purpose to put over its double-barreled program of federal ownership and control.

I have never lost my high regard for the ideals and the caliber of personnel in the Forest Service. I am sure that many of the extreme political and socialistic designs read into service policy are not there. On the other hand, there is no question that government propaganda in the past ten or twelve years has sought to break down public confidence in the capacity or purpose of industry to rebuild its forests. The propagandists have done their best to sell the public on the necessity of federal ownership and regulation. This campaign has not always been fair to private forest ownership.

To an old warrior, with sympathy for both camps, it has seemed unfortunate that a federal organization for the conservation of natural resources should be drawn into the heat of political controversy. Inevitably its scientific accuracy has been attacked and its capacity for constructive leadership has been reduced. However, zest for a fighting cause is in the blood of the organization. The Forest Service was born in controversy and baptized with the holy water of reform. The crusading spirit of Gifford Pinchot lives on.

The steady barrage of propaganda became a sort of cold war. It threatened the industry directly with federal regulation of its operations. It attacked the industry also on its highly vulnerable standing with the public. It gave aid to the hue and cry of "devastation." A public-opinion poll made by a lumber group in 1941 showed that Americans generally had a very low opinion of lumbermen. They were almost in the class of Public Enemy No. 1.

All of this bludgeoning rolled off the backs of many lumbermen, who were of the old order and in process of cutting out and quitting anyway. But other men in all branches of forest industry, who were in business to stay, took the government attacks and popular disfavor to heart. Many an operator who had just drifted along took a fresh look at his hole card. There was more awakening to local public relations and obligations. Community pressures for a more assured future bore upon many situations. Consciously or not, timbermen of the old stripe of rugged individualism were swayed by public opinion. It put more force into the economic ground swell for industrial tree growing that set in again after the depression.

On the propaganda front, men in the business of cutting and using trees took the offensive. They were tired of standing up to the pillory in silence. The feeling spread that the best way to defeat federal regulation was to convince the public that it was unnecessary. Through trade associations and new regional or national alliances wood-using industries told the public of their good works and took the lead in tree growing and fire prevention. There was a new fighting spirit—in forestry and in publicity, a determination to show that private initiative would take care of the "forest problem." Tree farms appeared in the Pacific Northwest. The old war horse of protection battles, the Western Forestry and Conservation Association, rode again into the arena, as champion of western forest owners in conservation policies and public relations.

The collective voice of the timbermen, as far as they have one, is the Forest Industries Council. It speaks for the national lumber, pulpwood, and paper and pulp manufacturers. On the crucial forest questions of the day, the council, in 1944, declared for:

1. Practices on all forest lands which insure the continuous production of timber.

2. Private ownership (rather than public) of all lands which can be profitably managed for continuous production of forest crops.
3. Public regulation when necessary or desirable, to be administered under state law; the need and scope of such regulation to be determined by the people of each state.

This is the industry's riposte to federal propaganda. One of its strongest points is "public regulation when necessary . . . under state law." The Oregon Conservation Act of 1941 started a procession of state laws to control or influence private timber cutting. Forest industry has been active in promoting these laws. There has been much ironic comment in official circles that the industry wants state controls chiefly as a means of forestalling federal controls. It is partly true. The state laws are, to some extent, a political answer to the threat of federal invasion. But many people, in forest industry and outside of it, regard such legislation as the natural and constructive outgrowth of the experience of a state in its own affairs. It has no necessary bearing on policies enunciated in Washington. Forestry is becoming established in the land usage of the country; and, as we have seen, is generating many local methods of encouragement or control.

In the meantime, the effort to put federal policemen in the woods goes on. The latest regulatory bill, introduced by former Secretary of Agriculture Clinton Anderson, is presumed to be the last word in official planning for a regimented timber crop. Its adroitness in fitting the techniques of a police state to traditional institutions and processes of free America is worth a bit of study.

The proposed Forest Practices Act contains many terms and procedures customary in co-operation by the Department of Agriculture with the states, but the heart of it is the almost absolute authority vested in the Secretary. He alone will determine what should be done for the "conservation and proper use of privately owned forest lands." The rest of the bill is machinery to carry out the Secretary's will. Local and national forestry boards are provided, but they are purely advisory. There may be hearings and appeals, but the Secretary's final determination will always stand if "within the law." His decision on facts cannot be set aside. There would, under this bill, be one uncontested boss over all the privately owned forests in the United States.

In the management of private woodlands, the Anderson Bill would make the states administrative agents of the Secretary of Agriculture or block them out altogether. The Secretary may approve state regulatory plans which satisfy him, inspect their enforcement, and reimburse the state one half the cost of their administration. He would hold a tight rein on any "co-operating" state. If he should find its forest rules or their enforcement "inadequate," he could withhold financial assistance in forest practices. He could also cut off federal aid under "any other Act administered by him in furtherance of any other forestry program," or of erosion or flood control. The Secretary could thus withdraw co-operative funds for forest protection, farm forestry, Norris-Doxey wood-lot management, and soil conservation—any or all of them.

After three years of grace, the Secretary may move into any state which has not offered a satisfactory plan of forest control, set up his own rules of practice and enforce them through the federal courts. Every private woodland owner in such an "administrative area" would be subject to a fine of $5,000 for failing to keep the records prescribed by the Secretary, for violating any forest-practice regulation, for disposing of any wood cut in violation of a rule of practice or for disposing of any wood cut from any *other land in the same county* for three months after noncompliance with a regulation. The bill does not rank with the Nazi Gestapo in making a suspect's family hostage for his conduct, but its police arm could quarantine all of a malefactor's logging operations in the county.

II

In 1949 J. P. Weyerhaeuser, Jr., told the U. S. Chamber of Commerce that the very industries under the most fire in the early days of conservation are now our greatest assurance of perpetual forests. The timber survey of '45 put the large holdings well in the lead of forest management. On the propaganda stage, there was a sudden shifting of villains. The spotlight of regulation, long aimed at the timber barons, was thrown on the peasantry of little owners. The men with wood lots were discovered to be the "hard core" of the problem, and for the little sinners particularly policemen were demanded in the woods.

Fifty-six per cent of all our commercial forest land is broken up into small ownerships. There are four and a quarter million of these little forests. Nine tenths of them are on farms. The last tenth is mostly in the absentee ownership of the butcher and baker, the doctor and banker, the widow and the estate. There are many fine examples of painstaking and profitable forestry on these little woodlands. It is difficult to believe the hard facts of the official census, but only one out of every twenty-five is rated "under good management." On over two thirds of the wood lots, cutting practices are classed as "poor" or "destructive," with no provision for regrowth.

Many foresters have been keenly disappointed by the seeming failure of our years of effort to change the forest-cutting habits of the American farmer. They think something is lacking in our approach or technique, and they turn to the forest controls of Europe for the answer. My friends in the federal regulation camp like to draw beautiful pictures of the settled forest culture of old world countries, where generation after generation has grown up under required woodland practices. I have seen enough of the French and German countrysides to get the feel of a settled rural economy and of a forest system long established and never questioned. I have seen the lines of sturdy country women coming out of the woods with great fagots of branches on their backs. Certainly not a scrap of wood is wasted.

But it does not follow that a system developed for the crowded lands and disciplined peoples of Europe can be transplanted successfully to the soil of the United States. We still have much of the free land tradition in our blood. We still hold strongly the right of every man to hew his own way to success or failure, to use what he has as he sees fit in a free economy.

We do well to learn the forest technologies and skills of other lands, but a successful forest policy cannot be borrowed. It must be home-grown. It must reflect more than a need for wood or flood control. It must express the national psychology, accustomed relations between government and citizen, the incentives which make the economy tick, and accumulated experience in dealing with natural resources. There is a good, homemade pattern right at hand—in American agriculture. Our agriculture and our forestry have the same background of free land. Agriculture, like forestry, has been forced to change from the wasteful exploitation of a virgin resource to its intensive use and culture. Farming in this country owes its progress to the land-grant colleges, the experiment stations, the extension and soil conservation services, and to public co-operation with the man on the land. Its regimentations are voluntary. It has apparently been too successful in creating great surpluses of food. Whatever its imperfections, it is an American land policy of outstanding power and success. It was not borrowed from the Old World; it was created at home to fit our own people. And its progress, step by step, has rested upon the education of the men who work the soil.

The job of getting full timber crops on our little woodlands is tough enough, and there are many larger holdings whose management also leaves much room for betterment. Yet it remains my conviction that more trees will be grown by working with the forces and incentives indigenous to the United States than by copying the police controls of other countries.

We have a native genius for educating people to know on which side their own bread is buttered. It is part of our democracy. Furthermore, while we have been at it for many years, our wood-lot schooling has not yet had a fair chance. It has had mostly an uphill fight against a national oversupply of forest products. Only since Pearl Harbor has the law of supply and demand been pulling with the forester. The forest-earned dollar has become the most persuasive teacher in our school of the woods.

Last year a group of New England wood lots realized an average stumpage value of $11.18 per thousand board feet on the timber they sold. Many a small farmer in the South has learned that his pulpwood is worth $2.00 a cord standing in the woods, aside from what he may earn from the labor of extraction. One of my Puget Sound neighbors awoke to the fact that he could harvest $60 worth of stumpage to the acre from his second-growth forest and, by careful marking, have as much more to cut in another ten or twelve years. The No. 1 theorem in today's schoolbook is that it pays to grow trees.

Every year more federal and state foresters are teaching in the woods and helping with the marking ax. The regional experiment stations of the Forest Service and their outlying pilot plants have developed remarkable skills in teaching the A B C's of wood-lot management. The best demonstration of paying forestry I have ever seen is the "Farm Forty" operation of the station at Crossett, Arkansas. To maintain the maximum growth rate and show that forestry can pay off as often as a cotton field, some cutting is done on this little tract every year. Each fall the year's crop is piled out by the highway and the whole county comes to see it. There is a pile of pine logs, ricks of pulpwood, some gum and oak logs, hickory-handle bolts, and a few cords of rough stuff suitable for fuel. Every pile is marked as to its quantity and value. The Crossett "Farm Forty" is returning a net yearly profit of $8.00 an acre.

The federal Soil Conservation Service brought a new and very practical form of forest education to the farms in 1935. Any farmer within a soil conservation district may obtain a working plan for handling all the soils and crops on his property, including his woodlands and needed tree plantings. Expert help is available in carrying out his forestry operations. Much of the best farm forestry we have yet seen in the Pacific Northwest was under the planning of soil conservation men. There are 120 million acres of woodlands in the soil conservation districts of the country. The current work upon them includes twelve million acres of cutting to improve the growing stock and 700,000 acres of tree planting. From its own forest nurseries and purchases from state nurseries, S.C.S. distributes some thirty-five million trees and shrubs annually throughout its districts. This is certainly getting forestry down into the "hard core" of the problem.

The domain of the Tennessee Valley Authority includes some fourteen million acres of forest land. The authority has pushed hard for organized protection from woods fires and for better management of the thousands of wood lots in the Tennessee Valley. Two hundred and fifty continuing demonstrations of good cutting practices have been established. One hundred and seventy-three million seedlings from T.V.A. nurseries have been provided for replanting. Educational work among the little sawmills aims for better use of the low-grade woods. Here is probably the most intensive center of practical teaching for forest owners in the United States.

The better economic climate for tree growing has brought a new educational force to the support of state and federal agencies. This is the commercial interest of wood-consuming industries in future supplies of raw material. As good wheat crops are matters of business to flour millers, many wood-using industries are making regional tree growing part of their business. One hundred and fifty foresters industrially employed give much or all of their time to advice and timber marking on neighboring wood lots.

The young profession of consulting foresters is moving in on woodland management at many points. Often a consulting forester takes over timber marking, planting, and other parts of a job after a public forester has sold an owner on the idea.

Wood pulp and lumber companies are preaching forestry to their wood-lot neighbors, distributing trees for them to plant and gathering groups together in the woods for tree-cutting demonstrations. In 1939 the Southern Pulpwood Conservation Association made this an industry-wide undertaking for the South. Its conservation foresters are educating southern landowners to crop their wood lots carefully and grow more pines. Training camps are conducted for farm boys in co-operation with the state forestry departments. In 1949 the southern pulp mills planted forty-five million trees on their own lands and put another eleven million in the hands of small owners nearby. These plantings reforested a total of sixty thousand acres.

In many instances, pulp company foresters discuss methods of cutting with an owner or contractor before closing the deal for his wood. Often they decline to buy if the lot is destructively slashed. Such buying policies may be tempered to the size of the stock pile at the pulp mill or the practices of a competitor. But, by and large, the educated dollar in a buyer's hands, backed by free marking service and free planting stock, is a powerful curb upon destructive cutting. A pulp company on the Olympic Peninsula has covered its supplying woodlands with illustrated circulars about its wood-buying program. No wood will be taken from second-growth stands of Douglas fir, seventy years or younger, unless a good growing stock is left on the land.

An outstanding enterprise in neighborhood forestry was launched by paper mills on the Wisconsin River in 1944. "Trees for Tomorrow" distributes planting stock among the small woodlands and assists their owners in practical forest management. It also has a long-range educational program, including forestry scholarships, school forests, and memorial forests. Its Conservation Camp at Eagle River, operated in co-operation with the Forest Service, is the summer headquarters for groups of teachers and for educational conferences on forestry.

The education of all potential growers of trees is the purpose of American Forest Products Industries. This organization was set up in 1941 by lumber companies who resented their relegation to the doghouse by public opinion. They were determined to spell down the dreaded word, "devastation." In a few years, A.F.P.I. became the grand lodge of the principal wood-using industries, with a program of "Trees for America." It took over the "Keep America Green" movement, which lumbermen had started in the Northwest and spread it through twenty-eight states. It has led a national campaign for American tree farms as a means of promoting and popularizing good woodland management. In 1946 "Cash Crops from Your Woods" became the slogan for a series of intensive state campaigns which sought to arouse the interest of woodland owners in growing trees and to strike vigorously with county demonstrations and forestry know-how on the ground. These educational drives have set an encouraging pattern of co-operation among all the public and private agencies who want more and better forestry.

New and diverse movements and organizations with a common tree-growing purpose have sprung into life since the start of World War II. The education of small woodland owners has become the most virile phase of American forestry. Distinctive and true to the traditions of its birthplace is the New England Forestry Foundation. This is a nonprofit corporation formed by industrial and conservation leaders and designed to give expert technical service to private woodlands at cost. It establishes management centers, employs foresters, and serves clients owning 125,000 acres of woodlands. Trained forestry crews will log for any client under foundation supervision.

Most of the ten Forest Products Co-operatives in the United States add their bit to the education of the owners of little forests. The Otsego Association of Cooperstown, New York, serves its members all the way from woodland management to the milling and sale of lumber. I have never witnessed a more interesting example of the infinite possibilities of American rural organization than the deal between the Washington Forest Co-op and the Cascade Mills Association. The co-op wanted a stable market for the wood grown and cut on its 22,000 acres of farm lots; the mills wanted an assured supply of saw logs, peelers, and pulpwood. Accordingly, a twenty-five-year contract was signed: at the current market prices, the mills take all the wood, of all sizes and species, that the co-op has to sell, and the co-op is expected to give the mills first call on what it wants to cut. This setup and the co-op's forest management are real assets to the small timber owners of northwestern Washington.

Forestry is flowing out through the woodlands of America like ripples on the surface of a pool. The strongest currents are the expanding educational undertakings of the public services. New flow lines have been set in motion by the devices of many states to induce or prescribe better forest practices. Other strong currents spring from the commercial interests of businessmen. It is all very characteristic of our habitual ways of doing things.

The outward spread of forestry is ragged and uneven; there are gaps and duplications and misspent energies. The movement has no goose-stepping. It lacks the orderliness and precision of a Prussian forest regime, but it is full of vitality and energy. And it is getting results. Every year it is making us a more forest-conscious people. While we argue over teachers versus policemen in the woods, we are growing more trees.

CHAPTER XIV

The Forest Balance Sheet

WASHINGTON Irving draws a delightful picture of Wouter Van Twiller, the rotund governor of New Amsterdam, sitting as a magistrate and hearing charges brought by one merchant against another for unpaid accounts. The worthy governor demanded the ledgers of the contending tradesmen. He counted their pages and balanced the weight of one against the other in his enormous fists. After smoking several pipes of stout tobacco in unhurried reflection, the verdict was rendered. Since the ledgers contained the same number of pages and were of the same weight, the accounts therein were held to balance each other and the case was dismissed. Great was the fame of Wouter Van Twiller as a lawgiver; so great, says the historian, that not again was the peace of his administration disturbed by litigation.

For two and a half centuries the forest ledger of the United States carried no entries save withdrawals from capital account. And, with the exception of half a dozen farsighted men like Carl Schurz, the country took it with the imperturbability of Van Twiller himself. But from the days of Gifford Pinchot's clarion warning of "timber famine," the state of the nation's woodpile has aroused lively public interest. Six official inventories of our forests have been made during the last forty years. Each of the last four tree tallies has brought a fresh round of debate and new demands for federal legislation.

The last census of standing timber and of the growth and drain taking place in our forests was made by the Forest Service in 1944 and 1945. About two thirds of America's woodlands were then covered by the official timber survey of the service itself. From these areas the returns are reasonably uniform and accurate. For the rest of the forests, the enumerators drew upon the best estimates to be had from state, county, or company cruises. Attempts to estimate the growth of trees in the infinitely varied and complex forest types of the United States are difficult and at best involve many calculated assumptions and approximations. Our figures on wood supply are not exact, but they are trustworthy within reasonable limits and give us a fair picture of the situation. We are still supplying our needs for wood in part by deficit spending from the capital account of virgin timber, but the balance sheet is the best in forty years.

The standing saw timber, supposed to have exceeded five trillion board feet in the primeval forests of the continental United States, has now shrunk to one and six tenths trillion feet. But the timber crop, the yearly income from growth, has risen steadily since the taking of forest inventories began. The last count puts the crop at thirty-five billion feet of saw timber and sixty-six million cords of polewood, pulpwood, and fuel. Converting every kind and size of wood to a common measure in cubic feet and including everything down to five-inch trees —fuel, poles, pulpwood, and pit props, along with the saw timber and peeler logs, the forests are growing over thirteen million cubic feet of wood a year. The rate of replacement has more than doubled since the reckoning in 1920. The total drain upon the forests from every source exceeds the growth by only 2 per cent. We have almost balanced this page of the ledger. Eighty-nine per cent of the cubic-foot drain, or outgo, from our forests is the consumption for innumerable uses of wood. The other 11 per cent is the toll paid to forest fires, tree-destroying insects, and disease.

The entries on the ledger page headed "saw timber" are less favorable. They cover only the trees large enough to be used currently in milling lumber. The present growth in trees of this size, nine inches and up in the South, twelve inches and up in the West, and so on, is nearly three times the growth estimated in 1920. However, against the saw-timber crop of thirty-five billion board feet, the saw-timber drain is nearly fifty-four billion feet. We cut or lose three board feet of milling wood for every two board feet of new growth. Ninety-two per cent of this take is for commodity uses; eight per cent feeds the forest fires, the bugs, and the fungi.

The gap between drain and growth of saw timber has been narrowed substantially since the surveys of 1920 and 1933. Both of those estimates put the drain at five times the current growth. But the gap has not yet been closed. The country can still be charged with deficit spending. Nineteen billion feet of saw timber in an average year are taken from capital assets. This red-ink entry is the inspiration for persistent warnings of a worsening forest economy and repeated recommendations for national control of cutting practices. "Timber famine" is still the most potent slogan of propaganda for a more drastic forest policy.

We can look upon the forest balance sheet as a still picture, as a portrayal of cold facts and their implications. That seems to be the viewpoint of official Washington. It is an interpretation strongly colored by the crusading zeal for reform, which was Gifford Pinchot's greatest legacy to the service he created. It is colored also, consciously or not, by an avowed policy to build up public support for conferring greater powers upon the federal foresters. The Forest Service has always had new worlds to conquer. That is one reason why it has been such an effective organization. It must carry the public with it. It would never do to let the people quit "viewing the situation with alarm."

On the other hand, we can look upon the last forest accounting as one film in a moving reel. We can relate it to the pictures which went before it. We can interpret it in the light of what is going on in the length and breadth of this big country. I have worked in the forests of America for nearly fifty years, and their condition during that time certainly has not been static. I have seen the best estimates of the day change from an over-all drain four times the calculated over-all growth to the present finding that drain and growth are almost in balance. About twenty-five years ago our forest growth started on an upward curve. It has maintained that rise to the present day, and there is every reason to believe that the annual crop of wood will continue to increase.

There is a constant spread, in acres and in effectiveness, of organized protection from forest fires. And only now, within the last four or five years, are we beginning to put a comparable curb upon the huge cut in the timber crop taken by forest insects.

The steady progress of logging has converted millions of acres of virgin timber in which growth is at a standstill into young, growing forests. Forestry begins with the ax. A typical section of virgin Douglas fir is like a heavily stocked lumber-yard. It is storing up wood in vast quantities but it is growing little or none. One of the tough problems of West Coast lumbermen is what to do with overripe timber, with the big trees that have stood far too long on the stump and are shot with decay in various stages. When such timber is cut, a stagnant woodpile becomes a growing forest.

Then we have the remarkable increase in timber planting, the restocking today of 400,000 acres every year. Our nurseries produced 380 million forest seedlings in 1949. Probably the most important of all factors in the long-range productivity of American forests is the economic incentive to grow trees. The very agencies which at the start of the great crusade seemed most to threaten forest destruction have now become foremost in forest renewal. American industries and commercial technologies want more wood. It has become their business to produce it.

Much of our increasing growth of wood from these sources was not measurable in the calculations made by the official tabulators in 1944. Still, on the record, they reported an over-all timber crop 19 per cent greater than in 1939; 50 per cent greater than in 1933, and 123 per cent greater than in 1920. This "record" is not accepted at face value in federal circles. The official view holds the earlier returns as too inaccurate for acceptance and, for practical purposes, starts from scratch with the tree count of 1945. Admittedly, the last census was more accurate than any of the earlier surveys, although its figures are now being revised in many states. Yet all of these accountings do reveal the trend of what is happening in our forests. They show broadly how the forests have responded to twenty-five years of protection, tree planting, and forest-wise cutting. Most important in the whole series of estimates is the finding, in 1945, that all forest growth, in cubic feet, now practically replaces all the drain. That is the meat of the coconut.

Many things now going on in the woods show that the annual yield of our forests is increasing. The slow-growing stands of ponderosa pine in the Dakota Black Hills have been under forest management since 1898. The first cycle of selective cutting in those ragged and fire-swept forests is about completed and the second cycle is beginning. The hills no longer have the open, parklike stands of pine I rode through as a young forester. Their woods have thickened up. As the result of cutting the overmature veterans and stopping fires, there are many more young, vigorous trees than in the early days and the average acre is growing a third more wood.

One of the large southern pine operations which is working its forests for continuous production makes a detailed inventory of timber volume and growth every ten years. Their inventory in 1949 showed an average increase in the growth on this property of 20 per cent over the rate figured ten years before. In the Douglas fir country we have been rather smug over the *net growth* in our junior forests. It runs up to 1,500 board feet per acre on the best sites and in the best growing years; it will average 500 to 750 board feet on most of the tree farms. Of late, however, we have awakened to the *gross growth* of these lusty young stands, to the vast numbers of trees which are killed out from decade to decade in nature's process of thinning. The next great advance in West Coast forest management, now beginning at several points, will be the recovery of these surplus young trees by systematic thinnings. When we can get at all of our seven or eight million acres of overcrowded young forests in the systematic fashion of the Finns and Swedes, we will add another billion feet of wood a year to the harvested growth of the region.

In 1944 the Forest Service hazarded a guess that the forest growth of the country could be doubled by general application of the best methods of management then in effect on some private lands. The present spread of intensive forest practices and more complete use of wood is moving steadily in that direction.

The physical measure of our forest resources, in units of use and consumption, is also changing. It cannot be computed for the future like barrels of petroleum or tons of iron ore in place. The minimum sizes of saw timber used in the last official survey doubtless reflect the American practices of 1944, but for the long range they are quite arbitrary. The sawmills of Sweden and Finland process many trees below the minimum sizes on which the government estimates the American supply. Much Finnish and Swedish plywood is made from logs that America's veneer peelers would reject today. Yet our lumber and plywood manufacturers have always been ready to adapt their machines and techniques to smaller and lower-grade logs whenever it will pay to do so. If the 1944 estimates were recast on the minimum sizes used in Europe, they would show our growth and drain of saw timber more nearly in balance.

European foresters, looking over western American logging and milling, marvel at the little trees we do not use and at the cutting of big trees into little pieces of lumber. One of them reported last year to the Danish Forestry Association: "There is no danger of a lumber shortage in the United States. The forests will be able to increase their production simultaneously with increased market requirements."

This is another way of saying that wood supply is not static. Increasing uses and demands for wood, and the cutting practices they make possible, react directly upon the quantities of material our forests can supply. As a matter of fact, American mills have been sawing smaller and smaller trees for the last fifty years. In every manufacturing region the size and grade of logs put into lumber have dropped as its saws turned from virgin to second-growth stumpage. West Coast loggers are now bringing in 15 or 20 per cent more footage from similar stands of timber than before the last war. New cruises of many properties bring home the fact that we now have more wood, measured by use, than we ever had on these acres before.

Another changing phase of the American picture is the

shifting of large consumption requirements from the sawn board to fiberboard or paperboard. Many wooden boxes have been replaced by paperboard containers. A large part of the lumber formerly used for sheathing the frames of buildings has been displaced by fiberboards or composition boards of many makes. The hardest and densest fiberboards are even used for the floor under our feet and the finish on our walls. How far such changes in use may go, under the aggressive technologies of pulping, fibering, laminating, and plasticizing wood, is unpredictable. Equally unpredictable are the shifting battlefronts of competition between lumber and other materials fighting for the construction market—steel, cement, brick, glass, tile, and the rest. These kaleidoscopic puzzles simply emphasize that wood consumption is no more static than forest growth. Neither growth nor drain fits the assumption of any formula out of calculus.

In stressing the overcutting of saw-timber growth and its dangers, the conservation accountants have largely passed over the national requirements and emergencies which have pushed the sawmils in their pursuit of gain. They forget what the lumber was used for. There has been waste, to be sure, but essentially our forests were overcut to build the westward expansion of the nation, to win two world wars, to provide materials for every periodic upswing in home building, to house the veterans returning from World War II. In every emergency of war or peace, after every great flood, after the Chicago and San Francisco fires, the country has demanded immediate and unstinted supplies from its forests. In World War II the energies of the Forest Service were largely concentrated upon obtaining the maximum possible production of wood. The service was an arm of the War Production Board, searching out every little idle sawmill and finding timber for it to work on.

In specific situations, like a war or a fire-swept city or a national housing shortage, doubtless no one would dare to question the practical wisdom of taking what we need from our forests as we have always done. It is not hard to imagine the fate of a forest marshal who halted truckloads of logs or pulpwood rolling out of our woodlands during the last world war. Picture him in the middle of the road with upraised hand shouting "Stop! You are overcutting this forest!" He would probably have swung from the nearest tree himself, as a Nazi fifth columnist.

Yet it has become the fashion to blame the lumber industry for "timber shortages," real or imagined. This was a popular theme five years ago when millions of new homes for veterans were demanded overnight. They must be built and ready as fast as the boys got off the transports. The sawmills and lumberyards had scarcely recovered from unprecedented war demands for wood. The resumption of long-deferred civilian construction was in full cry. But many people, not excluding business analysts, naïvely assumed that additional billions of feet for G.I. homes should pour out of the lumber mills like white rabbits from a magician's hat. This was the third "national lumber shortage" I have witnessed. Like its predecessors, it disappeared in a few months. From V-J Day to 1949, the sawmills provided lumber for over three and a half million new homes and for repairing or remodeling four million old homes. The total of home construction was greater than in any other like period of American history. While this was going on, the mills also put 8.5 billion feet of lumber back into the depleted stock piles of the country. The lumber industry met all postwar demands and, in early 1949, had a production in excess of the current orders.

At one of the Clarke-McNary hearings in Minnesota, a Congressional witness referred to the impetuosity of American public opinion, in its swings from pole to pole. "The typical American," he said, "wants to put a setting of eggs under the old hen in the evening and have broilers for breakfast." We are a very impatient people and we have been reared under an economy of superabundance. Our standards of living continually advance. President Truman has recently held out alluring prospects of a further fabulous increase in the earning power of the American people during the next fifty years. People of such a temper and such a background will most certainly have lumber shortages in the future. They are due for shortages in any basic commodity, depending upon the current whims of popular consumption and the latest turns of industrial technology and sales promotion.

In the face of the unpredictable demands of the American people and the seemingly limitless expansion in their technologies and consuming habits, it is futile to attempt a forest balance sheet of the precise structure of European models. That sort of statistical strait jacket simply does not fit the United States. We must blunder along and work it out as we go. We are, however, making headway on fundamentals. We are passing from the timber mine to the timber crop as our chief source of wood. Timber is no longer a *wild land* crop. It is constantly receiving more care and more technical culture. The principal force behind the expanding timber crop is economic. It pays to grow trees, and movements in forest industry and forest education are giving that incentive constantly wider application. The net result of all these factors is an increasing timber crop, an over-all forest growth that has risen steadily for twenty-five years.

We have not yet balanced the forest budget, but we are moving in the right direction.

CHAPTER XV

The Ways of Democracy

WHEN I was a small boy, I attended several town meetings at Gilmanton, New Hampshire. Village road improvements came up for discussion at one of them. My grandfather, who was a clergyman, offered a suggestion. The debate warmed up. One of the neighbors, well known for a short temper, advised the "reverend to stick to the Scriptures and leave the ruds to them as built ruds."

The vigor of the town meeting was never more forcibly demonstrated than at the Fourth American Forestry Congress. It was called by the American Forestry Association in 1946, at the end of World War II. Although the association offered a comprehensive platform built around education and the responsibility of the states for local progress, every shade of forest thinking in the United States had its inning.

Attorney Anthony Smith of the C.I.O. general staff demanded immediate and complete regimentation of forest use by the federal government. Under Secretary Chapman of the Department of the Interior championed national regulation of forests in somewhat general and Utopian terms. Secretary of Agriculture Clinton Anderson advocated his specific plan of federal control. To the attending lumbermen there seemed to be a policeman behind every tree. During the congress, however, much was said for state controls of destructive logging and the merits of leaving this problem to the public agencies closest to the trees. The Agricultural Extension Service and other apostles of education in the wood lots had a field day, and struck some sparks over the technical issue of where education ends and management on the ground begins. The industry men held stoutly to their trilogy of free enterprise, education at the grass roots, and co-operation without dictation from national agencies. The congress endorsed the Forestry Association's program of education and state action, but it sharply reflected the contending schools which are trying to shape our future course. Some of us want forestry from the top, and some of us want forestry from the bottom.

Lyle F. Watts, head of Uncle Sam's foresters, has often talked it out with industry groups in down-to-earth terms. His favorite symbol for the policy he advocates is the *tripod*. One leg stands for the national forests. He believes there must be a larger backlog of forests in public ownership to stabilize the situation. The second leg is *co-operation*. The government should bring about as much of the needed betterment as it can by co-operation with the states and private owners. The Clarke-McNary policy is here to stay. The third leg in Mr. Watts's tripod is *regulation*. He believes the program will be incomplete and inadequate without federal power, as a last resort, to stop destructive cutting.

At one of his talks, the lumbermen were crowding Mr. Watts on when and how drastically, in the development of his whole program, he would use the club of regulation. He replied with the story of the Mormon bishop who was trying, on a Sunday morning, to wean an obstinate and contentious bull calf. The calf would not drink the milk from his bucket, but kicked it all over the cowshed. The bishop did not lose his temper—oh no! But he looked grimly at the obstreperous beast and said: "If this wasn't Sunday morning and I didn't have on my best black meeting coat and if I wasn't late for Ward Sunday class, I'd jam that bucket down your confounded throat."

Forest regulation has been a national issue for thirty years. It has been steadily and forcibly propagandized for the last fifteen years. A succession of regulatory bills have paraded before Congress, but there is still no federal coercion in the woods. Meantime, state activities, industrial timber cropping, and wood-lot education are spreading over the map. The greater the stress for regimentation from the top, the greater seems to be the aggregate force working from the bottom. In the meantime, also, the law of supply and demand is taking the lead as the best forester we have.

For the actual situation, the wheel is perhaps a better symbol than the tripod. We seem to be spinning in a state of tension between centripetal and centrifugal forces.

It is a fortunate American trait that we fight like wild men over ideologies and at the same time work heartily together in doing what must be done right now. Behind the skirmishes over regulation, Mr. Watts's second plank, "co-operation," is carrying more and more of the actual load. Many government and industry men learned to co-operate in the hard school of forest fire fighting and prevention. On many sectors of the front, day-by-day relations in the protection of adjoining woodlands, in practical application of the Clarke-McNary Act, in selling and logging federal timber, in a thousand projects of research and mutual community interest, have kept co-operation very much alive.

Teamwork is developing effectively between government, state, and forest owner in the vital field of insect control. The Forest Pest Control Law of 1947 gave us a good working harness. Technical leadership and organization come from the Bureau of Entomology and Plant Quarantine. Project costs are divided between public and private ownerships in the ratio of their acreages. Legislation to require cost sharing by all private woodlands threatened by insects on the warpath is spreading from state to state, like the fire laws after passage of the Clarke-McNary Act. The success of recent aerial attacks in destroying tussock moths and spruce budworms on thousands of acres at one sweep ranks among the great victories of men over their natural enemies. The bureau is organizing a scouting and warning service to locate the first sign of invasion, like our radar screens along the northern latitudes. This is federal co-operation at its best.

Chief Forester Watts relaxed the commandment of his predecessor, that there should be no major expansion in federal forest policy without regulation as a *sine qua non*. In 1944 the Forest Service joined western lumbermen in asking Congress to enact Public Law 273 for the unified management of adjoining public and private woodlands.

The idea back of this legislation and its potentialities in stabilizing forest industry are well illustrated by the first contract under its terms. This was made in 1946 between the Forest Service and the Simpson Logging Company on Puget Sound. An industrial tree farm of 160,000 acres and 110,000 acres in the Olympic National Forest were put under unified management for one hundred years. The Forest Service will determine the rate and method of cutting on the entire area. The logging company will maintain the growing stock on its own holdings, as well as on the national forest lands as required by the federal foresters. This obligation includes replanting in case of fire. In return the company has first call on whatever timber the service decides to cut from the government's acres at the prices determined by federal appraisers. All of the timber taken from the combined unit is to be manufactured at two forest-borne communities, Shelton and McCleary. The wood-manufacturing pay rolls of some 1,500 workers are thus underwritten for the next century.

One of the most revolutionary features of this unique contract is a commitment by the company to install any conversion facility deemed necessary by the Secretary of Agriculture for effective utilization of the timber. Again we run upon the outstanding characteristic of America's forestry in this day and age. It is the union of wood-using technology with intensive timber culture. By virtue of an assured one hundred years' supply of raw material, the Simpson Logging Company, Uncle Sam's partner in this enterprise, has expanded its facilities to manufacture fiberboard, plywood, doors, and many factory items as well as lumber. Its hemlock pulpwood supplies another plant in the same community. It has become one of the industrial leaders of the Northwest in the number of employees kept at work by each thousand feet of stumpage.

Under the discretionary authority of Public Law 273, the Forest Service has set aside another large area in the South Olympics as a *community-sustained yield unit*. It will supply the lumber, plywood, and pulp mills on Grays Harbor. This setup includes no private lands and involves no contracts with individual operators. Every sale of stumpage from some two hundred thousand acres of National Forest requires that the logs be manufactured in one of the harbor communities. Thus all the wood which now stands or may be grown on a substantial section of the federal timberlands, is committed to the perpetuation of this old West Coast lumber center. This means an assured sixty million board feet of logs per year at the start and much more in the future.

This foresight in forest administration is far removed from the routine bureaucracy so common to public affairs. Co-operative sustained-yield contracts involve the weighing of monopoly in selling government timber against public benefits in forest management, closer utilization of wood, and community security. The administration of Public Law 273 requires an exceptional degree of skill and judgment. It marks the high tide of resource and community planning in the public land policies of the United States. And it opens up limitless horizons of constructive co-operation between the government and its citizens.

The co-operation of the Forest Service with state forestry departments and with industry men and forest owners extends into innumerable local enterprises. The service is an active collaborator in any number of regional movements like "Trees for Tomorrow" in Wisconsin or the State Institute of Forest Products in Washington. Around every regional forest experiment station is a circle of study and demonstration areas. Most of them are private forests leased to the government for $1.00 per year. These are proving grounds for thinning, growth measurements, sample tryouts of ideas for better regional forestry.

Almost overnight the wood demands of World War II threw the West Coast industry into large-scale cutting of second-growth forests. There was an immediate demand for facts on the growth and management of these junior stands of Douglas fir and western hemlock. The Pacific Northwest Experiment Station of the Forest Service summoned us all to a council of war. Everyone turned in whatever information or experience he had on second growth. It was all put into a "quickie" bulletin, published within a year after the first call. A second more detailed edition is now in print. The station is making case-history studies of all the second-growth cutting jobs it can locate, and is establishing permanent experimental areas for tests of thinnings and growth rates. When men work together like this on practical needs, we need not worry too much about conflicting ideologies. Whatever eventually boils up in our melting pot of forest policy, we may be confident that Mr. Watts's second ingredient, co-operation, will savor the whole brew.

To many Americans the forest story has seemed only a long funeral dirge over our virgin woods. It is only human to deplore the passing of the "tall, whispering pines," especially when they made a sanctuary of a favorite fishing stream or formed the familiar and well-loved horizon for a valley of farms and towns running through the hills. But let us not be too hasty in writing off the virgin forest. Every flight across the midwestern states by airplane leaves me with a feeling that the pine forests of the Great Lakes are with us still. The virgin trees have been transmuted into homes, churches, and factories, into solidly built farmhouses and big barns. The forest has kept right on living with its people. I always get the same feeling when I look at a New England homestead built in the seventeenth or eighteenth century. Those superb pines and oaks are still part of the community, like the creativeness and craftsmanship of the men who built homes from them.

The cycle of forest growth, harvest, and use goes on. The wilderness has disappeared, but forests still surround and sustain us. The carpenter and his woods still build most of our homes. The new techniques of every generation multiply the gifts from the forest to our advancing standard of living. The loss of much of the virgin timber of my grandfather's day has been repaid many times over, not only by what it has built but by the appreciation and understanding of the forest which we have gained from its use. The first old trees are gone, but the forest is here to stay.

One day last fall I visited the forest pilot plant of the International Paper Company on Phillips Brook in the northern Connecticut Valley. I had tramped through some of these spruce and maple woods with Austin Cary, the forester from Maine, forty-five years before. At that time planning for continuing supplies of pulpwood was just beginning. Now I found a paper company devoting an entire watershed of twenty thousand acres to experimental cuttings. Here it will determine the number of crop trees and cords which should be left standing to yield the greatest possible annual growth of the prized pulping woods—spruce and balsam fir.

The baked beans and mince pies in the camp had the old flavor of the north woods, but the painstaking felling and logging had no counterpart in any forest work I have seen in America or in France. The marking of the trees to be cut is an exact engineering job, calculated to take out 35 per cent of the softwood volume—no more and no less. Measurements of the growth and windfall losses every five years, on this and other fellings left with different stocking, will give the company the facts it needs to grow a maximum timber crop in its northern forests.

My mind ranged westward to the studies of fire control which set up the first tree farm in Washington and south to Crossett and Bogalusa and a hundred other industrial centers where the most effective growing of wood is the first object of business management. I thought too of the farm wood lots where timber crops are growing, of the federal experiment stations in every forest region, and of the far-flung examples of good cutting on the national forests. I thought of the helicopters reseeding the Tillamook Burn in Oregon and of the forest seedlings pouring out of public and industrial nurseries by the hundreds of millions, and I wondered if there is any undertaking to which the American people have set their hands more unitedly than the growing of trees.

The forest story is very largely the story of America. The zest for free land and the almighty dollar, the creation of new skills and industries, the craze for bigger and better sawmills—all are part of our people and temperament. The commercial genius of the country was never displayed more dramatically than in the industrialization of the depleted pineries of the South or the teaming up of wood technology and forest management since the outbreak of World War II. And just as true to the American heritage are the challenge and idealism of conservation. From our own experience and leadership we have created an enduring philosophy of restraint and forethought in the use of natural resources.

My first interest in forestry was aroused in 1900 by a long chat with Bernard E. Fernow, then dean of the forest school at Cornell. The kindly, gray-haired German told me I would make a good forester because my long legs would take me through the woods and help me scramble over logs. For fifty years of scrambling over logs I have witnessed the battle between "free land" and "conservation." Forces grouped under these opposing banners have fought and compromised, cooperated like practical Americans in doing first things first and then fought again. There have been the inevitable conflicts between earnest reformers who cannot wait and conservatives who hold to the status quo. And there has been the characteristic and fruitful reaction of the American people to so many contentions—a growing understanding and a getting to work at the grass roots. The greatest change in fifty years is the spread of forestry outward and downward. Forestry is now the common possession of us all. To me the fifty years have brought a rich experience in the democratic process at work, hammering out a national problem.

ACKNOWLEDGMENTS

American Tree Association, *Forest Directory,* 1949: for data on state and community forests.

Butler, Ovid, *American Conservation:* for many historical facts.

Coman, Edwin T., Jr., and Gibbs, Helen M., *Time, Tide and Timber:* for data on the early history of Port Gamble and Seattle.

Cotterill, George F., *The Climax of a World Quest:* for Captain Vancouver's observations on the forests of Puget Sound.

Glesinger, Egon, *The Coming Age of Wood:* for much about wood chemistry in World War II.

Hagenstein, W. D., *Oregon Business Review,* February 1950: for data on tree farms.

Holbrook, Stewart, *Burning an Empire:* for the historical record of great forest fires.

———, *Ethan Allen:* for the rum requirements of the American Revolution.

Horn, Stanley F., *This Fascinating Lumber Business:* for historical backgrounds of the lumber industry.

Hunter, Dard, *Paper Making*: for many facts about paper.

Jenks, Cameron, *The Development of Governmental Forest Control:* for the story of Santa Rosa and other reserves of live-oak timber.

Kinney, J. P., *The Development of Forest Law in America:* for the references to early state laws.

Kirkland, Burt P., *Forest Resources of the Douglas Fir Region:* for data on northwestern forest management.

Marquis, Ralph W., "Employment Opportunities in Full Forest Utilization," *Journal of Forestry,* Vol. 46, No. 5, 1948: for data on the employment of labor per unit of raw material.

Municipal or Community Forests, state of New York, 1941: for data on community forests.

Pinchot, Gifford, *Breaking New Ground:* for many phases of the Pinchot-Roosevelt period.

Publications of the U. S. Forest Service, particularly: *Some Public and Economic Aspects of the Lumber Industry,* U.S.D.A. Report No. 114, 1917.

Forests and National Prosperity, U.S.D.A. Miscellaneous Publication No. 668, 1948: for data on timber supply, drain, and growth.

Puter, S. A. D., and Stevens, Horace, *Looters of the Public Domain:* for data on the Oregon timberland frauds.

The Russell Town Forest, Massachusetts Forest and Park Association, 1946: for data on community forests.

Wood, A. D., and Linn, T. J., *Plywood:* for many facts about plywood.

Woods, John B., Jr., assistant state forester of Oregon: for data on the Tillamook Burn and state plans for its rehabilitation.

Index

This book is a reprint of the edition originally published by the Stackpole Company and the Wildlife Management Institute in 1951. The text is a direct reproduction of the original. Page numbers remain the same. Great care was taken to locate all the original photographs and because of improved printing techniques the photographic reproduction is much improved in the reprint edition. All the photographs are identical to the original except in a few cases where similar photos from the author's files or from Yellowstone National Park were used.

Gene Downer, Publisher

ISBN: 0-933160-02-X Hardback edition
ISBN: 0-933160-03-8 Paperback edition

Library of Congress Catalog Number 79-83649

Photo of the author taken in 1952
Cristof photo

About the Author

Dr. Olaus J. Murie was a field naturalist of tremendous experience, and was recognized as the foremost authority on elk and caribou in North America. Born in 1889, his first scientific work was undertaken in 1914 when he made an expedition to Hudson's Bay to collect specimens for the Carnegie Museum. He made another expedition to Labrador three years later. After service in the Army Air Force in World War I, he joined the U.S. Biological Survey, now the Fish and Wildlife Service, and conducted an extensive study of the caribou of Alaska and the Yukon Territory. The results of this six-year study were published by the Department of Agriculture as a part of the North American Fauna Series. From 1920 to 1923 he also served as the fur warden for Interior Alaska, while continuing his studies of caribou, brown bear, and waterfowl.

For 36 years he made his home in Jackson Hole, Wyoming, in the heart of America's most famous elk range. Here he conducted the painstaking research for this book, **The Elk of North America,** living and working in intimacy with his subject. His studies of the Jackson Hole elk herd brought him international fame, and in 1948 he was called to New Zealand to advise the government of that Dominion on the management of its introduced elk.

An accomplished artist, he illustrated ten full-length books and countless articles on wildlife. He was the author of more than 75 popular articles on natural history and innumerable scientific papers. A leading champion of the preservation of our few remaining wilderness areas, he was Director of The Wilderness Society from 1946 until 1962, the year before his death.

FOREWORD

THIS BOOK on the elk, monarch of the West, is the first comprehensive treatise on this noble American animal based on an adequate field study. It is the result of years of tireless field work by Olaus J. Murie, one of the world's top field naturalists. No other technician has ever devoted as much time or effort to acquiring a knowledge and understanding of this magnificent game animal, and the thoroughness and quality of his work is revealed on every page of the manuscript. This volume is, and will be for a long time to come, the last word on the behavior and characteristics of the elk. Doctor Murie's basic studies were with the Yellowstone herds, but he has checked his findings there by field studies on all other major herds.

It was my good fortune many years ago to be associated with the author in the field on the Olympic Peninsula, home of the Roosevelt elk. Those days are remembered with pleasure, for they were some of the most pleasant in a lifetime of interesting outdoor experiences.

Murie has produced an extremely significant and worth-while volume which will be useful to hunters, wildlife administrators and technicians, students and educators for many, many years. It represents a real achievement in a life already marked by many other outstanding accomplishments. The Institute is proud to present this interesting and enlightening report as one of its series of books on practical wildlife management.

Ira N. Gabrielson

President
WILDLIFE MANAGEMENT INSTITUTE

Washington, D. C.

CONTENTS

ILLUSTRATIONS

INTRODUCTION

THE HISTORY of pioneer times in North America records a life and environment that have gone, and we read the accounts of those days with regret and envy as we reflect on the disappearance of the great wilderness with its teeming wildlife and its opportunities for exploration and adventurous living. That same history continues, though, as a source of inspiration and enjoyment as we vicariously relive the stirring events of our glamorous frontier.

In the early accounts of the abundance of game, the wapiti, now generally known in North America as the elk, figured prominently. It was sought as game and was one of the most widely distributed of our deer. Early travelers spoke enthusiastically of the hordes of "buffalo, elk, and deer" or of "buffalo, elk, and antelope."

As man advanced westward, however, the elk disappeared from the regions settled until, about the close of the nineteenth century, they had vanished from most of their range, and the herds that remained found refuge chiefly in the Rocky Mountain region and parts of the Pacific coast.

During the early explorations by field parties of the United States Bureau of Biological Survey [1] valuable information on the distribution of elk was obtained. The field reports of C. Hart Merriam, Vernon Bailey, Merritt Cary, H. E. Anthony, and other biologists are rich in data that are of great value today in our attempts to reconstruct the original natural game environment.

With the range of the elk so greatly restricted, the animals' plight had become obvious, and for a number of years the American public had been conscious of the "elk problem." Attention

[1] On June 30, 1940, the Bureau of Biological Survey was consolidated with the Bureau of Fisheries to form the Fish and Wildlife Service.

was focused on the Yellowstone-Jackson Hole areas of Wyoming, where the greatest herds remained, and in 1911, the Biological Survey detailed E. A. Preble to investigate conditions in Jackson Hole. His findings were published the same year (Preble, 1911).[2] Vernon Bailey and E. A. Goldman made a number of trips into the elk country to study various phases of the situation. Then in 1926 the President's Committee on Outdoor Recreation created a special Elk Commission to give attention to the elk problem, and this body requested the Biological Survey to undertake a comprehensive study of the life history of the elk to be used as a basis for efficient elk management. The writer was assigned to the proposed study and began work in July 1927, with Jackson, Wyo., as headquarters. He continued his field studies for about 5 years, making only minor interruptions for other duties. Subsequently, while engaged in field work in Montana, Washington, Idaho, California, Oklahoma, and parts of Canada and Alaska, he made studies on other elk ranges and while working in still other states examined ranges where elk had been exterminated.

The present report is an attempt to assemble the principal facts in the distribution and life of the elk of North America and to offer suggestions, based on the studies made, for future elk management. Chief reliance has been placed on original field studies; but supplemental information has been derived from perusal of the voluminous literature on the subject. In the work of compilation, especially as to the extent of the primitive ranges, the files of the Fish and Wildlife Service have been invaluable.[3]

---

[2] Publications referred to parenthetically by date (alone or with colon and specific page) are listed in the Bibliography, p. 333.

[3] The writer is indebted to the members of the Fish and Wildlife Service, Forest Service, and National Park Service who so heartily cooperated in the conduct of this work; to the Game and Fish Commission of Wyoming for the many helpful courtesies extended; to Childs Frick, of the American Museum of Natural History, for kindly placing fossil bones in his extensive collections at the writer's disposal; to Almer P. Nelson, in charge of the National Elk Refuge at Jackson, whose cooperativeness played an intimate part in the studies; to the two late veteran wardens of Jackson Hole, Fred Deyo and O. A. Pendergraft, who contributed vitally with information and help in the field; and to Adolph Murie, who spent several seasons in the field with me and assisted directly with the research program. Other persons who aided greatly in the work, including members of the organizations mentioned and residents of Jackson Hole and of communities in other states, are too numerous to list here; but anyone who has undertaken, singlehanded, a life-history study in a wilderness environment will appreciate the importance of the cheerful help of rangers and other citizens, without which the work would lag interminably.

1

THE NAME

ALTHOUGH EARLY COMERS to the American continent were undoubtedly familiar with the red deer *(Cervus elaphus)* of Europe, for some reason the closely related American animal *(C. canadensis)* became known as "elk," a name that properly belongs to the European moose. The misnomer has persisted, however, and today is firmly established in general usage. It is reported that in Ceylon the sambar is also known as "elk" (Donne, 1924: 129).

Ordinarily a change in nomenclature could readily be accepted as permanent, but in this instance there are complications. In scientific usage, especially in publications dealing with world distribution, the name "wapiti" designates a large group of animals of the genus *Cervus,* of which our American forms are only a part. Even in more popular writings "wapiti" is often used in preference to "elk." In Asia are to be found a large group of species of the genus *Cervus,* some of which differ little from our American "elk." European writers refer to this group as "wapiti" because they cannot use the word "elk," already applied to the European *Alces* (Brooks, 1910).

Thus it appears best to recognize both names, "elk," which is already in general usage in America, and "wapiti," which is often used in written accounts and which is preferable when writing on aspects of international significance.

2

ORIGIN OF THE AMERICAN WAPITI

THE DEER FAMILY has been traced back as far as the Oligocene. Matthew (1908: 535) stated: "This genus [*Blastomeryx*] proves to be a very primitive deer, approximately ancestral to the American Cervidae, and derivable in its turn from the Oligocene genus *Leptomeryx*, whose relationship to the Cervid phylum had not been suspected. We are thus enabled to trace the ancestry of the American Cervidae back to the Oligocene, . . ." (Matthew, 1934).

Representatives of the modern American deer, however, generally are thought to have come to this continent more recently, not before the Pleistocene at the earliest; though some doubt is cast on this by evidence found by Frick (1937: 189, 191)—a ramal fragment from the Uppermost Pliocene of the Eden Beds of southern California that bears dentition indicative of a true Cervine form, which was named *Procoileus edensis*, new subgenus and species. The place of origin and center of dispersal of the recent deer, more particularly of the genus *Cervus*, has been thought to be in Asia, where several groups have been developed. Lydekker (1898) designated the following groups under *Cervus:* Elaphine, Sikine, Damine, Rusine, and Rucervine. The Elaphine group contains the red deer *(C. elaphus)* and the wapiti *(C. canadensis)* and related species and subspecies. Some writers

3

THE AMERICAN FORMS OF ELK

CERVUS CANADENSIS CANADENSIS ERXLEBEN

Eastern Elk, Canadian Elk or Wapiti

Type locality. Eastern Canada

This name was generally applied to the elk of the East as well as the Rocky Mountain region until Bailey (1935) described the Rocky Mountain animals as distinct. The indigenous eastern elk are extinct; in fact they disappeared early in the period of the settlement of America, and apparently only one skin has been preserved in an American museum, although there are a number of skulls available, most of which have been excavated.

CERVUS CANADENSIS MANITOBENSIS MILLAIS

Manitoba Elk or Wapiti

Type locality. Manitoba and eastern Saskatchewan, Canada

The race *manitobensis* is thought to range chiefly in Manitoba and Saskatchewan, although the limits of its range have not been very well determined.

This form is darker than the typical subspecies *canadensis* and has smaller antlers.

CERVUS CANADENSIS ROOSEVELTI MERRIAM

Roosevelt Elk or Wapiti, Olympic Elk

Type locality. Mount Elaine, near Mount Olympus, Wash.

For a number of years this form was known as *C.c. occidentalis;*

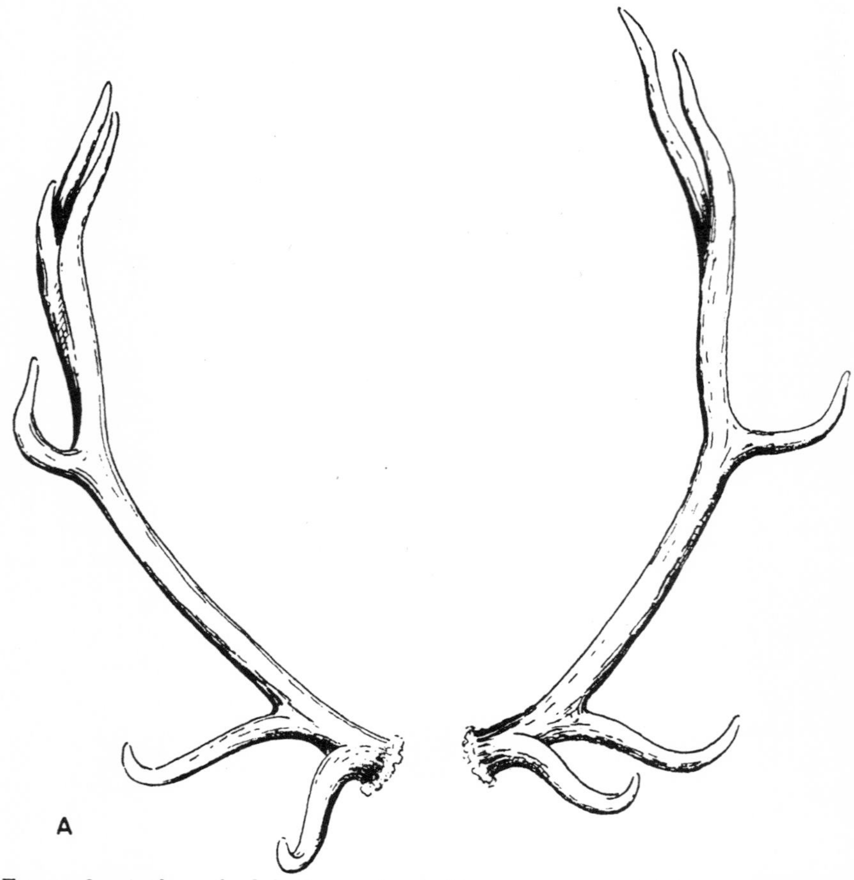

FIGURE 2a. Antlers of adult Roosevelt elk *(Cervus c. roosevelti)* from Hoh River, Olympic Mountains, Wash. Note the tendency toward crowning. Length of beam, 45¾ inches; widest spread, 33 inches;

but as brought out by Bailey (1936: 81, footnote 4) there appears to be considerable doubt as to the origin of the material on which Smith (1827) based that designation, so that it seems proper to use Merriam's name *roosevelti,* inasmuch as it is based on an authentic specimen.

The Roosevelt elk inhabits the rain forests of the Pacific coast intermittently from Vancouver Island south to northern California, and a small herd has been introduced on Afognak Island, Alaska. Formerly this subspecies had a practically continuous

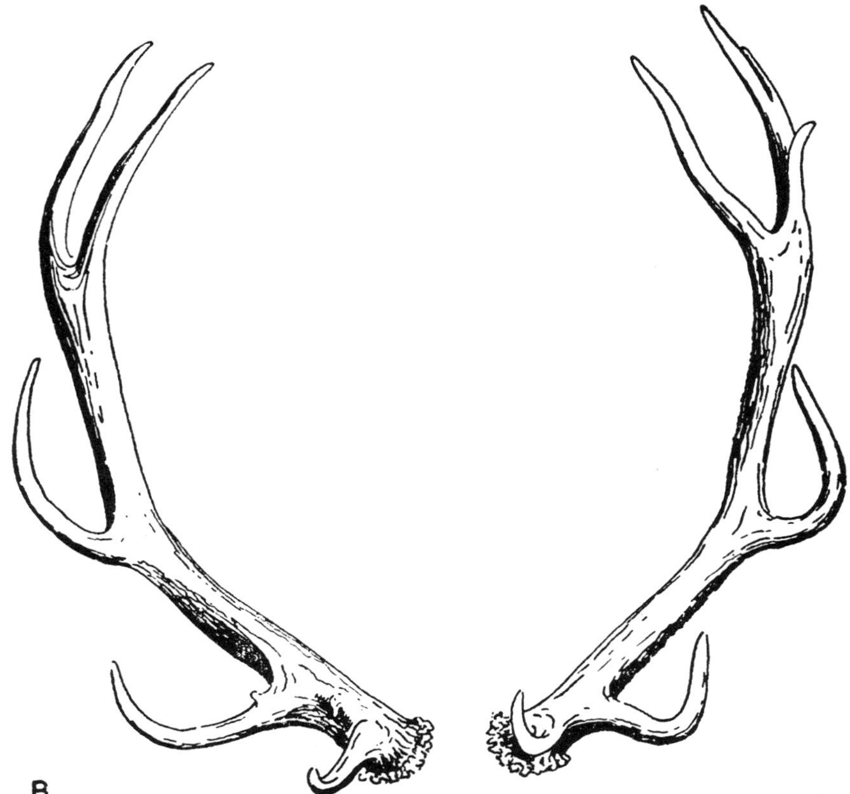

FIGURE 2b. Antlers of adult Roosevelt elk. Length of beam, 42¾ inches; widest spread, 33¼ inches.

distribution southward to the vicinity of San Francisco Bay and appears to have occupied the west slope of the Cascade Range.

This elk is distinguished from the Rocky Mountain elk by its darker coloration, a distinction that has been doubted by casual observers familiar with both the Rocky Mountain and the Pacific coast herds. The difference in color can be observed, however, by proper comparisons.

The Roosevelt elk has been characterized as larger than the Rocky Mountain elk; and measurements of a small series of skulls confirm this statement, as the skulls of the Pacific coast animal were all consistently larger than those of the Wyoming elk. Presumably *roosevelti* is heavier, too, than *nelsoni;* and field examinations of dead animals support this assumption, but unfortunately precise weight records are not available.

The antlers of *roosevelti* (Figure 2a and 2b) are rugged and heavy but on the average do not have the length of beam or the spread of those of *nelsoni* (Figure 3). They have a greater tendency to "crowning," and among the antlers examined the basal tines

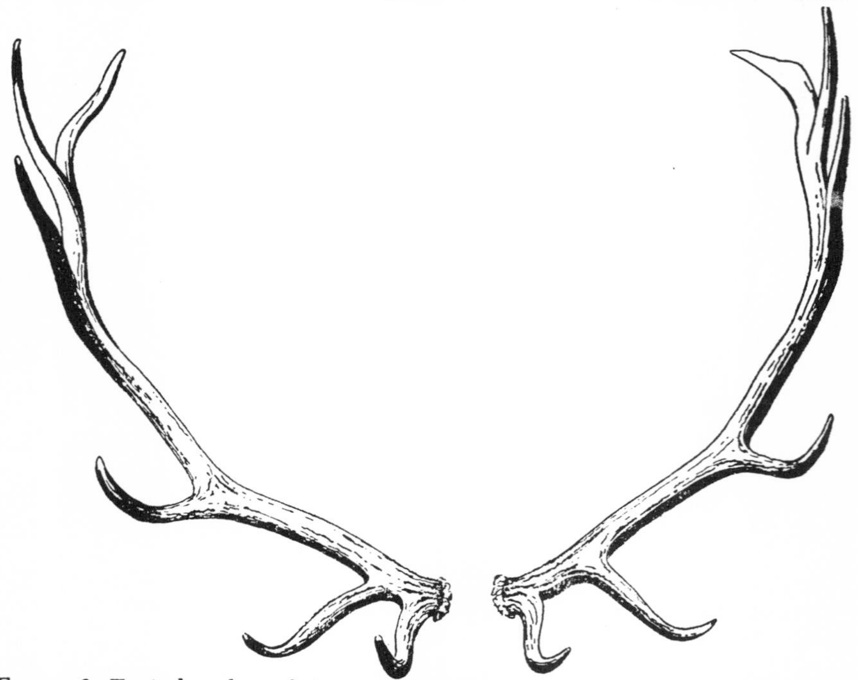

FIGURE 3. Typical antlers of 4-year-old Rocky Mountain elk *(Cervus c. nelsoni)* from Jackson Hole, Wyo. Length of beam, 49½ inches; greatest spread, 43¾ inches.

averaged heavy and long, seemingly at the expense of the upper terminal tines, which averaged smaller, an arrangement that is more or less true of the antlers of other elk but that is accentuated in those of *roosevelti.*

CERVUS CANADENSIS MERRIAMI NELSON—

CERVUS MERRIAMI, NEW SPECIES

Merriam Elk or Wapiti, Arizona Wapiti

Type locality. Head of Black River, White Mountains, Apache County, Ariz. Altitude, about 9,000 feet.

"*Type*, No. 11639, ♀ adult, U. S. National Museum, collected August, 1886, at head of Black River, White Mountains, Arizona, By E. W. Nelson."

The Merriam elk, now extinct, apparently ranged only through a few mountain areas of Arizona and New Mexico, where it was more or less isolated by surrounding arid territory.

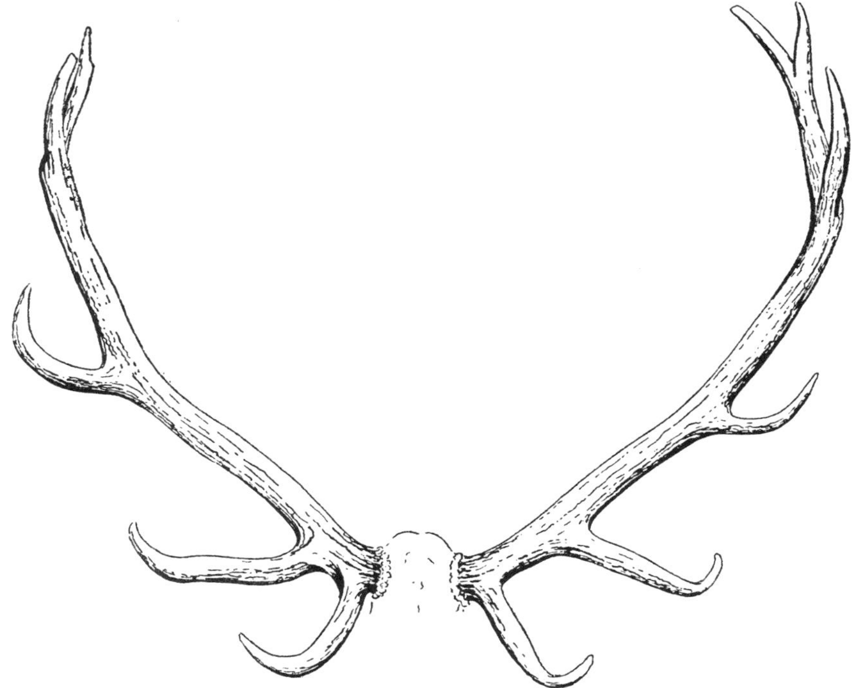

FIGURE 4. Antlers of Merriam elk *(Cervus c. merriami)* found in 1910 on the Apache National Forest, Ariz., by Jesse Burke. Length of beam, 51 inches; greatest spread, 53 inches.

Available descriptions of a few specimens indicate that *merriami* was larger than *nelsoni* and *roosevelt* and more uniformly colored, perhaps paler, with less contrast in color pattern, and it had "more massive" antlers than *nelsoni* (Figure 4).

The maxillary tooth row of the *merriami* specimen measured by Nelson (1902: 12) is longer than the average tooth row of 21 *nelsoni* males from Wyoming measured by the writer, longer, in fact, than the maximum measurement for the Wyoming series. It is longer also than the corresponding measurements for two *roosevelti* males but shorter than the average tooth row of a series of nine *nannodes*. In other cranial measurements, too, Nelson's specimens exceeded in size the other forms here considered. It is possible, of course, that it was an exceptional specimen, but it is the only one available.

CERVUS CANADENSIS NELSONI BAILEY

Rocky Mountain Elk or Wapiti

Type locality. Yellowstone National Park, Wyo.

"Type from Yellowstone National Park, ♂ adult (8½ years old) No. $\frac{49722}{124656}$, U. S. Nat. Mus.: died September 21, 1904, in the National Zoological Park (No. 671¼); tanned skin in fresh early fall pelage, not the short summer nor the long coarse winter coat; complete skeleton and skull with antlers sawed off." (*From* Bailey, Vernon: A new name for the Rocky Mountain elk. Biol. Soc. Wash. Proc. 48: 187-190. 1935 [p. 188].)

The original range of the Rocky Mountain elk included the Rocky Mountain region, but its limits are entirely unknown, as the primitive range was not materially interrupted to the east.

The taxonomic status of the elk of the extreme east, long extinct, and of the elk of the Rocky Mountain region has been a problem. In 1935, Vernon Bailey (1935) named the more western group *Cervus canadensis nelsoni*, basing his action on the long-recognized assumption that the eastern elk undoubtedly differed from the elk of the Rocky Mountains, a perfectly logical procedure. It is unfortunate, however, that adequate material was not available for precise characterization of the eastern elk. Old published descriptions of pelage colors are hardly reliable. The coloration of an animal such as the elk is not as readily or as precisely described as that of the plumage of birds, for example, and permits of great latitude of expression. There are, too, individual variations and variations owing to sex and to age. Bailey refers to Audubon's painting of the eastern elk. The animal, in the summer pelage, is not readily distinguishable from the Rocky Mountain elk in summer. Audubon painted the animal as the artist sees it, in its bright colors as contrasted with the environment (Audubon and Bachman, 1846-54: vol. 2, facing p. 82).

The usual technical descriptions of the summer pelage of necessity do not convey the striking difference between it and the winter pelage. In summer, in its red coat, the live Wyoming elk in its environment would look much like Audubon's painting. The old skin of the eastern elk now in the Philadelphia Academy of Sciences is too faded to be very useful.

On the whole, the form *nelsoni* has not been satisfactorily differentiated by precise characters, so that its validity must rest on the probability that one form could not extend all the way from the Rocky Mountain region to the Atlantic coast. It was properly designated as a subspecies, for elk distribution was interrupted very little in all that distance. A fact in support of the new form is the relatively sedentary habit of the eastern elk, as contrasted with the mobility of the western animals, which indulged in long migrations.

In the present study there has been no opportunity to examine specimens of *manitobensis* for comparison with *nelsoni*.

In due time it should be possible to assemble enough skulls and antlers of the eastern elk, from material that may be excavated, to ascribe definite characters to *C. c. canadensis*.

CERVUS NANNODES MERRIAM

Tule Elk, Valley Elk, California Wapiti, Dwarf Elk or Wapiti

Type locality. Buttonwillow, Kern County, Calif.

The tule elk formerly was apparently confined to the interior valleys of California; it is now restricted to Owens Valley and a fenced preserve at Tupman, near Bakersfield (Plate 2).

This is the smallest of the North American elk. It is also the palest and has the least contrast in color pattern. Although decidedly smaller than other elk in general body measurements, as well as in cranial measurements, the tooth rows of nine male specimens examined are actually larger than those of the Wyoming and the Pacific coast elk, and some of them exceed in size the tooth row of *merriami* as described by Nelson (1902: 12). The antlers of *nannodes* are much less developed than those of the other elk. Among eight specimens of males in the Museum of Vertebrate Zoology, University of California, three bore supernumerary prongs on a brow tine; in one specimen, there were three small ones on the right brow tine and one on the left.

CERVUS WHITNEYI ALLEN

Type locality. Blue Mounds, Wis.

This form is based on fossil bones described by Allen (1876: 48). Just what relationship it has with more recent wapiti is uncertain.

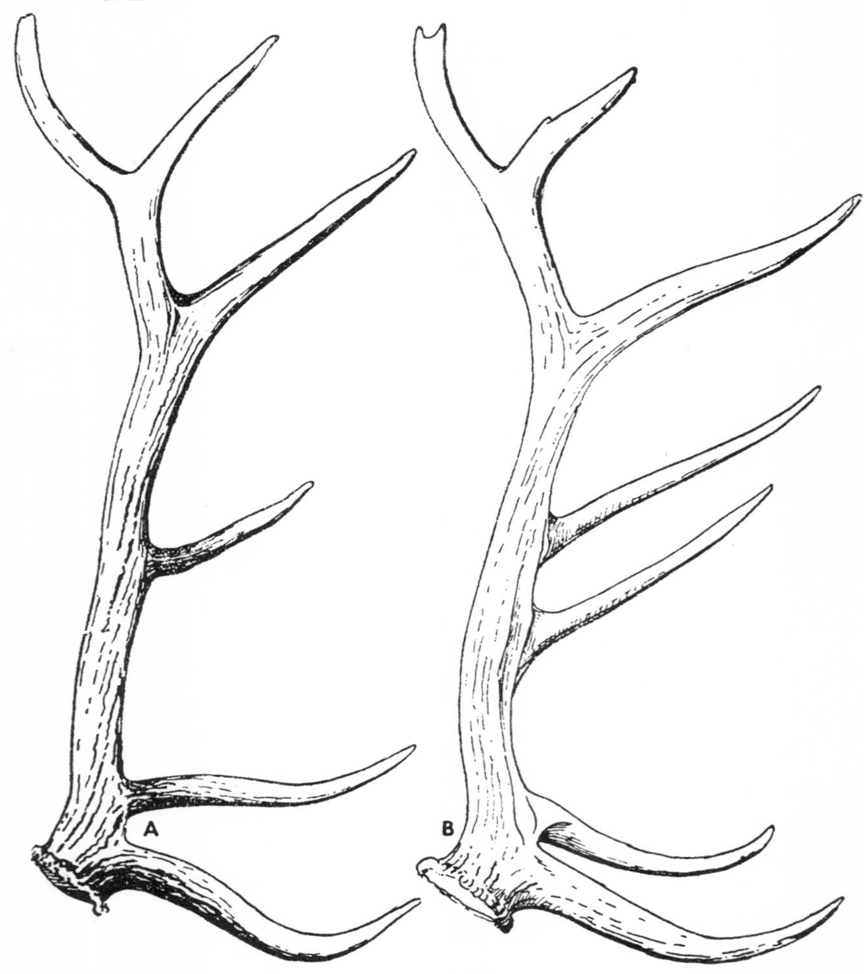

FIGURE 5. Antlers of wapiti from the Pleistocene of the Fairbanks district of Alaska—Childs Frick Collections, American Museum of Natural History. *(A)*, Antler from Little Eldorado Creek, quite comparable to that of present-day wapiti. Length of beam, 47½ inches. *(B)*, Unusual antler from Cleary Creek, one of two specimens that have two tines in place of the usual single trez. Length of beam, 50¾ inches.

CERVUS C. SUBSPECIES: FOSSIL WAPITI OF ALASKA

For several years Childs Frick, of the American Museum of Natural History, has supported field work in the Fairbanks area of Alaska in cooperation with the University of Alaska. In the course of this work, fossil bones brought to light by placer gold operations have been preserved, among them a good series of wapiti remains that Mr. Frick kindly placed at the writer's disposal.

Measurements of mastoid width indicate that this fossil animal was slightly larger than *Cervus c. roosevelti* and definitely larger than *C. c. nelsoni.* The antlers cannot be distinguished from those of modern wapiti if two unusual specimens are ignored. These two antlers may indeed represent a second form from Alaska. Under the circustances, in view of the very slight difference in size of the one significant skull measurement, it appears unwise to name this fossil wapiti. The work is continuing and may bring to light material of more significance.

At any rate, it is certain that a wapiti quite similar to the living animals occupied interior Alaska in the Pleistocene (Figure 5).

CERVUS LASCRUCENSIS FRICK

Frick (1937: 196-197), who described this animal from the Pleistocene at Las Cruces, N. Mex., stated: "As seen in the metatarsus, the stature of this species was intermediate between the wapiti and the European red deer."

CERVUS AGUANGAE FRICK

This is another Pleistocene form of the genus Cervus from Aguanga, southern California. What relation it may have had to our elk is not known.

4

EARLY ELK DISTRIBUTION IN AMERICA

WHEN WHITE MEN first came to this continent, the American elk, or wapiti, was the most widely distributed of the deer, as it was found across what is now the United States from the Pacific Ocean almost, if not quite, to the Atlantic, northward far into parts of Canada, and south almost to what is now Mexico (Figure 6). Elk were missing or were scarce, however, in some parts of the Atlantic coast. There are no actual records of their occupying the areas now included in Maine, New Hampshire, Vermont, Massachusetts, Connecticut, Delaware, and Florida, nor the eastern parts of what are now the States of New Jersey, Maryland, Virginia, and the Carolinas. Nevertheless, though there are no available precise records showing Atlantic seaboard distribution, early writers expressed the opinion that elk did at one time reach the sea, and it is at least probable that at some early date elk occupied Atlantic coastal areas. Wyman (1868: 574) noted that elk bones were found in shell-heaps on Mount Desert Island, Maine; Stone (1908: 55) reported bones from aboriginal refuse heaps near Trenton, N. J.; and Bailey (1896: 94) stated that elk formerly occurred in the District of Columbia.

There were other gaps in elk distribution. In the Pacific coast region, elk occupied the Coast and Cascade Ranges from Vancouver Island southward through Washington, Oregon, and into the southern part of California; but curiously enough, much of the Great Basin of the West, including most of Arizona and Nevada and parts of Utah, Oregon, and Washington, appears to

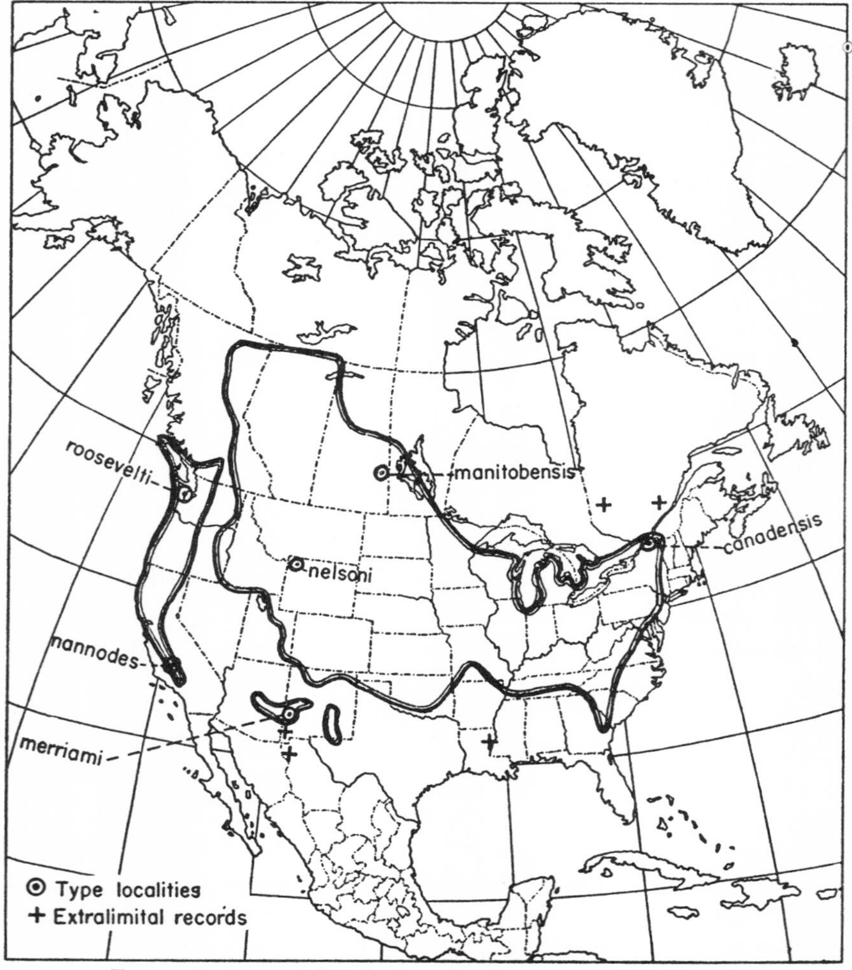

FIGURE 6. Original distribution of wapiti in North America.

have been unoccupied by elk, thus leaving an apparent gap between *Cervus c. roosevelti* and *C. nannodes* on the west and *C. c. nelsoni* of the Rocky Mountain region to the east.

The discussion of primitive elk range given here, as well as the accompanying map (Figure 6), is based on written or printed

into the State were made, Colorado being one of the few States in which the animals were not completely exterminated.

Lloyd Swift (1945: 114) has furnished an excellent summary of the Colorado elk remnants and restoration. He comments that in 1910 there were about 10 small bands, and that according to Forest Service records these totalled probably between 500 and 1,000. To restore the elk, 14 introductions were made over the period 1912 to 1928, totalling 35, mostly in areas where the original animals had been extirpated. "From about 25 foci of native and planted elk, totalling not much more than 1,000 head, the elk are now common over most of the wild and mountainous country of the state. The 1943 estimate for the state was 24,000 elk."

According to this account the native elk had persisted, sometimes in very small numbers, in the Estes Park country and adjacent areas; upper reaches of Saguache Creek and the Rio Grande drainages (especially Goose Creek); head of San Juan River (probably recruited from Goose Creek); Hermosa Creek region; Dolores River drainage; drainage basins of Coal and Soap Creeks; head of Gunnison River; probably the Leroux Creek area; Middle Park (about 50 head); headwaters of White River (where the largest group of native elk survived); and the area now known as the Routt National Forest.

The history of elk in Colorado illustrates the value of wild country for preservation of native animals. Swift comments on the White River country: "In this large roadless area, natural factors of isolation and ample food and cover combined to give the elk more protection than was afforded on other portions of their ranges."

GEORGIA

The records do not indicate general distribution of elk in Georgia in early days. Evidently most of the State was unoccupied by them, as Bartram (1928: 62) in recounting his travels of the year 1773 wrote: ". . . there are but few elks and those only in the Appalachian mountains."

IDAHO

Elk occurred throughout Idaho, but it is doubtful that the

entire country was elk range. Apparently the animals were seldom found on the arid plains of southern Idaho but roamed largely in and near the mountain ranges, having been specifically recorded from the Salmon River, Crags, Pahsimeroi, Lemhi, Sawtooth, Blackfoot, Bruneau, and Elk Mountains. Early reports frequently mentioned that the Henry Lake area and other localities were occupied by elk.

ILLINOIS

In early days elk ranged throughout Illinois, where, according to Allen (1871b: 5), the prairies were once preeminently their home; but they disappeared early from this State also. Michaux (1904: 72) recorded one killed by his guide in 1795 in the vicinity of Kaskaskia, Randolph County, and Cory (1912: 70) wrote that elk "are claimed to have been common about 1820" in southern Illinois. There are records of elk "tracks in the forest north of Peoria in 1829" (Caton, 1877: 80), of an elk killed near Mount Carmel in 1830 (Wood, 1910: 515), and of "several elks shot" in Cook County (Kennicott, 1855: 580).

INDIANA

Elk were found at one time throughout Indiana but had become relatively scarce even at the time of the first settlement of the country by white men. Elk persisted in the early part of the nineteenth century. Butler (1895: 83) stated on the authority of E. J. Chansler that a Mr. Stafford said he saw an elk that was killed on Pond Creek, Knox County, in 1829, that Brad Thompson told of seeing a wild elk in the same county in 1830, and that a Mr. Bruce reported seeing elk horns in Daviess County as late as 1850. Writing in 1934, however, Butler indicated (p. 248) that the last elk in Indiana disappeared as early as 1818.

IOWA

Among data obtained from pioneers of Sac County, Iowa, Spurrell (1917: 275) recorded the following:

"All the earliest settlers united in saying that elk were plentiful. They were found from solitary individuals to 500 in a herd. . . .

"The elk were an important source of meet [sic] of the earliest settlers . . ., their place being taken by deer later. Elk horns could be picked up by the wagon load in 1856. . . . The last elk in Sac County was a herd of about forty, which was seen in October, 1869, and went from east of Storm Lake, south through Sac County, crossing the 'Goosepond' at Wall Lake."

Inasmuch as the elk in Iowa necessarily lived in open country, they were easily exterminated. It is said that the early residents could have elk meat any time they wished. With reference to Greene County, Spurrell (1917: 276) quoted the following from *Biographical and Historical Record of Greene and Carroll Counties of Iowa,* published in 1887 by the Lewis Publishing Co., of Chicago:

"Game such as deer and elk was in great abundance until the winter of 1855-56. The snows of that winter were so deep that it was impossible for them to escape the pursuit of men and dogs, and the number destroyed seems almost incredible. It is said that they were overtaken by men, boys, and even women, and beaten to death with clubs. Since then there has scarcely been an elk or deer seen within the county. Their rapid and sudden disappearance astonished everyone."

Concerning elk in nine counties southwest of the center of the State that he investigated in 1867, Allen (1871a: 185) reported as follows:

"During the early settlement of this part of Iowa, they were of great value to the settlers, furnishing them with an abundance of excellent food when there was a scarcity of swine and other meat-yielding domestic animals. But, as has been the case too often in the history of the noblest game animals of this continent, they were frequently most ruthlessly and improvidently destroyed. In the severer weather of winter they were often driven to seek shelter and food in the vicinity of the settlements. At such times the people, not satisfied with killing enough for their present need, mercilessly engaged in an exterminating butchery. Rendered bold by their extremity, the elk were easily dispatched with such implements as axes and corn-knives. For years they were so numerous that the settlers could kill them whenever they desired to, but several severe winters and indiscriminate slaughter soon greatly reduced their numbers, and now only a few linger where formerly thousands lived, and these are rapidly disappearing."

The writer does not have record of the last elk in Iowa, but the date above marks approximately the disappearance of the elk from the State.

KANSAS

Donald Hoffmeister (1947: 75) has given us a good account of elk in Kansas. Apparently elk roamed pretty well throughout the State. In his catalogue of mammals of Kansas (Trans. Kansas Acad. Sci., 4:20, 1875), Hoffmeister quotes Knox: "Quite common in the west part of the state."

Three letters written in 1892 by J. R. Mead, of Wichita, Kansas, to Professor L. L. Dyche, are of special interest, coming from an enthusiastic observer of big game. The following three extracts are quoted, in part, from Hoffmeister's paper.

"In 1859 to 64," Mead writes in his letter of March 12, 1892, "the Eastern range of Elk in Kansas would be a line drawn north and south through Eldorado [,] Butler Co [unty]. All country west of that in Kansas was ranged over by them and I presume occasionaly [sic] east of that line."

On March 13, 1892, Mead wrote, "Elk followed the timbered creeks, probably for the browse they seemed to prefer. For instance, a herd would cross the Solomon [River] coming from the North. At or near the mouths of some stream followed that stream up to its head. Cross the divide to say the head of Spillman's Creek, follow it down to its mouth. Then follow the valley of Saline River down a few miles feeding as they went to the big ford at the narrows where Lincoln now stands [,] cross there, and feed along up Elkhorn Creek to the head of Alum Creek and follow that down to the Smoky Hill River, and from there went I do not know where . . . I only saw them [large herds of a '1000 more or less'] in summer or fall. But old Bulls were found in the broken hills between Saline and Solomon [rivers] all winter . . . [The elk] preferred broken country with timbered canons and streams and were not plentiful in the level treeless country adjacent to the Arkansas [River] . . . Elk were much more numerous north of the Smoky Hill River in Kansas than south of it . . ."

In describing the abundance of the elk in the 1850's and 1860's, Mead writes on March 11, 1892, "I have seen 1000 more or less in one drove, and they crossed the Saline [River] at the ford right where the town of Lincoln now stands . . . I have killed Elk on the Solomon, Saline, Smoky Hill, and Arkansas rivers and their tributaries . . . Have seen 50 to 100 in a drove in the Indian Ter [ritory] just over the line 10 miles below Kiowa, Barber County [thus, near Burlington, Alfalfa County, Oklahoma]. Have known them killed in Butler Co. [Kansas] on the Walnut [River]."

"In what is now Lincoln, Mitchell, Osborne, Smith, Phillips, and Rooks counties," writes Mead on March 13, 1892, ". . . I frequently saw where large herds had recently [stopped?] and occasionaly [sic] saw the animals themselves."

In western Kansas, elk were still fairly abundant in 1871, for J. A. Allen, in reporting upon the mammals observed chiefly in the vicinity of Fort Hays, wrote: "More or less common near the streams, especially on Paradise Creek, and occurs as far east at least as Fort Harker [Ellsworth County]" (Bull. Essex Institute, 6:48, 1874).

KENTUCKY

Elk were once common in Kentucky and were utilized for food by travelers. Shoemaker (1939: 22) says:

"Daniel Boone is always pictured as encountering vast herds of Wapiti and Bison in Kentucky, where he first went in 1769, but John Strader and James Yager, who were there several years earlier, had already described the myriad heads of game about the salt licks. Colonel Thomas Walker of 'Castle Hill,' who made an expedition over the Cumberland Mountains to Kentucky from his estate in Virginia in 1769, killed many wapiti and Bison. A century after Strader's first appearance, followed by Boone, Kenton, Wetzel, Girty, Colonel Crepps and other great killers, big game was gone from the dark and bloody ground . . ."

Elk were still present, though in reduced numbers, in 1847, according to Audubon and Bachman (1846-54). Apparently elk disappeared from the State about the middle of the nineteenth century.

LOUISIANA

Dr. Milton Dunn wrote in a letter from Colfax, June 2, 1918: "We had elk here when this part of Louisiana was sectionized by surveyors in 1829"; and in Porter's Spirit of the Times (12: 555, Jan. 21, 1843) is recorded the killing of a bull elk with a gross weight of 704 pounds near Mound, Madison Parish, on Walnut Bayou in December 1842. Apparently elk did penetrate parts of Louisiana but probably were not widespread or abundant.

MICHIGAN

In the early part of the nineteenth century elk were common in parts of the Lower Peninsula of Michigan. In 1856 they were fairly plentiful at Sand Point, Huron County, and according to Wood (1911) one was killed there that year. Vernon Bailey (field notes 1894-95, 1905) recorded that in 1861 they were abundant in the heavy pine forest in Tuscola County and that his father, from 1856 to 1863, saw many of them there in small droves of 16 or fewer and said that they were as common as deer. A number of reports show that elk were present in Huron, Tuscola, Sanilac, and Gratiot Counties until 1870 or 1871 (Wood and Dice, 1924: 466). Some still remained in the northern Lower Peninsula as late as 1877 (Archer, 1877: 192; Caton, 1877: 80), but undoubtedly they disappeared soon after that date. Available records do not show that the Upper Peninsula of Michigan was ever occupied by elk.

MINNESOTA

Elk were once abundant in Minnesota and are said to have ranged throughout the State. Surber (1932: 74-75) stated:

"At the time of the first settlement of the state this noble deer ranged over the entire southern half of the state as far north at least as Aitkin County, thence northwest to the extreme northwestern corner of the state . . . It ceased to exist in the southern part of its range previous to 1860. In the northwestern part of its range it survived much longer. A herd of 16 of the animals was seen in the southwestern part of Roseau County in 1896 but these disappeared shortly thereafter."

In a letter of July 16, 1888, E. L. Brown reported that he and others killed a number of elk near Warren, Marshall County, in October and December 1887 and saw many more. Herrick (1892) wrote: "Mr. W. W. Cooke of Moorehead informed me that both moose and elk are always found near lakes Itaska and Caribou while in 1885 they were common about Red Lake." Johnson (1916: 2) wrote: "Probably extinct as a wild animal in Minnesota at present though reported for the Lake of the Woods region about the year 1900."

No doubt, the elk lingered longer in Minnesota than in some other states because some of them occupied the less accessible forested areas of the north woods.

Swanson, Surber, and Roberts (1945: 12) have furnished a thorough resume of the history of elk in Minnesota. Quoting from "Reminiscences," in *Minnesota Historical Collections*, 3: 265, they say in part:

"That it was still fairly numerous in southeastern Minnesota as late as 1841 is shown by Sibley's statement that a hunting party of Sioux Indians killed fifty or sixty that winter in the region that is now Dodge and Mower counties and adjacent portions of Iowa and that, the fall before, while returning across country from the same hunting ground to Fort Snelling, he and a companion 'fell in with two herds of elk on the route, numbering at least five hundred in each . . . .'"

Old settlers and others had reported to them that elk "still lingered along the west bank of the Mississippi as late as 1850 or 1860."

From the manuscript of Dr. Evadene B. Swanson, the following events are recorded by these authors:

"Samuel Pond believed Elk were becoming rare along the Minnesota River in 1834. Two were killed near Bloomington about 1840, an event he considered unusual.

"In 1847 several seen near Crow and Saux rivers.

"Several stragglers seen on Sunrise River in November, 1854. Near St. Peter in September, 1856, and July, 1857, 'some.' A few seen near Glencoe in August, 1857, and 15 miles from that town in January, 1872.

"Crow Wing county in 1874 and 1897."

These same authors have also assembled the following valuable historical data:

"Chief Game Warden Andrus asserted in 1894 that the area near Thief River Falls was the only place that they might still be found in Minnesota. 'During the final 20 years of the last century the growing scarcity of both elk and caribou was recognized generally' in Minnesota.

"Turning to the western part of the state, Alexander Henry II, the fur trader, made record in his diary in August and September, 1800, that Elk were very abundant about Fort Pembina on the Red River of the North. The long expedition met with a band of fifty or sixty not far north of Lake Traverse on July 28, 1823 (Keating, vol. 2, page 11). But they were more common in the sparsely timbered, park-like region to the eastward, and were present, though in dwindling numbers, until shortly after the middle of the last century in Becker, Otter Tail, Grant, Pope, Stevens and other adjacent counties.

"In the diary of E. L. Brown (in Minnesota Historical Society) he states that he killed a cow Elk on October 31, 1896, and preserved the skin. It was one of a band moving eastward in the Mud Lake region. He saw the calf next day. Brown also says that he saw the hide of an Elk killed by a settler in the winter of 1896-7 and that others were seen at the same time. Brown records that he saw signs of Elk in the Roseau River country in October, 1898. This is the last Elk entry in his diary though he was hunting in the same region in 1899. When the writer was in eastern Marshall and northern Beltrami counties in 1900, with Brown as guide, no evidence of Elk at that time could be found. So apparently they had left the Roseau-Thief Lake country about 1898.

"Mr. P. O. Fryklund of Roseau says that so far as he knows the last native Elk taken in northwestern Minnesota was killed in Roseau County in 1896 by a French hunter, Charles Desjarles.

In a letter dated May 18, 1921, to Mr. Carlos Avery, in the files of the State Game and Fish Department, Mr. Stephen H. Withey, of Crookston, furnishes the following information gleaned from hunters and trappers: "The last Elk that was killed in Minnesota was in 1908; the last Elk seen in Minnesota was four years ago (1917), there were three of them seen about 18 miles southeast of Roosevelt in Beltrami County." (Now Lake of the Woods County.)

"Dr. Gustav Swanson, while in the Northwest Angle in 1932, gained the following information in regard to Elk in that isolated region:

"'While in the Northwest Angle, I learned that Mr. Julius Lofgren, who spends his winters in the Angle on a homestead near the Manitoba line, had seen a band of Elk there in February, 1932.' Mr. Lofgren is an experienced woodsman and well acquainted with the larger game animals. Dr. Swanson adds to the above note, 'It is not to be wondered at that the animals should roam into the Northwest Angle, however, because the part of Manitoba adjacent to the Angle is wild and uninhabited and Elk or any other big game could exist there practically unmolested by man' (Journal)."

MISSOURI

Missouri was elk country, and the animals were common in the State at one time. They were reported still present, in herds, in the southeastern part of the State near the Arkansas border in 1834 (Featherstonhaugh, 1835: 56) and abundant near New Madrid as late as 1859 (Old Timer, 1898).

Shoemaker (1939: 26) gives a much later date for their disappearance, saying: "One of the wonders of wild life tenacity was the way in which the last of the eastern Wapiti held on in Missouri. If an effort had been made to protect them they would be affording grand sport today, but though they lasted until about 1898, they are all gone now."

MONTANA

Elk once roamed in great numbers over Montana and together with deer, antelope, bison, mountain sheep, grizzlies, and smaller mammals and birds, presented one of the most remarkable wildlife spectacles on this continent. To judge from the journals of Lewis and Clark (1893) and the accounts of other early travelers, the herds of elk were distributed in greatest numbers in the eastern and central parts of Montana, being especially numerous on the plains in wooded bottomlands and islands of the rivers, in brushy ravines or river breaks. Elk were also in the mountains, where they sometimes persisted in limited numbers, as in Wyoming, until the restocking program got under way and the elk population once more took an upward turn.

It appears from the records, however, that elk were very scarce or absent from the mountain region of northwestern Montana and northern Idaho.

NEBRASKA

Elk occurred at least in fair numbers, throughout Nebraska, but they could not survive the influx of travelers and settlers in this prairie country and largely disappeared in the 1870's. There are reports of their presence as late as 1880.

NEVADA

Probably the original elk distribution was greater than records indicate, but in the light of existing information, it may be said with confidence that elk did not occupy Nevada in any considerable numbers. The State is outside the normal elk range, yet there are at least two early records. On July 20, 1859, Simpson (1876: 121) wrote: "An elk was seen for the first time yesterday in Stevenson's Canyon, and one today in Red Canyon; . . ." This was in the Snake Range a little north of the 39th parallel. In a field report of June 7-21, 1898, A. K. Fisher wrote:

"Elk occur north in the Wild Bruneau Mountains. Last winter, I am told, seven were seen by cattlemen, and of these a small herd was making inroads on a haystack from which they were driven only with difficulty." This referred to the vicinity of Mountain City.

In 1940, on an occasion when he had been perusing some of his early field diaries, the late Dr. C. Hart Merriam informed the writer that elk were formerly found in the Charleston Mountains of Nevada. Some of these mountains are high, with pine forests.

NEW MEXICO

The status of elk in New Mexico, appraised on the basis of all the available information, is fully discussed by Bailey (1931: 39-45) in his *Mammals of New Mexico*, in which he wrote (p. 42) about the occurrence of *Cervus merriami* as follows:

"Merriam's elk is now probably extinct; certainly it no longer occurs in New Mexico. Forty years ago it was common in the Sacramento, White, and Guadalupe Mountains east of the Rio Grande, in the Mogollon group of mountains west of the Rio Grande and in the White Mountains of Arizona. . . . There are old records for the Datil and Gallina Mountains of Socorro County and a doubtful record for the Manzano Mountains. To the north there are no more elk records until the Jemez and Pecos River Mountains are reached, where the Colorado elk comes down from the north."

Allen (1875: 65) stated: "Mr. Mecham has seen them [elk] as

far south as the Mexican boundary, and speaks of having met with droves of two thousand individuals in southern New Mexico."

NEW YORK

Positive records of elk in New York are not numerous, but they are sufficient to show that the animals at one time ranged through the State. There has been some doubt concerning their occurrence in the Adirondacks, but Merriam (1884: 143) stated as follows: "That the American elk . . . was at one time common in the Adirondacks there is no question. A number of their antlers have been discovered, the most perfect of which that I have seen is in the possession of Mr. John Constable." Merriam recorded several other finds. Rowley (1902: 32), too, stated that the elk was "known to have existed in the Adirondack region. One of the last elk reported seen in the State apparently is the one mentioned in the following excerpt from De Kay (1842: 119):

". . . Mr. Beach, an intelligent hunter on the Raquet, assured me that in 1836, he shot at a stag (or as he called it, an elk) on the north branch of the Saranac. He had seen many of the horns, and describes this one as much larger than the biggest buck *(C. virginianus)*, with immense long and rounded horns, with many short antlers. His account was confirmed by another hunter, Vaughan, who killed a stag at nearly the same place. . . . in 1834, I am informed by Mr. Philip Church, a stag was killed at Bolivar, Allegany county. My informant saw the animal, and his description corresponds exactly with this species."

Shoemaker (1939: 19) says:

"In the broken, densely wooded regions of southwestern New York, Wapiti were numerous. There is a fine set of antlers at Letchworth Park, New York, the stag having been killed, it is said, in the Genesee Valley, about 1843; and another was formerly in the Albany Museum, taken in Cattaraugus County the same year. It is likely that the range of the Wapiti extended to the western edge of the Adirondack wilderness, but they disappeared about 1847."

NORTH CAROLINA

Elk were reputedly abundant in the Carolinas as late as 1737, according to Hays (1871), and their presence in those States was mentioned briefly by other writers. Oberholser (1905: 4) stated that elk occurred in North Carolina "in colonial times, at least until about the year 1750."

NORTH DAKOTA

Bailey (1926: 33) summed up the early elk distribution in this State as follows: "Originally elk ranged over all of what is now North Dakota, and were equally at home in the timber and over the open prairie."

In the same paragraph Bailey quoted entries from Alexander Henry's journal made on September 5, 1800 (Henry and Thompson, 1897, vol. 1: 84, 85), while on a trip up the Red River, to the effect that the red deer [i.e., elk. Ibid.: 2, footnote 3] "being now in the rutting season, are heard in every direction excepting toward the plains" and that "large herds were seen at every turn of the river," and added, with reference to Henry:

"During the next 6 years he frequently mentioned them, and next to the buffalo they seem to have been the main source of meat supply for him and his parties of trappers in the Red River Valley and adjacent country."

The Missouri River was the convenient highway of travel for the early explorers, and Lewis and Clark (1893) in 1804-05, Wied (1839-1841) in 1833, and Audubon (1897) in 1843, in their accounts of their expeditions repeatedly refer to the numbers of elk along the Missouri in North Dakota, where they were often found in the bottomlands and wooded islands as well as on the adjacent plains. Theodore Roosevelt (1893: 167) found elk along the Little Missouri, where, he said, they were very plentiful until 1881.

Apparently the elk had disappeared pretty generally from the State by 1881. Bailey stated (1926: 34, 35) that in 1887 he was "told by an old hunter at Larimore of two elk killed near there in 1881 or 1882, and at Devils Lake there were said to be still a few"; that "in 1915, Remington Kellogg was told of six elk killed in 1883 near Elkton [Elkwood], in Cavalier County"; and that in 1919 (field report, Oct. 1-4) "Mr. Eugene D'Heiley tells me that an old hunter killed the last he knew of not far north of Walhalla about 1889."

These dates probably mark the passing of the original North Dakota elk, though in 1913 Bailey (field report) wrote of a cow elk killed near Flaxton in 1906 and in 1915 Remington Kellogg (field report) stated: "A man named Hart killed one 2 years ago in the summer near Plaza. No one knows where it came from." These are curious occurrences that took place long after the elk generally had disappeared from the State.

OHIO

This State falls within the former range of the elk, although recorded data are not conclusive as to its abundance. Brayton (1882: 80) quoted from an 1838 history of Ohio as follows: "When Circleville was first settled the carcasses, or rather skeletons, of 50 individuals of the family of Elk lay scattered about on the surface." Brayton (1882: 185) presented also a letter from Judge Emory D. Potter of Toledo stating that elk disappeared from the State about 1828, whereas Kirtland (1838: 177) stated in 1838:

"The Elk was frequently to be met with in Ashtabula County, until within the last 6 years. I learn from Col. Harper of that county, that one was killed there as recently as October of the present season." These dates evidently mark the disappearance of elk from Ohio.

OKLAHOMA

From the sketchy records available, the early distribution of elk in Oklahoma appears to have been very scattered. Bailey (1905: 60) mentioned that elk had been reported in the Wichita Mountains in 1852 and that one was killed at Rainy Mountain, west of Lawton, in 1881. Apparently the Wichita Mountains were their favorite range.

OREGON

Oregon was well supplied with elk, chiefly the coastal form, *Cervus canadensis roosevelti,* which roamed all through the Coast Ranges and eastward to the Cascade Mountains. Some of the early narratives report on abundance of elk in the Williamette Valley, and others record the animals as occurring throughout the Cascades. The Rocky Mountain elk *(C. c. nelsoni),* which entered the State chiefly in the northeast corner, occupied the Blue Mountains, the Wallowas, and neighboring localities and were recorded as far west as the headwaters of the Deschutes River.

There are two reports of elk in southeastern Oregon—one in

the Steens Mountains in 1914 and one on the Blitzen River in 1876. These animals were probably stragglers, as the arid districts evidently were not normal elk range.

PENNSYLVANIA

Records indicate that elk at one time wandered over nearly all parts of Pennsylvania and that their favorite ranges were in middle sections of the State, particularly the Allegheny Mountains. Referring to Monroe and Pike Counties, Rhoads (1895: 389) wrote:

"The 'Elk' was probably never as numerous in this region as in the central Allegheny Mountains, those individuals taken in former days being considered by the natives as stragglers from the main body. The last capture in Pike County was probably not later than 1840 or 1845."

The same author (1898: 207) stated:

"The former range of this animal in Pennsylvania was closely coextensive with that of the Bison, both species using the same trails, feeding grounds and licks among the western Alleghenies and passing thence eastward by the same routes to the Delaware valley. The elk was most numerous among the elevated mountain glades and eastern tributaries of the Allegheny and Monongahela Rivers. It was also fairly abundant in the early part of the century in Clinton, Potter, Tioga and Lycoming Counties. The latter named regions formed the hunting grounds of my veteran friend, Seth I. Nelson, whose diary between 1831 and 1837 shows that he killed 22 elk during the period . . .

"The range of the elk and buffalo into the south central counties of Pennsylvania, east of Fulton County, is very improbable, if, indeed, they ever wandered that far. The main line of their eastern range on Mason and Dixon's line was probably along the valley of Castleman's River in Somerset County and the main ridge of the Allegheny mountains near that place, which formed a continuous trail of safety between their haunts in West Virginia and the Keystone State. North of this region their range probably spread northeastward as far south as the Juniata valley, but by far the largest number did not come south of the east and west branches of the Susquehanna . . ."

In a brief summary of the history of the elk in Pennsylvania, Gerstell (1936: 6) stated:

"The opening of the last century found the wapiti exterminated in southeastern Pennsylvania; rare west of the Allegheny River and in the Blue Ridge and Cumberland Mountain Ranges; numerous on the Pocono Plateau; abundant throughout the Allegheny Mountain and Plateau sections.

"By 1830, the archives show that the animals had disappeared from the southwestern section of the State, an area once among their favorite haunts. Extermination in the Pocono Plateau district in the diagonally opposite corner of the Commonwealth was completed between 1835 and 1845.

"During the 1840's and up until the early 1850's, a fair number of elk yarded and were annually hunted in those sections of Elk, Cameron, and McKean Counties lying between the headwaters of Bennett's Branch of the Susquehanna on the southeast and the Clarion River on the north and west. It was due to the presence of these animals in the region that Elk County received its name when established in 1843."

Continuing his account, Gerstell reports that in the winter of 1852 a herd of 12 elk were said to have yarded near the present town of Ridgway but 7 of them were killed and "a band of native elk apparently never again yarded within the State"; that from 1855 to 1865 elk were found in various parts of north-central Pennsylvania; and that in 1867 the last known native Pennsylvania elk was killed along the headwaters of the Clarion River by an Indian, Jim Jacobs. Gerstell noted that this animal killed by Jacobs appeared to be "also the last individual of the species to live in the vast Allegheny Mountain section of eastern North America."

Shoemaker (1939: 19) reported:

"In northern Pennsylvania, the Wapiti made one of the last stands in the east; unconfirmed reports tell of two bulls killed as late as 1878; but live specimens were seen by John G. Hamersley, a careful student of wild life, in 1869 and 1870. Colonel James Duffey, a leading Democratic politician of Marietta, Lancaster County, became alarmed at the threatened extinction of the Wapiti of Pennsylvania and, during the Civil War, collected a fine herd in his game park overlooking the Susquehanna. These animals were captured in the northern part of the state . . .

"A fine set of Pennsylvania Wapiti antlers mounted with the

hide is preserved in the Academy of Natural Sciences, Philadelphia. This Wapiti, according to Dr. Witmer Stone, was shot by Sherman Devens near Teutonia, McKean County, in 1853; according to others at Cherry Springs in Potter County, in 1858; but in either case 'Sherm' is credited as the killer."

Between 1912 and 1926 Pennsylvania imported or purchased locally 177 elk, of which 98 bulls were killed legally from 1923 until 1931. In 1940 about two dozen live elk remained in the State (Anonymous, 1936; Leffler, 1940).

SOUTH CAROLINA

Hays (1871) reported that elk were said to be plentiful in the Carolinas as late as 1737, and other writers referred briefly to their occurrence there. True (1883: 212, footnote), stated that the last elk in South Carolina was killed in Fairfield County.

SOUTH DAKOTA

South Dakota appears to be still another state from which the original elk have entirely disappeared. Available records, though not voluminous, are sufficient to show a state-wide distribution in primitive times and to indicate that in the early part of the nineteenth century, the animals were abundant especially in the Black Hills.

Visher (1914: 87) said of elk in Harding County: "The last were killed in 1879 when a large crew cut trees for the N. P. R. R. in the Long Pines." There are reports of three males killed near Box Elder Creek in 1874 (Winchell, 1875: 53) and of an elk killed 20 miles from Yankton in September 1881 (West, 1881), and a few elk were said to occur in the Black Hills in 1887 (Bailey, 1888: 434).

TENNESSEE

Ashe (1808: 267) stated that the elk was still present in Tennessee, chiefly in the mountains, in 1806; and Rhoads (1897: 180) reported that at the beginning of the nineteenth century it was probably a visitant to every county in the State, abounding in the high passes and coves of the southern Alleghenies, frequenting the licks near the present site of Nashville, and roaming through the glades and canebrakes of the Mississippi bottoms. He stated (p. 181) that a Mr. Miles wrote to him: "The last elk killed in West

Tennessee that I can learn of was at Reelfoot Lake about 1849 . . . In 1865 I heard that an elk was killed in Obion County."

TEXAS

The few available reports for this State indicate that elk at one time occurred in the Texas Panhandle. Bailey (1905: 60) was told by several old ranchmen that elk were originally found down over the Texas line in the southern part of the Guadalupe Mountains. Most of Texas, however, apparently was not in the original elk range.

UTAH

Records show that in primitive times elk were found in largest numbers in the northern Wasatch and the Uinta Mountains, yet they evidently occupied other parts of the State. Orange Olsen informed the writer that elk formerly ranged on the Mt. Nebo Range, but disappeared there about 1880. Presnall (1938: 18), writing of Zion-Bryce and Cedar Breaks, says:

"Elk at one time were native to this region, as shown by the Paiute name Paria (meaning 'elk') for the stream which drains from Bryce south to the Colorado; and also by the finding of a badly weathered elk antler in Willis Creek, Bryce, 1932. It is reported also that an elk was killed at Willis Creek by a Mr. Johnson, of Cannonville, sometime between 1900 and 1910. The subspecies is thought to have been *nelsoni*."

VIRGINIA

Beverly (1722: 135) stated that elk were found on the frontier plantations. Hale (1886) said:

"Formerly, the elk or wapiti, *Cervus canadensis* L., ranged over most of the United States and southern Canada, and is known to have occurred throughout most of Virginia, although it was found more abundant in the mountainous regions of the Allegheny and Blue Ridge Ranges. In 1666, twelve years after the discovery of the New River, Henry Batte, with fourteen Indians, started from Appomattox, near the present site of Petersburg, and in seven days reached the foot of the mountains. On crossing them they came to level, delightful plains with an abundance of game, deer, elk, and buffalo."

In 1670, Lederer wrote that he ". . . traveled through the

savannah among vast herds of red and fallow deer which stood gazing at us, and a little after we came to the promontaries or spurs of the Appalachian Mountains." He no doubt referred to white-tailed deer and elk.

According to Wood (ms.), Johnston (1906) wrote that on a hunting expedition in 1753, James Burke, one of the Draper's Meadows settlers, wounded an elk and followed it through what is now called Henshaw's Gap into the beautiful body of magnificent land which has since borne the name of Burke's Garden, Tazewell County. Burke moved into the "Garden" in 1754, but had to leave the area after the Indian uprising. He took with him large numbers of tanned hides from deer, elk, bear, and other game which he had killed while residing in the area.

According to Hale, the elk travelled over the country of the New River, making well-worn trails to mineral licks. Roy Wood (ms.) points out that the Ellwood Estate, 15 miles west of Fredericksburg, was originally called Elkwood because of the large numbers of elk previously found in the nearby forests. He also points out other Virginia place names, such as Elk Creek near Bedford, Elk Garden, Elk Hill, Elko, Elkton, Elkwood, and Elkcreek.

McAtee (1918) states that elk persisted in Virginia until 1844. According to Hays (1871) there were a few elk in the mountains of Virginia in 1847. In the Smithsonian Institution, there are two specimens collected by Colonel Gos Tuley of Clarke County during the years 1854 and 1855 (Baird, 1859).

WASHINGTON

Records of the Rocky Mountain elk *(Cervus canadensis nelsoni)* in Washington are scarce and sometimes obscure. Hallock (1892: 402) stated that there were elk in the Okanogan district, West of Spokane, in 1892; Symons (1882: 118) wrote of elk in the mountains near the present site of Spokane in 1881; and Dice (1919: 20) found evidence of former occurrence in the Blue Mountains in the southeast part of the State. Apparently elk range extended westward from Montana to include parts of eastern Washington. The plains of the Columbia were not occupied by elk, however, so far as can be told by published records, and there must have been a gap in elk distribution between the Rocky Mountains and the Cascades.

The Pacific coast of Washington, however, was occupied by *C. c. roosevelti,* which has persisted there to the present time. The Olympic Peninsula was the stronghold for these animals, but they ranged over the entire coastal forested area and also inhabited the Cascade Range to the east.

Although the original ranges of *C. c. nelsoni* and *C. c. roosevelti* were separated by territory unoccupied by elk, there was a good chance for limited intermingling of the two races through parts of southern British Columbia and adjacent parts of Washington. There may have been some contact, too, by way of the Columbia River Valley. This intermediate country, however, was not ideal elk range.

WEST VIRGINIA

In West Virginia, the elk was "once of rather common occurrence in our higher mountain regions" (Brooks, 1911: 12). Some of the early Indian tribes called Elk River, in West Virginia, "Tiskelwah," meaning River of Fat Elk (Hale, 1886). According to Hale, the last elk killed in the valley of the Kanawha was taken by Billy Young on Two Mile Creek or Elk River about 5½ miles from Charleston, West Virginia, in 1820. However, Audubon and Bachman (1846: 93) stated: "On a visit to Western Virginia in 1847, we heard of the existence of a small herd of elk that had been known for many years to range along the high and sterile mountains about forty miles to the west of the Red Sulphur Springs . . ." Brooks (1932: 1) said that there were records in "Border Settlers" of the killing of an elk at Elk Lick, Middle River, Pocahontas County, in 1867 and of a fresh elk track on Cheat River as late as 1873. Shoemaker (1939: 22) declared that "As late as 1875 Wapiti were found at the headwaters of Tygart and Greenbrier Rivers." These were undoubtedly the last occurrences in the State.

WISCONSIN

Elk occurred throughout Wisconsin and were present in the State until almost the close of the nineteenth century. Jackson (1908: 15) stated:

"The elk is without doubt now extinct in Wisconsin, but cast-off antlers scattered throughout the lakes, marshes and woods of northern Wisconsin attest of its former occurrence there. I have examined antlers of *Cervus canadensis* found in Ashland and Iron Counties."

There are reports that the elk was present on Hay River in 1863 (Hoy, 1882: 256) and in the vicinity of Green Bay in 1878 (Brayton, 1882: 80); but during the geological survey of 1873-79, according to Strong (1883: 347), it occurred "very rarely in northern and central Wisconsin." This probably marked the end of the original elk distribution in the State.

WYOMING

Wyoming was probably more densely populated with elk than any other state. At any rate the herds, which were widely scattered over the State, persisted longer in its mountain ranges and in greater numbers, remaining until public sentiment brought about protection and restocking of many of the early ranges, thus assuring the preservation of the species and recovery of total elk population to the present high level.

It is worth while to consider with some care the early distribution in this important elk range. As centers of abundance may be mentioned the Black Hills of northeastern Wyoming and part of South Dakota; the Big Horn Mountains; the Laramie and Medicine Bow Mountains; the foothills of the Uinta Mountains; the Shirley, Ferris, Green, and Granite, and Rattlesnake Mountains; and finally the extensive mountain ranges about Jackson Hole and Yellowstone National Park, including the Salt River Range extending south along the western boundary.

BLACK HILLS

According to field reports of Vernon Bailey (1913) and Merritt Cary (1912), elk were at one time abundant in the Black Hills and Bear Lodge Mountains of northeastern Wyoming but by the middle 1880's had become very scarce. Cary's report stated that in the winter of 1884-85 a few elk ranged in the limestone country of the Black Hills and thence northwest to Sundance Mountains, and that in 1885 one elk was killed on Sundance Mountain and another in the region at the head of Stockade. Beaver, and Inyankara Creeks. Elk were reintroduced in this region in 1912 and 1913.

BIG HORN MOUNTAINS

Writing from Fort McKinney of a hunting party organized to supply the garrison with game, Hatch (1886) stated:

"The Big Horn Mountains contain large game in abundance, though the frontier hunters consider it scarce and have given up hunting for a living . . . The heads of the many tributaries of Powder River were first visited, when it was apparent the game had been driven out of the high ranges to lower ones."

Hatch's party then proceeded southward to the Casper Mountain, where good hunting was found, 72 elk, 102 blacktail deer, and 45 antelope being killed for the garrison. From this it would appear that elk customarily occurred along the highlands from the Big Horn to the Casper Mountains.

In 1894, Kellar (1894) reported elk high in the Big Horns and Bailey recorded (field report) that elk horns were common all along the route from Belle Fourche Valley to Sheridan, particularly in the valleys. Two years later elk were reported at Ten Sleep.

In 1909, Cary (field report) stated that elk at one time had been common in the Casper Mountains but that he was informed none had been seen there since about 1904.

By 1910 elk had become scarce. Apparently none remained in the No Wood River Valley (D. D. Streeter, Jr., field report); but a few, at least two bands, were reported in the Big Horns west of Buffalo, one of these in the Cloud Peak region (Cary, field report). At that time, however, elk had been reintroduced in the northern part of the Big Horns.

LARAMIE MOUNTAINS

According to accounts of old-timers, Cary reported (1909 field notes), the Casper and Laramie Mountains were at one time among the best elk ranges in Wyoming. The large heaps of elk antlers that Cary found at most of the ranches from Laramie Peak north to Douglas and from Laramie Peak west to Toltec testify to the truth of the assertion.

Wyeth (1899: 156) reported an elk killed along the North Platte River, above Laramie River, in 1832; Parkman (1924: 126) saw a herd of about 200 elk near Fort Laramie in 1846. Bailey (field notes) recorded that in 1888 elk were said to be quite common in the Laramie Mountains between Douglas and Rock Creek but that in 1890, during a trip through the same region, he was informed that they were then scarce there but had been very numerous several years earlier; he found a great many antlers scattered about. In 1909, Cary reported (field notes) that the elk had disappeared from most of the Laramie Mountains many years before but that a very few scattered elk still remained in the mountains at the head of Labonte Creek, and that in the winter of 1902 a lone individual had wandered into the hills north of Guernsey. He also noted that the few elk that had inhabited the Pole Mountain vicinity near Laramie had disappeared and that those formerly in the region about Rawhide Butte, near Lusk and other neighboring localities, had been absent for many years.

MEDICINE BOW MOUNTAINS

Elk were at one time common in the Medicine Bow and Sierra Madre Mountains. Ludlow (1870) stated that they were "abundant" in the Elk Mountains about 1864, and Allen (1875: 60) also reported them "abundant, particularly about Elk Mountain, and neighboring portions of the Medicine Bow Range" in the early 1870's. As early as 1888 Vernon Bailey recorded in field notes that the elk had largely disappeared but were still to be found on Elk Mountain. In 1911, H. E. Anthony wrote in a field report that the elk were extinct in the Medicine Bow Mountains but that cast off antlers, as well as statements of old trappers, testified to their former abundance there.

FOOTHILLS OF THE UINTA MOUNTAINS

Crampton (1886) reported that elk come down into the northern foothills of the Uinta Mountains "when the snowfall in the mountains is great"; but in 1911, there was no evidence of these animals there. No doubt the headquarters of these elk was in the Uinta Mountains of northern Utah.

GREEN MOUNTAINS AND NEIGHBORING RANGES

These more or less isolated groups of mountains were formerly occupied by elk. The high Ferris Mountains were at one time a favorite elk range. Ingersoll (1883: 138, 139) wrote that during a trip in June 1877 his party noted bands of elk at the foot of the Seminole Mountains, and he stated as follows:

"From Camp 4 to 5 the road led over sage-grown ridges along the southern foot of the Sweetwater hills, the direction being westward, and the distance about 15 miles. Elk in small bands, several accompanied by young calves, were visible all along, and antelopes were constantly in sight . . ."

The Shirley, Rattlesnake, and Green Mountains were once elk range, but by 1909 Cary (field report) found the animals scarce in all these places. At that time he was informed that until a few years earlier small bands of elk had passed through the Granite Mountains each winter on their way to the Green and Ferris Mountains, and he found that a very small band of perhaps six or seven still ranged along the upper slopes and crest of the high Ferris Range. Apparently this very small group was the last remnant of the herds that once roamed through these mountains.

JACKSON HOLE REGION

Under Jackson Hole Region are discussed the various mountain ranges surrounding Jackson Hole, including Yellowstone National Park, the Tetons, Wind River Range, and all other tributary areas involved in the ancient travels of the Jackson Hole elk herd. Of these the Yellowstone National Park-Jackson Hole area is the center of abundance of the great elk herds of the region.

On the east the Absaroka Range, the Owl Creek Mountains, and the Wind River Range were all occupied by elk that today are still represented by the Shoshoni herd on the crest and eastern slope of the Absarokas, a small band in the Owl Creek Mountains, and a herd in the Wind River Range on the headwaters of Green River.

The rugged Teton Range, west of Jackson Hole, was long occupied by elk, which in 1910 Cary (field notes) reported "abundant" at the northern end of the Tetons. In Webb Canyon and the neighboring slopes Cary found their well-worn trails crisscrossing in all directions "from the dense willow thickets along Moose Creek up to the limit of trees at 10,000 feet."

In early days elk ranged through the Caribou Mountains of Idaho and the Wyoming and Salt River Ranges of western Wyoming and occupied the Hoback Mountains and the Hoback Basin.

This discussion has been concerned chiefly with mountain

ranges. Elk also frequented the lowlands, especially in winter. They were known to winter on the flats about Pinedale, the sagebrush plains at Big Piney, and the Red Desert itself. In 1868, Lexden (1914: 333) saw a large band of cows and calves near Bitter Creek; and in 1911, H. E. Anthony reported (field notes) that elk were said to winter near Fontenelle.

ALASKA

It is of interest to note also that fossils prove that wapiti occupied interior Alaska at a time when the mammoth, horse, and ancient bison ranged the north country. Apparently elk were not present in Alaska after the Pleistocene until reintroduced on Afognak Island.

CANADA

The Roosevelt elk *(Cervus canadensis roosevelti)* ranged north into the coastal part of southern British Columbia, especially Vancouver Island. Records do not indicate that this elk went farther north on the mainland.

The Rocky Mountain elk *(C. c. nelsoni)* had a much greater distribution. It ranged north through eastern British Columbia at least as far as the upper Liard River, and through much of Alberta, as far as Fort Smith on Slave River. In Saskatchewan, it reached the headwaters of Churchill River and Cumberland House, and in Manitoba the north end of Lake Winnipeg. From that point the northern boundary of its range swung sharply southward through eastern Manitoba to the United States boundary in northern Minnesota. It ranged through the southeastern part of Ontario, between Lakes Huron and Erie, as far east as Ottawa, and there is a curious record for Ontario of elk 200 miles northeast of Lake Nipigon.

In Quebec there are some indefinite records of elk north of the St. Lawrence River.

5

ELK HABITAT

ELK AS A GENUS are tolerant of diverse environments, as shown by their original wide distribution over the American continent and by their varied habitats today. The remaining herds, of several described forms, occupy the highlands of the Rocky Mountain area, the more humid and hotter country of Oklahoma, the arid mountains of Arizona, the dry valleys of California, and the cold more northern parts of western Canada, as well as the dense, cool, rain forests of the Pacific coast. Some of these habitats, it is true, are occupied by different subspecies, but many localities diverse as to climate and vegetation have been restocked successfully from the Rocky Mountain herds about Yellowstone National Park. It is well, however, to discuss habitat separately for the individual forms.

CERVUS CANADENSIS NELSONI

It has been assumed that the Rocky Mountain elk were different from the eastern elk, but the exact location of the region of intergradation between these two subspecies cannot be known now. For the purpose of defining the habitat preferences of the Rocky Mountain form, it is safe to include the States of Wyoming, Montana, Idaho, Utah, and some neighboring areas. Special reference is made to the Jackson Hole and northern Yellowstone, Wyo., elk herds, on which the basic studies were made.

Today these elk are primarily mountain dwellers. Practically nowhere do they occur on the plains. Yet records of the early days state that at times elk were noted on the plains in great numbers. Those records so commonly mention elk associating with bison that the thought has developed that the elk is primarily a plains animal which in early times did not inhabit the mountains but has been driven there, to an unnatural home, in comparatively recent years by advancing civilization. To support this contention is the undisputed fact that formerly hordes of elk lived on the plains. Moreover, many early travelers failed to find elk, or at any rate failed to mention them, in certain mountain areas; and some even stated positively that game was scarce.

Unfortunately, the fact of migration often was overlooked, or at least was not discussed, and it is not always clear whether the plains animals were already on winter range or were in transit to or from it. Now it is known that the elk went from the mountains to the plains each winter in great numbers, sometimes on long, well-organized migrations. It seems natural that in the early days of great abundance of game an animal so widely distributed as the elk and tolerant of such a diversity of habitats would occupy the plains to some extent even in summer.

Skinner (1927:169) discussed the early dearth of elk in the Yellowstone National Park, quoting other writers who agreed that big game was scarce in the mountains in primitive times, and he and Rush (1932) both quoted many reports of park superintendents that showed a steady increase of elk in the late 1870's and 80's, the increase presumably resulting from protection after the elk had been driven to the mountains.

This matter of supposed change of environment from plains to mountains is of sufficient importance to deserve special consideration. There is much evidence to show that in early times the elk was more generally a dweller of woodlands and mountains than has been supposed.

Seton (1929, vol. 3, pt. 1: 8) stated:

"After this date [1653-54], the numbers of travelers increased in America, and their accounts frequently included descriptions of the 'Great Stag that was of the bigness of a Horse,' and whose numbers were so great, in the high country, that their trails through the woods were convenient ways of travel."

Lewis and Clark (1893, vol. 3: 845) stated of elk: "They are common to every part of this country, as well the timbered lands as the plains, but are much more abundant in the former than in the latter." They also recorded (1893, vol. 2: 537) that on August 25, 1805, several miles below the confluence of the Lemhi with the Salmon River, a herd of elk was seen among the pines on the mountainside.

Washington Irving's "Captain Bonneville," recording Bonneville's travels between 1832 and 1835, contains a number of references bearing on the mountain habitat of elk. Time and again in writing of the game of a country he grouped the elk, deer, and bighorns together. Speaking of a camp somewhere on Bear River (p. 190) he stated: "They remained, therefore, almost starving in their camp, now and then killing an old or disabled horse for food, while the elk and the mountain sheep roamed unmolested among the surrounding mountains." On December 19, 1833, on the north fork of Salmon River Captain Bonneville's party found "numerous groups of elk" and "large flocks of the ahsahta or bighorn." In this instance, finding elk in the mountains, possibly in lower portions, even in winter, parallels present-day conditions in the Jackson Hole region, where many elk winter in the mountains.

In 1834, Bonneville found the "Grand Ronde Valley of Oregon a good pasturing ground for his horses in winter, "when the elk come down to it in great numbers, driven out of the mountains by the snow" (p. 5, pt. 2). Again, the author said (p. 189, pt. 2): "By hastening their return they would be able to reach the Blue Mountains just in time to find the elk, the deer, and the bighorn."

Osborne Russell (1921) related in diary form numerous observations pertiment to elk habitats. On December 20, 1840, on Weaver's River, near Great Salt Lake, Utah, he recorded: ". . . we also found large numbers of elk which had left the mountains to winter among the thickets of wood and brush along the river."

On January 10, 1841, while camped near Great Salt Lake, he went into the mountains for several days of elk hunting and while at "Ogden's Hole" on Ogden Fork, where the snow was 15 inches deep, wrote: "Towards night the weather cleared up and I discovered a band of about one hundred elk on the hill among the shrubbery." He killed a "very fat doe."

In his diary of 1834, Russell wrote of Ham Fork of Green River:

"The face of the adjacent country was very mountainous and broken, except the small alluvial bottoms along the streams. It abounded with buffalo, antelope, elk and bear and some few deer along the river."

In 1835, at Pierre's Hole (now known as Teton Basin, Idaho), he wrote: "This was a beautiful valley, consisting of a smooth plain intersected by small streams and thickly clothed with grass and herbage and abounding with buffalo, elk, deer, antelope, . . ."

He stated again:

"We stopped there [Fort Hall] until the 1st of January, 1839, when we began to be tired of dried meat, and concluded to move up the river to where Lewis's Fork leaves the mountain and there spend the remainder of the winter, killing and eating mountain sheep." On January 20 he wrote: "We found the snow shallow about the foot of the mountain with plenty of sheep, elk and some few bulls [bison] among the rocks and low spurs." On a trip up Snake River Canyon to hunt elk, his party found a band half a mile up the mountain. Here he found elk and sheep on the same winter range, partially in the rugged Snake River Canyon.

Russell visited Jackson Hole several times and on July 2, 1835, wrote: "This valley, like all other parts of the country, abounded with game." Again, July 28, 1836, at Jackson Lake, he wrote: "Game is plentiful and the river and lake abound with fish."

These last few statements indicate clearly that elk, bison, and antelope sometimes occupied the same or adjacent ranges even in summer, further evidence that elk did occupy the lowlands in summer also.

Russell's references to what is now the Yellowstone National Park proper—which he visited year after year, and which has been considered gameless prior to the 1870's—should be examined.

On August 19, 1836, while at the north side of Yellowstone Lake he wrote:

"This valley was interspersed with scattering groves of bull pines, forming shady retreats for the numerous elk and deer during the heat of the day . . . It being very difficult to get around these places, we concluded to follow an elk trail across it for about half a mile." He recorded killing a fat elk here.

On August 5, 1837, as this trapper sat viewing the scenery from a mountain a few miles northeast of Yellowstone Lake, he soliloquized in part as follows:

". . . these stupendous rocks, whose surface is formed into irregular benches rising one above the other from the vale to the snow, dotted here and there with low pines and covered with green herbage intermingled with flowers, with the scattered flocks of sheep and elk carelessly feeding or thoughtlessly reposing beneath the shade, having Providence for their founder and Nature for shepherd, gardener and historian."

A week later, August 12, he made the following journal entry: "We hunted the branches of this stream [Shoshone River], then crossed the divide to the waters of the Yellowstone Lake, where we found the whole country swarming with elk. We killed a fat buck for supper and encamped for the night."

In the summer of 1839 he again visited the Yellowstone Park area and said:

"Vast numbers of black-tailed deer are found in the vicinity of these springs and seem to be very familiar with hot water and steam, the noise of which seems not to disturb their slumbers, for a buck may be found carelessly sleeping where the noise will exceed that of three or four engines in operation."

In August of the same year at Yellowstone Lake, he wrote: "Game was very scarce on the west side of the lake," but he did kill a cow elk there. At another part of the lake, though, while lying wounded and watching a band of hostile Blackfeet Indians, he observed: "Then they [the Indians] began shooting at a large band of elk that was swimming in the lake, killed four of them, dragged them to the shore and butchered them, which occupied about three hours."

At a later date, 1870, Hayden (1872: 129-130) found elk quite abundant in the Elk and Sheephead Mountains of Wyoming, especially in the valleys of small mountain streams. In reporting a tour to North Park made in August 1868, he stated (p. 126) that aspen groves in the moist ravines of the lower mountain ridges of the Medicine Bow Range were the favorite resort of elk, deer, and other game, a statement which is largely true for elk today. In September 1872 Bradley (1873: 254) saw two herds of about 20 each in the groves on the top of a ridge along the upper Snake River, just above Mount Hancock. On September 6, 1870, according to Doane (1871: 21), at the southeast angle of Yellowstone Lake "the ground was trodden by thousands of elk and sheep." On September 1 he had seen elk feeding in small bands in the Yellowstone valley above the canyon.

These early accounts reveal conditions similar to those of the present, except that today the numbers of animals are greatly reduced in most localities. In earlier times, on the prairies and in the mountains, whites and Indians alike sometimes went hungry because of scarcity of game. Probably this was a local circumstance, caused by movements of the animals. Even today, in the very heart of the elk country the animals may be temporarily scarce in certain localities, so that to one unfamiliar with seasonal and local travels of the herds the country may appear largely devoid of game. Also to be considered are possible fluctuations in numbers, the extent of which among our larger mammals is little understood.

It has often been stated that in primitive times elk did not winter in Jackson Hole, but there is good evidence that they did (Plate 3). A. P. Nelson, in charge of the National Elk Refuge, some years ago interviewed the late Emil Wolfe on this question. Wolfe had first come into Jackson Hole some time in the 1870's and camped one winter in the upper part of the valley. He found elk wintering at various points, including parts of the Snake River bottoms and the vicinity of Deadman's Bar. The greatest numbers were to be found near the south end of Jackson Hole. In fact, Wolfe said his impression was that the elk were distributed then about as they are today in this area.

There is a somewhat parallel situation in the case of the bison, which is even more of a plains animal than the elk. Early hunters spoke of the "mountain buffalo," and Dodge (1877: 144) wrote:

"These animals are by no means plentiful, and are moreover excessively shy, inhabiting the deepest, darkest defiles, or the craggy, almost precipitous, sides of mountains inaccessible to any but the most practiced mountaineers." Fryxell (1926, 1928) accumulated much evidence to show that bison inhabited the Rocky Mountains, even to elevations above timber line. He quoted (1928: 138) a communication (dated October 7, 1926) from Dr.

PLATE. 3. *Above.* Habitat of introduced Rocky Mountain elk in Guadalupe Mountains, N. Mex., May 8, 1939. At one time this was probably occupied by Merriam elk. *Below.* One type of winter range in Jackson Hole, Wyo. In such situations are found grasses, rabbitbrush, and other short browse species relished by elk. (U. S. Fish and Wildlife Service photographs.)

George Bird Grinnell, which said in part: "The bison commonly ranged in and occupied the mountains, and even the high mountains, all through the year." Bison are known to have ranged in Jackson Hole, at least in summer, and bones found in cut banks indicate that such occupancy antedated the "great slaughter" on the plains.

To return to the elk—In Idaho, too, the elk were distributed in mountain ranges, according to Merriam (1891: 80), who wrote:

"Common [1890] in the Saw Tooth and Pahsimeroi Mountains, and not rare in the Salmon River Mountains; occurs also in the Brunneau and Elk Mountains in extreme southern Idaho, and is said to inhabit the Blackfoot Mountains.

"In 1872 we found Elk in abundance in the region about Henry Lake."

Apparently these animals had not been driven to the mountains from the plains, either, because when some of Bonneville's party traveled through the Snake River plains of southern Idaho in the 1830's, they found no elk (Irving, 1843). In fact, there are no records available to show that these plains were ever inhabited by elk.

From the foregoing it may safely be concluded that the elk have always been at home in the mountains as well as on the plains. It is reasonable to suppose that the favorable ecological niches of the high mountain meadows with their luxuriant forage would be occupied by big game, and records show this to have been the case. The plains elk, both those that spent the whole year in the open country and those that only wintered there, naturally would be destroyed first, as they were so accessible; and the destruction of mountain elk while on the winter range on the plains could very well account for the relative scarcity of these animals even in the high mountains in the few years immediately after the so-called great slaughter. Furthermore, the evidence here presented does not preclude limited migration of even the plains-loving elk to the mountains. The latter animals were often found in timbered bottoms or in brushy draws in the vicinity of streams or hills, and they did not restrict themselves to the open to the same extent as the pronghorn. Whatever the sequence of events was, the herds now living in the mountains are undoubtedly the descendants of elk that were originally mountain dwellers.

CERVUS CANADENSIS CANADENSIS

The eastern elk disappeared very early. The available records show that in habitat choice eastern elk resembled the Rocky Mountain animals in that the former occupied mountain ranges as well as lowlands. In early times elk were particularly common in the Allegheny Mountain area in Pennsylvania and occurred in the uplands of the Appalachians in West Virginia and in the forests of Michigan. They were widespread throughout Indiana and Illinois, where they certainly occupied the prairies as well as available woodlands, and occurred in the canebrakes of the Mississippi bottoms in Tennessee. They were numerous on the prairies of Iowa and were distributed over the other Prairie States of the Midwest. Thus the eastern elk also were tolerant of a variety of habitats, although they apparently preferred the mountains and were more plentiful in woodlands than on open prairie.

CERVUS CANADENSIS MANITOBENSIS

Boundaries between the ranges of *Cervus canadensis canadensis*, *C. c. nelsoni*, and *C. c. manitobensis* cannot be drawn. Apparently, from what is known of habitat and distribution, *manitobensis* shares with the other forms their choice of environment and tolerance of varied terrain.

CERVUS CANADENSIS MERRIAMI

The now extinct Merriam elk ranged through the Sacramento, White, Gaudalupe, and Mogollon Mountains of New Mexico and in the White Mountains of Arizona and appears to have been primarily a mountain species. (See quotation from Mearns given on p. 21.). Even this far south, it seems, the Merriam elk still clung to the general habitat preference of other elk (Plate 3).

CERVUS CANADENSIS ROOSEVELTI

The Roosevelt elk is an animal of the heavy rain forests of the Pacific coast. It, too, seeks the mountains but is not confined to them and thrives in lowland forests as well. In Oregon in early days it inhabited not only the western slopes of the Cascade Range but also the valleys west of the range, its distribution in the State thus corresponding closely to that of the coast black-tailed deer (Bailey, 1936: 81-83, 87).

CERVUS NANNODES

The so-called valley elk, or tule elk, of California, occupied the San Joaquin and Sacramento Valleys and apparently did not penetrate the chaparral or woodlands of the adjacent mountain slopes. It chose the arid plains and was probably more truly a plains animal than any other American elk. According to Van Dyke (1902: 168), when the elk were being slaughtered during the gold rush days of California, those that escaped sought refuge in the tule marshes along the streams and lagoons. The valley elk may be considered the most specialized of all elk in habitat choice and the one, apparently, with the greatest tolerance for desert conditions.

OLD WORLD DEER OF THE GENUS CERVUS

It is of interest to note here the habitat choice of some of the Old World members of the genus *Cervus* that are still found in wild country. Sowerby (1937: 250) stated that the Manchurian wapiti, which he designated *C. canadensis xanthopygus,* was formerly very plentiful "in the forested areas of Manchuria, the Amur, Transbaicalia, the Ussuri, the Primorsk and Northern Korea" and that the North China wapiti *C. c. kansuensis* is "found in the mountains of Suiyuan Province on the Mongolian border of the Shansi, westward . . ." On the other hand, a north African race of the red deer, *C. elaphus barbarus,* is said to occupy open, treeless country in Tunisia and Algeria. It appears that the genus as a whole is highly versatile and adaptable.

6

LIFE ZONES

CONSIDERED on the basis of their continental range as a whole, the early elk may be said to have occupied mainly the Boreal and Transition Zones. In the mountainous parts of their distribution areas they sometimes ranged well up through the Hudsonian and even occasionally into the Arctic-Alpine Zone, a few even wintering there. But at the northern fringes of their distribution, in Canada, they barely approached the edges of the Hudsonian in a few places and in eastern Canada did not penetrate far into the broad, transcontinental Canadian Zone.

Within the United States particularly, early elk range lay largely in the Transition, although in the West the animals ranged to timberline and sometimes above. In some states elk occupied the Sonoran Zones, particularly in those Middle-western and Eastern States in which the Boreal or Transition Zones are not available over large areas. Southern habitats evidently were not favorable for elk. The animals did not thrive, apparently, in desert or semidesert areas. Most of Texas, New Mexico, Arizona, and Nevada had few elk, and there are no records of any for the hot plains of southern Idaho and eastern Oregon and the dry plains of the Columbia in eastern Washington.

Even in such southern localities as parts of New Mexico and Arizona, the Merriam elk confined itself largely to the Boreal and upper Transition Zones, coming down to the Upper Austral only in winter. Elk are much more tolerant of cold than of warm climates and push north until the physical barrier of snow stops them. It is of interest to note that wapiti antlers were found in the Pleistocene of interior Alaska. The animals entered this continent from the north during what must have been a favorable climate but later were cut off in that direction by climatic change.

The elk have reached their highest development in the temperate climates and have not yet become adapted to the desert, but that there has been some tendency toward desert adaptation is evident in the valley elk of California. The result has been smaller size and a pale, suncurled pelage, a somewhat inferior animal by ordinary standards.

Caton (1877: 282) made a good general statement concerning elk distribution:

"They do not confine themselves to a limited range, but are liable to roam over extensive districts of country: now high up the mountains, again in the deep canons or fertile valleys, and again, far out on the plains along the borders of some water-course."

7

MIGRATION

ON THE WHOLE, the elk is a migratory animal, but not much can now be known in that respect about the eastern elk. It is probable, however, that in many parts of the eastern United States there was little drift of elk from summer to winter range, although undoubtedly there were local wanderings. Where altitudinal changes were slight, there would have been little occasion for well-defined migration to any considerable distance.

Consider Iowa, for instance. Because of the uniformity of its terrain and the fact that its average winters are not too severe, normally there would be no reason for migration. In the unusual winter of 1855-56, as quoted from Spurrell (p. 24), the snow was so deep that the people of Greene County killed great numbers of elk with clubs. This would indicate that the animals were not in the habit of migrating and so were caught at a disadvantage when a severe winter occurred. Based on information obtained from pioneers of Sac County, Spurrell (1917: 275) reported as follows:

"The elk scattered out in summer time but in October herded together, remaining in herds until spring . . . In case of storms in winter they took refuge in reed and rush grown ponds, where the reeds and rushes were 10 feet or more in height."

It was in the Rocky Mountain region that elk developed well-defined migrations. This is perfectly logical, as the high mountain ranges attractive in summer are unavailable under the deep snows of winter and the favorable lowlands are within easy reach.

The most striking and best-known migration occurred in the Yellowstone National Park-Jackson Hole region. The so-called Southern Herd, which in recent years has numbered some 20,000 animals, spent the summer in the highlands about Jackson Hole, most of them about the southern boundary of Yellowstone National Park but some north of the boundary, about the headwaters of Snake River and the Pitchstone Plateau, and still fewer in the Upper Yellowstone area.

On the approach of winter some of these elk moved into Idaho, into the Henry Lake area at the north end of the Teton Range; over the range by way of Mosquito Creek farther south; and out into Snake River Canyon. According to early reports, the principal migrations into Idaho took place by way of Conant and Bitch Creeks, north of the Teton Range. Some drifted into Star Valley. A large herd left the ranges at the north end of Jackson Hole, some working south along the Teton Range, some down the valley itself, and many south through the highlands bordering Jackson Hole on the east. A few remained behind on summer range, wintering in small numbers here and there—on a windswept mountain where snow did not accumulate or in some other favorable location. Many drifted into the upper Gros Ventre Basin and remained there. Many others went into southern Jackson Hole in what is now South Park, where abundant willows and grasses furnished winter forage.

Large bands proceeded farther south—some through the Gros Ventre Basin into the Green River country; others down Granite Creek, or up Cache Creek, then down Granite, especially through South Park, then up the Hoback, through the Hoback Basin (Fall River Basin) and on to the sagebrush plains. By these various routes hordes of elk moved down to the sage plains in the Green River drainage just southeast of Jackson Hole, and usually they remained on these northern plains and in the willow bottoms of Green River unless deep snow forced them to move farther south. In such severe winters some of them actually reached parts of the Red Desert in southern Wyoming. As stated on page 6, in 1868 Lexden saw a band of cows and calves near

Bitter Creek and in 1911 H. E. Anthony reported that elk were said to winter near Fontenelle. Many old-timers testify to the fact of this early migration of elk into the Red Desert region.

Other elk drifted onto these same plains, particularly near Big Piney, from the Wyoming Range on the west; and from the southern end of the Wind River Range elk came into the lowlands near Pinedale and others passed on through the Granite, Green, and Ferris Ranges. There must have been considerable mingling of herds in winter.

Some old-time residents of Wyoming recalled that the elk preferred the foothills at the edge of sage plains, the aspen ridges and favored slopes, and did not necessarily seek the open plains in great numbers. On the east the Absaroka Range, the Owl Creek Mountains, and the Wind River Range were all occupied by elk herds, which in winter ranged on the eastern lower slopes or the lowlands from the vicinity of Cody and Meteetse to Dubois and Lander.

The so-called Northern Herd of Yellowstone National Park migrated northward, finding winter range in the Yellowstone Valley and adjacent mountain slopes.

In the Big Horn Mountains, at the northern boundary of Wyoming, elk move northward and go into the low Garvin Basin of Montana for the winter. In many other parts of the country, too, there were formerly local migrations, most of which have been unrecorded because of the early disappearance of elk. Captain Bonneville, in 1834, found good pasturage in the Grande Ronde valley of Oregon in winter, "when the elk come down to it in great numbers, driven out of the mountains by the snow."

THE CAUSE OF MIGRATION

The reason for migration is apparent after one observes the animals over a period of years. The elk simply move to find new feeding grounds. All the evidence obtainable, including all field observations, indicates that in the fall snow is the chief causative factor of elk migration.

The Jackson Hole herd is an excellent one to study because it makes the most striking migration on record. During summer these elk graze the mountain meadows and parks but even then are not entirely confined to limited areas; there are local movements of the various bands. As changes in the vegetation occur, especially after the first frosts, there are decided shifts to new feeding places. Then in September comes the rut and the movements become more animated and widespread. These are not, however, migratory movements as the term is usually understood, for the elk are still on summer range.

Normally fall migration does not take place until after the rutting season. The date varies greatly, being dependent on meteorological conditions. In fact, migration is so dependent on the weather that hunters await the autumn snows that drive the elk down into open territory. Sometimes in October there is a snowfall in the mountains heavy enough to affect the availability of the forage. Then the elk begin to move promptly. If the snow does not continue to accumulate, however, only a few elk trickle over the migration route. Usually the heavier snows do not come until some time in November, and occasionally there is an open winter when the elk remain in the mountains until a much later date. At any rate, when the summer pastures are irrevocably snow-buried, the herds string along in scattered bands over their ancestral routes to the lowlands or to other favorable winter pastures where forage is available.

The spring migration of the Jackson Hole herd is largely the reverse of the fall migration. In spring, however, there is an apparent eagerness to get back to summer range, as many elk will follow the snow line back into the mountains at a time when the valley bottom is green with abundant new growth and new vegetation in the mountains is still scarce. Quite a few of them even wade through deep snows over extensive areas to reach favorite mountain slopes beyond. All the elk do not move so promptly. Evidently there are individual differences, for the migration extends over a period of some weeks and bands of elk may be found in the lowlands long after the main herd has gone.

Possibly there is greater palatability in the very newest vegetation—difficult for man to measure but detected by the elk—that lures the animals upward in the wake of the retreating snow; or perhaps there is a stimulant in the early spring atmosphere that creates an impulse to travel—and travel would naturally be over accustomed routes; or maybe there is actual nostalgia for remembered summer pastures.

The movements of another race in another locality are somewhat different. The Roosevelt elk in the Olympic Mountains, Wash., do not make an extensive migration—for instance, from the high mountains to the lowlands bordering the sea—although this would undoubtedly be to their benefit in some winters. There is some migration, though, from the higher mountains to the valley bottoms and lower slopes. This is dependent on the snowfall, and if the snow disappears early, many of the elk again scatter over the range. Here again, however, not all the elk move down when the heavy snows come; some, particularly old bulls, seek out favorable bare, wind-swept mountain ridges and winter there.

The writer spent the fall and winter of 1916-17 in the Olympics. In the autumn elk were scattered through the mountains, but when the permanent snows came most of the elk forsook the higher country about the Elwha River drainage and shifted to the vicinity of the snow line. Later a similar movement was noted on the Hoh River. Once they had gained the river bottoms, however, these elk apparently failed to go downstream any great distance though in some winters they could readily have escaped severe and fatal snow conditions by doing so and thus getting out of the heavy snow belt.

There is some evidence that in earlier times longer migrations downstream took place. Suckley and Gibbs (1860: 134), for example, stated:

". . . Near the last locality [Sekwim Bay] they are very abundant during the winter, being driven down by the snows on the mountains. They run in large droves, following well beaten trails, and at that season are an easy prey to the hunter. In January, 1857, two men in the vicinity of Sekwim bay killed eleven fine elk in one day."

J. G. Cooper, in the same report, says:

"In severe winters, also, when the elk leave the mountains, and in large herds descend to the warmer prairies along the coast, they are tracked in the snow to their lairs, and shot. Many frequent these prairies every winter, returning in early spring to the mountains."

The Merriam elk of Arizona and New Mexico had a vertical migration between the summer range in the Boreal Zone of the mountains and the winter range, at lower elevations, in the lower Transition and Upper Austral Zones.

THE MIGRATORY HABIT

It is difficult to specify the psychological bases for animal actions, yet there are many suggestive circumstances. It is clear from the above discussion that the time of elk migration is governed not so much by instinct as by changes in the physical environment in fall and by the coming of new vegetation in spring. These events occur more or less regularly on the average, but there may be great variation from the average at times. It is noteworthy that in some years, when snow is very late in coming (as it was in the autumn of 1944) and the main body of the elk herd lingers back in the mountains, a few elk come down to the lowlands more or less at the usual time, even though the highlands are still bare of snow.

Aside from the time, though, why do not all elk migrate? Why do some settle for the winter after a short migration? Why do some stop in the Gros Ventre Basin in Wyoming and others go down into the lowlands farther on? It seems worth while to attempt an explanation based on available observations.

Over a period of years elk have wintered regularly on the government Elk Refuge at Jackson, Wyo., and in the Gros Ventre Basin, and on many occasions these elk have been counted by districts. The counts show that, roughly, about the same proportion of either herd winters in each well-known winter range every year. In other words, a given herd presents essentially the same distribution pattern each winter.

Moreover, people who have fed the elk on the refuge have been able to recognize certain animals by some abnormality year after year. Evidently some individuals returned to the same spot winter after winter.

Considering all the facts, one cannot escape the conclusion that the migratory movements of the elk are conditioned by habit. Having once spent a winter in a given locality, the animals will return to it; and having once used a given route, they are likely to go the same way again. The calf follows the mother the first time;

the next time it will be familiar with that same route. The herds mingle more or less throughout the summer and become broken up in fall during the rut, but in due time the force of habit segregates the animals and they arrive at their respective winter areas by way of the "traditional" routes.

That, it would seem, is why only a few of the elk winter on certain elevated ridges or other highland ranges where a large number cannot be accommodated. If the numbers in such areas increase, it is likely that a severe winter will reduce the number to the carrying capacity of the range. Thus in each area is maintained the proper number of animals that habitually go there. That is why the present Jackson Hole elk no longer go out on the desert; they know nothing about the desert. That *may* be why the Olympic elk do not migrate far down the river valleys to the lowlands; it "has not been done" by members of the present population.

This conservatism among animals in acting along traditional lines is probably more rigid among our deer, but it is a well-known and widespread trait. J. Wong-Quincey, in "Chinese Hunter," p. 358, speaking of wapiti and other big game in China, says:

"Furthermore, game are perhaps like human beings in the matter of being conservative about their habitats. They are in a district because their ancestors were there before them. They will not move until starvation and great danger threaten their very existence. They will remain in a poor environment which used to be good simply because circumstances or sheer inertia prevent them from seeking better surroundings. Sometimes they may even stick to one district until extermination overtakes them."

Coupled with this more or less rigid habit, other traits must be considered. There appears to be a logical tendency for elk herds to go to lower elevations when there is a heavy snowfall and to gravitate to the lower parts of a given valley. At times certain groups have broken away from the main herd on the Elk Refuge in Wyoming and have wandered still farther down the valley, where they have annoyed ranchers and sometimes have been driven back up by a group of wardens. Among some of the elk there is a venturesome spirit that tends to break established custom. This appears to be prevalent enough to insure the spread of the species into new territory, so far as favorable environment can be found.

Figure 7. "Such a two-directional migration may be noted also in the Upper Yellowstone Region."

It was stated that the Jackson Hole elk do not move south to the desert any more because that is not in the experience of the present elk population. Yet, as noted, there is an exploratory tendency in some of the animals and this seems to be borne out by recent developments in elk distribution. Elk have been observed venturing to the plains from the Wyoming Range. Some have been working south from the upper Green River district. Others have appeared in the vicinity of Pinedale, having come from the Wind River Range. There is some evidence that some of the Gros Ventre River elk have been moving into upper Green River valley, over the traditional route. In 1944, the State of Wyoming transferred a number of elk from Jackson Hole to some of the sage plains areas farther south. Given enough time and oportunity, the present elk herds would undoubtedly develop again the primitive migratory routes.

Referring again to the Olympic Mountains herd, local information bears out other observations to the effect that if a group of elk calves born in the high country experience a series of open winters, they do not learn to go down. They do not know any other range and do not migrate. In a subsequent hard winter many such elk—not only bulls, but also cows and calves—will winter high up. They may descend a short distance during a storm and then later go back up. During the rutting season they may became scattered, with some exchange of individuals between two watersheds; but after the rut these individuals will often cross a high divide to return to their accustomed ranges, so that while some are crossing in one direction, others are crossing in the opposite direction. Meantime, if a heavy snow has fallen, in their determination to reach their accustomed winter range on the other side of the divide, the elk work up gradually as the snow packs hard enough for them to travel on it.

Such a two-directional migration may be noted also in the Upper Yellowstone region of Wyoming, where elk moving westward over Two Ocean Pass as they head for their winter ranges in Jackson Hole pass small groups of mule deer traveling eastward toward the eastern slopes of the Absarokas (Figure 7). Yet members of other deer populations winter with the elk in Jackson Hole.

8

PHYSICAL CHARACTERS

SIZE AND WEIGHT

MEASUREMENTS of elk necessarily vary in the taking according to the conformation of the ground on which the animal lies, or to the individual who does the measuring. In general, however, the total length from tip of nose to tip of tail of an adult Rocky Mountain elk is about 80 to 100 inches; the length of tail, approximately 4 to 5 inches; the length of hind foot, 24 to 28 inches; and the height at shoulders about 49 to 59 inches. Caton (1877: 81) recorded a 5-year-old male "southern elk" that was "over 16 hands high" (more than 64 inches) at the shoulder, the largest he ever had owned. The Roosevelt elk of the West Coast is said to be an even larger animal, but recorded data are not sufficient to determine the point satisfactorily. The valley elk or dwarf elk of California is decidedly smaller.

The elk is a large deer, second only to the moose in weight. Various writers have recorded old bull elk weighing as much as 1,100 pounds live weight, and it may be safely stated that the maximum live weight of old bulls in their prime does not exceed that. A large male of Wyoming, weighed by the writer in August (when elk are at their best), came to 1,032 pounds live weight and 657 pounds dressed weight. Caton (1884: 82) thought that 600 pounds would "exceed the average live weight of the full grown buck," but Seton (1927, vol. 3, pt. 1:6) gave "about 700 pounds" as the average for this class. Judging by the weights obtained during the studies in Wyoming, the writer considers the latter a fair figure when the animals are at their best.

Elk usually lose weight during the winter. One may expect them to be heaviest during the late summer, while on green food, and least in spring. On good winter range, however, elk may still be fat as late as January.

Various specimens of Rocky Mountain elk of Jackson Hole were

TABLE 1. WEIGHTS OF JACKSON HOLE ELK AT DIFFERENT SEASONS

Males				Females			
		Weight in pounds				Weight in pounds	
Date	Approximate age	Live	Dressed	Date	Approximate age	Live	Dressed
July 22, ?	2 months	130	98	Dec. 2, 1927	adult	...	312
Nov. 17, 1927	2 years ..	...	310	Dec. 8, 1927	do	...	312
Dec. 2, 1927	6 months	...	150	Dec. 21, 1927	do	603	366
Aug. 19, 1928	old	1,032	657	Aug. 3, 1928	do	545	365
Nov. 1928	6 months	...	156	Sept. 16, 1928	do	608	361
Mar. 11, 1931	21 months	475	...	Sept. 16, 1928	52 months	440	276
Mar. 17, 1931	9 months	285	...	Mar. 17, 1931	adult	532	...
Feb. 13, 1941	old	680	...	Mar. 17, 1931	do	500	...
Feb. 13, 1941	mature ..	600	...	Feb. 13, 1941	do	565	...
Feb. 13, 1941	do	750	...	Feb. 13, 1941	do	450	...
				Feb. 13, 1941	do	543	...

[1] Dressed weight as here used is the weight of the carcass with the head, feet, and entrails removed.

weighed at different times of year, with the results shown in Table 1.

A number of elk were weighed at Yellowstone National Park in January and February 1946 (Skinner, 1946). These appeared to be slightly heavier on the average than those weighed in Jackson Hole. With some exceptions, the elk weighed in Jackson Hole were fed hay. Those weighed in the Yellowstone had been foraging on the open range. But differences in some groups are very slight, and may have no significance (Table 2).

Many of the animals weighed on February 13, 1941 were marked and weighed again on March 20, 1941. Among the calves there was an average loss of 29 pounds between these two dates, and an average loss of 54.6 pounds among the adult elk, both sexes combined.

The weights of female and male calves overlap considerably, though the average male is slightly heavier than the average female. The series is so small, however, that the difference shown may not be significant. The weights of newborn calves vary

TABLE 2. COMPARISON OF ELK WEIGHTS BY LOCALITIES

	Jackson Hole		Yellowstone National Park[1]	
Class of animal	Number weighed	Average weight	Number weighed	Average weight
Mature male	15	620	15	642
Mature female	29	510	9	499
Yearling (male spike) ..	2	394	2	406
Yearling female	2	386		
Calf male (8-10 mo. old)	12	255	9	268
Calf female (8-10 mo. old)	21	247	11	251

Average Weights of Elk, Jackson Hole and Yellowstone National Park, Combined

	Number weighed	Average weight
Mature bulls	30	631
Mature female	38	520
Yearling male (spike)	4	400
Yearling female	2	386
Calf male	21	260
Calf female	32	249

[1] Skinner, Curtis K. Live elk weights from small group weighed during winter of 1946. Nature Notes Yellowstone National Park 20(3): 1-12. May-June 1946. (Processed).

greatly. Hornaday (1904: 122) recorded one weighing 30.5 pounds, and Rush (1932: 31) gave 37 pounds as the average weight of calves ranging from 23 to 45 pounds in weight. In the present study the weights of three fetuses obtained in June, practically at the time of parturition, were 24.5, 37, and 39.5 pounds, and a newborn calf weighed 30 pounds. It is safe to say at birth most elk calves weigh between 30 and 40 pounds, although some weigh less than 30 pounds.

It will be noted that there is great variation in weight of animals even at the same age. This begins at birth, with recorded weights from 23 to 45 pounds, and is obvious during the first winter, when calves may range from 205 to 320 pounds on the same date in

February. Though it is difficult to correlate exact ages in the older animals, the indications are that some variation in weight continues through life. During the first year there is little if any difference in weight between male and female, but the males gradually become heavier as the animals mature.

The Roosevelt elk is often referred to as a larger animal than the Rocky Mountain elk, but figures are not available for comparison. The tule elk of California is decidedly smaller, however, as is shown by the live weight of eight bulls collected for the Museum of Vertebrate Zoology at Berkeley, which were as follows: one spike, 356 pounds; five young adults, 328, 383.5, 385, 434 and 441 pounds; and two old adults, 493 and 522.5 pounds.

PELAGE AND COLORATION

Winter Pelage

The winter pelage of the Rocky Mountain elk consists of a woolly undercoat largely concealed by the usual cervine guard hairs that give this animal its characteristic color, a brown varying from grayish on the sides to very dark, sometimes almost black, on the neck and legs. In the areas of transition from light to dark, as on the lower sides of the body and the upper parts of the legs, there are rich shades of brown, most pronounced in old bulls. The rump patch, bordered on the sides with a black line that fades away toward the dorsal line, is usually tawny. In old bulls the sides are almost whitish, in great contrast with the dark head, neck, legs, and under parts. In cows, calves, and young bulls there is less contrast, although the same pattern of color intensities is followed; the sides are not so light as those of old bulls, and the edge of the dark areas is not so rich a brown.

The head is usually dark brown. A lighter ring, sometimes not very noticeable, generally surrounds each eye. The ears have a black spot on the lower edge of the anterior surface. The chin is light brown, or tawny, with a black spot near the angle of the mouth. Below the chin is a contrasting streak of lighter brown that is quite conspicuous when the animal stands with raised head facing the observer. On the neck and throat is a coating of very long hairs that form the dark mane so characteristic of the elk. Even calves in their first winter have this mane.

In many animals, there is a somewhat lighter patch on the throat, involving the light streak mentioned above, but paler, and spread out fanlike, faintly suggesting the white throat patch of the mule deer.

The hoofs are normally black, but an occasional animal has one or more hoofs that are light horn color, usually with a few black streaks. The tips of the hoofs and dewclaws have a tendency to show a lighter color.

Within all age classes there is considerable variation in color among the animals, some being much darker than others. The rump patch will vary from a pale whitish or diluted shade to a very rich tawny, this variation being observable even among calves. These differences are apparent to hunters, who sometimes refer to the "blue" cows, for instance, as being darker than the others. As the winter progresses there is a gradual but decided fading or bleaching until in early spring many of the older animals appear almost white at a distance. Even as early as February the pelage of some of the elk acquires a roughness, and during March it begins to show decided wear. At first a few ruffled spots appear, which may be caused by the elk scratching or biting at its sides when the ticks begin to be felt. More and more such spots appear until there is a general irregularity of the surface of the pelage (Plate 4). During March and April, as scratching for ticks becomes more common, the winter coat deteriorates rapidly. In many cases the hair is rubbed off in a ring around the base of the neck and later most of the lower neck becomes bare, giving the animal a ludicrous appearance. The elk scratches with a hind hoof, which reaches the base of the neck most conveniently, and this probably accounts for the odd pattern of wear.

The Roosevelt elk is similar to the Rocky Mountain elk in essential pelage color pattern but on the average is decidedly darker even though many specimens show little difference in color. In some specimens examined the neck was actually black. A spike buck in the Hoh River, Wash., area had a coal-black mane, the black extending to the roots of the hairs. The rump patch is often a deeper tawny, and this same richness of tone appears also on the chin and the ears.

The tule elk of California has a less differentiated color pattern than other elk and is grayer and strikingly paler, even on the under parts, which in some specimens are a "light pinkish cinnamon" or buff. The sides are light buff and the legs pale orange yellow, cinnamon on the anterior edge, in great contrast with the dark brown in the other elk species. The neck is largely "clay color." The hoofs and dewclaws have more extensive light horn-colored tips, bearing out the general scheme of much lighter pigmentation. The rump patch is pale, the black border indistinct. Furthermore, even in November and December the pelage is rough or "frizzled" in appearance, somewhat similar to an ordinary elk hide that has lain in the sun until the tips of the hairs have curled.

Dr. E. A. Mearns' description of the winter pelage of Nelson's specimen of the Merriam elk (Nelson, 1902: 9-10), being a mere enumeration of colors, does not satisfactorily distinguish the Merriam from the Rocky Mountain elk. Nelson (1902: 8) stated, however, that *merriami* had a "nose darker and head and legs more reddish than *Cervus canadensis*" and Bailey (1931: 42) that it had a "color paler and more reddish" than that of the Rocky Mountain elk.

The winter color of the Manitoba elk follows the general type of the Rocky Mountain elk, but specimens were not compared.

Summer Pelage

The following description of the summer pelage and its changes is based on the studies of the Rocky Mountain elk, but with slight deviations caused by somewhat different coloration it is doubtless applicable to most other elk forms as well. The distinctions are likely to be most noticeable in the tule elk, but its pelage changes have not yet been studied.

As early as April a little of summer pelage may be found under the old coat. In May, the new hair becomes noticeable, first about the head. During this month the old tattered remnants of the winter coat give the animal a most unsightly appearance; yet often, while the old coat still clings in patches on the body, the head has acquired the smooth, sleek summer aspect. The legs also shed their hair early. By the first of June about half the elk have completely shed the winter coat, and by July practically all of them have done so, although on July 5, 1929, a spike bull was seen that had not quite rid itself of the last vestiges of the old hair. Thus the summer pelage is dominant during June but is most characteristic in July.

The summer coat consists of short, stiff, sparsely distributed hairs that lie close to the skin; there is little, if any, underfur. At close range the color is somewhat tawny, but at a distance, when the animals are viewed in a green landscape, it appears distinctly reddish or light bay. When examined closely, the hairs appear glossy, in contrast with the dull hairs of the winter coat. The legs and head remain dark, but the neck does not bear the long mane so characteristic of elk in winter. When the change is complete, the strong and bulky-appearing winter elk is transformed into the trim and graceful animal of the green summer pastures. The finely modeled musculature is apparent through the thin summer pelage.

The new born calf, tawny like its mother but even richer in hue, has a very pronounced yellowish-brown rump patch and is speckled all over with light spots. In some specimens a blackish dorsal line and a dusky cast overlie the brown. The spots are arranged much like those of the fawn deer but are somewhat more diffuse and not so clear-cut. Down the neck and back, there are two rows of spots, one on either side of the median line. On the sides the spots are more irregular but tend to form rows. In July, as the hair lengthens, the spots become much less distinct and at a distance are not readily discernible.

Early in August a change in the elk pelage takes place. It is first noticed as a slight gray cast on the red. The earliest recorded date of an animal showing this gray suffusion is July 23 (1931), when other elk were still "red." Another instance was recorded on August 1 (1929). By the 10th of August the gray tone has become somewhat more prevalent, and by the 20th most of the elk are grayish although some still retain the summer pelage. By September 1 the summer coat in its purity has vanished and the elk may be said to have acquired the equivalent of the "blue" coat, or the short winter pelage that in some members of the deer family at this season has a blue or gray tone that later becomes a distinct brown.

The transition from summer to winter pelage entirely lacks the ragged, tattered stage so characteristic of the change in spring. This is because the summer hair that is dropping out is short and scattered and does not hang together like the shedding winter hair.

Albinism

Almost all dark mammals occasionally show some degree of albinism, and the following instances reveal that the elk are no exception. The head and neck of a cow elk brought to the local taxidermist at Jackson, Wyo., were liberally blotched with small areas of white; a letter from E. L. Brown to the Biological Survey in July 1888 stated that in 1887 a hunter in northern Minnesota "reported seeing one female in a band that was plainly spotted on the sides with white"; and Elmer Fladmark [4] recorded a young elk in Glacier National Park whose "body was a dirty white color with the legs and head a little darker."

On October 8, 1939, in a casual conversation with a deputy game warden in Oregon, the writer was informed that there was an ablino female elk at Goldburg, Oregon. In winter the pelage was described as fawn color, with white rump patch, and the animal had pink eyes.

ANTLERS

Form and Succession of Antlers

The following discussion of antlers is based on studies in the Jackson Hole, Wyo., region, but examinations of antlers of other elk groups and observations in the field on the Pacific coast and elsewhere indicate that with some minor variations the facts presented apply fairly well to all North American elk. Diagnostic differences in elk antlers are discussed under each elk form (pp. 9-17).

The typical antler of the normal adult Rocky Mountain elk bull is characteristic in form (Figure 3 and Frontispiece.) It consists of a long, round beam sweeping up and back from the skull and bearing six points or tines without palmation. Although the beam extends back in a fairly straight direction, it has not only several successive undulations, or curves, each culminating upward in a

---

[4] Fladmark, Elmer. Albino elk. U. S. Nat. Park Serv. Glacial Drift 7: 29. 1934. (Processed).

tine, but also an outward torsion, which in some specimens is very apparent.

The brow tines and the bay, or bez, tines are close to each other near the base of the beam and extend out above the muzzle in a formidable array of four points, designated by the old-time woodsmen as "dog-killers" or "war tines" but today more commonly called "lifters." The tray, or trez, tine arises higher up on the beam. Still further up rises the fourth point, the royal, sometimes known as the "dagger point," a very heavy tine and generally the dominant one. Often the beam curves upward toward this fourth point so decidedly that the point appears essentially a continuation of the beam itself. The remaining two points, sometimes known as the sur-royals, are smaller and often appear to be merely a forking of the end of the beam. The sixth tine generally takes a decided downward slant.

The first antlers normally begin growth late in May, when the young bull is about a year old, but February 15, 1928, an elk calf was seen near Jackson with prominent knobs showing where the new antler growth was forming. This was a very unusual occurrence, however, and is the only instance of such early antler growth observed. Numerous male calves, or "short yearlings," have been seen as late as early May with no indication of antler growth, but in 1933 on May 25 many of the yearling bulls had plainly discernible knobs, and this seems to be about the usual date for such development. Normally the first antlers are simple spikes ranging in length from 10 to 20 inches, but occasionally they have small forks at the tips (Figure 8 F and Plates 5 and 6). Once in a while unusual antlers occur. On March 9, 1932, each of two spike bulls observed had a fork at the tip of one antler and three prongs at the tip of the other, in addition to several slight points and knobs near the base of each antler.

The second antlers, which attain their growth when the animal has passed his second birthday, appoach the typical six-point form but ordinarily carry only four or five points. There may be fewer than four or as many as six points, but even when there are six, the antlers are small and slender compared with those of adults (Figure 9 and Plate 4).

After a bull has passed his third birthday, his third antlers are

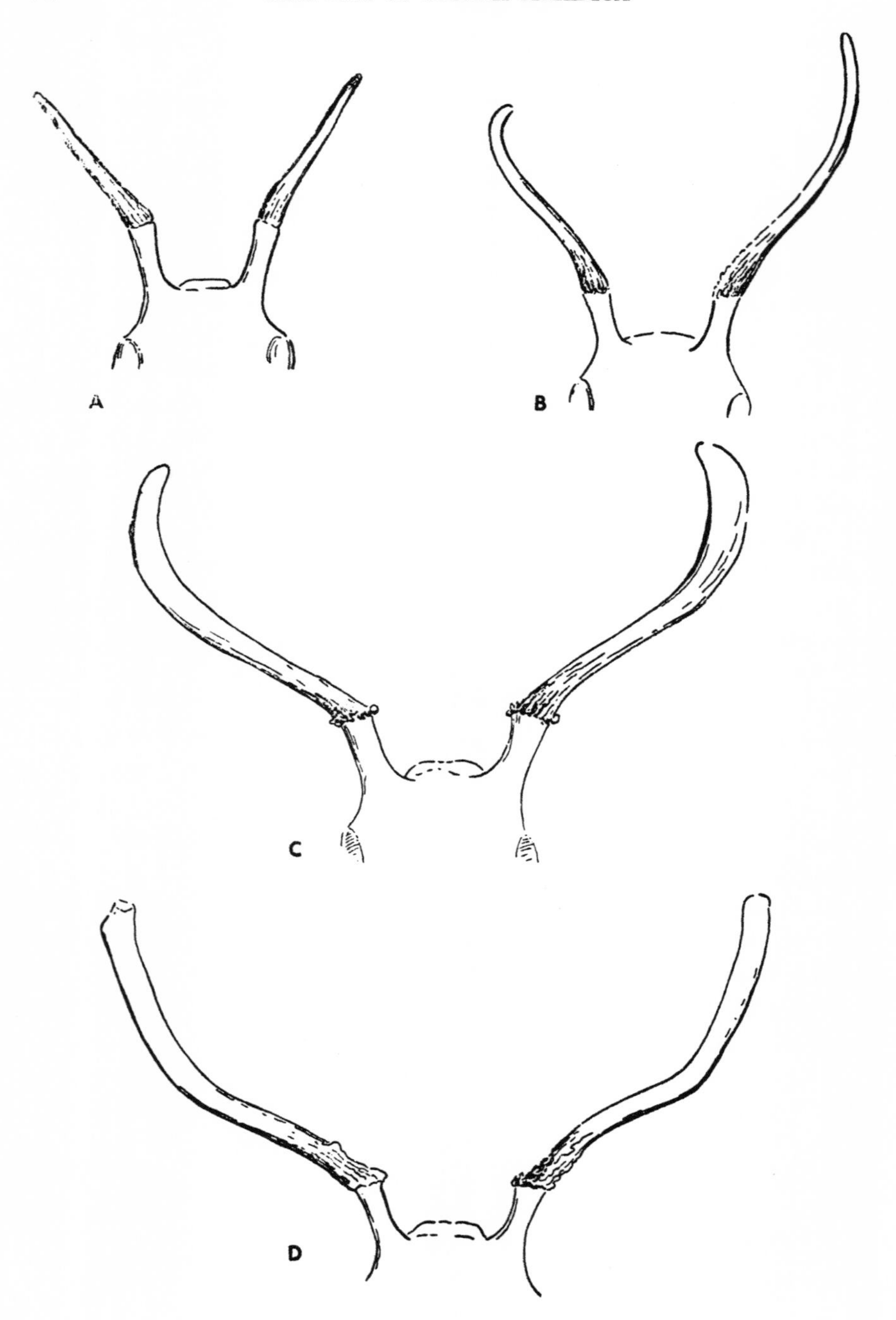
A
B
C
D

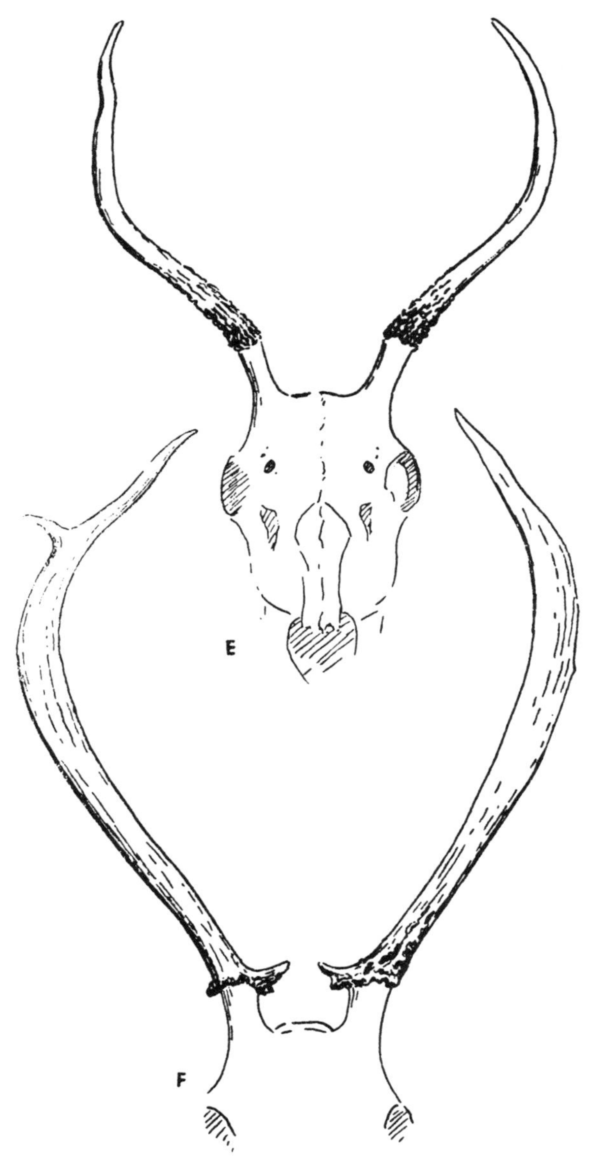

FIGURE 8. The first elk antlers are normally simple spikes. This series (drawn to scale approximately) from Jackson Hole, Wyo., show variations. *(A)* and *(B)*, Represent small types often seen, greatest lengths being 6½ inches and 14¼ inches, respectively; *(C)* and *(D)*, widespreading type; (*E*) and (*F*), narrower, perhaps the most common type; *F.*, being 24 inches in greatest length. Note the long pedicels in this group.

still very likely to have less than six points, usually four or five, and these are sometimes superficially hard to distinguish from those of the 2 year old. Closer scrutiny, however, reveals that the antlers of the 3 year old are heavier, that the pedicel is also heavier and definitely shorter.

Succeeding antlers are normally of the six-point form that is typical of the mature bull elk until he has passed his prime, after

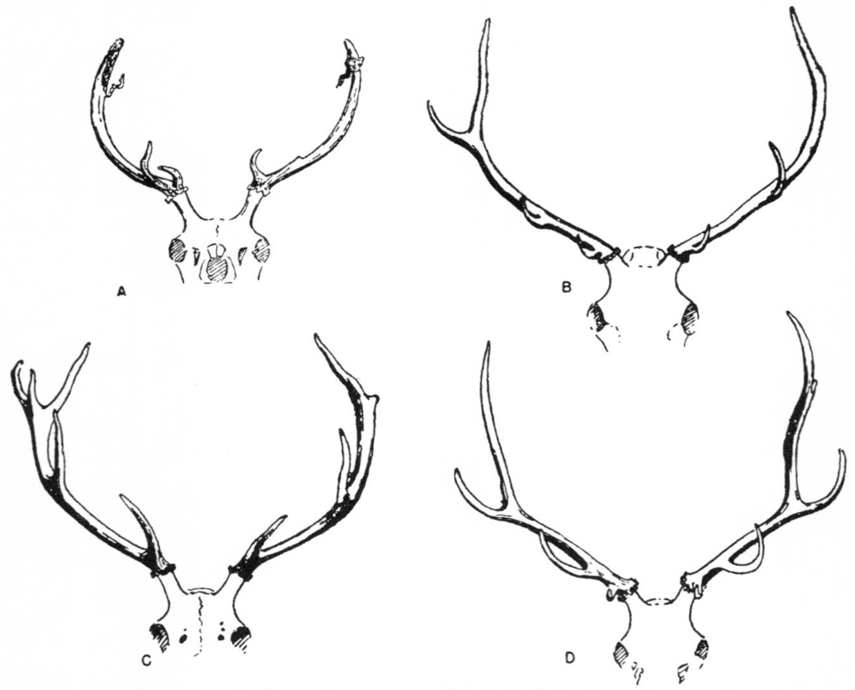

FIGURE 9. The second elk antlers approach the adult form but rarely carry six tines. This series from Jackson Hole, Wyo., is typical. On an elk of this age (approaching 3 years) the pedicel is still long. *(A)*, Antlers unusually small, length of beam being 15 inches, which is about the length of some spike horns. Length of beam in *(B)* and *(C)*, 25 inches and in *(D)*, 33 inches.

which they are likely to decrease in size. William Rowan (1923: 112) records the antlers of a captive bull elk in Assiniboine Park, Winnipeg. The antlers were at their best, with a spread of 61½ inches, in 1917, when the animal was about 12 years old. In 1919 the spread was still about 60 inches. The animal became senile, with poor teeth, in 1920.

There is so much variation in the size and form of the antlers, as well as in the number of points at some ages, that it is difficult to estimate the age of the elk by the antlers alone. No specific set of antler measurements can safely be used as indicative of any given age, but there is an interesting development of the pedicel of the antler that furnishes a general clue to the animal's age. In the spike bull the pedicel is very long and slender (Figure 8), but in succeeding years it becomes progressively thicker and shorter. Boas (1923: 4-5, figs. 1-3), speaking of deer in general, explained that at the time of separation of the antler from the pedicel there is resorption of bone at the end of the latter, thus shortening it each year.

The record lengths of beam, recorded in *North American Big Game* (1939) are 64¾ and 64½ inches. Twenty antlers measure 60 inches or over, and 108 others have one or both antlers between 55 and 60 inches in length. The record spread is 74 inches, on an animal taken in the Teton Mountains of Wyoming, with a length of beam of 58½ inches. Record circumferences of the beam between bez and trez tines are 9¾ and 9½ inches, while numerous others are from 6 to 8 inches.

These record heads are almost wholly from the Rocky Mountain elk, only one or two representing the Roosevelt elk of the Pacific coast. They illustrate the size and beauty of antler growth on this magnificent member of the deer family.

Shedding of Antlers and of Velvet

March is the month during which most of the bulls shed their antlers. The oldest bulls, however, begin to drop theirs even in February, when some of them begin sparring or rubbing with their antlers, stimulated probably by an "itching" about the antlers similar to what may exist when the velvet begins to dry up. On February 16, 1933, each of two mature bulls observed had dropped one antler, but as late as February 24 each was still carrying the other. From about the middle of February on, antler shedding increases in frequency until it reaches the peak in March. Some adult bulls carry their horns through the month of April, however.

Although some 2 year old bulls shed their antlers in March, others of that age have been seen with antlers as late as May 8.

Curiously enough, the spike bulls, which normally carry their antlers the longest, show the widest range of shedding dates, as indicated by the following observations. On January 20, 1928, one spike bull was seen that had shed both antlers and another that retained only one. In 1933, on February 16 one spike bull had dropped both antlers and on February 24 two others each carried only one horn. Again, in 1935 a spike bull seen had shed one antler on February 13 and there were reports of several others that had shed their antlers a little earlier. A number of spike bulls shed their horns in March, many carried them through April, and several in captivity shed theirs on May 5, 11, and 19. Even as late as June 10 a spike bull was observed still carrying both antlers.

It is probable that in many cases the immediate cause for dropping an antler is collision with some obstacle, such as a tree, or a bush. This is by no means a necessary condition. Many times shed antlers are found on open, bare ground, and often they are found concentrated at a point where elk are in the habit of jumping a fence. Apparently a sudden movement of the head at the right time is sufficient.

On several occasions the process of shedding was observed. Usually there was a distinct, audible snap as the antler let go, and in every case the animal was obviously startled. Once a young bull, that had shed only the left antler, gave a sudden leap, seemingly coincident with the snap and release of the remaining antler. The animal ran off in apparent fright, holding his head tilted up on the right side, where the muscles were not as yet adjusted to the sudden loss of the weight. On March 17, 1928, an older bull was observed shedding. One antler clattered to the ground, startling the bull and driving the neighboring elk away in a short stampede. The bull wandered about carrying his head tilted on one side and twisting it and his neck about with snakelike motions, seemingly unable to adjust his muscles at once to the uneven burden of the remaining antler. When last seen, he was standing as if in a stupor, his head gradually coming more and more on the level (**Plate 7**).

On March 15, 1941, a bull was observed dropping both antlers at once. This animal, also, moved its head about uncertainly, much as did the one mentioned above that had shed only one antler.

There is some bleeding as the antler falls off, but soon a scab forms and the new growth begins. As the swelling bud of the new antler takes form, the scab persists, but with the further growth and lengthening of the antler it contracts and finally disappears (Plates 6 and 8).

In April, the new antlers of many of the older bulls have attained very noticeable length although probably most of the actual antler growth, including the long beam and tines, takes place in May and June. On April 25, 1928, at a time when many younger bulls were still carrying the old antlers, three old males were observed with antlers more than a foot long that were beginning to divide at the tip, and undoubtedly there were others with still longer new antlers. On May 26, 1940, at Norris Junction, Yellowstone National Park, there were three adult bulls, all of which had the brow, bez and trez tines pretty well grown and the fourth tine budding out. Some time in July, the antlers complete their growth and even begin to get ready for shedding the velvet (Plate 9). It should be remembered that such events are greatly modified by variations in age and individuality, however, so that precise, all-inclusive dates cannot be given.

The old bulls are generally the first to shed the velvet, some of them beginning the rubbing of velvet in a small way in the first week of August. These are undoubtedly the individuals that had shed their antlers in the previous February or early March (p. 81). In 1931, several bulls were seen that had shed their velvet a few days prior to August 10. By September 1, most of the bulls have rubbed off the velvet and the rut has begun.

The younger animals, either because of the light weight of the antlers or because the breeding urge is less potent, apparently do not rub their antlers as vigorously as do the older bulls. Spike bulls appear to be the last to lose their velvet—although in 1930 one was seen rubbing the velvet on August 20—and some of them never do rub it off. On February 1, 1928, of 71 spike bulls observed 8 still retained the velvet, 1 of them on only one antler. Shed spike antlers with velvet intact or only partially rubbed are not uncommonly found in spring (Figure 10). At least one 2-year-old bull was seen in winter with a little velvet still clinging to the antlers, and an occasional shed antler of a 2 year old is found with velvet adhering.

In a personal letter under date of January 27, 1948, Dr. Philip

L. Wright, of Montana State University, has commented: "With respect to the antlers of spikes carrying velvet, the packer who has packed me into the upper South Fork of the Flathead drainage the last two seasons told me this fall that he had never seen a spike elk come out of the South Fork except with antlers in the velvet whereas in the upper Blackfoot drainage only 50 miles to the south he has never seen one in the velvet. I still have the feeling that elk living in rather cold areas where the food is not good may be retarded in this regard."

It could very well be that this phenomenon varies geographically and that it is dependent on nutritional influences, directly or indirectly.

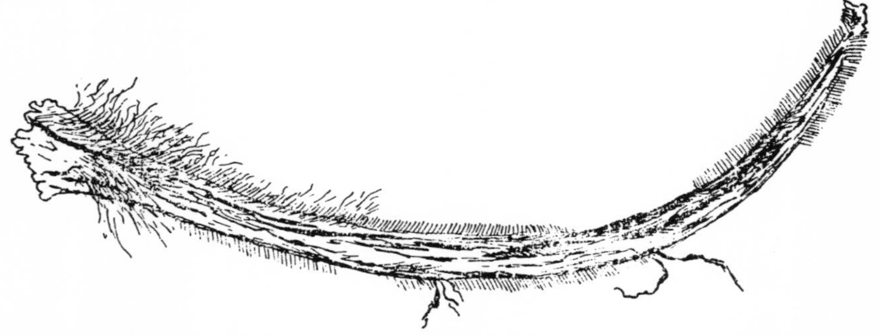

FIGURE 10. Shed antler of spike bull, which had retained most of the velvet. Jackson, Wyo., March 17, 1928.

The velvet is loosened and scraped away by constant rubbing and threshing by the restless elk against bushes and tree limbs, and during the process numerous young evergreens are rubbed bare of limbs and bark and are demolished. As the rut approaches, the bull prods the ground and fights the willow bushes with loud rattling and crashing. This constant rubbing eventually leaves the antlers cleaned, polished, and stained and a portion of the landscape more or less disfigured.

Immediately after the velvet is shed, the antlers are bloody; but the blood soon disappears, leaving the newly dried antlers a gleaming white. Continued rubbing against bushes and the trunks of small trees in a very short time stains the antlers a beautiful brown, and still further polishing produces an ivory whiteness on the smooth tips.

PLATE 4. Bull elk, Jackson, Wyo. *Above.* Yearling, beginning to shed the winter pelage, April 27, 1929. (U. S. Fish and Wildlife Service photograph.) *Below.* Bull nearly 3 years old, with second pair of antlers. These usually carry four to five points. (Photograph by E. A. Preble.)

PLATE 5. A Rocky Mountain elk, Jackson, Wyo. *Left.* April 27, 1929 at about 1 year of age. *Right.* November 20, 1929, with its first antlers. Occasionally the spike horns are forked, as in this case. (U. S. Fish and Wildlife Service photographs.)

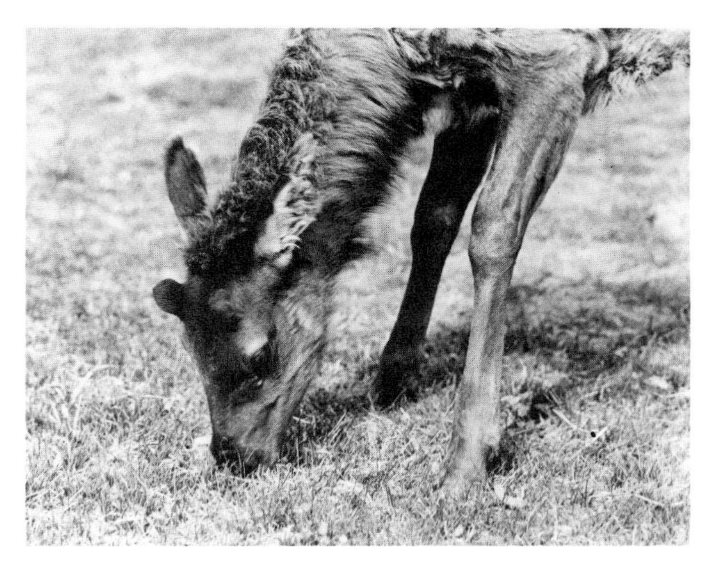

PLATE 6. A Rocky Mountain elk, Jackson, Wyo. *Above.* At 1 year of age, May 17, 1929, with spike horns budding. *Below.* April 1932, at nearly 4 years of age, with the fourth pair of antlers beginning their growth. Note the disappearing scab formed by the recent shedding of the previous antlers. (U. S. Fish and Wildlife Service photographs.)

PLATE 7. Elk at Jackson, Wyo., March 17, 1928. *Above.* Cow bearing a single velvet spike. *Below.* Bull elk that has just shed one antler and that because of the unaccustomed absence of weight on one side is having difficulty holding his head level. (U. S. Fish and Wildlife Service photographs.)

There has been some question about what produees the brown stain on antlers. One thought is that the stain is produced by the bark of the trees that are used in rubbing off the velvet. Adolph Murie informs the writer that in the Olympic Mountains a local opinion is that the elk in the lowlands, which are likely to rub on alders, have dark-stained antlers, whereas those in high country, which rub on conifers, have light-colored horns.

As an experiment an old bleached antler was rubbed vigorously against fir limbs and trunks in an effort to simulate the actions of an elk. A slight staining resulted, but the rubbing period was too short to be conclusive. It is true that in large elk antlers small bits of bark and other material often lodge among the rugosities

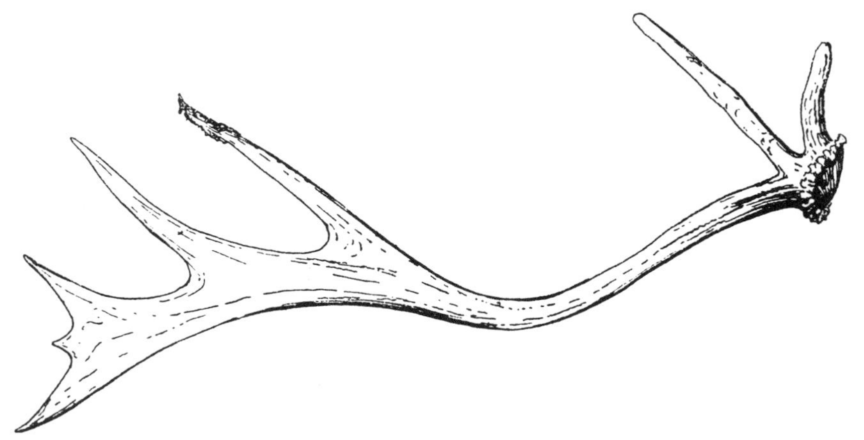

FIGURE 11. A palmated antler from Jackson Hole, Wyo.

of the rougher parts. It would appear that the antler actually is stained by the materials it comes in contact with, but more conclusive proof should be obtained.

Freak Antlers

One winter an old antlerless bull was observed on the Elk Refuge. The men who fed the elk were familiar with him, having observed him all winter, and they declared that he had never had antlers. Perhaps there had been an extremely early shedding, unnoticed by these men, or possibly the bull had been castrated in some manner.

On the other hand there are several reports of bulls with three antlers. Seton (1929, vol. 3, pt. 1: 28) and Bird (1933) each recorded the occurrence of a three-horned elk, and in the winter of 1931-32 a head was obtained in Jackson Hole that had two large, normal antlers and a supernumerary burr very close anteriorly to one of the antlers.

There are various abnormalities of the tines, too. Some tines may have a clublike form directed downward beside the head; others may bear numerous points or have palmations (Figure 11).

An old weather-beaten set of antlers was found on upper Flat Creek, Jackson Hole, Wyoming. The tines were broken off. Evidently the tips of the beams had been palmated. It was an unusually massive set. The beam midway between the bez and

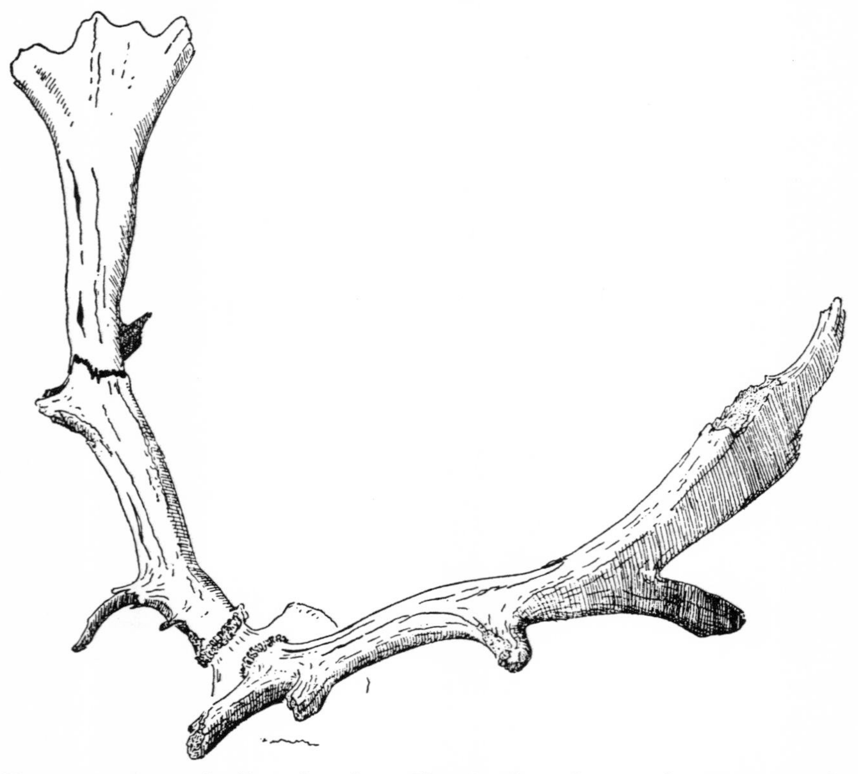

FIGURE 12. Unusual elk antlers from Flat Creek, Jackson Hole, Wyo. Lengths of broken beams, 35½ and 37½ inches; circumference of beam between bez and trez tines, 8½ inches. Note the tines projecting from posterior side of beam, as in the caribou.

trez tines measured 8½ inches in circumference, not far from the record. A peculiarity was the fact that there had been a pair of tines directly back from the posterior side of the beams, as in the caribou (Figure 12).

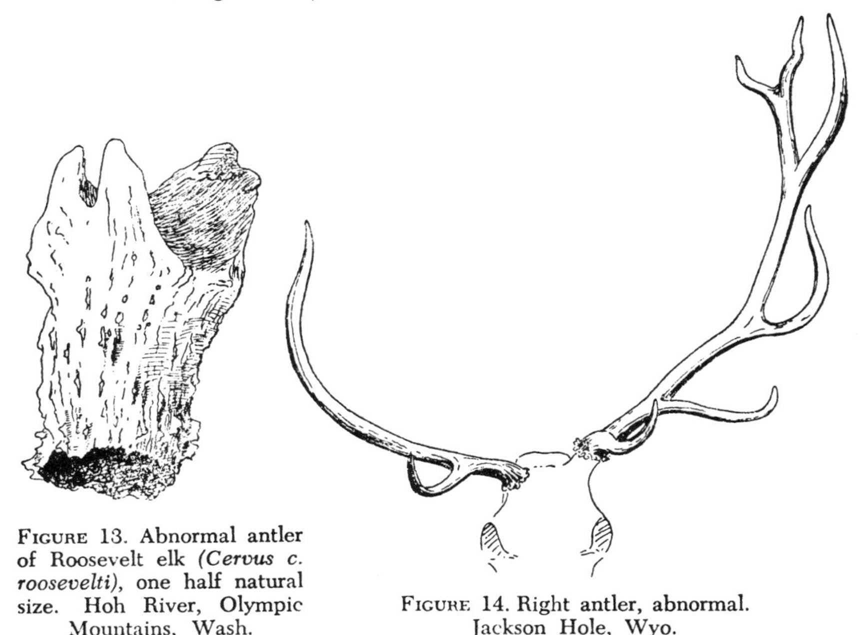

FIGURE 13. Abnormal antler of Roosevelt elk *(Cervus c. roosevelti)*, one half natural size. Hoh River, Olympic Mountains, Wash.

FIGURE 14. Right antler, abnormal. Jackson Hole, Wyo.

Sometimes as in the case of the spike bulls (p. 77) that had three points at the tip of one antler and some smaller projections at the base of both antlers, or as in the case of large bulls that bear seven, eight or more points on each antler, the abnormality appears caused by exceptional vigor or perhaps by an abundance of antler-building food. This type of freak should be distinguished from that caused by castration, disease, or milder afflictions that lower bodily vigor (Figure 13) or by accident (Figure 14).

A good example of the effect of physical debility on antler growth is illustrated by a tame bull that was raised from a fawn at the Elk Refuge. The first spike antlers were large and forked, indicating robust health and vigor (Plate 5). The second antlers, still denoting good condition, bore six points, which was unusual, as almost all second antlers bear only four or five. The following winter the elk became infested with scabies. He was treated repeatedly but was never completely free of scab all winter and remained in poor condition. In the spring the last of the scab mites disappeared and the green forage caused an improvement in the animal. Apparently the damage was already done, however, for the new antlers, the third pair, which should have been large and heavy, came out small and decidedly asymmetrical, with one bez tine missing entirely and other points projecting at irregular angles. Evidently the animal's vitality and perhaps

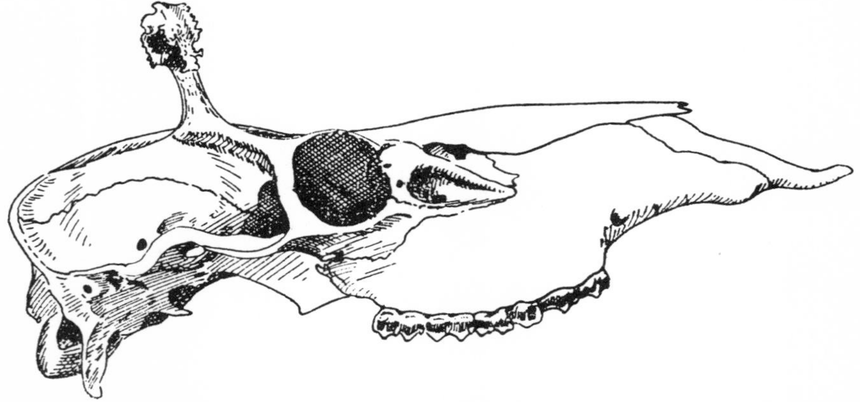

FIGURE 15. Skull of antlered female, fully adult. Jackson Hole, Wyo., Jan. 12, 1933 (one quarter natural size).

its reserve of mineral matter built up in the previous summer, had been drained away by the prolonged attack of scabies. The next winter this elk became afflicted with what appeared to be arthritis in a hind leg and remained emaciated. In spring the antlers again came out abnormally and were small. The following summer the animal became dangerous and was shot.

In the winter of 1942-43 there was an abnormal bull elk on the feedgrounds of the National Elk Refuge. Although it was a large, fully adult animal, its antlers were not of the typical form, being unusually straight with narrow spread and irregular tine formation. The dried velvet was retained. The animal was unusually fat for the winter season. Perhaps of great significance, it was noted that the abdomen was swollen in the area surrounding the sexual organs, suggesting some degree of accidental castration.

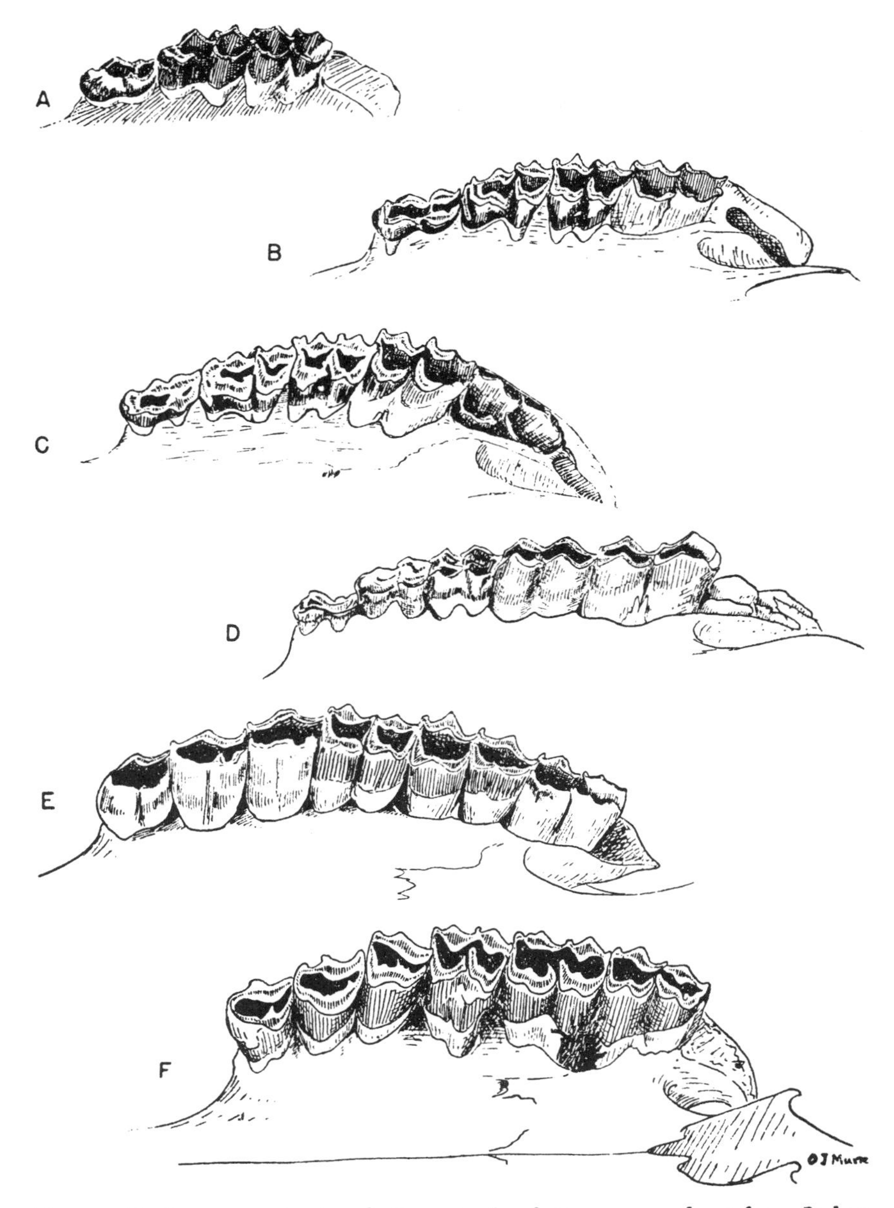

FIGURE 16. Development of molariform teeth of *Cervus c. nelsoni* from Jackson Hole, Wyo. Upper left teeth from: *(A)*, Elk calf, about 1 month old; *(B)*, male calf, about 7 months old; *(C)*, female, about 1 year old; *(D)*, spike bull, nearly 2 years old; *(E)*, adult male, nearly 3 years old; and *(F)*, old female, 10 years old.

It was learned from workmen that the antlers were shed in the latter part of the winter. In the winter of 1944-45 apparently the same animal was observed, again with abnormal antlers in velvet, this time with a wider spread, but with the tips of the tines dubbed off. Late in winter both were found broken off.

Antlered cows are not uncommon. At least five were observed during the course of these investigations, each bearing a single short spike (Figure 15 and Plate 7). In addition several mounted heads of females with more complex antlers were noted in the possession of residents of Jackson Hole. In every instance the velvet had been retained.

THE TEETH AND THEIR DEVELOPMENT

General Characters of the Teeth

The teeth are a vital part of the anatomy of any mammal, and their importance to an herbivorous mammal cannot be overestimated. Hooton (1937: 77) remarked that ". . . if the human dentition breaks down, it will carry with it in its fall the human species." Although he spoke in a phylogenetic sense, his remarks have a certain applicability here. The condition of the teeth is often the deciding factor in the length of life of an herbivorous animal and, as shown later (p. 103), it not only is a rough index to the status of the mammal in the vital statistics of its species but also plays a big part in the winter mortality of certain age classes. Therefore, the writer gave considerable attention to elk dentition, and its development is here discussed in some detail.

The formula for elk dentition may be expressed thus:

$$\frac{\text{I0—C1—PM 2-3-4—M 1-2-3}}{\text{I4—C0—PM 2-3-4—M 1-2-3}}$$

Possibly the fourth lower incisor is an incisiform canine, but for convenience and to follow the custom of most authors, it is here given as an incisor.

The teeth—of the brachydont form common to the deer—are relatively large. The premolars and molars are usually double semilunar, though the lower premolars are irregular in this respect. The incisors are large and heavy. The upper canine is unusually well developed in both sexes but is much more specialized in the male, its size thus possibly being a sexual character (Figures 16 to 20; Plates 10 to 12).

First Year Dentition

A number of fetuses were examined microscopically, and the early development of the teeth and jaws was noted (Figures 16 to 20).

In a fetus of March 30, 1928, weighing 6 pounds 10 ounces, none of the teeth had erupted, of course, but the deciduous teeth were fairly well formed and the tooth row (at this stage consisting of the premolars) was approaching its full length, being about 53 mm. in length in the upper jaw and 51 mm. in the lower. A molar had begun to form in the lower jaw, but there was little if any trace of an upper molar. (A microscopic examination would undoubtedly have revealed more extensive representations of teeth.) Incisors and canines had already formed but were not full size.

In a fetus of April 18, 1928, weighing 11¼ pounds, the row of premolars had reached practically full length, although the teeth had not come through the gums. There was a trace of the first upper molar, and the first lower molar appeared to be well developed.

In these two fetuses bony tissue had not completely filled in the spaces over the outside of the teeth, and consequently there was a row of three openings, or fenestra, in the thin bone of the outer surface of each mandible, opposite certain roots of the premolars.

In another fetus weighing 11¼ pounds the three fenestra were still evident on each maxillary, two opposite the roots of the second (or anterior) premolar and another opposite the third premolar, but those on the mandibles had disappeared.

In a 12-pound fetus the three fenestra of each mandible were still indicated but were closing, and two were still present opposite each upper anterior premolar. Apparently these openings begin to close at about this stage of fetal development.

In a 16¼-pound fetus the fenestra on each mandible were only indicated by three dark spots, but there were still two openings opposite each second upper premolar. Also, in this specimen an opening was present opposite each first lower molar.

Even in a 24½-pound fetus two or three mandibular fenestra were still represented by dark spots and the third (posterior) one

A

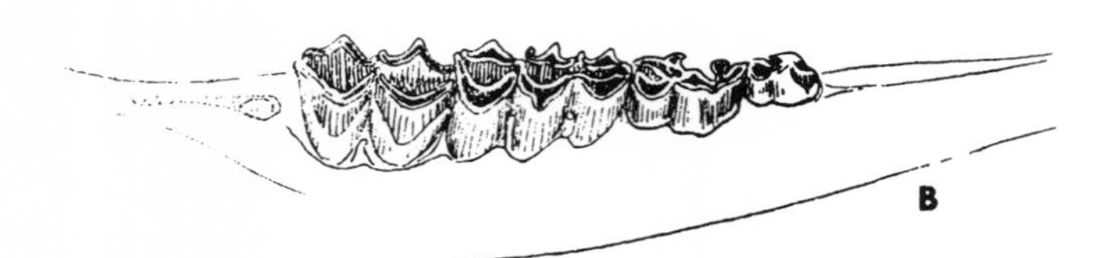

B

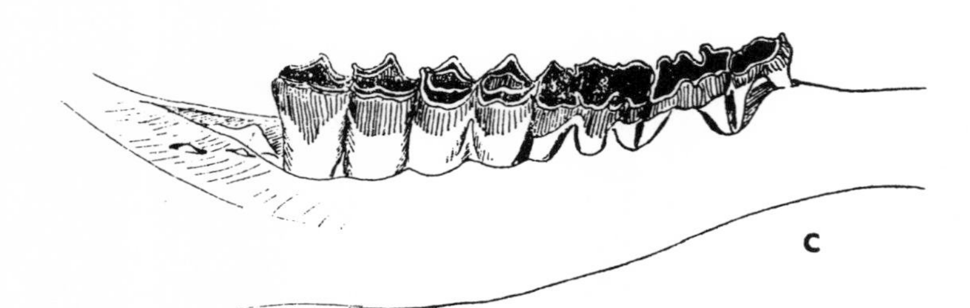

C

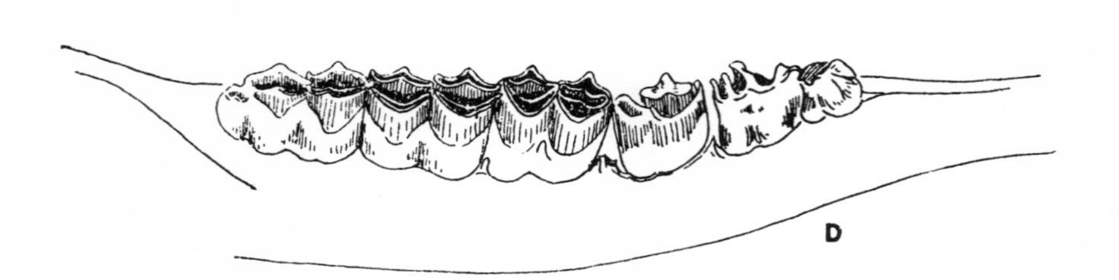

D

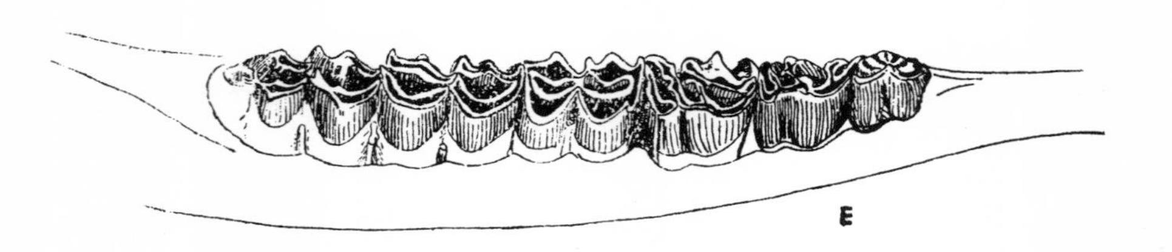

E

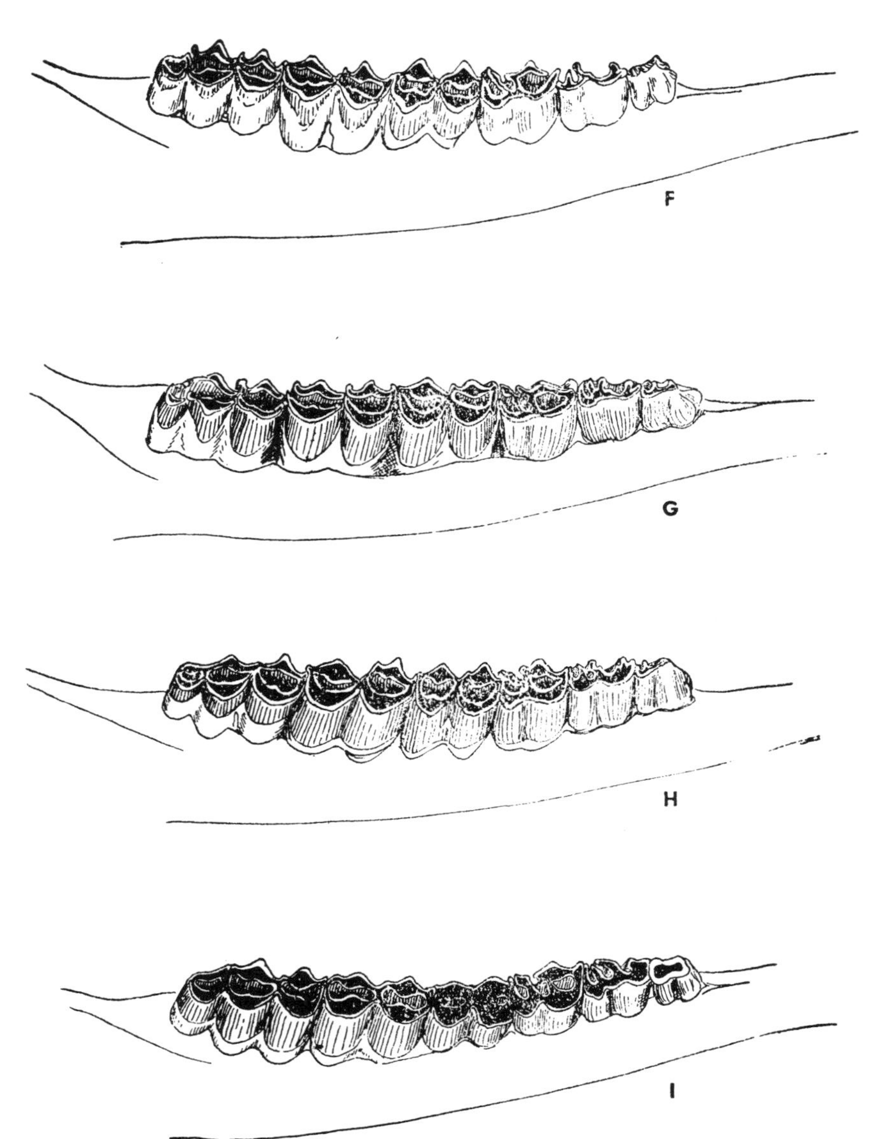

FIGURE 17. Development of molariform teeth of *Cervus c. nelsoni* from Jackson Hole, Wyo. Lower right teeth, from: *(A)*, A 2-month-old calf, with premolars only; *(B)*, an 8-month-old calf, with the first molar added; *(C)*, a spike bull nearly 2 years old, with two molars and with considerable wear on premolars; *(D)*, female, 2 years 5 months old, with all three molars and with permanent premolars unworn; *(E)*, male, 2 years 8 months old, anterior premolar and posterior cusp of last molar unworn; *(F)*, female, 3 years 2 months old, anterior premolar still unworn, but slight wear on posterior cusp of last molar; *(G)*, female, 3½ years old, showing some wear on both anterior premolar and posterior cusp of last molar; *(H)*, male, 3 years 8 months old, with anterior premolar and posterior cusp of last molar definitely worn; *(I)*, older male, with all teeth showing a heavy pattern of wear.

was not entirely closed. In addition, there was an opening opposite the posterior root of the molar. In the upper jaw the fenestra had disappeared completely and the maxillary had thickened.

In a 37-pound fetus, taken very near the time of birth, all

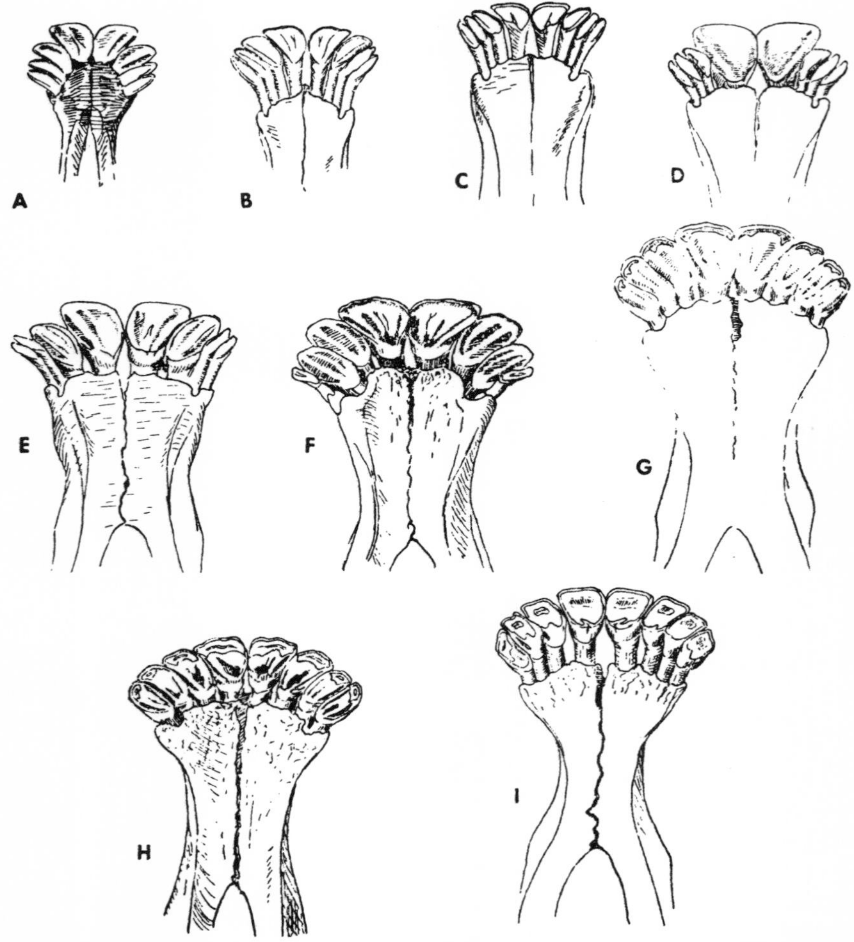

FIGURE 18. Development of incisors of *Cervus c. nelsoni*, Jackson Hole, Wyo., (all one half natural size). *(A)*, Deciduous incisors of female calf, 7 days old; *(B)*, of male calf, 7 months old; and, *(C)*, of female, about 14 months old; *(D)*, male, about 20 months old, which has acquired the first pair of permanent incisors; *(E)*, female, about 2 years old, four permanent incisors; *(F)*, female, about 2 years 5 months old, six permanent incisors; *(G)*, male, about 3 years old, eight permanent incisors showing only slight wear; *(H)*, female, about 4 years old, with wear a little more pronounced; *(I)*, female, 10 years old, teeth well worn.

openings had disappeared, but even in this case an occasional dark area on the bone marked the position of a premolar root.

After birth all signs of the fenestra have disappeared. It would appear that the rapid tooth development of the growing fetus takes place at the expense of the adjacent tissues, particularly the bony covering, and that only when the tooth has reached full size (about time of birth) does the adjoining bone fully develop and thicken. It is interesting to reflect that in old age some of these same bony areas have a tendency to disintegrate or be resorbed so as sometimes to expose parts of the roots of the molariform teeth.

To resume discussion of the growth of the teeth, in a 37-pound fetus only the milk incisors were in the process of eruption. As it was very near the time of birth for the fetus, the condition of its teeth may represent very well the condition of the teeth of a newborn calf.

Milk dentition in a calf 7 days old showed the upper third and lower fourth premolars appearing through the gums; the canines barely visible; and all the incisors erupted, but with their bases still mostly hidden in the gums. The incisors had attained practically full growth, except perhaps in the root portions, but because of their semi-circular arrangement the row of incisors was far from the length it would have been in an adult. Apparently only the first molars had developed within the jaw; they, of course, were not erupted.

In a calf about 5 weeks old all three milk premolars were fully erupted and were apparently full size. The first molars were still hidden within their sockets but were well developed, and small alveolar cavities posterior to them indicated the beginning of development of the second molars, although no hard structures were present.

In one November specimen about 5 months of age there was partial eruption of the first molars and in another, taken early in the month, the first molars were barely visible at one point. The milk canines had erupted.

In some winter specimens taken in January and February, when the animals were 6 to 8 months of age or even older, the first molars were fully erupted and the second well developed within their capsules; but in other elk of this age the molars had hardly advanced beyond the stage of those in the November specimens already described. The length of the row of the deciduous premolars in these winter specimens was practically the same as in a July specimen about 5 weeks old. The difference was so slight that it is safe to assume that by July the premolars of the elk calf have attained full growth.

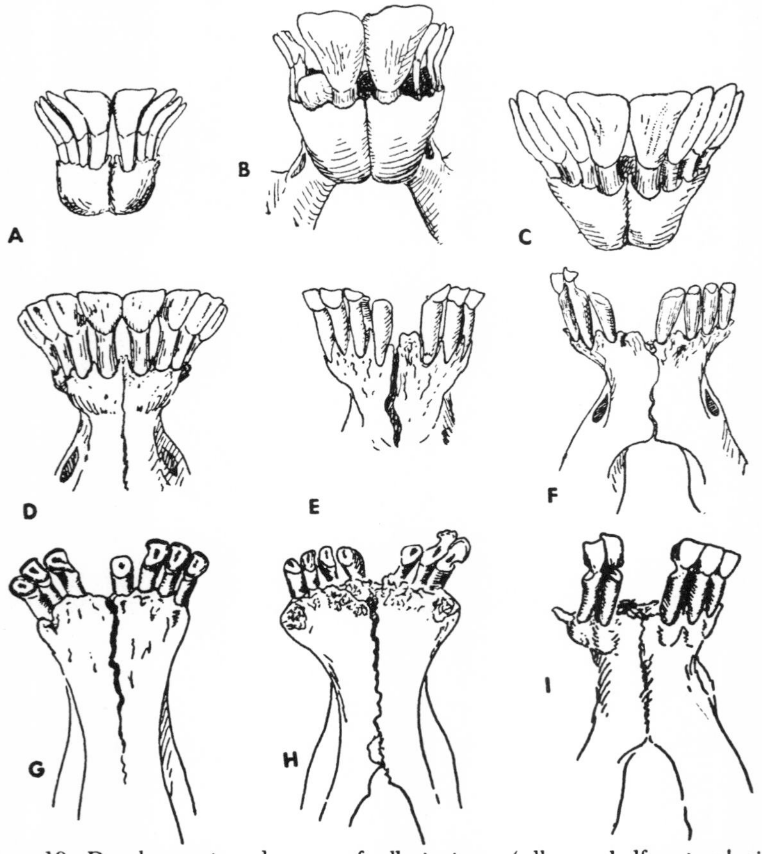

FIGURE 19. Development and wear of elk incisors (all one half natural size). *(A)*, Male, nearly 1 year old, all deciduous incisors; *(B)*, female, nearly 2 years old, two permanent incisors, and a third one ready to appear; *(C)*, female, nearly 3 years old, all permanent incisors; *(D)*, female 10 years old, incisors considerably worn; *(E)*, aged female, one incisor missing, the other badly worn; *(F)*, aged female, crowns of teeth practically all worn away; *(G)*, *(H)*, and *(I)*, aged female showing extreme wear, tending to destroy the central teeth first, and (as in *(H)* and *(I)*) with grooves in the sides which would eventually sever the tooth.

Elk dentition for the first year may be summarized briefly. As early as in a fetus of March 30 the milk dentition is formed but has not erupted and the first lower molars have begun to form. At birth of the calf only the milk incisors have erupted, but within about a week all the deciduous teeth have appeared and by November the first molars are coming through. During its first winter the calf normally acquires its complete milk dentition and the first molars (Plate 10).

Second Year Dentition

In a yearling female elk taken July 25 and in a male collected August 6, the second molars had not yet erupted. The female had dropped the milk canines, but the permanent canines had not appeared. The male still retained the milk canines, but the permanent ones were swelling in the gums. No permanent incisors had appeared.

A yearling male, taken September 10, 1943, during the hunting season, still retained all the deciduous incisors and the second molar had not yet appeared. This animal, however, had a diseased jaw, which may have retarded development.

In animals from 12 to 14 or possibly 15 months of age, taken during June, July, and August, the third molars were developing within their sockets. Apparently the second summer is a period of more rapid growth of the permanent dentition.

The deciduous canines are dropped some time during July and August of the second summer, when the elk is 13 to 15 months old; but apparently the permanent canines, or tusks, do not fully mature until autumn, when the elk is probably 15 or more months old. The permanent canines, not yet solidified within, are characteristic of animals in their second winter.

In a female taken in November, when about 18 months old, the second molars had erupted. In animals from this age to at least 2 years, the first and second molars are present.

Between the ages of about 18 to 24 months the elk acquires the two middle pairs of permanent incisors, but there is much irregularity in the time of their appearance. A female 18 months old (in November) already had these two pairs, whereas an animal approximately 22 months old (March 26) had but a single pair. The irregularity in the time of eruption of the incisors is indicated by the fact that in a series of 19 elk, 20 to 23 months old, 4 females and 4 males had one pair of permanent incisors each; one female and one male, 1½ pairs each; and seven females and two males, two pairs each. A spike bull killed about November 1, 1943, perhaps 17 months old, had already acquired two pairs of permanent incisors, the permanent canines, and two molars. An even younger spike, killed October 26, had two pairs of permanent incisors and permanent canines just emerging.

To summarize dentition during the second year, it may be stated that at the age of 18 months, possibly in some cases a little earlier, the elk acquires the second molars and thereafter until the animal is two years old (i.e., throughout the second winter), other permanent teeth erupt until the so-called yearling acquires the characteristic permanent dentition, which consists of the first and second molars, the four middle permanent incisors, and the hollow permanent canines (Figures 16 to 20 and Plate 11).

Third Year Dentition

Some time during the autumn of the third year, when the elk is about 30 months of age, the milk premolars are generally shed, the fresh, unworn permanent premolars are in place, and the third molar and the remaining permanent incisors also are usually fully erupted. There is a great deal of variation in the time of this change. In a female that died during the winter and was found in May, the fourth pair of incisors had not yet fully erupted. Another winter-killed female still carried the milk premolar caps, and the animal's third molars and fourth incisors were not fully erupted. Therefore, the complete permanent dentition may not be acquired in some individuals until they are approximately 33 months old, possibly older, indicating a spread of six months or more in which various individuals acquire the complete dentition. By the time the animal is three years old, the new teeth have begun to show a little wear, though this may not be readily apparent on the third and fourth incisors. The posterior cusp of the third molar may also show little if any wear at this stage (Figures 16 to 20 and Plate 11).

Fourth Year Dentition

During this year the canines become somewhat more flattened

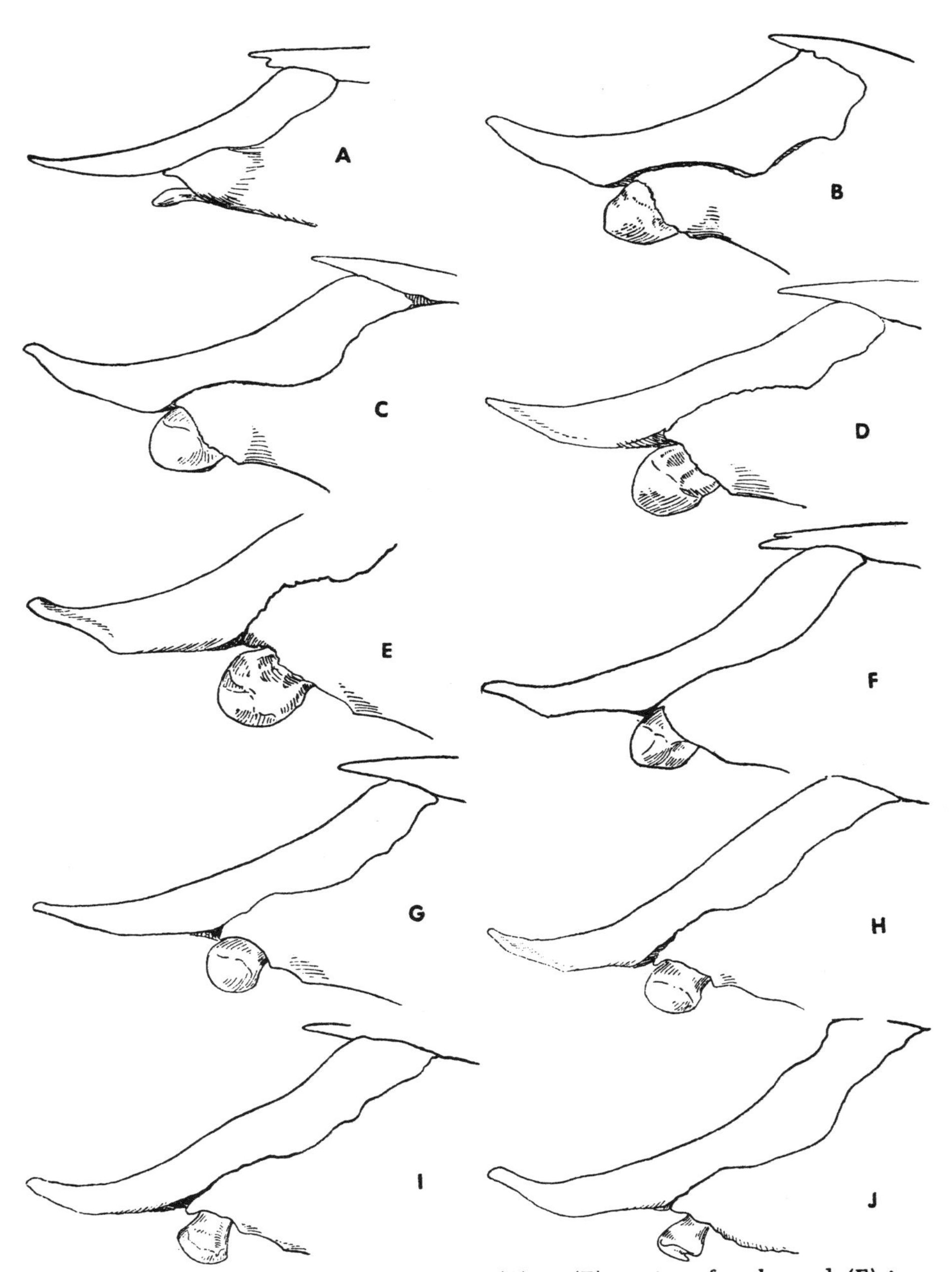

FIGURE 20. Development of elk canines. *(A)* to *(E)*, canine of males and *(F)* to *(J)*, of females (all one half natural size). The deciduous incisor of the female, not illustrated here, is similar to that of the male, as in *(A)*. *(A)*, about 8 months old, with deciduous canine; *(B)*, nearly 2 years old, with permanent canine; *(C)*, nearly 2 years old, with point of canine somewhat worn; *(D)*, nearly 3 years old; *(E)*, 4½ years old. Females: *(F)*, nearly 2 years old, canine somewhat pointed; *(G)*, nearly 3 years old, canine rounded; *(H)*, adult, canine well worn; *(I)*, old, canine much worn; *(J)*, aged, canine worn and grooved.

laterally, the third molars show wear all the way to their posterior margins, the anterior lower premolars are a little more abraded, the two outer incisors begin to wear a little, and the two middle incisors become noticeably worn (Figures 16 to 20).

Dentition of Older Animals

After its fourth year the age of an elk cannot be told with precision by the teeth, though one can estimate roughly for the fifth year. By reference to Figure 17H, it will be noted that at the age of 3 years 8 months, therefore approaching 4 years, the posterior cusp of the last molar shows decisive wear and a flattened surface. In a specimen 4 years 3 months old the wear is not noticeably greater. In two other specimens in the fifth year the wear is definitely more pronounced.

In estimating age of animals in the 3- to 5-year group, it will be noted that there is a particular type of variation. Wear on the posterior cusp of the last molar may be unusually heavy, and unusually light on the anterior premolar (or the reverse). This is no doubt due to the fact that the lower tooth row is longer than the upper, and wear at one end or the other may depend on the individual chewing habit of an elk. In fact, in one example, a slight difference in wear may be observed between the anterior premolar of the left and right jaws.

Beyond the fourth year aging becomes guesswork. In due time, with the accumulation of specimens of known age, a more comprehensive age key may be provided. But at the present time such material is inadequate.

Wear of the Teeth

Use of the teeth begins in the first month of the elk's life. Wear begins on the three milk premolars that are present, and during the first summer the sharp edges of these teeth are dulled. By the following winter, when the calf is 6 or more months old and the first molar has erupted, the premolars have worn down into a definite pattern (Figures 16 and 17) and even the new molar has lost the sharpness of its edges, particularly on its anterior half.

In the second winter, the first and second molars are in use and show moderate wear. The anterior milk premolar is very badly worn but the other two show less and less wear posteriorly

(Figures 16 and 17). This is, undoubtedly, due to the presence of the new large molars, which take the brunt of the chewing and somewhat shield the adjacent premolars.

In the third summer, after about 27 months of use, the milk premolars are practically worn out and remain for a time as flat caps covering the permanent premolars, which have then erupted.

To understand the tooth pattern caused by wear it is necessary to consider the morphology and mechanism of the jaws and dentition as a whole. As is well known, the tooth row of the lower jaw is considerably longer than that of the upper jaw, owing in part at least to the fact that the posterior lower molar is longer than the others and has three cusps instead of two. The lower teeth are individually narrower than the upper, however, and the lower tooth rows closer to each other than the upper. As a result, and because of the lateral movements of the lower jaw, the comparatively narrow lower row slides about on the broad receiving surface of the upper. The lower teeth receive the heaviest wear on the cheek, or buccal, side, while the upper teeth correspondingly receive most wear on the tongue, or lingual, side, and the grinding surfaces become slanted accordingly.

The profile of the lower row is concave; that of the upper row is convex. This relationship sometimes becomes greatly accentuated in old age. There is a tendency for an angle to form in the upper jaw, the projecting point being formed by the first molar, sometimes in combination with the adjacent premolar; and this angle wears a depression near the middle of the lower row. Sometimes the second molar forms the point of the protruding angle. Because of this feature the first lower molar may be worn out and have only the roots remaining or may be missing even while in the same jaw the anterior premolar is still sound and only moderately worn (Figure 21). As a matter of fact, the anterior premolar, PM no. 2 in the series, receives little wear, and that only on its posterior half, as it tends to project forward beyond the end of the upper row.

Sometimes two opposing teeth wear unevenly, as is shown in Figure 21 C. In this case the back cusp of the upper third molar has resisted wear and developed into a long fanglike snag that has worn away the posterior cusp of the opposing lower third

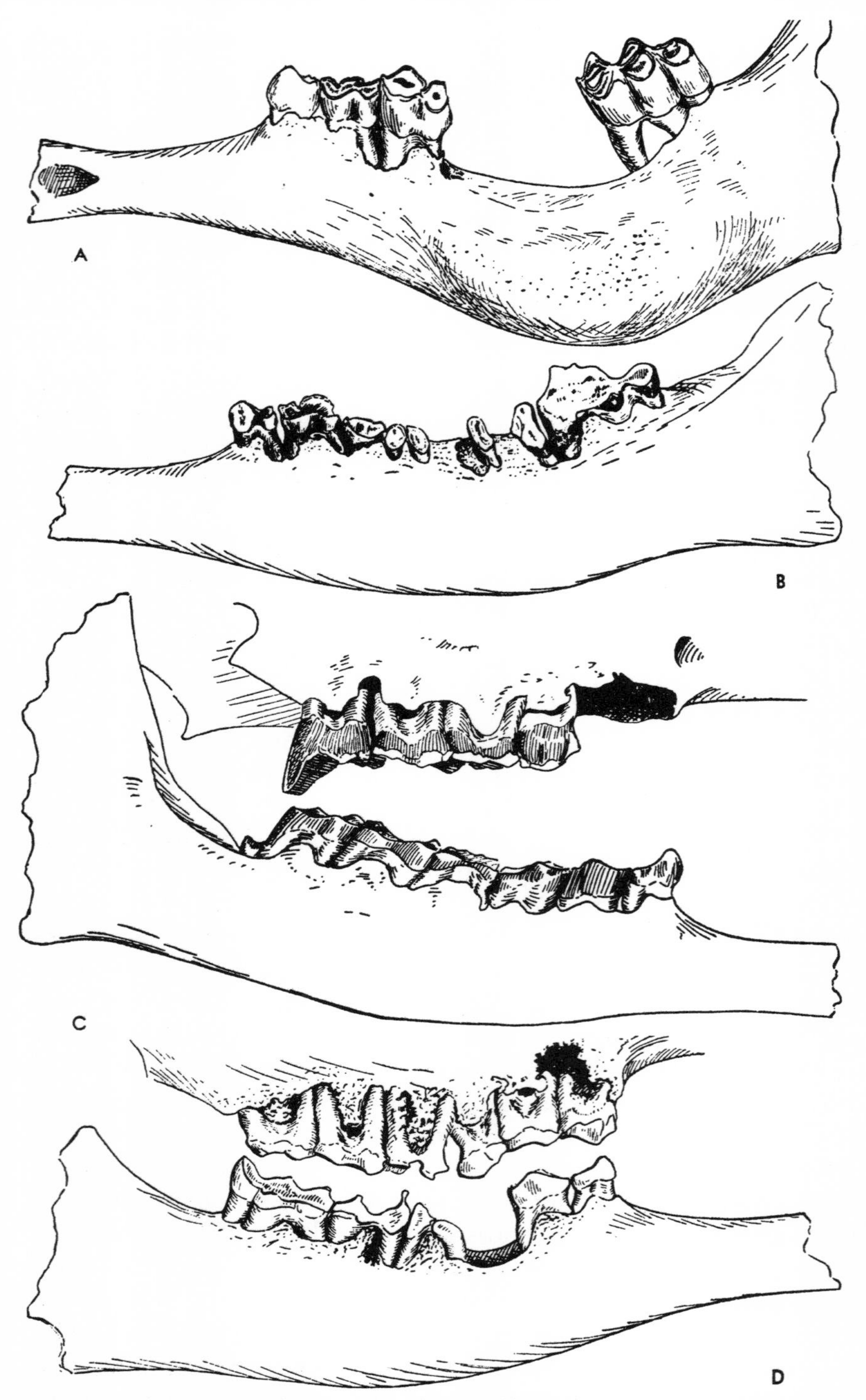

FIGURE 21. Teeth of old female elk (all one half natural size). *(A)*, Lower jaw of old female; middle teeth are lost and bone indicates previous diseased condition; *(B)*, lower jaw of old female. The center teeth have almost disappeared, though the anterior part of the anterior premolar is still unworn; *(C)*, teeth show uneven wear of upper and lower posterior molars; *(D)*, teeth show necrosis about the roots.

molar. A variety of such abnormal conditions were observed. Such uneven wear as noted probably hastens the time when the animal suffers from poor dentition, for when the tooth is worn down to the roots and gums at any one place, the animal is subject to injury from rough forage.

The milk incisors are often—although not always—shed before they have undergone severe wear. The permanent incisors wear slowly and keep a sharp anterior edge for many years. The middle ones persistently receive the hardest wear, the lateral ones lagging a little in degree of attrition. Perhaps this has brought about the large size and strength of the middle pairs and the diminishing size of the others laterally. Eventually, however, the incisors are worn out completely, the crowns may disappear, and all the wear be applied to what is left of the roots. The animal is then a "gummer" and has reached extreme old age (Figure 19).

When freshly erupted, the incisors are large and thin-edged, making an excellent tool for nipping herbage (Figure 19). Continued wear takes off the thin edge, and as the tooth wears down, the surface occluding with the dental pad becomes flatter until at the lower level of the tooth a large, squarish bearing-surface forms, with a sharp edge at its anterior border (Figure 18I). After the greatest diameter of the tooth is passed in the wearing process, the biting surface decreases in area and spaces may appear between the teeth. Finally, if the animal survives long enough, the crowns are completely worn away and in some cases even the roots of the middle incisors (which have borne the brunt of foraging) may disappear (Figure 19). The teeth bordering the gap left when a tooth is lost receive undue wear on their lateral surfaces, and sometimes deep grooves are worn at the base of the crowns. In Figure 19I note that the crowns of the second and third incisors have been almost severed from the roots by this lateral wear and that the outer surface of the third incisor shows lateral wear, probably because of the loss of the fourth incisor and the realignment of the remaining teeth.

It is at first a little puzzling to account for such severe attrition on the sides of the teeth, but undoubtedly it is caused by pulling at the vegetation. As the animal pulls up to sever the plants held between the incisors and the dental pad, plant stems are no doubt crowded into the gaps and as these stems are not severed at once, they tend to slip out, rubbing the adjacent tooth surfaces in passing. Repeated day by day, this process continues to file away and undermine the teeth from the side. This can be the only explanation. There may be, and probably is, a more or less lateral thrust of the muzzle sometimes when the animal is feeding that tends to accentuate the lateral wear.

The canines, commonly known as tusks, also pass through a cycle. The deciduous canine of the young elk (Figure 20A and Plate 10) shows little change and apparently experiences little if any wear. It is dropped when the permanent tooth, of large and apparently swollen form, erupts in its place. The new canine (Figures 20B, F; Plate 11), while almost rounded, comes to an obtuse point that may be somewhat worn off even before the animal is 2 years old (Figure 20C). During the elk's second year the pulp cavity of the canine has not yet filled with solid material and the tusk is spoken of as "hollow." During the elk's third year the canine becomes rounded (Plate 11). It follows approximately the same course in both sexes but is much better developed in the male. In the adult bull it becomes somewhat flattened laterally and often develops the brown-stained, polished surface so much admired by trophy hunters. The tusks of the female remain small and comparatively insignificant and soon wear down to small diameters (Plate 12).

At first glance the wear of the tusk presents a puzzle. It never comes in contact with another tooth or with the dental pad. It is situated opposite the gap of the tooth row of the lower jaw and only rests against the tissues of the adjacent part of the upper lip and the tongue, yet the wear on it proceeds as regularly as on the other teeth.

A clue to the cause of wear may be found in the fate of the incisors, described above. The passage of vegetation past the canines apparently produces attrition, and possibly a certain amount of the browse catches on this tooth, especially after the incisors have become less effective. Attention is called to Figure 20J, showing a groove and an actual notch worn into the posterior edge of the canine of an aged cow elk. This is not an isolated instance. Many such teeth were noted in the field, and in all of

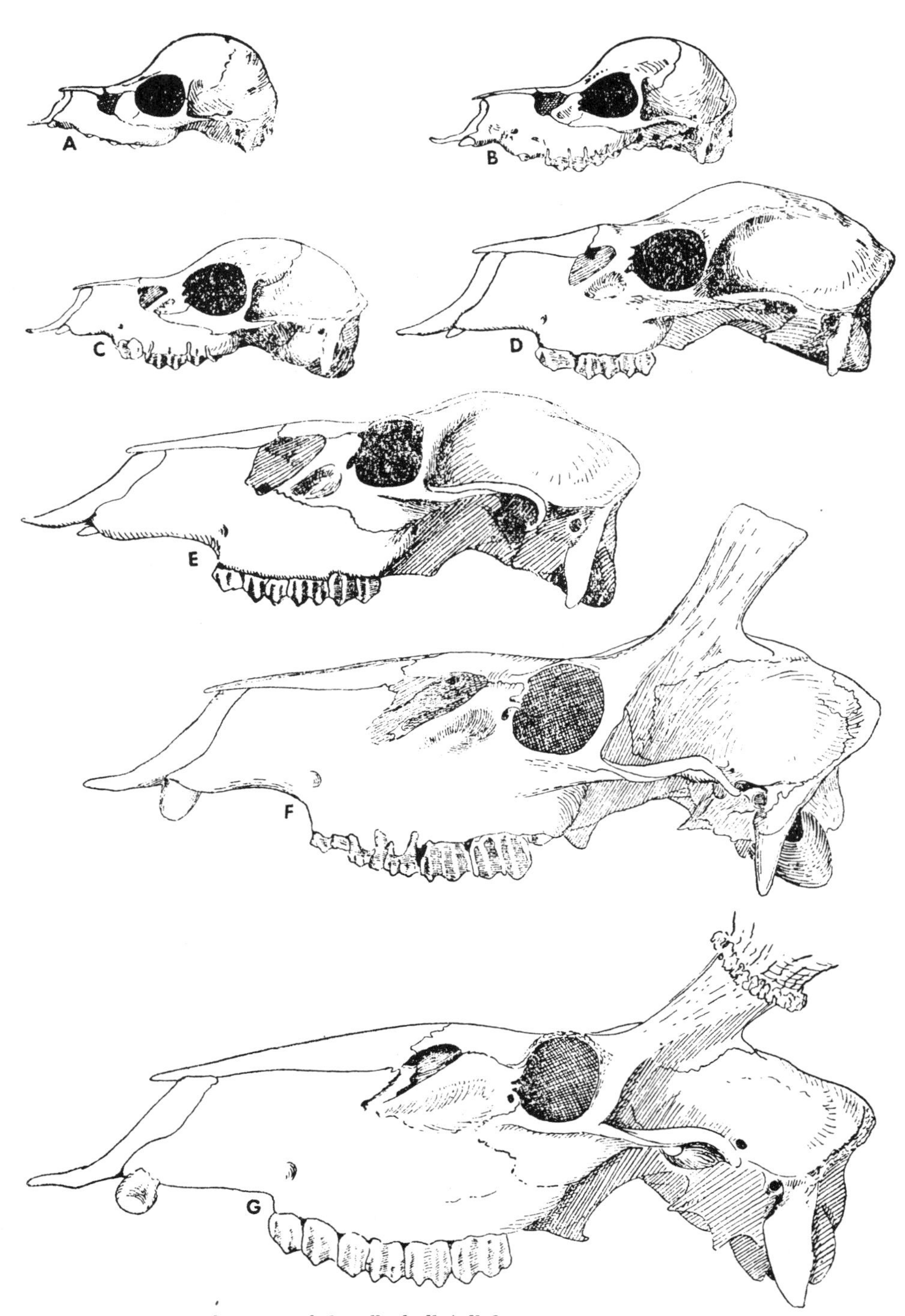

FIGURE 22. Development of the elk skull (all figures one quarter natural size). *(A)*, Skull of fetus weighing 6 pounds 10 ounces; *(B)*, of female fetus weighing 16 pounds 4 ounces; *(C)*, of female 7 days old; *(D)*, of calf, nearly 2 months old; ***(E)*, of male, about 8 months old; *(F)*, of male, nearly 2 years old; and *(G)*, of male 4½ years old.**

them the grooves were on the posterior border and to some extent on the sides. It would indicate that this wear resulted from pulling at vegetation and perhaps letting it slip a little. Probably the biting-off of vegetation is not clean cut or completely efficient.

GROWTH OF THE SKULL

Throughout the first year the sutures of the skull, with few exceptions, remain distinct. The fetal development of some of the cranial bones was observed briefly. For skull proportions see Figures 22 and 23.

The Interparietal and Parietal

In the elk fetus the interparietal bone is more or less distinct and in some calves persists a short time after birth. It is irregular in its development. A small fetus weighing 6 pounds 10 ounces showed this bone well defined but with its median suture practically absent, there being only a slight indication of it at the anterior border. In another fetus, 11¼ pounds in weight, the interparietal was distinct and had a median suture extending about one-fourth of the distance across from the anterior border; in a 12-pound fetus it was nearly divided by a median suture that extended at least three-fourths of the distance back from the anterior border; and in a fetus weighing 16¼ pounds and in another weighing 24½ pounds it was completely divided by a median suture.

A 37-pound fetus had nearly lost its interparietal. On the dorsal surface of the skull there was a short indication of the original lateral sutures at the posterior border. The median suture of the parietals extended all the way to the supraoccipital, having come in line with the median suture of the interparietals. Examination of the ventral surface of this bone group showed that the interparietal was coalescing with the parietals.

Even in a 7-day-old calf the interparietal was still present but had no median suture. The median suture between the parietals was still present, however. In a 2-month-old calf there was no trace of the interparietal and the median suture between the parietals had disappeared. In elk in all subsequent stages of development there was no indication of these sutures.

Occipital

All the elements of the occipital are distinct in the fetus, in the newborn calf, and in a calf 2 months old. In a calf 2 months of age the posterior portion of the basioccipital on each side is adapting itself more fully for taking part in the joint formation of the condyles. In an 8- or 9-month-old calf this suture of the basioccipital may be distinct but in many cases fusion has taken place, especially in the condylar surface. The exoccipital-supraoccipital sutures persist much later, being still present when the calf is 10 months old and even until it is 2 years of age.

During the third year in a great many specimens, in males particularly, the elements of the occipital bone are still separated by

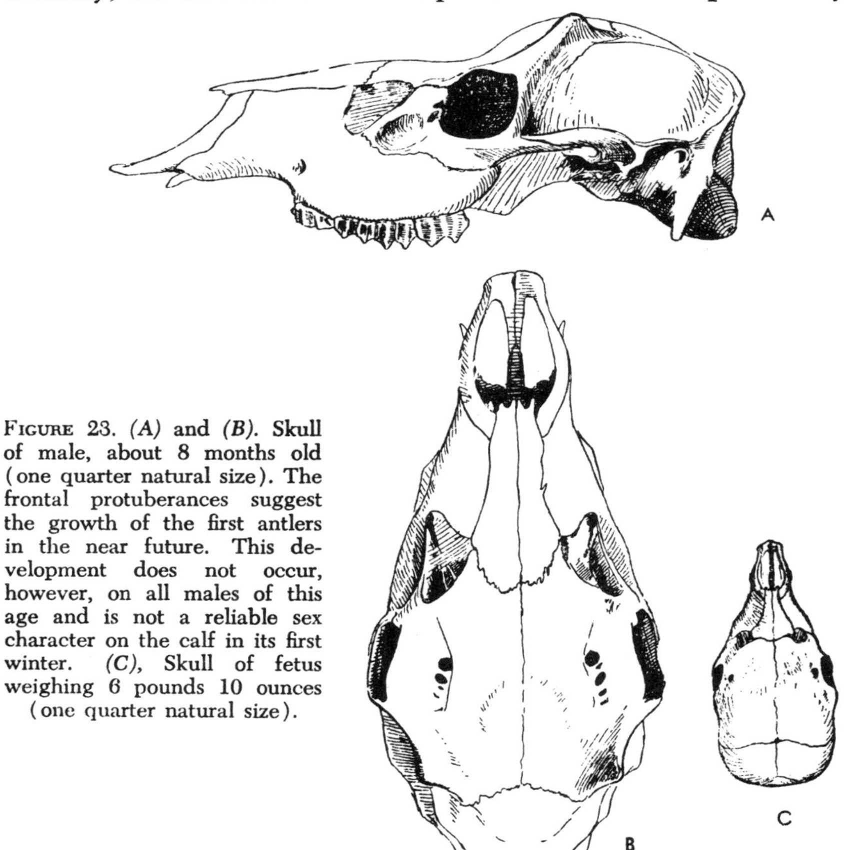

FIGURE 23. *(A)* and *(B)*. Skull of male, about 8 months old (one quarter natural size). The frontal protuberances suggest the growth of the first antlers in the near future. This development does not occur, however, on all males of this age and is not a reliable sex character on the calf in its first winter. *(C)*, Skull of fetus weighing 6 pounds 10 ounces (one quarter natural size).

sutures, although in a great many females and in some males they have fused, having no visible sutures, and the basioccipital has merged anteriorly with the sphenoid.

In 4-year-olds these elements invariably appear to be fused with the suture between the basioccipital and the sphenoid has completely disappeared. In older animals most of the skull sutures continue to tighten up but remain discernible, and the bones of the cranium become very thick. As mentioned in the development of the antlers, the pedicels of the horns become much shorter and of much greater diameter with the advancing age of the animal.

GLANDS

In the elk, as in many other deer, the lachrymal gland is well developed and the lachrymal pit is prominent in the skull.

A metatarsal gland also is present, situated only a short distance below the hock on the outside of the hind leg. The golden brown streak of color on the back of the leg widens out to encircle this gland, and a definite streak of white hairs marks its center. In the tule elk examined, the location of this gland was less conspicuous, partly because of the more uniform color pattern of the leg and also because the gland was entirely overgrown with hair. In the calf elk, the gland is marked by a dark brown patch on the leg and the streak of white hairs in the center.

No indications of the tarsal gland or interdigital glands of the foot have been found in the elk.

Surrounding the elk's tail vertebrae, but closely attached to the skin, is an extensive soft, glandular brownish mass, somewhat divided into lobes, one on either side of the tail bone. The division is only slightly indicated, however, and the tail itself appears swollen by the inclusion of this gland.

DUNG

"Elk sign," or droppings are useful as a means of identification in the field, and familiarity with their characteristics is important.

As in the case of other deer, elk droppings are in pellet form during winter (Figure 24, C and D) and more amorphous like those of cattle in summer (Figure 24, A and B), the form depending upon the relative succulence of the seasonal foods. About the first of May, although still in pellet form, the droppings are softer and tend to be larger than in winter, and soon thereafter they become somewhat flattened. By the middle of May the pellet shape begins to disappear, and by the first of June the droppings are soft and resemble those of domestic cattle, in miniature, although they remain somewhat firmer. The transition here described indicates accurately the increasing abundance of green food in the elk diet through the spring.

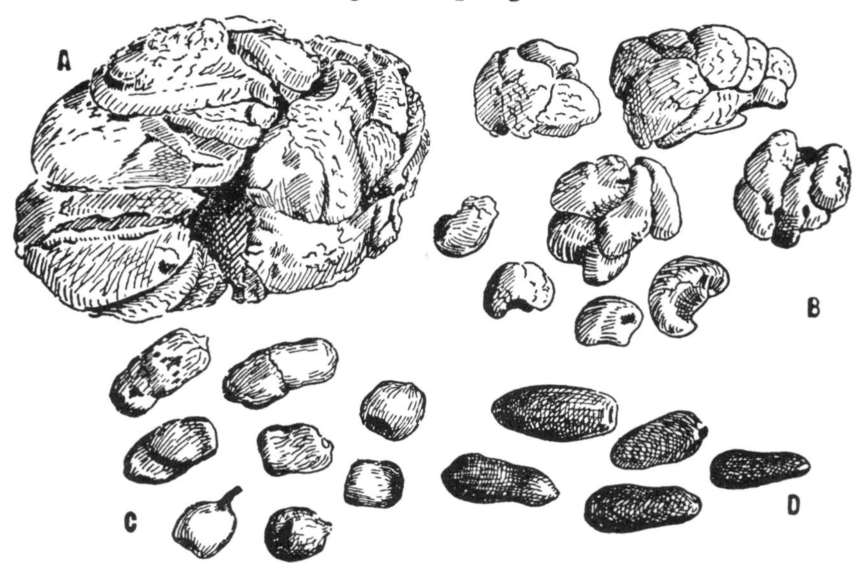

FIGURE 24. Elk droppings: *(A)*, The amorphous type characteristic of the season of green forage; *(B)*, somewhat more pelletlike but soft and also common when animals are on green feed, Sun River, Mont., July 21, 1934. *(C)*, Characteristic pellet form assumed when elk are on dry feed, as in autumn and winter, Sun River, Mont. *(D)*, Elongated pellet of the Roosevelt elk, form taken when animals are on a browse diet. These often resemble moose droppings. Hoh River, Wash., March 1, 1935. (All figures one half natural size.)

In fall, the process is reversed. Large, soft pellets have been observed as early as August 17, but a comparatively amorphous type of dung ordinarily prevails until the end of August at least and occasionally occurs until the middle of September. Shortly after that the pellet shape is assumed, and the droppings become harder as the winter advances.

A confusing feature of elk droppings is their variation in shape. Normally the form is roughly globular, with two more or less flattened surfaces or "ends" and other irregularities (Figure 24C), but often it is elongated and resembles that of some moose pellets. Apparently this is the result of a preponderant diet of browse. In the Olympic Peninsula, where the winter diet is more largely browse than in the Rocky Mountain area, the elongated form of dung pellet is prevalent. It is also noted in winter among the deer, which are heavy browsers.

Figure 24C reveals a curious process by which two normal pellets tend to fuse, end to end, to form an elongated single dropping, and it suggests that possibly some of the variations are produced in this way. Perhaps, after all, the form is determined by the length of time the material is held in the colon—the longer it is held the more consolidated and the more elongated the pellet. The type of forage used may determine the length of this period, undigested dry browse doubtless being retained longer than the unused parts of succulent food.

TRACKS AND GAIT

The foot of an elk is heavier and blunter than that of a deer, but smaller and much less pointed than that of the moose. Elk tracks are easily distinguished from those of deer or moose but may readily be confused with those of cattle, especially with those of young cattle. In general the elk track is trimmer, or less "blocky," than the average cattle track yet this does not wholly describe it. There are intangible differences difficult to convey in words, and experience is a necessary guide. Figures 25 and 26 illustrate elk tracks.

If alert, the elk walks with head held high; otherwise, with head and neck extended forward and held rather low. It produces a trail similar in pattern to those of related animals. When trotting, the elk holds the muzzle horizontal and sometimes thrusts the nose even above the horizontal (Figure 27). There is nothing unusual about the gallop, in which the actions are similar to those of many other hoofed animals. On a few occasions elk were observed to use a peculiar bouncing gait for a short distance, a gait not unlike that of the mule deer in which the four feet are swung forward in unison. Young and Robinette (1939) ascribed this gait to young animals in a playful mood and apparently observed it chiefly among the calves in summer. This is no doubt the proper interpretation, though the writer has observed it in late winter also.

The writer has no extensive reliable tests of the speed of elk. On a few occasions he raced a group of elk in open country with a saddle horse. Some were overtaken, but others kept ahead of the horse. The speed attainable by these animals has been stated to be 40 to 50 miles an hour, presumably for short distances. Such estimates appear too high. Cowan (1947: 160) timed elk running

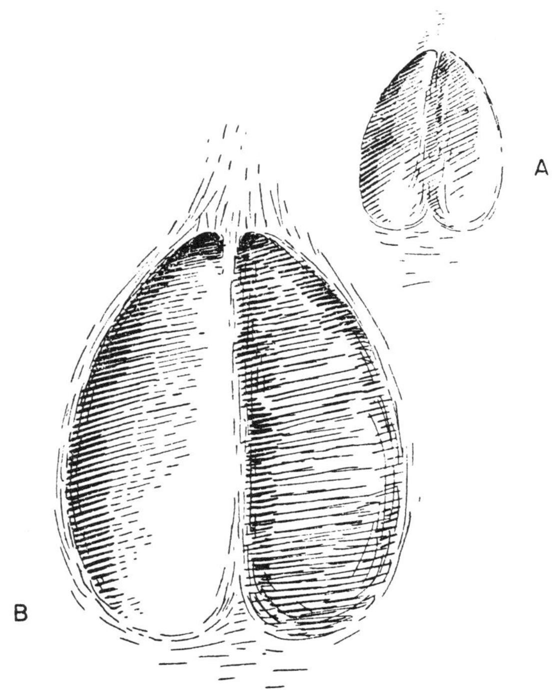

FIGURE 25. *(A)*, Track of young elk calf. Jackson Hole, Wyo., June 13, 1932. *(B)*, Track of adult cow elk on hard, sandy soil. Jackson Hole, Wyo., Aug. 23, 1928. (Both one half natural size).

FIGURE 26. Track of adult cow elk, *Cervus c. roosevelti,* on sand bar. Hoh River, Wash., March 6, 1935.

28 and 29 miles per hour, and considers their speed somewhat greater than that of the timber wolf.

FIGURE 27. When trotting the elk holds the muzzle horizontal and sometimes thrusts the nose even above the horizontal.

JUMPING

The ordinary cattle fence is no obstacle to elk; the animals cross such enclosures at will. They do not jump the least bit higher than necessary, however, and as a consequence often scrape the top pole of a buck and pole fence or sometimes become disastrously entangled in a wire fence. Where elk habitually enter a fenced field or a hay corral, the fence is usually more or less broken at the points of crossing.

In the Olympic Mountains of Washington the Roosevelt elk will sometimes scrape the top of a big log when leaping over or will even land on top before going down the other side.

That an elk can jump much higher, though, than is necessary to clear the average pole stock fence, is illustrated by a number of observations, such as the following from February 1935 records of the Biological Survey:

"Geno A. Amundson, in charge of the Niobrara Game Preserve, Nebr., reports an incident that is of interest in showing how high an elk can jump . . . Mr. Amundson relates that while elk at the preserve were being separated with a view to cutting out some cows for shipment, two bulls were placed in separate pens. Later in the day it was found that one of them had jumped the plank fence 7 feet 6 inches high surrounding the pens and had joined the herd in a holding pen. He had knocked a splinter off the fence, but joined the herd apparently in good shape."

Ordinarily, though, a fence that high, whether a drift fence or one enclosing a large pasture, is high enough to hold elk.

SWIMMING

Elk are good swimmers, being quite similar to other deer in swimming ability (Figure 28). They readily cross rivers, even small fawns following their mothers. One elk was seen swimming across Two Ocean Lake in Jackson Hole, Wyo., and when elk were introduced on Afognak Island, Alaska, one of them swam across to Whale Island, a distance of a mile, and several crossed to Deranof Island.

VOICE

The Calf Call

The newborn elk calf has a lusty voice (O. J. Murie, 1932 b). When disturbed or alarmed, as when temporarily separated from its mother, it emits a loud squeal, high-pitched and often piercing that may be written "Eee-e-e-e-e-uh!" Sometimes it is short, sometimes long, depending on the feeling of the animal. In the normal course of the calf's calling to its mother, with no danger present, the sound is more or less as indicated above, moderately short; but if the calf is seized as it lies in hiding, it may give vent to its terror in a startling medley of squeals and prolonged screams, which sometimes bring the mother on the run. Again, when beside the mother and undisturbed, the calf may utter a somewhat different version of the call, more in the nature of a low bleat. Such calls are difficult to detect in wild herds.

A young bull calf, taken June 6, 1928, was raised at the Elk Refuge at Jackson, Wyo. As the animal matured, the baby calls persisted but became deeper in tone. As late as August 21, 1932, when the animal was well over 4 years old and had its fourth pair of antlers, this low bleat or modified squeal was still heard, an incongruous sound coming from so large an animal. It seemed to

Figure 28. Elk are good swimmers, being quite similar to other deer in swimming ability.

express mild annoyance, as at being approached too closely when not in the mood to be handled or as at having some tidbit of which it was very fond, such as a cigarette, withheld.

Adult cows reply to their offspring with a note of essentially the same form as the calf squeal, but stronger and more mature. When a calf has been disturbed and has wandered from its original hiding place, the returning mother will call to attract the little one's attention and guide it to her.

In summer, a traveling band of cows and calves may be traced in its course by the noisy squealing, with which they maintain a more or less continuous communication that probably tends to keep the animals together, although in rapid flight mothers and offspring must become mixed. The vocalization of traveling herds varies from an occasional call in a slowly moving group to a confusion of excited sounds in a herd fleeing from danger. In the latter case the greatest volume of calling occurs when the animals slow up after a preliminary burst of speed. Such rapidly moving bands were repeatedly described in field notes as accompanied by "squealing, barking, roaring and bugling, making a big racket."

It is interesting to note that other herd animals have comparable calls when banded together, in each case characterized chiefly, it would seem, by the calling between mothers and offspring. Domestic sheep have their monotonous bleating; cattle, their characteristic bellowing; and caribou, their babel of grunting calls.

The calf squeal appears to be a primitive sound which issues from the throat with a minimum of modification by the vocal organs. It can hardly be considered a means of direct, intentional communication between individuals, particularly when it is a squeal of pain or terror. It is more likely a spontaneous expression of feeling or emotion, which, however, may be correctly interpreted and appropriately responded to by other animals chancing to hear it.

It may be concluded that the various forms of the calf-elk call indicate terror, anxiety (as when separated from the mother), or desire (as perhaps when the calf seeks its mother for milk), or may be more or less minor acknowledgments of attention from one calf to another, in which case they are low, intimate vocal expressions. Companions hearing these sounds appear to understand the emotion involved. Sometimes, as when the mother calls to the calf to guide it to her, one cannot escape the thought that after all here may be a rudimentary form of conscious effort to convey an idea to another individual.

The Alarm Call

When alarmed or suspicious of lurking danger or possibly when in a state of excitement and confusion, all elk over 1 year old, male or female, use a better defined call: a sharp, startling explosive "Eough!" a hint of squeal mingled with a loud, hoarse bark. This danger call, or, perhaps better, this call of mental disturbance, may be used at any hour, in any season.

Although the call serves as a warning to other members of the herd, the writer is convinced that there is no conscious desire on the part of the elk that gives the alarm to convey that idea to others. Rather, the call appears to be a spontaneous outburst expressing the emotion of the individual. Just as human beings are able to interpret correctly a cry of pain from a dumb animal, so the other elk instinctively recognize the emotion involved in this call and become alert to detect the danger. Perhaps the nature of this call may roughly be compared with that of the human's cry of sudden fear or spontaneous "Oh!" of surprise, pleasant or otherwise, which is readily interpreted by bystanders.

An incident in June 1932 will illustrate the process. The writer had reached the top of a high hill, in Jackson Hole, Wyo., and was looking into the valley on the other side, where he spied a small herd of elk, chiefly cows and calves, peacefully grazing in a small open meadow, the calves nosing their mothers or frisking about in casual play. Suddenly one of the cows came to attention and uttered a sharp bark of alarm as she gazed steadily down the valley. Instantly all the other cows froze into alert attitudes, peering into the surroundings for the source of danger. After a time the bark was repeated. The cows turned their heads this way and that, searching in all directions. The calves sensed the feeling, crowded around the adults, and looked about. From time to time some of the animals shifted their positions, and some seemed inclined to begin grazing. However, even though the danger had not materialized so far as could be seen, the nervous tension was too great and the animals gave way to their fears and fled. The elk calls in this case clearly indicated alarm.

Another instance also is worth description. Late in May, in an aspen grove in the same locality, the writer spied a lone cow, whose calf, he suspected, was hidden nearby. Creeping up close to the cow, the writer watched her feeding for a long time. Then, as an experiment, he gave the coyote howl. Up came her head and she stood still, apparently looking for the coyote but searching in the wrong direction. Presently the writer repeated the howl at a distance of less than 50 yards. The cow started toward the sound at once but was passing by, on the run, having missed the exact direction, when the writer stepped into view. Instantly she was startled and ran off into a willow patch, barking in alarm. There she took her stand at a supposedly safe distance and continued barking. Obviously her calf was hidden somewhere in the vicinity; a little later the calf was found 200 yards away.

In this case, at the sound of the coyote call the mother had silently sought the animal, which she apparently regarded as a potential danger to her calf but not as a serious threat to herself. She felt she could cope with the situation, and she did not give the danger call. Upon seeing the human being, however, she recognized an hereditary enemy too powerful for her to attack or drive away and so at once gave the bark of alarm signifying her fear and concern for the safety of her calf, as well as her own helplessness. There were no companions with whom to communicate, and the call here was clearly an impulsive expression of emotion, although directed, to be sure, toward the intruder.

Near camp on the evening of August 16, 1928, several cow elk spied the writer's tent and after barking a little disappeared in the woods. One cow, however, continued barking at frequent intervals. Darkness came, but still the sounds continued to come for at least an hour and three-quarters longer. This prolonged performance surely did not involve sudden fright or continual warning to other elk. Rather, the animal seemed very much puzzled and curious; perhaps also she vaguely sensed possible danger. In such a case, curiosity and fascination for the lighted tent could easily be controlling forces, comparable to the excited chatter of the red squirrel under certain conditions.

PLATE 8. A 2-year-old Rocky Mountain elk, Jackson, Wyo. *Left.* May 17, 1930, with new antlers forming. *Right.* June 5, 1930, with brow and bez tines nearly full grown and the trez tine about to branch off. (U. S. Fish and Wildlife Service photographs.)

PLATE 9. Bull elk on Teton National Forest, Wyo., with velvet antlers nearly full grown. Old bulls are inclined to be solitary in summer, feeding and resting in the luxuriant vegetation. (U. S. Fish and Wildlife Service photograph.)

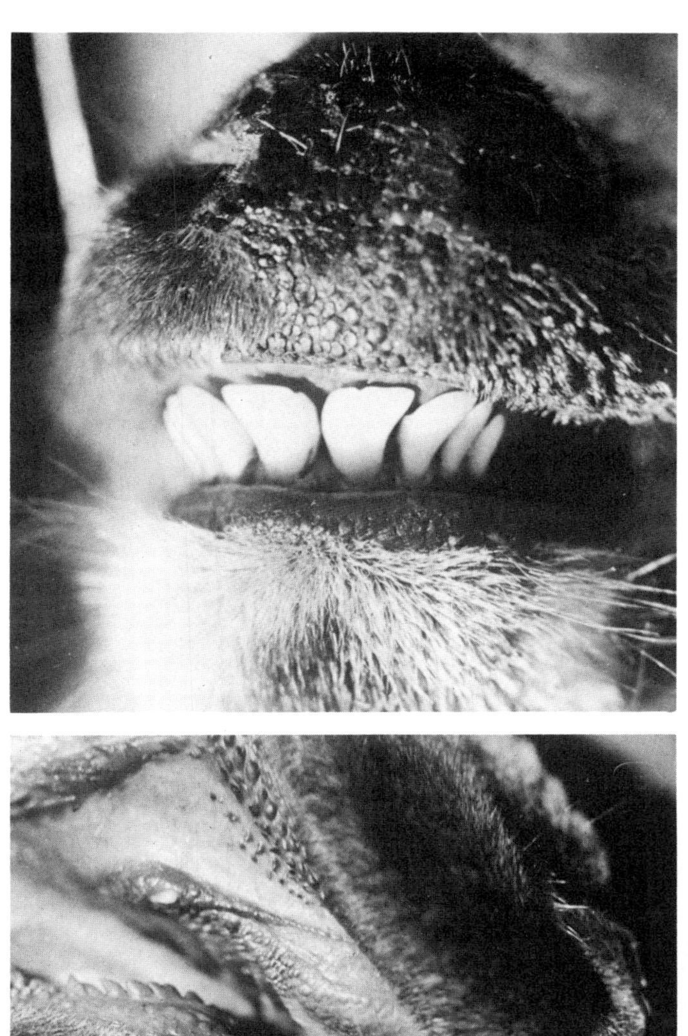

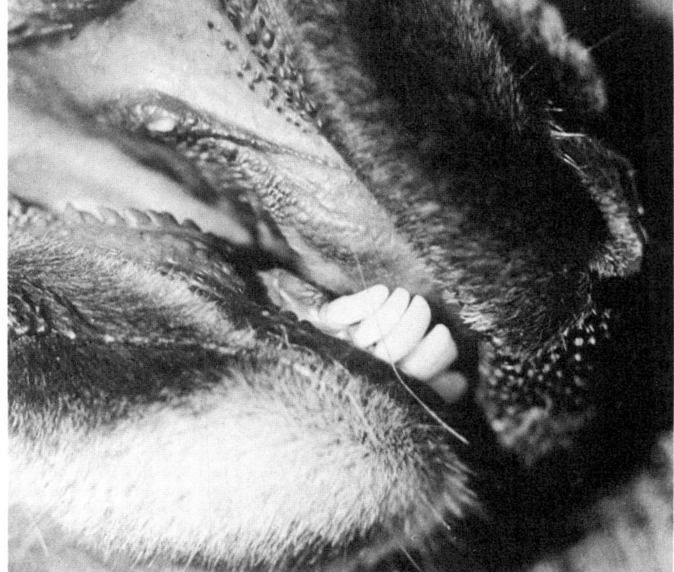

PLATE 10. *Above.* Deciduous incisors of calf elk nearly 1 year old. *Below.* Deciduous incisors and canine of calf elk nearly 1 year old. (U. S. Fish and Wildlife Service photographs.)

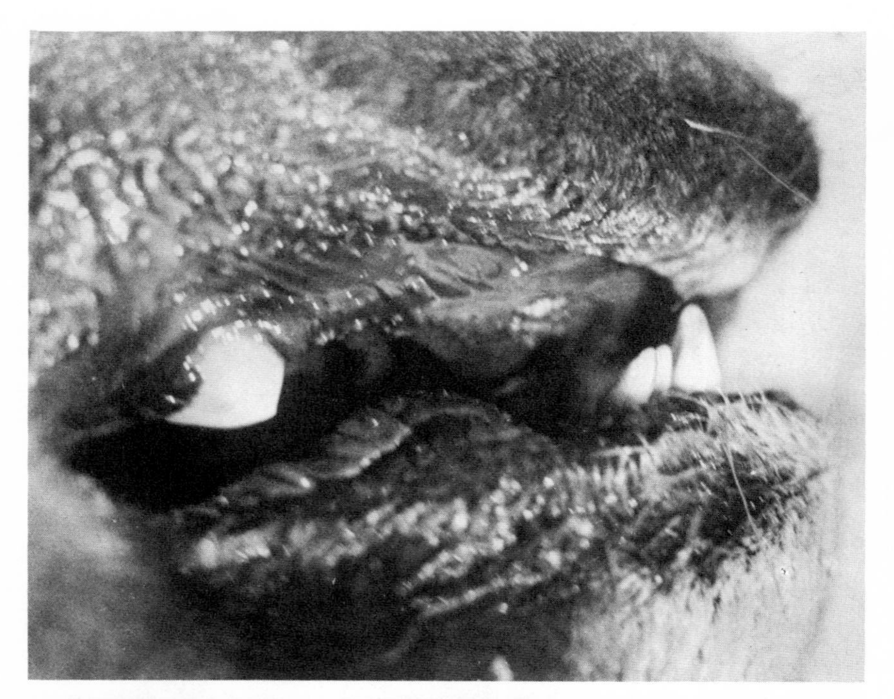

PLATE 11. Upper canine of female elk. *Above.* Of a female nearly 2 years old, with the point still unworn. *Below.* Of a cow nearly 3 years old, with the point worn off into a rounded surface. (U. S. Fish and Wildlife Service photographs.)

Bugling

Under the head of bugling are included several calls that have a general resemblance to each other and appear to have a common background. Bugling proper has generally been understood to take place in the fall, during the rutting season, and to be confined to adult bulls, but this conception must be broadened.

One old woodsman stated that he had heard bull elk bugling in practically every month of the year, and similar statements have appeared in the literature. When questioned, the woodsman admitted that he had not actually seen the animals that he had heard making the sounds in spring and had only assumed that they were young bulls.

Each spring during the present study considerable bugling was heard in May and early June, a puzzling circumstance—inasmuch as bugling was considered characteristic of the mating period—until it was discovered that cows were doing much of it. It is possible, of course, that some of the bulls also bugle in the spring, but they were not seen in the act. Moreover, it should be explained that although this spring call of the cows is termed "bugling," it is by no means so loud and full-chested as the resonant calls given by the old bulls in autumn.

On the evening of June 7, 1929, the writer observed a cow elk giving voice to peculiar hoarse sounds—a "combination of squeal and groan, difficult to describe," as it was entered in the notebook. She grazed along hurriedly, frequently emitting the peculiar squeals that sometimes dropped to low, short whines. It was suspected that she was near parturition. Another cow spied the observer and gave the alarm. The first cow stood for a long time at attention, looking for the danger. When she finally spotted the human being, she ran a distance and began barking. When circling about, she deliberately picked up the writer's trail. She continued to act very much as if she had a calf nearby, and she was barking when the writer left at darkness.

Cow elk have been heard bugling in July and August, also. This is not very common and is rather unaccountable.

Elk calls reach their finest development in the bugling of the old bulls in the rutting period. This sound varies greatly, justifying the many descriptive terms applied to it in literature, among them roaring and whistling. Sometimes the rutting bull produces a roaring sound that at a little distance reminds one of the hoarse bellowing of a domestic bull. It is as though his voice, wearied by frequent use, stuck on a low pitch and failed to rise to the high, clear notes. Normally, however, the call begins on a low note, glides upward until it reaches high, clear, buglelike notes, which are prolonged, then drops quickly to a grunt, frequently followed by a series of grunts. The call may be very roughly represented thus: *"A-a-a-a-ai-e-eeeeeeeee-eough! e-uh! e-uh!"* At close range the low notes are clearly heard. They possess a reedy, organlike quality that changes rapidly as the high note is struck. A deep resonance from the capacious chest accompanies it all. At a distance the low, hoarse notes are lost to a great extent and the high bugle notes are especially clear, giving rise to the impression of "whistling." There is much individual variation in the pitch and character of the bugling. This is especially noticeable when three or four bulls in the same vicinity are heard in close succession.

Sometimes one hears only the grunts—"Euh! Euh! Euh!" or "Euh-uh! Euh-uh! Euh-uh!"—the second syllables of the latter series being caused by the intake of the breath. More rarely another sound is heard, a peculiar "tapping" or clucking, with a liquid quality, possibly merely the mechanical clicking of the vocal organs.

During these investigations spike bulls were never observed bugling. As for 2-year-olds, there is fairly good evidence that they bugle occasionally, although most of those seen during the rutting season were silent. In some cases identification of 2-year-olds in the field is a little doubtful; but on several occasions it was practically certain that such animals were bugling, some at least in a higher-pitched voice that lacked the depth of the call of the old bulls.

Bugling of bull elk was heard as early as August 21, but in 1949 Dr. David Love noted bugling July 15, and many times July 25 to 31, indicating an unusual season. Bugling lasts at least through October though Adolph Murie noted it often as late as November 14. Normally August bugling does not mean active pursuit of cows; even in early September full fervor has not developed. Late in October bugling wanes, and is intermittent in early November.

The bugle of the bull elk is usually interpreted as a direct challenge to a rival. It is difficult to interpret mental activity of animals; but after close observation of rutting bulls and study of the circumstances it was concluded that the bugle is not a direct challenge, as such. True, the master of a harem of cows will give voice to the finest, loudest call within the power of his great chest and lungs. Undoubtedly he feels aggressive and is perhaps in a challenging mood. A real rival, lingering on the outskirts of the band, will reply in kind, and back and forth the vocal battle is waged, figuratively speaking. However, the bull begins bugling early in the season in the presence of male companions, often with no sign of ill feeling among them. At a later date, if there is no rival present, the bull bugles just as energetically and regularly. Moreover, a wandering, unattached bull bugles at intervals even though he is not conscious of the presence of other bulls. Bulls have been observed lying down in the heat of the day, tired and listless, with eyes half shut, yet at intervals uttering a lazy bugle call. In many instances like this it would be extremely difficult to attribute to the animal a direct challenge to unseen rivals; to be consistent, all circumstances should be weighed in reaching a conclusion.

During the rut the bulls are in a turmoil of unrest. The physiological development at this period has produced a swollen neck and other sexual changes. The tremendous sexual urge and intense emotional state of the animal require definite expression. The elk cows are very evasive. The bulls are under terrific strain, and the bugling appears to be but a partial outlet for their pent-up feelings. Among birds the male often gives voice to song in the mating period. The bugling of elk may be, after a fashion, comparable to bird song as a natural expression of emotion.

It is true, there is rivalry among bulls and there are definite attempts at intimidation, but never has the slightest indication been seen that an intruder has been frightened off by a rival's bugling. Even 2-year-olds watch their chance to steal a cow when the herd master is not looking. On the other hand, the slightest offensive gesture on the part of the bull in charge, such as a move in the direction of the intruder, does have an immediate effect. It would seem that the elk bugle is to be compared with the howling of a dog rather than with the dog's intimidating growl.

Referring again to the bugling of cows in spring, it is noted that this takes place during the general period of parturition. This bugling involves no rivalry, and the conclusion seems inevitable that it is an expression of feelings derived in some way from important sexual development attending the near approach of parturition or possibly from maternal feelings toward the young calf. An instance has been cited showing that a cow elk that was probably about to give birth to a calf gave voice to her feelings with peculiar calls. The bugling of cows, though but an echo of the powerful call of the rutting bull, appears to be of the same pattern as that of the bull and seems to be associated with the reproductive process.

It is of interest to note that the call of the related red deer of Europe, often referred to in the literature as "roaring," is quite different from the call of the American elk; whereas some of the Asiatic wapiti that are so closely related to the American forms are said to have a bugling call that corresponds somewhat with that of the American elk.

LONGEVITY

It is, of course, impracticable to determine the age of old elk observed in the field. The tagged skull of a 10-year-old elk obtained in Jackson Hole, Wyo., contained teeth that were not unduly worn, which indicates that animals with much abraded teeth are very much older, possibly 20 or more years of age. Roosevelt elk kept on the Olson Ranch in the Olympic Mountains years ago reached ages of 12 to 14 years, and one female that was sent to a zoo was 18 years old when last heard from and had had a calf when she was 17. Rowan (1923: 112) told of a bull elk in Assiniboine Park, Manitoba, that in his fifteenth year had antlers with a spread of 60 inches. Two years previous its antlers had had a spread of 61½ inches. The bull was mated for the last time in its fourteenth year and proved fertile. In its sixteenth year it showed signs of senile decay, with loss of teeth and semiblindness. In November 1937, south of Winslow, Ariz., a hunter took a large bull elk that bore a Biological Survey ear tag showing it to be one of the elk sent to Arizona from Wyoming in 1913 when nearly a year old. It had therefore reached an age of at least 25 years. Audubon and Bachman (1846-54: 92) wrote:

". . . The elk lives to a great age, one having been kept in the possession of the elder Peale, of Philadelphia, for thirteen years; we observed one in the park of a nobleman in Austria that had been received from America twenty-five years before."

As to the closely related red deer, in reporting on longevity of mammals in the Philadelphia Zoological Garden, Brown (1925: 267) listed a male that lived for 19 years and 2 weeks and a female that lived for 12 years and 4 months after being received at the zoo.

9

REPRODUCTION

SEXUAL DEVELOPMENT AND BREEDING

NORMALLY a cow elk breeds for the first time in the third rutting season after birth, an average age of about 2 years 4 months. Although a large number of so-called yearling elk cows have been examined in winter for disease and indications of breeding, one with calf was never found, and only one published record of such an occurrence has been noted. While examining elk slaughtered in Yellowstone Park, Mills (1936: 250) found an elk cow "approaching 21 months of age" that carried a fetus.

Indications are that the male becomes sexually mature at an earlier age than the female. On August 11, 1931, in a large band of elk a spike bull, probably 14 months old, was seen mounting a cow elk. This was in advance of the normal rutting period. Possibly the sexual development of spike bulls is erratic, for a tame spike bull attempted to mount wild cow elk before he was 7 months old and in 1931, when a little less than 3 years of age, bred with domestic cattle early in March, a time of year when elk breeding in the wild would be very abnormal; perhaps the semidomestication of this bull influenced its reactions.

On the whole, although spike bulls occasionally may be found somewhat active sexually, they are not so normally. Facts tending to show their sexual immaturity are the relative indifference of most of them during the rut, when ordinarily they do not actively seek the cows, and the retention of velvet by many of them. Although spike bulls are driven away from the cows by the herd master, presumably as possible rivals, the actions of the spike bulls indicate sociability, or a herding instinct, rather than rivalry with older bulls. Small bands of spike bulls have been seen off by themselves, evidently temporarily ostracized from mixed bands.

The 2-year-old bull presents a different picture. He is sexually mature in the rutting season, ardently seeks to steal cows from the herd master, bugles occasionally, and undoubtedly does a certain amount of breeding. He has not reached full size or development, however, and evidently does not experience the extreme ardor that seizes old bulls during the rut.

It is difficult to determine the ages of the older bulls seen in the field, but there is evidence that the 3-year-old bull is sexually mature, and is definitely seeking the cows. It is doubtful, though, if the 3-year-old is able to hold a herd of cows long in the face of stronger competition. In succeeding years, the bull becomes heavier and stronger and for a period is a formidable rival.

As in the case of the caribou, another gregarious animal, there is a question as to whether the young 2-year-old males or the old bulls do most of the breeding. Many woodsmen and other observers are of the opinion that the old bulls are kept so busy dashing about chasing rivals that much of the actual breeding is done by the young bulls, by stealth. On the other hand, it is very evident that the old bulls are exhausted by the ordeal of the rut. To learn which view is correct, there should be voluminous field notations of activities, and these are extremely difficult to obtain on any great scale under natural conditions. The data acquired in the present elk studies are inconclusive.

THE RUT

There are several records of bulls bugling in the latter part of August—apparently an indication of the approaching rut—and on August 15, 1944, on Astringent Creek in Yellowstone National Park, the writer observed a bull elk with a group of two cows and two calves. The bull's antlers were cleaned of velvet, and he showed his interest in the opposite sex by pursuing one cow for a short distance. This was the earliest that such indication was observed. Some bulls shed the velvet as early as the first week in August, and by September 1 most of them have done so and the rut has definitely begun. At this time bugling is general and the old bulls have begun herding the cows into small bands. The breeding season has not yet come into full swing, however, and there may even be several bulls in the presence of cows without a show of rivalry. A little later there is distinct competition. The active rutting season extends from the first part of September to the latter part of October, though the ardor has greatly diminished by October 10.

Evidently adult bulls have a period of sexual activity more or less corresponding to that of the females, although there are exceptional incidents. It has already been noted (p. 123) that a tame spike bull bred with cattle at various times of year and when less than 7 months old attempted to mount cow elk when presumably they were not in heat. Ignar Olson, whose family raised elk for the market many years ago on their ranch in the Olympic Mountains, Wash., related that tame elk used to go off on the range with the cattle and return with them. At one time there were two domestic bulls, yet it was noticed that for several seasons there were no calves in the cattle herd. Thinking there was "something wrong" with the bulls, the Olsons disposed of these bulls and purchased two others. Still there were no calves. Then it was discovered that when a domestic cow came in heat, she was promptly served by an elk bull, that would not let a domestic bull come near. The elk bulls were dehorned, whereupon one of the domestic bulls was able to breed but the other one was still intimidated by the elk. Mr. Olson said a bull elk would serve a domestic cow whenever she came in heat, at any time of year, including even the time when the bull's antlers were in the velvet.

Should it be reasoned, therefore, that the breeding period of elk is established by the sexual readiness of the female, which serves as a stimulus to sexual development in the male? Or does the sexual impulse in the male normally reach full development in the rutting period, in a physiological rhythm, perhaps being awakened at other times also in response to the proper stimuli, at least enough for occasional breeding?

Observations on wapiti and red deer introduced in New Zealand are important in the study of the sexual cycle. It is known that those introduced deer adapted their reproductive cycles to the seasons of New Zealand. Donne (1924: 287) reports:

"On the 4th October, 1907, two stags and four hinds, all four years old, were shipped from Warnham Court Park for New Zealand. Two of the hinds died on the voyage, the other deer reached New Zealand in mid-November and were placed in the Paraparaumu Game Farm.

"In April following the imported stags came out of the bush with clean antlers and began roaring on the 2nd May, about two or three weeks after the New Zealand rut had finished and about six weeks after it began. This shows that they came "in season" and that their antlers had been shed and reproduced twice in the one year. Their first antlers grown in New Zealand were, however, not fully developed and smaller than those they had shed. In 1909 they roared in the New Zealand season, and it is stated by the keeper of the Game Farm that their antlers were normal. The two hinds dropped calves in April, 1908, seven months after the September rut in England. In 1909 they again both calved in February, two months earlier than the previous year and about two months later than the normal period of New Zealand. In 1910 they had adapted themselves to local conditions."

Adaptation of wapiti and red deer to a reproductive cycle suitable for New Zealand in such a short time signifies the importance of environmental factors in shaping the function of reproduction into harmony with the prevailing seasons of the habitat.

During the summer most of the cows, calves, and young bulls band in herds of 20 to 300 or more individuals, but there are also various small, straggling groups of 5 to 20 (Plates 13 and 14). As the rut begins and the old bulls start to seek the cows, the big herds divide, probably because one bull cannot hold together a very large band of cows when other bulls also are eager to acquire herds of their own. A herd bull may have under his charge a band ranging from 6 to 30 or more in number, including cows, calves, and yearlings. Commonly there are 10 to 15 cows, most of them of breeding age.

The bull endeavors to "round up" his cows and direct their

movements. The tame bull elk at the refuge, previously mentioned, herded some of the domestic cattle on a neighboring ranch, as observed by a number of cowboys, and the same practice was noted among wild elk. On Big Game Ridge, Wyo., on the evening of September 28, 1932, half a dozen elk cows were feeding along in the direction of the writer, by the edge of a wood. With them was a young bull, 3 years old or possibly younger. Presently, across an open grassy vale between two slopes, another group of cows, followed by a large bull, came downhill on the run into the grassy bottom in the direction of the first group. There they stopped to feed. The young bull with the first group gazed intently at the approaching large bull, then turned and meekly walked away into the woods. Meanwhile the large bull came on alone in a direction to one side of the first group, seemingly about to pass them by. After reaching a certain point, however, he veered back toward them. As he turned, the group of cows promptly reversed their direction and quietly fed away. Although the bull was still at some distance, the moment he changed direction, with uncanny precision they had divined his intention and moved away. The bull then went directly up to the group. Threatening one cow with his antlers, he put her on the run, and soon all the cows in the band were moving toward those grouped in the bottom. In a few moments all 15 or so cows were together and the big bull was in charge.

Interpretation of the incident would be as follows: During the middle of the day—when old bulls, as well as cows, are more or less quiet and the bulls are listless and inclined to rest at frequent intervals—the big bull had not troubled a great deal to keep his band assembled and some of the cows had drifted away, feeding along slowly, and one of the young bulls, alert for such an opportunity, had joined these strays. As evening approached—when bulls are again filled with energy and resume great activity—the old bull, no doubt aware of the straying cows a short distance away, proceeded to assemble his band. He drove those with him down the hill toward the others. Having reached the meadow, he made a wide detour and skillfully directed the wanderers back to the others. All this was accompanied by frequent bugling, which signified his awakened interest. The incident not only illustrates how the bull maneuvers to keep his cows together but also shows how the younger males fare in competition with their older and stronger rivals.

On another occasion the writer suddenly came upon a bull with his band of cows and several other bulls in the vicinity. As the elk fled, in the panic the old herd master ran off by himself and one of the other bulls promptly joined the cows, more or less accidentally, and went off with them. This was an unintentional quick change of leadership. As actual possession is a strong point in the art of intimidation among elk, the new master was probably able to maintain his position for a time.

A good indication of rutting activity is the intensity of the bull's bugling. During the middle of the day the bull may bugle a little but usually at long intervals. About 3:30 or 4 p.m., bugling becomes more common, and in the evening and again from the break of day until a short time after sunrise, rutting bulls are particularly noisy. There may be bugling throughout the night, especially on clear, moonlit nights; but the greatest intensity occurs in the evening and early morning. Early one morning during the height of the bugling activity an old bull kept bugling at intervals of 7 to 75 seconds for a considerable time.

Typical of the activities of a group of elk during the rut are those observed near Enos Lake, Wyo., on October 11, 1931. En route to a favorite meadow late in the afternoon, the writer had glimpses of elk at the edge of a wood. An old bull clashed antlers with a 2-year-old, but the battle was not savage and soon ended. A little later the same old bull, with a peculiar undulatory motion of his head and neck that appears to be a characteristic threatening mannerism, drove off another 2-year-old, which promptly fled a short distance. A little beyond, in a meadow, was a typical band of cows and calves and probably some yearlings, all in charge of an old bull; several other large bulls were at a short distance; and at the edge of the group there were two spike bulls and four or five 2-year-olds.

Most of the animals were busy feeding, but one old bull held the center of the stage. Bugling incessantly, he stamped back and forth, pursuing first one cow, then another, and then stalked off aimlessly, only to come back after another cow in a frenzy of excitement. The cows appeared indifferent and evaded his advances. In the background two big bulls restlessly threshed the lower limbs of a bushy tree with their antlers. Once or twice was heard the clashing of antlers as two bulls fought, but as there was no serious battle, the activity probably represented jealousy between two bulls that possessed no cows. One bull prodded the ground with his antlers. The 2-year-olds and the spike bulls in the outskirts of the band were all outcasts, not tolerated by the rampant old bull. Apparently these younger males were not interested in the cows but were intent only on grazing. The extra old bulls showed more interest; but their ardor was not sufficient, it seemed, to make them take issue with the bull in possession. It is probable that at this date (October 11) the rutting heat was waning somewhat among many of the bulls and that earlier in the season their behavior would have been more strenuous. It is interesting to note that clashing of antlers among younger bulls was noted in Yellowstone Park by Adolph Murie as late as November 21, and bugling as late as November 14.

THE WALLOW

An accompaniment and sure indication of the rutting season among elk is the act of wallowing, which was observed then and at no other time. Its purpose seems to be the soothing of the rutting fever, a cooling for the body, an outlet for pent-up energy. The habit should not be confused with the bathing indulged in by all classes of elk in summer (p. 262). The earliest date a fresh wallow was found was August 22, in 1930. In September fresh wallows are common. On the second of that month in 1936 the writer saw a number of them on Afognak Island, Alaska, where the Roosevelt elk has been introduced. The bulls were rutting at the time.

The wallow is made in a swampy spot where seepage trickles from a slope, in a wet sedge meadow, or even in the edge of a mineral lick. Where the ground is saturated, an old wallow often remains as a permanent pool of clear water and may be recognized several years later (Plate 15). The old bull elk make wallows and often appear with their hides caked with mud, but how often they use them and whether all old bulls use them could not be ascertained. Apparently 2-year-old bulls wallow only occasionally, and no evidence was found that spike bulls or cows ever do so.

The formation of a wallow is illustrated by the following incident. In northwestern Wyoming, on the evening of September 26, 1932, a small band of cows and calves, accompanied by a small bull with rather straight narrow antlers, came drifting through a thin forest growth. The bull, bugling at intervals, presently descended a steep bank and at a wet, sedge-covered place proceeded to prod the ground with his antlers. He dug the ground and threw sod high in the air over his back. After a little he pawed the ground with his hoofs, and again the dirt flew. Presently, he lay down in the wet mud and continued prodding with his antlers, reaching out here and there. Two or three times he bugled while lying down. In 3 or 4 minutes he got up and commenced belaboring a young fir tree. Twisting his head this way and that, he fought the tree from all angles, moving around it occasionally for a new point of attack. This lasted several minutes. Then he stopped, looked around, climbed the bank, and went back into the woods, bugling. The cows had drifted on, calmly feeding.

THE BATTLE

Much has been written on the subject of battles among the various species of deer, and everyone is more or less familiar with the general process. Interlocked elk antlers have been found, and battles have been described by various observers; but during the entire time of the present investigations not a single really serious battle among the elk was observed. A number of times 2-year-olds were seen sparring, sometimes clashing their antlers together with some spirit, but such jousts never lasted long and were but the expression of the prevailing hostility of one male toward another.

It appears that when a bull gains possession of a group of cows, he thereby acquires an advantage in that other bulls recognize the fact of possession and are the more easily intimidated. The herd master maintains his position largely by bluff, by offensive gestures rather than by bugling. Two bulls will be bugling in each other's faces, so to speak, without battling; but when the "outside" bull comes dangerously near, the herd owner gives chase, and in no case observed did the intruder stand his ground. The chase is not long, and the fugitive is soon back. Although the intruder is easily driven off and may even be impressed by the bugling of a rival, there appears to be no great fear. As recorded above (p. 128), even a 2-year-old may become engaged in a fight with an older bull, with no serious consequences.

There is no doubt that at times a vigorous battle occurs, sometimes with fatal results. One autumn during the hunting season a large bull was shot whose thigh bore a deep wound that had every appearance of having been inflicted by a tine of an elk's antlers. Yet even the fact that an elk is seriously injured or killed in battle, is not a sure indication of an intense fight. Antlers may become interlocked in a slight sparring match, or fatality may result from a single stroke; for, although antlers are employed in combat chiefly during the breeding season, they are also used during the winter in minor quarrels. At least seven instances were recorded in which elk were killed by the thrust of a bull's antlers in a petty quarrel over a bit of hay. Two of these victims were adult cows that were pierced in the abdomen and died later as a result, and two were calves, each killed outright by an antler thrust. Two other cows were killed in the same manner, though in these instances an antler tine had penetrated through the forehead to the brain. An adult bull, in a dying condition, had been gored in the side.

In using their antlers bulls have been seen striking in a forward-downward arc, which admirably brings into play the various tines. Once in early spring a spike bull, threatening a companion, swung downward in the same manner, although his most effective method would have been a forward-upward thrust.

Not all fighting is done with the antlers. In minor quarrels a more common method among cows and bulls, as well as among younger animals of both sexes, is to strike with the front hoofs. (See pictures in Leek 1918: 359.) Apparently there is a well-defined technique. Both disputants rise on their hind legs and, with muzzles held in the air, rapidly punch at each other with both front feet, using them alternately. Adult bulls sometimes indulge in such punching matches in winter, when the animals are well equipped with sharp antlers, and so, the method is not merely substituted for antler combat at a time of year when the antlers may be shed or when they may be in the velvet.

These sparring matches occur among other members of the deer family. In *Nature Magazine* (Anonymous 1935: 253) is a photograph of two white-tailed deer in this attitude. In *Deutsche Jagd* are similar photographs of the red deer (Heck, 1934: 390) and the fallow deer (Rheinfels, 1937: 398). Two cow moose in this attitude are described and illustrated by Ben East (1937). In the *National Humane Review* for March 1944, is shown a photograph of two horses on their hind legs in a similar sparring position, and we know that this is characteristic of horses in certain moods.

In defense of her young a cow elk will use her front hoofs to strike at a dog or coyote and probably at other similar foes. Without rising high on the hind legs, and sometimes even while running swiftly in pursuit, she will quickly rise a little and strike downward with a front hoof. This is a very effective attack and a dangerous one for her opponent, as she can deliver a powerful blow.

PHYSICAL EFFECT OF THE RUT

When the elk enter the rutting period, they are fat, the bulls especially, undoubtedly in the best condition of the entire year. During the rut the bulls do not eat so much as formerly, and the actively breeding ones indulge in such tremendous expenditures of energy that when the rut is over, the surplus fat is gone and winter finds them in poor condition. The cows, on the other hand, feed regularly during the rut. They do not lose flesh noticeably and enter the winter in fairly good condition.

GESTATION PERIOD

The writer has had no opportunity to determine experimentally the gestation period of the elk. Lantz (1910: 26) gave it as 249 to 262 days (about 8 to 8½ months), which is fairly well substantiated by field observations. It is difficult to observe actual breeding of elk on a scale sufficient for statistical treatment; but assuming that bugling is indicative of breeding activity, the period from September 20 to 25 may be taken as the height of the breeding season. As June 1 is considered the peak of the calving period, a gestation period somewhat longer than 8 months—possibly as much as 8½ months—is indicated.

CALVING

The period of calving is generally given as May 15 to June 15, which is fairly accurate, although most of the calves are dropped sometime late in May or early in June. There is much variation in the time of birth. Rush (1932a: 29) gave May 13 as the earliest date he recorded in Yellowstone National Park, but two earlier dates were noted in this study. On the Elk Refuge at Jackson, Wyoming, a calf was born on May 11, 1929, and it was reported that on April 14, 1932, a calf estimated to be not more than 2 days old was seen following its mother up the butte west of Jackson. There are records also of later birth dates. On July 3, 1929, two elk calves only a few days old were found hiding. In the winter of 1932-33 the men engaged in feeding the elk observed a young calf whose pelage still showed faint spots, which meant, that this animal must have been born late in summer. A news item of October 3, 1937, at Cody, Wyo., reported the birth of a calf elk in the zoo. Here, of course, captivity was involved as an upsetting factor. But Seton once found an elk calf, probably less than a week old, on October 15 (1929: 34).

On the Selway district in Idaho calving is predominantly on winter range (Young and Robinette, 1939: 11). In Yellowstone National Park, Wyo., it occurs to some extent on winter range but also on summer range. In the Jackson Hole, Wyo., region very few calves are born on winter range, some appear during migration, and still others on strictly summer range, but perhaps the majority are born on what may be termed intermediate or spring range.

Almost invariably the cow elk gives birth to a single calf, though twins occasionally occur. During the reduction program on the National Elk Refuge in Wyoming in 1935, when over 500 elk of all ages and both sexes were killed, two pairs of twin fetuses were found. In similar operations in Yellowstone Park in January 1943, of 156 adult cows examined there was not a single instance of twin fetuses. There have been some verbal reports of twin elk calves from time to time, and Caton, who kept many of the deer family under domestication over a period of years, wrote to the effect that twins were not uncommon, and that even three are sometimes produced. However, the evidence is overwhelming that in the present day elk herds a single elk calf at birth is the rule.

THE ELK CALF

At first the elk calf is unsteady on its feet and rather inactive, spending much time lying down. When no danger is near, it lies at ease, head raised, looking about; but at the first sign of peril, fancied or real, it crouches at once, the head and neck extended on the ground (Plates 16 and 17). This habit of lying still in the presence of danger is well known. Sometimes when a group of elk is traveling along, a calf may drop prone on the appearance of danger, although some of the older calves usually flee with the adults. Once when a band of elk was fleeing, one of the calves stumbled and fell, whereupon it lay perfectly still and was subsequently handled without trying to escape.

Elk calves attempt to hide in a variety of situations, often beside a log, tree, or bush, in the sagebrush, or not infrequently in the open, on green grass, where they are rather conspicuous. Many times calves are found in or near aspen groves, no doubt because the adult elk are finding good feed there; and often calves are discovered well hidden in the brushy growth of a patch of willows where the adult elk are feeding. There appears to be no regularity as to the kind of hiding place and no attempt to find a situation that harmonizes with the coloring of the pelage, but there may be a natural tendency to lie down beside the nearest object.

The calves are perfectly free to move about; and while the adults are feeding, the little ones may stray off for some distance and then lie down. In the absence of danger the cows usually show little concern for the resting calves. They may graze along and wander far from them, several yards or possibly a quarter of a mile. In numerous instances, however, the mother remains quite near her offspring. The writer found no evidence that the cow deliberately placed her calf in concealment.

At the approach of danger to the calf the cow acts differently on different occasions. In the case of human intrusion she may flee with little demonstration, stopping at frequent intervals to look back. Sometimes she will stand at a distance and bark repeatedly, retreating very slowly. If the calf is handled and caused to make an outcry, the mother will often reappear and sometimes come close, although she probably would not actually attack a human being in such a situation. If the intruder is a coyote or dog she will promptly attack, as described on page 116 and 154. A cow elk probably will not attack a bear ordinarily although as related on page 155 one did venture into the water after a bear that was swimming off with her calf in his mouth; and another cow mustered enough courage to strike a black bear that was making away with her offsping.

When a calf has been disturbed and caused to flee, it may run 200 or 300 yards, possibly more, depending on its age, and then lie down again. Later it may return to the vicinity of the original hiding place unless the mother finds it in the meantime. Sometimes, after an interval, the mother will return to the place of concealment and, with nose to the ground, seek the scent of the little one. The actual tracking-down of an elk calf by its mother was not observed, but a doe mule deer was seen returning to the point where her two fawns had fled in a direction different from her own and then carefully following their trail for a considerable distance, eventually disappearing at the exact spot where the fawns had been lost to view. In this instance the fawns were fairly well grown.

This brings up the question whether young fawns or calves in the deer family emit a scent. The view has been generally expressed that they do not scent, and sometimes precise observations are cited to confirm this belief. It is a question difficult to determine ordinarily, and the evidence has been confusing. Certainly one might argue that any flesh, young or old, should have scent that could be detected by keen-nosed animals.

On June 7, 1933, the writer produced positive evidence on the subject by the use of an Eskimo dog (of the Siberian type from St. Lawrence Island, Alaska) whose nose, presumably, was not so sensitive as that of a hound or bird dog. The dog had shown some momentary interest in elk trails but appeared to depend on his eyes to detect game. The writer had a companion to help watch the dog's reactions. An elk calf, judged to be 5 or 6 days old, was found lying flat in a patch of willows; and as it lay very still, the dog did not see it. The dog was led slowly by a chain to leeward of the calf. When 9 yards from it, the dog raised his muzzle and, with nostrils working, extended it horizontally toward the animal, very plainly interested in its scent. The observers noted the exact moment when the dog caught the scent. Then a young calf, possibly 2 or 3 days old, was found at the edge of a lodgepole pine woods. This time it was difficult to determine the lee side, as the wind was extremely erratic and air currents were confusing. At a distance of 11 yards, however, the dog definitely caught the scent, as his actions attested. He was led away, and the process was repeated. Again he reacted very positively at about the same distance.

Adolph Murie (manuscript, 1935) recorded the following pertinent information collected in the Olympic Mountains, Wash.:

"Olson said he had trained dogs to hunt calves by seeking the body scent. An untrained dog would often pass by a calf and follow the mother, but a trained dog would not do this. Some of his dogs became quite expert at finding the calves.

"On one occasion three hounds passed close to a deer fawn in plain sight. None of the dogs noticed the animal, although they were good trailers. On the way home the trail led past a trap which had been set for a bobcat. A bobcat was in the trap under a log. The first two dogs passed the bobcat without seeing it. As the third dog came opposite, the cat suddenly pounced upon him. Olson explained the incident by stating that the dogs were 'thinking about something else' which may not be far from the truth."

It is true that the three dogs passed by the deer fawn, but they also passed close by the bobcat, and it cannot be argued that the bobcat gives no scent. Direction of wind, preoccupation of the dog, perhaps other factors, may play a part. At any rate, the hunter quoted above had made use of dogs to hunt elk calves at a time when he was obtaining elk for domestication.

In his studies in Yellowstone National Park, Adolph Murie (1940) observed a cow elk walking down a slope. Suddenly, at her feet, an antelope fawn leaped up with a cry. The elk was startled and wheeled to one side in a big jump, probably having stepped on the fawn. The elk appeared too startled to have been aware of the fawn and thus must not have caught its scent. Later the antelope mother came to the spot where her fawn had been but was unable to follow its tracks—or at least she did not—and kept hunting for her fawn at intervals until dark. This incident is at variance with the observation on the mule deer and her fawns (p. 135), but those fawns, it should be remembered, were much older animals.

Thus there is varied evidence, positive and negative, but the positive evidence of the reactions of dogs in the presence of young elk calves cannot be ignored. It is probably justifiable to assume that the calf, because of the comparatively undeveloped state of its glands, and perhaps because of other physiological conditions as well, gives off less scent than the adult and is correspondingly more difficult to detect by smell. The facts show, however, that there is some scent, and no doubt, with due allowance for direction of wind and distance, it may be detected by the keen-nosed coyote and other mammals.

A newborn calf will lie close, passively allowing itself to be handled or lifted from the ground, but a calf a few days older, particularly one 5 or 6 days old, is apt to flee when closely approached. By the middle of June there are few calves that will allow close approach, and very soon they all follow their mothers in flight at the appearance of danger. Their short period of reliance on concealment or "freezing" has then definitely passed.

Young elk calves are playful, just as other young animals are. They skip about, bunt at each other, and in other ways exhibit their baby exuberance. In summer they particularly like to run and splash around the shallow waters of a lake or pond, a pleasure that is enjoyed by the older elk as well. When nearly a year old, after a hard winter, in the thawing days of March the "coming yearling" (as well as many older animals) will scamper and kick up its heels in the exuberance of the early spring.

The calf's period of dependence on the mother's milk has not been precisely determined. The premolars are in the process of eruption when the calf is about a week old, and an elk calf 2 or 3 weeks of age has been seen picking at the grass. Probably most elk calves begin to feed on vegetation when less than a month old. In June the calves are often seen browsing or mouthing the vegetation slightly. On July 22, a calf that was probably 1½ to 2 months old had more than trebled its birth weight and had its paunch filled with vegetation. No doubt the calf had fed on vegetation for some time. The premolars had already begun to show wear (Figure 17A).

Suckling continues throughout the summer, however. The men who fed the elk on the refuge reported that suckling was observed as late as December 27, 1930, and even in January, although the specific dates were not recalled. In the course of the elk-herd reduction of 1936 a number of elk were slaughtered at the refuge on January 20. A. P. Nelson made notes on each animal, and from his records it appears that 27 mature cows were dry, 8 were with milk, and all but 2 were pregnant. Many of the dry cows may have lost their calves earlier, so that these figures do not necessarily indicate the relative numbers of weaned and suckling calves.

On May 30, 1944, near Yellowstone Lake, a young spike bull with velvet spikes about 2 inches long was observed getting down on his knees and suckling a cow elk presumably his mother, butting her in typical fashion. The cow elk did not object but stood quietly.

Although milk is a welcome and desirable addition to the elk calf's diet, it probably becomes unnecessary in autumn, when the calf has attained considerable size and has begun to acquire some permanent teeth. The mother exhibits a growing impatience with nursing during the fall months.

The calf remains with its mother during the rut, probably throughout December, but during the winter when the elk band up considerably, it is difficult to observe whether the calves still associate with their own mothers. Certainly the characteristic calling back and forth between mother and calf that is so common in the summer is no longer so noticeable, but the incident of the year-old spike that still nursed is proof that some of them remain with their mothers throughout the winter.

RATE OF INCREASE

Of vital interest to game management is the rate of increase in the herd. This is one of the most difficult facts to determine, and probably in no other phase of game problems is there greater discrepancy in estimates. During the study of the Jackson Hole, Wyo., herd many pertinent factors were noted and an attempt was made to place them on a quantitative basis. Whenever elk counts were made from time to time, at intervals of a year or more, a classified count was obtained on the Elk Refuge proper. Seven of these, from which pertinent facts may be singled out, are given in Table 3.

In addition to these organized counts there are other statistics relating to the proportion of calves in a given elk herd. During two summers the writer compiled classified counts on Teton National Forest, south of Yellowstone Park, on summer range, securing a total of 1,192 cows and 458 calves, or a cow-calf ratio of 38 per cent. In the summer of 1937 Adolph Murie made similar classified counts in Yellowstone National Park, securing a total of 931 cows and 385 calves, or a cow-calf ratio of 41 per cent. Under such circumstances it was impractical to obtain ratio of calves to entire herd because of the distribution of bulls at that time of year. It is interesting to note that for the Selway Game Preserve in Idaho, Young and Robinette (1939: 46) reported a calf-cow ratio of 74 per cent based on summer counts, and estimated that it would actually be 80 per cent.

In the winter of 1943-44 a herd of 354 elk were being fed at Black Rock Ranger Station in Jackson Hole. Of this number 16 per cent were calves.

In the spring of 1938, while checking on winter losses in the Gros Ventre River basin, on Teton National Forest, Wyo., a field party noted that 20 per cent of the live elk seen were calves.

In January 1943, in the course of a reduction program in Yellowstone National Park, on a herd-run basis, with no selection of sex or age, 691 elk were disposed of: of these 57, or approximately 9 per cent, were calves.

In December 1935 to January 1936, in the course of a similar nonselective reduction program, 541 elk were killed, of which 435 were classified. Of this number 16 per cent were calves.

There is another method of computing calf crop during the hunting season, in Montana and Wyoming, as there are no restrictions on age or sex of animals that may be killed, and because modern hunting is virtually nonselective.

Data compiled in 1943 by the Protection Division of the Yellowstone National Park staff show that among the animals killed

TABLE 3. CLASSIFIED ELK COUNT ON THE NATIONAL ELK REFUGE, WYO., AND NEARBY FEED GROUNDS

Age class	1928			1932			1933			1935			1936			1938			1941		
		Percentage			Percentage			Percentage			Percentage			Percentage			Percentage			Percentage	
	Number	Of total elk	Calf-cow ratio	Number	Of total elk	Calf-cow ratio	Number	Of total elk	Calf-cow ratio	Number	Of total elk	Calf-cow ratio	Number	Of total elk	Calf-cow ratio	Number	Of total elk	Calf-cow ratio	Number	Of total elk	Calf-cow ratio
Cows:																					
All ages	4,835	64.0	32.2	4,305	62.6	28.2	6,975	62.1	30.3	6,142	68.5	30.3	656	56.0	17.5	4,552	68.4	23.5	5,876	61.8	32.6
Bulls:																					
All ages	1,161	15.4	....	1,354	19.7	....	2,149	19.1	....	961	10.7	....	399	34.1	....	1,032	15.5	....	1,712	18.8	....
Adult	679	9.0	....	1,160	16.9	....	.....	....	....	646	7.2	....	.....	....	....	768	11.5	....	1,182	12.4	....
Spike	482	6.4	....	194	2.8	....	.....	....	....	315	3.5	....	.....	....	....	264	4.0	....	530	5.6	....
Calves	1,555	20.6	....	1,214	17.7	....	2,115	18.8	....	1,860	20.8	....	115	9.8	....	1,071	16.1	....	1,916	20.2	....
Total elk ..	7,551			6,873			11,239			8,963			1,170			6,655			9,504		

over a period of years, north of Yellowstone Park, the following proportions were calves:

TABLE 4. KILL NORTH OF YELLOWSTONE NATIONAL PARK

Year	Total number of elk killed	Percentage of calves
1935-36	2,609	19.1
1936-37	793	34.0
1937-38	3,867	16.0
1938-39	2,570	16.0
1941-42	1,997	15.9
1942-43	7,000	13.9

The classification summary of 18,836 elk from Yellowstone records is: Bulls, 6,913 or 32.9 per cent; cows, 9,561 or 50.8 per cent; and calves, 3,082 or 16.3 per cent.

Data for Jackson Hole, Wyo., in 1932, show the classified hunter kill of elk totalled 1,749, of which 11.5 per cent were calves. In 1943 the hunter kill, of which sex and age were known, totalled 4,742, of which 963, or 20.3 per cent were calves.

From data such as these an attempt is made to estimate the percentage of annual increase in the herds in and near the Yellowstone National Park. The first point that attracts attention is the great variation in the percentages of calves in the various computations.

Considering, first, the northern herd that winters in and near the northern part of Yellowstone National Park, there are no classified winter counts comparable with those made possible on the feed grounds in Jackson Hole. If guided by the hunter kills in that herd, however, an average of 16.3 per cent calf crop over a 6-year period is found.

The data on winter counts for the southern or Jackson Hole herd for eight of the winters from 1927 to 1944, show a total of 52,309 elk counted, of which 9,904 were calves. This points to an average of 18.8 per cent that were calves. Classified data on calf percentage in hunter kill in Jackson Hole are more meager. The three records available vary between 11.5 and 20.3 per cent with an average of 16.7 per cent.

From these various computations percentages are found between the extremes of 9 and 20.8 per cent. Some of this variability is unquestionably due to differences in the herd composition and breeding success from year to year; part must be due to errors in faulty sampling. But if one excludes the extremes and considers the recurring figures in the intermediate range, the average arrived at is probably fairly reliable. It will be noted that on the basis of hunter kill alone the herd increase for the northern Yellowstone herd would be in the vicinity of 16.3 per cent.

For the southern herd, on the basis of more extensive data, the annual increase averages 18.8 per cent.

Inspection of the figures given above shows that, roughly speaking, about one third of the ideal calf crop (one calf to each cow of bearing age) reaches the age of about 10 months.

What has become of the other two thirds?

One autumn an attempt was made to obtain a large number of uteri of elk shot by hunters in Jackson Hole, but the response was not great and the age of the animals was sometimes doubtful. On the basis of that material more than 90 per cent of the cows had bred. In a herd-reduction program conducted by the Wyoming State Game Department in December 1935 and January 1936, the writer examined 334 cow elk of breeding age. Of these, 89.2 per cent were pregnant. In the course of a similar reduction program in Yellowstone National Park in January 1943, the writer examined 156 cow elk of breeding age, 90.4 per cent of which were pregnant.

It is obvious then that normal breeding takes place and that the sex ratio is satisfactory. On the other hand, in the examinations of hundreds of elk that died in late winter it was found that about 50 per cent of the adult cows were without calf and that one winter the figure was even higher. It is true that these were animals that had died, and that the same proportions may not have applied to the rest of the herd, yet this pregnancy ratio may have been significant. If so, then it is a fact that about half the cows do not produce calves in the spring. This is affirmed to some extent by field observations in spring.

It has been determined that Bang's disease is present in the Jackson Hole herd, and abortions have been noted repeatedly (p. 176). Old age may be a factor accounting for some barrenness, too, though it may be a rather limited factor, for some very old cows were with calf and some young ones were without.

Referring to the recorded calf-cow ratios, which suggest that roughly one third of the cows produce calves that survive into the following winter, it is tentatively assumed that about 65 per cent of the cow elk more than 3 years old are not represented by calves in late winter, 50 per cent because they failed to produce calves in the spring. That leaves 15 per cent not represented by calves for reasons yet to be determined.

At the time of the elk census, usually in March, when the percentage of herd increase was determined, the winter losses have already begun, and because the loss of calves far exceeds the loss of cows, that would account for a part of the unexplained 15 per cent failure. (Of the usual winter losses, from about 65 to 70 per cent have calves.) Furthermore, some elk calves die at or shortly after birth, in numbers and from causes yet to be determined, and some succumb to accidents and natural enemies, factors that cannot at present be accurately appraised but whose combined effect presumably might account for the 15 per cent of calf crop failure not definitely explained.

The number of elk calves killed by hunters in the open season and the number of cows killed in the same way appear to bear the same relation to each other as they do in the winter elk census, so that deaths caused by hunters may be eliminated as a factor on the percentage basis.

Thus it will be seen that the most important limiting factor in the annual increase appears to be the failure to produce a calf in the spring, that about one third of the cows present in the herd in March are represented by calves, and that the annual increase of the elk herd, as represented by the hunter kill in the Yellowstone herd and by surviving animals in March in the Jackson Hole herd, varies from 9 to 20.8 per cent, averaging 16.3 per cent for the Yellowstone and 18.8 per cent for Jackson Hole.

10

NATURAL ENEMIES

THE NATURAL ENEMIES of the elk, past and present, include the mountain lion, bobcat, wolf, coyote, dog, bear, possibly the wolverine, and perhaps to a slight extent the golden eagle. The pressure of each of these on the elk population has varied greatly, depending on the composition of the fauna and many other circumstances.

MOUNTAIN LION

Norris reported (1881: 41-42) that in 1870, when he first explored the Yellowstone-Jackson Hole region of Wyoming—the greatest center of elk population in the country—mountain lions, or cougars, were "exceedingly numerous" but that "now *[1880]* the comparatively few survivors usually content themselves with slaughter of deer, antelope, and perhaps elk, at a respectful distance from camp."

In the winter of 1888-89, Pierce Cunningham found a mountain lion track in the Gros Ventre area, Jackson Hole, which at that time was considered a notable event. However, the mountain lions undoubtedly increased in numbers subsequently, for early settlers observed them commonly in the Jackson Hole country. During the period of their abundance they were known to prey on elk, but with increasing settlement of the country they were hunted aggressively until they disappeared from the elk ranges. A track was seen on the east side of Jackson Hole in 1928, and a few years later a dead mountain lion was reported found at the south end, but this predator has now been practically eliminated from the Yellowstone-Jackson Hole fauna. An occurrence of this animal is still being reported from time to time.

The cougar still occurs on Vancouver Island, British Columbia, in the Olympic Mountains, Wash., and at some other localities where elk herds are present. Observations were made by the writer in the Olympic Mountains in 1916, 1934, and 1935. In 1916 he noted cougars feeding on the carcass of an elk that had been shot, which indicates their willingness to partake of carrion. They are fully capable of killing adult elk, however, and frequently do so. Two such kills were given post-mortem examinations. One, a fresh kill examined on the Bogachiel River on April 5, 1934, was a female elk about 1 year old and in good condition. The other, examined on March 20, 1936, by Dr. Adolph Murie, was a cow elk killed by a cougar on the Hoh River. This animal was judged to be 5 or 6 years old and was not diseased except for a number of lungworms. Other cougar kills were reported but not examined.

Ignar Olson, who has hunted cougars for many years, stated that sometimes snow remains deep on the mountain so unseasonably late in the spring that the elk calving season arrives before the high land is free of snow. At such times the elk may be concentrated in the snow-free bottoms, thus furnishing the cougars better opportunities to find the newborn calves.

During the study in the Olympic Mountains an attempt was made to supplement other cougar observations with data obtained by examination of feces, but the cougars had been so greatly reduced in numbers by bounty hunters that opportunity for such study was reduced to a minimum. Table 5 enumerates the contents of the 22 cougar droppings and 2 cougar stomachs collected.

According to one informant, the cougar is able to capture the marten under favorable circumstances and preys on pine squirrels and mice. Several cougar hunters interviewed had noted pine squirrels in cougar stomachs.

These data are far too meager to form the basis for defining the food habits of the cougar, but they suggest that this animal does not confine its diet exclusively to deer and elk, though possibly these furnish the most important items by volume, if not

TABLE 5. PREY SPECIES REPRESENTED IN CONTENTS OF 22 DROPPINGS AND 2 STOMACHS OF COUGARS COLLECTED IN THE OLYMPIC MOUNTAINS, WASH.

	Number of occurrences in–				
Prey species	10 droppings: North Fork, Quinault River, May 7, 1934	9 droppings: Hoh River, May 1934 and Feb.-Mar., 1935	3 droppings: Calawah River	2 stomachs Bogachiel River	Total
Elk	2	6	..	..	8
Deer	2	1	3	1	7
Marmot	1	..	..	..	1
Pine squirrel ..	2	..	..	..	2
Mountain beaver	1	1	..	1	3
Snowshoe hare .	2	2	..	..	4
Ruffed grouse .	1	..	..	..	1

Schwartz[5] has presented a table showing the contents of 28 cougar scats collected in the Olympic Mountains and containing the following food items: Snowshoe hare, 11 (32.35 per cent); deer, 8 (23.55 per cent); elk, 3 (8.82 per cent); pine squirrel, 2 (5.88 per cent); mountain beaver, 2 (5.88 per cent); white-footed mouse, 2 (5.88 per cent); woodrat, 2 (5.88 per cent); meadow mouse, 1 (2.94 per cent); flying squirrel, 1 (2.94 per cent); and a slight percentage of indeterminate material. He comments that the cougar apparently prefers deer to elk, and this is upheld by the opinion of many cougar hunters and local residents. He remarks further that these data "indicate reliance upon a rather large variety of so-called economically neutral species."

in numbers. The smaller animals serve to ease the pressure on deer and elk, and predation is divided between these two species. A thorough-going study is needed, however, to appraise these ecological relations properly.

BOBCAT

The bobcat has been almost eliminated from the Yellowstone National Park-Jackson Hole region, and there has been little opportunity to study this animal on other elk ranges. In the Olympic Mountains of Washington, where both elk and bobcat were common, 90 bobcat droppings were collected, chiefly on the Hoh, Bogachiel, and Quinault watersheds. The collection does not cover the critical calving season of elk, as most of the droppings were obtained from mid-April to mid-May 1934, 8 were found early in March 1935, and 1 on September 17, 1934. Table

[5] Undated mimeographed report, U. S. Forest Service.

6 presents a list of prey species in droppings and stomachs of bobcats.

These data clearly indicate that in the Olympics rodents constitute the food most sought by the bobcat. The snowshoe hare leads in importance—as it does in the diet of the lynx in the north—but it is closely followed by the pine squirrel, mountain beaver, and field mouse. Birds apparently furnish very little of the food. Bobcats are said to kill deer occasionally, but in the present study

TABLE 6. PREY SPECIES FOUND IN 90 BOBCAT DROPPINGS COLLECTED IN 1934-35 BY OLAUS J. MURIE AND 6 STOMACHS AND 99 DROPPINGS COLLECTED IN 1935-38 BY JOHN E. SCHWARTZ, IN OLYMPIC MOUNTAINS, WASH.

	Olaus J. Murie		John E. Schwartz	
Prey species	No. of individuals	Percentage	No. of individuals	Percentage
Washington varying hare *(Lepus washingtoni)*	27	25.71	65	43.91
Pine squirrel *(Sciurus douglasii)*	18	17.14	27	18.24
Mountain beaver *(Apolodontia rufa olympica)*	17	16.19	2	1.33
Meadow mouse *(Microtus* sp.)	15	14.28	6	4.05
Marmot *(Marmota olympus)*	9	8.57	....	....
Deer *(Odocoileus columbianus)*	3	2.85	8	5.40
Woodrat *(Neotoma* sp.)	1	.95	7	4.73
Salmon *(Salmonidae)*	..	....	8	5.40
Fish (unidentified)	1	.95	1	.67
Flying squirrel *(Glaucomys sabrinus)*	..	....	7	4.73
White-footed mouse *(Peromyscus* sp.)	3	2.85	3	2.02
Elk *(Cervus canadensis roosevelti)*	3	2.85	..	....
Jumping mouse *(Zapus* sp.)	..	....	2	1.33
Chipmunk *(Eutamias* sp.)	2	1.90	..	....
Western mole *(Scapanus* sp.)	..	....	1	.67
Sooty grouse *(Dendragapus obscurus)*	3	2.85	1	.67
Ruffed grouse *(Bonasa umbellus sabini)*	1	.95	..	....
Winter wren *(Nannus hiemalis pacificus)*	..	....	2	1.33
Small bird (unidentified)	2	1.90	..	....
Feathers	..	....	4	2.70
Grass	..	....	2	1.33
Bones	..	....	1	.67
Fir needles	..	....	1	.67
TOTALS	105	100	148	100

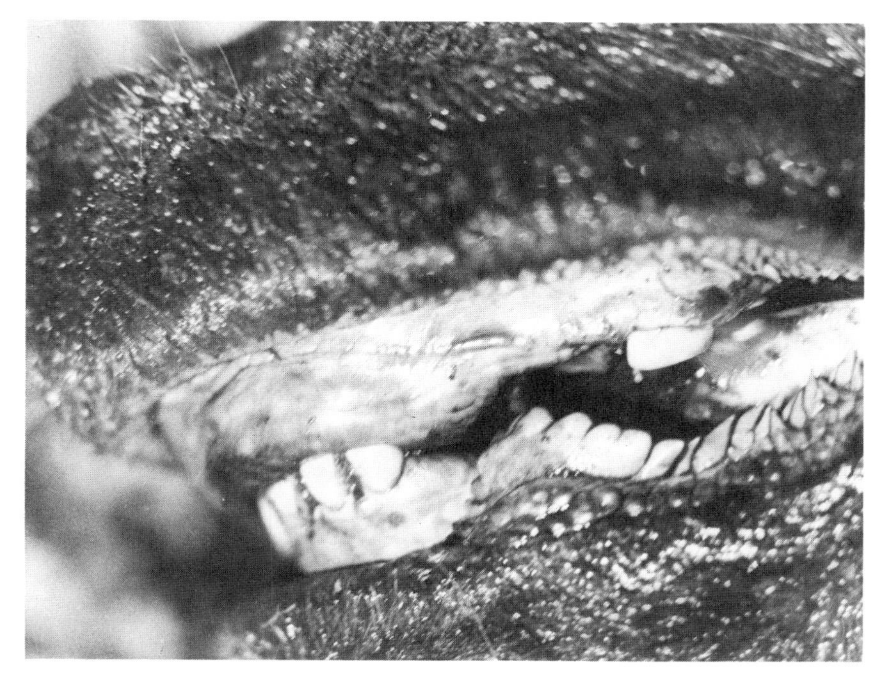

PLATE 12. *Above.* Canine and incisors of old cow elk. *Below.* Upper jaw of cow elk in which there are no canines and no indication that any were ever produced. (U. S. Fish and Wildlife Service photographs.)

PLATE 13. Band of elk on summer range on Teton National Forest, Wyo., July 26, 1928. The young calves are grazing with their elders. (U. S. Fish and Wildlife Service photograph.)

PLATE 14. Elk in Yellowstone National Park, Wyo. *Above.* A band on summer range in Pelican Meadows, in May. (Photograph by Adolph Murie.) *Below.* Animals feeding on new grass in early spring. (Yellowstone National Park photograph.)

PLATE 15. Jackson Hole, Wyo. *Above.* Wallow made by a bull elk in wet ground at head of Wolverine Creek and which remained as a permanent pool. *Below.* Typical elk lick on summer range, not to be confused with the wallow. (U. S. Fish and Wildlife Service photographs.)

PLATE 16. Elk calves, *Cervus c. nelsoni,* in Yellowstone National Park, Wyo. *Above.* At ease, when undisturbed. *Below.* In hiding attitude that is assumed at the approach of danger. (Photographs by Adolph Murie.)

PLATE 17. Elk calves. *Above. Cervus c. nelsoni,* cow and calf in Yellowstone National Park, Wyo. (Yellowstone National Park photograph.) *Below. Cervus c. nelsoni,* in hiding attitude, Jackson Hole, Wyo. (U. S. Fish and Wildlife Service photograph.)

it was impossible to determine whether or not the deer remains found were carrion. In two of the droppings the elk material was undoubtedly carrion, as it represented adult elk; that found in the third dropping was the foot of a calf elk. This specimen was collected on May 1, which date is a little early for calves although not impossible. It may have been old carrion or an unborn young from a dead cow.

It has been reported that bobcats kill elk calves in the spring, and it is entirely possible that, like coyotes, they do kill newborn elk calves left unguarded by the mother. This would be the only way in which the bobcat could prey upon the elk population. Intensive work during the calving season is required to determine the bobcat pressure at that time of the year.

WOLF

The wolf is gone from the Wyoming elk ranges, and little can be said about its relations with the elk, for the early records made only general statements. There is no question that wolves did kill elk in early days, but there is no measure of their pressure on elk population. When the writer visited the Olympic Mountains in Washington in 1916, there were still a few wolves left. Now they are gone and with them the opportunity to learn definitely about their status as enemies of elk.

In a recent study by Cowan, however, pertinent facts are reported on wolf predation on elk under conditions that obtained in Jasper and Banff National Parks in Canada (1947: 160). Dr. Cowan found that wolves prey commonly on elk, both on calves and on adults, and states that a single wolf can kill an elk. Apparently the wolves seize their prey by the throat or flank. Hamstringing was not observed.

Elk and other big-game species may escape—in deep soft snow, or by taking to the water. One cow elk, backed up a narrow promontory which protected her on three sides, was able to successfully beat off a band of seven wolves. But when elk are driven out on glare ice, they fall easy prey to wolves.

The study here referred to involved an area overpopulated with elk, of which 50 per cent were considered surplus, and Dr. Cowan says: "It is concluded that under existing circumstances the wolves

TABLE 7. MAMMALS FOUND IN 714 DROPPINGS AND 64 STOMACHS OF COYOTES COLLECTED IN JACKSON HOLE, WYO., LISTED ACCORDING TO NUMBER PRESENT

Species	Individuals exclusive of carrion		Carrion	
	Number	Percentage of total items	Number	Percentage of total items
Field mouse (*Microtus* sp.)	504	29.61	..	..
Pocket gopher (*Thomomys fuscus*)	379	22.27	..	..
Snowshoe hare or rabbit (*Lepus americanus bairdii*)	167	9.87	..	..
Golden-mantled marmot (*Marmota flaviventris nosophora*)	150	8.81	..	..
Elk (*Cervus canadensis*):				
Adult	..	..	80	4.70
Young	76	4.46	6	.35
Jumping mouse (*Zapus princeps princeps*)	96	5.64	..	..
Porcupine (*Erethizon epixanthum epixanthum*)	78	4.58	..	..
Horse	..	..	27	1.58
Uinta ground squirrel (*Citellus armatus*)	24	1.41	..	..
Mule deer (*Odocoileus hemionus macrotis*)	20	1.17	..	..
Beaver (*Castor missouriensis*)	12	.70	..	..
Coyote (*Canis lestes*)	..	..	12	.70
White-tailed jack rabbit (*Lepus townsendi campanius*)	10	.58	..	..
Buffalo (*Bison bison bison*)	..	..	10	.58
Deer mouse (*Peromyscus maniculatus osgoodi*)	8	.47	..	..
Chipmunk (*Eutamias* sp.)	8	.47	..	..
Domestic cattle	..	..	1	.06
Mantled ground squirrel (*Citellus* sp.)	7	.41	..	..
Pine squirrel (*Sciurus hudsonicus ventorum*)	6	.35	..	..
Moose (*Alces americana shirasi*)	..	..	5	.29
Muskrat (*Ondatra zibethicus osoyoosensis*)	4	.23	..	..
Antelope (*Antilocapra americana*)	..	..	3	.17
Wood rat (*Neotoma* sp.)	2	.12	..	..
Rocky mountain marten (*Martes caurina origenes*)	2	.12	..	..
Black bear (*Euarctos americanus*)	..	..	2	.12
Mountain sheep (*Ovis canadensis*)	1	.06	..	..
Flying squirrel (*Glaucomys sabrinus bangsi*)	1	.06	..	..
Dwarf weasel (*Mustela cicognanii leptus*)	1	.06	..	..
TOTAL	1,556	91.45	146	8.55

are not detrimental to the park game herds, that their influence is definitely secondary, in the survival of game, to the welfare factors, of which the absence of sufficient suitable winter forage is the most important."

COYOTES

An extensive study over a period of years was made in the Yellowstone National Park-Jackson Hole region in Wyoming to determine the effect of the coyote population on the elk supply. A total of 714 droppings and 64 stomachs of coyotes were examined. This material was collected chiefly in the southern part of Yellowstone National Park and at the north end of Jackson Hole, in the heart of the elk calving area. A report on the findings has been published (Murie, 1935b), so only a brief resume is given here. Table 7 presents the essential data on the mammals consumed by the coyotes. Percentages are based on the total number of individual mammals eaten and not on volume. In addition, there were found 72 birds representing 4.11 per cent of the diet, as well as some miscellaneous invertebrates, fish carrion, and a few vegetable and nonfood items.

In a similar study in Yellowstone National Park Adolph Murie (1940: 43) analyzed 5,086 coyote droppings. It is interesting to note the percentages of some of the more important food items there as compared with the percentages in Jackson Hole:

TABLE 8. COMPARISON OF FOOD ITEMS IN COYOTE DROPPINGS FROM YELLOWSTONE NATIONAL PARK AND JACKSON HOLE, WYO.

	Yellowstone Park	Chiefly Jackson Hole
	Per cent	Per cent
Field Mouse	33.93	29.61
Pocket gopher	21.61	22.27
Adult elk (carrion)	12.85	4.70
Calf elk	3.34	4.46
Snowshoe hare	3.40	9.87
Marmot	1.33	8.81
Jumping mouse	.07	5.64
Porcupine	.39	4.58
Uinta ground squirrel	.51	1.41
Mule deer	1.03	1.17
Muskrat	1.09	.23

These two independent studies show a striking similarity indicating fairly uniform coyote food habits throughout this region.

There can hardly be much question that adult elk are represented in the coyote diet by carrion only. There have been a few unverified reports of coyotes killing younger elk in winter. Coyotes could, of course, kill the animals weakened by starvation or disease—individuals already doomed—but even such attacks are certainly not common. It has definitely been shown, too, that some of the elk calves eaten by coyotes are carrion. A certain percentage of calves die shortly after birth and are later found by carnivores. Thus the 4.46 per cent that elk calf supplied of the coyote diet given in Table 7 would be modified if it were known just how much of the item was carrion. In his analysis of the 5,086 Yellowstone National Park coyote droppings, in which elk calf was represented by 3.34 per cent, Adolph Murie (1940, p. 43) also found good evidence that an unknown number of calves became available to coyotes as carrion.

The coyote's chief opportunity to prey on elk is among the newborn calves in spring, before they follow their mothers regularly, and there is pretty good evidence that calves are sometimes taken by coyotes then. Nevertheless, the coyote must find the calf when the mother is at a distance, for the female elk is determined and effective in defense of her young, as shown by the instances given on pages 116 and 154.

In notes on his observation in Yellowstone National Park, Adolph Murie recorded that in the spring of 1938 he saw a coyote traveling over grassland, obviously intent only on hunting mice. Each time the coyote inadvertently came near a cow elk with a calf, however, the cow immediately charged and drove it off. Thus the coyote mouser was given no peace so long as it happened to approach a calf.

McCowan (1926: 11) told of a band of cow elk in the Canadian Rockies trampling a coyote to death in January. In Wyoming, on the other hand, coyotes often have been seen moving about among elk entirely unmolested in winter, when presumably they could do no harm and the maternal instinct was dormant. On October 17, 1930, in a high meadow in Jackson Hole, Wyo., a big bull was observed feeding while a coyote nosed about only a few yards away. Neither paid the slightest attention to the other.

Mr. Almer P. Nelson, manager of the National Elk Refuge in

Wyoming, made an unusual observation on January 30, 1948. An adult cow elk had been attacked by a band of coyotes. When Mr. Nelson arrived on the scene, 11 coyotes were seen at some distance from the elk, which was then lying down. The coyotes had been feeding on the rump area of the living animal, and occasionally one would approach and tear at the flesh. Previously, workmen had seen this elk fighting off the coyotes, some tearing at the hind quarters, while others were being fought off in front. At Mr. Nelson's approach, she was barely able to get on her feet. This incident suggests one of the occurrences described in the following account of dogs as elk predators.

Studies of the rate of increase of the Jackson Hole elk herd and the factors affecting it, such as hunting, disease, and predation, indicate that coyote pressure on the elk herd is not excessive, as the herd is holding up well, even with an annual kill of 2,000 to 5,000 by hunters (p. 141). Similarly, the Yellowstone herd is easily maintained at numbers that continually tend to overstock the available winter range.

DOMESTIC DOG

Although dogs are not consistent killers of elk and as a rule have little opportunity to prey on them, dogs occasionally harass elk in winter. This is particularly noticeable on the Elk Refuge in Jackson Hole, Wyo., where thousands of elk congregate in severe winters. Once in a while a dog will run out on the fields and readily stampede the elk. Usually this is the only effect—the elk running here and there until the dog tires of the chase and trots away. Now and then an elk will advance toward the dog with threatening manner, but the spirit of the herd stampede generally prevails, causing the elk to turn and run with the rest.

Sometimes the consequences are more serious. Elk have been killed by dogs. Early one morning several shots were heard near the town of Jackson. Five dogs had attacked a cow elk and had her down, when the game warden appeared and killed three of the dogs. The elk was bitten about the head and neck but managed to stagger away. The question arises whether this was a weakened animal, as a healthy adult cow elk can strike a terrific blow and one would expect some injury to the dogs.

Once when the writer was riding down a trail with two young people from a neighboring ranch, two elk calves were spied some distance away at one side. At that moment two cow elk appeared just ahead, and a dog that had been trotting along with the horses ran forward. The two elk rushed out to meet the dog, and for a few moments the dog's life was in danger as rapid hoof-beats barely missed him by a few inches. Then the elk became aware of the proximity of human beings and fled. Had one of those powerful blows struck the dog, he would surely have been killed. The elk calves still lay, unmolested, when the incident was over.

Ignar Olson, who kept domesticated elk on the Quinault River, Olympic Mountains, Wash., told of elk attacks on dogs. Once a dog entered an enclosure where Olson had two bull elk being fattened for an Elk's Convention. Injured by one of the bulls, the dog managed to crawl under a sleigh. The bull elk struck again but that time hit the side of the sleigh and broke a front leg. Then the dog began to shift its position, whereupon the enraged bull struck again and broke his other front leg and so had to be shot. Several other tame elk were lost by breaking their legs over mangers under similar circumstances.

In the winter of 1937-38 near the Elk Refuge in Jackson Hole, two Chesapeake retrievers were observed attacking a young elk. An adult cow elk, presumed to be the mother, counterattacked and injured one of the dogs. Human interference terminated the incident.

On a visit on May 13, 1936 to Afognak Island, Alaska, where a number of Roosevelt elk had been released, it was learned that at first these elk had been very tame and had frequently entered the village, sometimes pulling washing off the clotheslines and otherwise being a nuisance. When dogs went after them, the elk would whirl and attack; several dogs were killed.

BEARS

It is very doubtful that bears ever kill adult elk except under very unusual circumstances, even though there are occasional records of grizzlies killing adults in spring. Both grizzly and black bears prey on elk calves, however, a fact that has been established in Wyoming and in the Olympic Mountains of Washington. During the short period in spring when elk calves are "lying low," a bear occasionally is able to seize one. When this happens, the mother elk is, of course, unable to defend her offspring as she does against the coyote. Once in Wyoming a black bear was seen swimming across a stream with an elk calf in its mouth. The distressed mother ran up and down the bank and even started into the water after the bear, but she did not attack it. Human interference made the bear drop the calf in this instance.

In June 1944 two Yellowstone Park rangers watched a black bear seize and start to carry off an elk calf in spite of the threatening actions of the cow elk. Finally the mother mustered enough courage to strike the black bear with her front hoofs, knocking the bear over and causing it to abandon the calf temporarily. Eventually, however, the bear succeeded in getting away with the young elk.

Both black bears and grizzlies are aware of the food possibilities during elk calving time, and they have been seen searching a likely area for a hidden calf. On one occasion a grizzly was observed to find two elk calves hidden in the same vicinity.

No systematic study of bear food habits was made, but field examinations of bear droppings in the Yellowstone Park-Jackson Hole area over a period of years did not reveal an excessive percentage of elk calf remains, and the elk herds in these areas continued to increase.

In the Olympic Mountain region observations were not so extensive, but as the ranges on several watersheds there are overstocked with elk, the indications are that bears and elk may readily be maintained on the same range, with a surplus of each for the sportsman.

EAGLES

Eagles should be mentioned here because of several interesting observations, even though their depredations on elk are so rare as to be of no significance in the rate of increase of an elk herd. No evidence has come to hand to show that eagles habitually prey on elk calves; in fact, there is no instance of it in the areas studied. A newborn elk calf will weigh as much as 40 pounds. The golden eagle is capable of striking a heavy blow, however, and could probably kill a calf, even though the bird could not carry the calf away.

One winter day a local warden, in company with several other men, observed a small band of elk traveling up an open slope not far from the town of Jackson, Wyo. Two golden eagles appeared, and one of them swooped and struck an elk calf in the neck, knocking it down. The calf got up, whereupon the other eagle struck it in the neck, again knocking it down. This was repeated four or five times, until the young elk crawled into a thicket of bushes. The following day the warden returned and found the animal dead. He said it was a weak animal, apparently one of those diseased individuals that succumb in winter. At that season a calf would weigh at least 150 pounds alive.

Once in the fall another observer noted an eagle swooping at a band of elk. Time after time the eagle swooped, forcing the band to change its course, but it did not strike. Other instances of eagles apparently playing with large game species have been reported, and the purposes of such antics, unless they truly be sport, is difficult to understand. It is known, of course, that on occasion eagles are capable of killing adult deer and antelope. Ordinarily the elk would be too large for successful attack. Can the eagle recognize a disabled or weakened animal? This is possible. If so, by harassing a band the eagle discovers a weakened animal that can be successfully attacked.

11

ACCIDENTAL DEATHS

NATURALLY, a certain percentage of mortality or of injury among wild animals is due to accidents, not often observed but probably occurring more frequently than is generally supposed.

DROWNING

Although elk are good swimmers, numbers are drowned under certain circumstances. A dangerous period is the time of ice formation on streams. Elk were often seen plunging boldly into a stream covered with thin ice and later having great difficulty getting out, when the somewhat thicker ice at the point of egress could not readily be broken and thus drove the animals back repeatedly. Actual drowning was not observed under such circumstances, but in the fall of 1930 when a band of elk attempted to cross Snake River in the northern part of Jackson Hole, 60 head were drowned at a place where elk had drowned before and that is known to be a dangerous crossing for them.

During the spring migration elk must cross a number of streams in some districts. If the spring is late and migration has been delayed, young calves following their mothers must also negotiate the crossings. Young calves are good swimmers but sometimes hesitate to plunge into a bad stretch of water. It was reported that years ago, owing to this reluctance, considerable congregations of elk calves had occurred at certain crossing places on the Gros Ventre River, in Wyoming, and that some calves were undoubtedly lost, though whether by drowning or desertion was not clear. Ordinarily the actual number of calves lost under such circumstances is not great.

In *Natures Notes* of Yellowstone National Park for July-August 1946 (pp. 9-10, Processed), Ranger-Naturalist Herbert T. Lystrop records the death of an elk calf on July 4, 1946. The animal had accidentally fallen into the hot Calida Pool, where it disintegrated in the intense heat. The mother elk was seen in the vicinity, "nervously sniffing the air and pawing the ground."

MIRING

In winter and early spring a number of elk become inextricably mired at places in Flat Creek in Jackson Hole, Wyo., where warm springs maintain some open water. It has not been demonstrated, but it seems likely that many of these are individuals weakened by disease; it is possible, of course, that healthy animals may sometimes become so deeply bogged that they are unable to get out.

From a vantage point on the top of a small butte, where all details could be observed clearly, the writer witnessed the process on the evening of March 11, 1939. A considerable herd of elk, numbering more than a thousand at least, was traveling southward on the Elk Refuge in Jackson Hole. On their arrival at the bank of Flat Creek there was some hesitation and massing of the animals, but presently a thin line of animals began crossing the stream cautiously and soon the entire herd was once more under way. Suddenly violent splashing and a stampede interrupted the crossing. A spike bull was mired and struggling desperately, his head under water part of the time. After some hesitation, when the bull became quiet, the herd resumed its crossing, but several times the struggles of the mired animal stampeded the herd. Presently the elk were using a new crossing farther downstream, but again there was a stampede when three more elk—a bull, a cow, and a calf—became bogged. The calf finally got out, but the other two remained. A raven lit on the back of the cow, evidently considering the elk a doomed animal. Eventually the herd completed the crossing and went south to a hay feeding ground, leaving behind the three unfortunate members mired in the creek.

Two low bridges have been built at strategic places on Flat

Creek for the use of the elk in crossing. Probably these have helped minimize the losses, but the elk do not always use the bridges and some losses still occur.

Vernon Bailey reported (ms. notes, 1905) that in 1857 his father, while hunting in Tuscola County, Mich., shot at a bull elk lying down at an elk lick. The animal did not move, so Bailey ran up, killed it with his belt ax and "then found that the elk was fast in the mud and that he had missed it with his rifle."

FIGHTING

Perhaps mortality from fighting should not ordinarily be classed as an accident, yet there are times, even in fighting, when the death of the elk is accidental, as, for instance, those relatively rare occasions when the antlers become interlocked and both combatants perish. Then, too, as already described (p. 131), in several minor quarrels an animal was killed by the single fatal thrust of an antler tine that struck just right.

BARBED WIRE

Perhaps barbed wire should be given special mention as a hazard to elk. A rail fence ordinarily is no obstacle to an elk, nor is it dangerous. Barbed wire, however, has caused the death of many elk, as well as other members of the deer family; and numerous such instances have been recorded both in America and in Europe. Two strands of wire sometimes become twisted about an elk's foot, holding the animal entangled until death occurs. One elk, passing through a barbed wire fence, cut its throat on the wire and went only a few hundred yards before it bled to death; and a young moose was examined that had met a like fate. Another young elk became so entangled in a snarl of loose wire that a loop formed tightly about a hind leg and cut deep into the flesh above the hock. The wound was gangrenous and the leg useless when the animal was found.

When it became evident that barbed wire was a serious hazard, all such fencing was removed from the National Elk Refuge at Jackson, Wyo.

MISCELLANEOUS

Gunshot wounds do not come under the head of accidents, but sometimes they are the cause of accidental death even after they have healed. For instance, a cow elk jumping over a pole fence tripped and fell in such a way as to break her neck. Examination showed that earlier a bullet had broken a front leg and that the bones had knit in an overlapping position. Although the wound had healed perfectly a long time before, the crippled leg doubtless was responsible for her awkwardness in jumping a fence that other elk had cleared easily.

A. C. McCain, at one time supervisor of Teton National Forest, Wyo., estimated that during a hunting season one elk is wounded and lost for every ten that are shot and retrieved. Many of these wounded animals wander off and die, and during these studies a number of such dead elk were found. Of course, some of the wounded recover, especially those shot in the legs. One old bull and one young cow had a hind foot completely severed and the stump healed.

Numerous post-mortem examinations have revealed occasional broken ribs or similar injuries. That a broken rib may cause death is indicated by a dead moose examined one winter. The heart showed pronounced subacute endocarditis, undoubtedly brought about by an adjacent broken rib.

One spring a yearling elk cow was found dead, partly eaten by

FIGURE 29. Right hind foot of female elk, showing abnormal hoof obviously resulting from an injury. Jackson Hole, Wyo., Jan. 17, 1936.

coyotes and ravens. On dissection it was found that a wood splinter 11 inches long and about 1½ inches in diameter had been driven into the thoracic cavity at the base of the neck and had pierced a lung. Evidently it was the tip of a snag against which the animal had run in sudden flight.

An incident told to the writer shows that an elk may lose its footing with serious consequences. One evening in August 1933 a party camped on a tributary of the North Fork of Sun River heard a crash nearby. Next day a few hundred yards away they found a dead six-point bull elk. It had fallen off a cliff, a height of 200 feet, then rolled down the talus. There had been a rainstorm, and slippery footing might have been the cause of the animal's fall.

On January 17, 1936, at Jackson, Wyo., a female elk was found with an abnormal right hind hoof, obviously caused by an injury. Such an accident is a handicap that may result in death under certain conditions (Figure 29).

In primitive times prairie fires in the Red River Valley sometimes blinded bison, and possibly elk too were caught. Today forest fires are a danger to game, including elk. Occasionally deer and elk succumb in snowslides.

12

PARASITES

IT IS difficult to draw the line between parasitic maladies and other diseases, as small organisms are concerned in both types. For convenience in this discussion parasites are treated as comprising insect pests, such as flies and mosquitoes, in addition to external and internal parasites as popularly understood.

The Jackson Hole, Wyo., elk are not seriously troubled with endoparasites. The late Dr. Maurice C. Hall, who visited the writer's base camp in 1927, described some elk he examined in the field as "unusually clean." Intensive work was not done in other areas. Various parasites were reported in the Yellowstone National Park elk herd by Rush (1932a) and Mills (1936: 251), and a few were found in elk in the Olympic Mountains, Wash.

Cowan (1949) and Green (1949) record the occurrence of hydatid cysts, *Echinococcus granulosus,* from elk in Jasper and Banff National Parks of Canada, and indicate the wolf, coyote, and mountain lions as terminal hosts.

FLIES

During summer, notably in July and part of August, the actions of a band of elk resting in a meadow plainly indicate that the animals are considerably annoyed by flies—horseflies, deer flies, and others. The elk shake their heads, flap their ears, and otherwise try to get rid of their tormentors. Frequently an elk gets up, walks about, and then lies down again, unable to keep still in the presence of these biting pests.

Under similar conditions horses find some relief by going into shade, and to some extent the elk appear to follow that practice. Often in midsummer they go into the hollow clumps or circles of firs that are characteristic of the Hudsonian Zone in the Rocky Mountain region. There, in the shady interior of thickets that the flies appear to shun to some extent, the elk obtain a degree of freedom from the flies. Yet many bands of elk are seen lying down in open meadows, exposed to sun and flies, patiently flapping the ears and tossing the head in routine fashion.

The elk seem resigned to the flies and take them as a matter of course. In fact, the animals do well and become fat during the fly season, a period of abundant and nourishing food. Possibly when the elk seek the shelter of fir groves, they do so as much to avoid the heat as the flies. Certainly elk do not move into the larger forests in numbers at this time, although there are always some elk in the forests. The general movement to woodland occurs later, after the worst of the fly season, when the frosts have made open meadow vegetation less desirable. There is no migration, even on a local scale, that can be attributed to flies.

In contrast, it is notable that horses seek the interior of barns or sheds, or the shade of a clump of trees, seemingly for the express purpose of getting away from the flies. It is doubtful if a group of horses would lie in an open meadow, fully exposed to both the sun and the flies as the elk so often do.

In the areas where these studies were made the mosquitoes have been of less importance, but at times a variety of larger flies have been of great annoyance to humans, horses, and game. The intensity of the fly pest varies greatly and seems worse in very dry periods. In some years the flies are not bad and may be confined to a large extent to lower elevations, especially valleys. On the other hand, in a season like that of 1931, which culminated a series of dry years and was itself hot and dry, the fly outbreak in the Yellowstone National Park-Jackson Hole area was unusually severe and no relief could be found at any elevation.

It is probable that the same species of flies attack elk, horses, and man. Among a number of flies collected near the elk and at fresh elk carcasses on various occasions, the following species were identified: *Tabanus hirtulus, T. rhombicus, T. affinis, T. punctifer,*

Chrysops carbonaria (the familiar deer fly that is well known by its "three-cornered" shape and the spot on its wings, as well as by its bite), and *Symphoromyia hirta.* Of course, there would be other species of the Tabanidae in many localities beside the particular ones here actually identified as elk parasites. It is difficult to estimate which of these is the most annoying and persistent—as each has a painful bite—but probably it is the abundant *S. hirta,* of the family Leptidae, a small, dusky, insignificant-appearing fly that has a surprisingly painful bite.

TICKS

The winter tick *(Dermacentor albipictus)* is a never-failing scourge that visits the elk each spring, its numbers varying from year to year. It is difficult to determine just when the ticks first become noticeable to the elk, because only dead animals can be examined closely, but these pests probably become markedly bothersome in the latter part of March although they attach to the animals long before that. The infestation lasts into May, about the time most of the winter pelage is shed. After that the ticks no longer annoy the elk for the season, as the females have dropped off to lay eggs. In the Olympic Mountains, however, Schwartz (1943) reported that elk are infested with ticks in all months of the year, including the hunting season in the fall, but they are more troublesome in winter and spring.

Usually the ticks are attached in large numbers on the animal's neck, along the back, about the base and under surface of the tail (where much bare skin occurs), and on the inside of the thigh. As stated earlier (p. 73), when an attack of ticks becomes severe, the elk does much scratching with the hind hoofs, so that on many an elk there develops around the lower neck (the point most conveniently reached by the hind foot in scratching) a more or less hairless band that gives the animal a comical appearance.

The elk also bite at the ticks around the hindquarters, and there are indications that the engorged ticks may be palatable to the elk. No wild elk were seen actually eating ticks, but a tame elk ate them with relish. In fact, unpalatable powdered medicine, given to a bull elk over a period of time in an effort to cure suspected necrotic stomatitis, was readily administered by wrapping the paper of powder in a ball of hair and engorged ticks taken from the animal's body. The bull obviously relished the mixture and picked up ticks that dropped to the ground.

Much confusion has existed with regard to the effect of ticks on elk. Many elk found dead are heavily infested with ticks, and it is often presumed that the ticks were the cause of death. In hundreds of examinations of dead elk it was found, however, that death invariably could be traced to other causes and that the presence of ticks was incidental. In fact, during the period of these investigations not one elk death could be attributed to ticks alone, although ticks were suspected of being the chief agent in the death of one young moose. Evidences of disease were present in so many of the elk examined that it appeared that the ticks seldom if ever were the direct cause of death. It is possible that a heavy tick infestation may at times be fatal, but it would be exceedingly difficult to prove this. Nevertheless, ticks are a heavy drain on an animal's vitality and may be a contributing factor in the death of some diseased elk.

Bishopp and Wood (1913: 178) reported that they found the winter tick an important pest of horses and cattle in California, Oregon, Montana, and elsewhere; and the writer observed winter ticks on a horse in the Olympic Mountains, Wash., in 1916. Such infestation of domestic animals may be common; yet curiously enough, in the vicinity of Jackson, Wyo., where thousands of infested elk assemble to feed each winter, the horses and cattle examined by the writer, particularly those on the Elk Refuge, were with one exception free of ticks or at least were not covered extensively enough for ticks to be readily noticeable. Apparently the domestic animals in the area are more or less immune to this pest.

Winter ticks do not readily accept the human host, although they will temporarily swarm over a man when he is handling an infested elk carcass. At any rate, during a test period of one year all ticks found on human beings were *Dermacentor andersoni*, the spotted fever tick, a species that was not found on the elk.

SCABIES

Scabies has been reported among elk as far back as records are available. Specimens of the causative mites occurring on the elk have been definitely determined to be *Psoroptes communis* var. *cervinae.* Dr. E. W. Price, of the Bureau of Animal Industry, commented in a letter as follows:

"This mite is a variety of the species which occurs on sheep and other animals. So far as definite evidence has been obtained, psoroptic scabies cannot be transmitted from one host species to another in spite of the fact that the mites causing this type of scabies are so similar that it is difficult to separate one variety from another morphologically."

Cattlemen have been concerned about the possibility of infection of cattle with scabies from game animals; but in view of the information given above, in which leading authorities appear to concur, it seems unlikely that scabies on elk will be transmitted to domestic animals. As a matter of fact, even transference from one animal to another of the same host species appears to be accomplished with difficulty. A pertinent observation was made one fall when the tame bull elk on the National Elk Refuge in Jackson Hole, Wyo., became badly infected with scabies, which persisted throughout the winter. This animal was closely associated at all times with a few head of milk cows, went into the barn with them and fed with them; yet none of these ever showed any sign of scabies. The elk had become infected while roaming about in the swamp near Jackson among domestic cattle. No scabies was known to be present among the cattle, and the wild elk had not yet arrived in the area from the summer range. Apparently, the only possible explanation of the source of this infestation was that there were available in the swamp the remains of dead scabby elk on which mites had survived from the previous winter.

Every winter a number of Jackson Hole elk become scabby, but in the average winter scabies does not figure largely in elk losses there. Although the degree of infestation varies, apparently about the same proportion of the herd is affected each year. The winters of 1938-39 and 1939-40 were exceptions, as both the proportion of scabby elk and the number of scabby females were unusually high (Plate 18).

Mature bulls are the chief sufferers, but quite a few adult cows are at times afflicted. The younger animals and calves are rarely scabby. Late in October the bulls have just completed the strenuous period of the rut, with tremendous expenditure of energy and scanty feeding. They appear more or less listless and have lost the fat accumulated during July and August on nutritious green forage. This, too, has largely disappeared, and the lean, winter months are at hand. All these circumstances tend to lower vitality and resistance. In this weakened condition the bull elk is probably very susceptible to scabies, especially if he has been injured, either in battle with a rival or by a bullet in the hunting season. Frequently a lone scabby bull has been seen in the valley lands far in advance of the migration of the main herd.

Probably many scabby elk recover. The infested tame bull mentioned was treated periodically with nicotine spray, but the treatment was not thorough enough as mites still remained on parts of the body in spring. In May, about the time the new pelage was acquired and when the elk was obtaining new green forage, the mites disappeared. This animal had no further attacks.

Some of the scabby animals were found to be otherwise diseased also, and it is by no means certain that all dead scabby elk have succumbed to scabies alone. It appears likely, though, that scabies aggravates the effect of other diseases. On the other hand, because of a slight weakness or disability that in itself would not be fatal, scabies may gain a hold and become so severe that it alone causes death. A scabby cow elk with the posterior part of the body, including the sides, so completely denuded of hair that she was obviously suffering from cold was observed in January 1939. She was lying down. When approached she arose but stood in a crouching attitude, backing up toward the slight breeze with mincing steps. She soon lay down again, curling up as completely as she could to reduce the skin surface exposed to the cold. A few days later she was dead (Plate 18).

It is notable, also, that scabies occurs in winter when the resistance of the animals is lowest and disappears in spring when rich, new forage becomes available. Is it possible that the disease indicates, in a measure, the condition of the range?

The only remedy that has been suggested for reducing scabies among free-ranging wild elk is to shoot scabby individuals and then burn the carcasses, or at least the hides, or else otherwise treat them to destroy the mites. As stated above, the scabby tame elk described evidently became infected from the mites that were on carcasses of dead elk on the winter range. The destruction of hides of infected animals would involve constant watchfulness on the part of game officials, though, and after all there may be factors that would nullify such efforts. Probably the best precaution would be the avoidance of overstocking, thus maintaining a good, productive elk range.

LICE

Although lice occur on various game animals, they are not present on the elk of Jackson Hole, Wyo., commonly enough to trouble them and so do not constitute a problem there. In the Olympic Mountains, Wash., more evidence of lice was found. One very old cow elk, examined March 12, 1935, shortly after death, was literally covered with lice. The teeth of this animal were worn to the gums in places, some were loose, and there were some necrotic lesions. Mills (1936: 250) found three cow elk in Yellowstone National Park infested with lice, one of them heavily so. Lice collected on elk in the Olympic Mountains have been described by H. H. Stage, of the U. S. Bureau of Entomology, as *Bovicola americanum.*

TAPEWORMS

Tapeworms have been recovered from the Yellowstone National Park elk, and they should be expected in elk on other ranges. In some elk of the Jackson Hole herd not a single tapeworm could be found, but in a number of others tapeworms occurred more generally and were identified by Dr. Maurice C. Hall as *Thysanosoma actinoides,* the fringed tapeworm. They were frequently found in the fourth stomach and small intestines and once in the paunch, which is unusual. The bladder worm *(Cysticercus tenuicollis)* was found in the liver of a number of elk. In the Olympic Mountains Schwartz found 7 out of 32 elk examined carrying tapeworms.

NEMATODES

The writer has not observed nematodes, except the lungworm, in elk; and Dr. Hall did not find any in the animals he examined in Jackson Hole, Wyo., in 1927. Mills (1936: 251) recorded a few nematode larvae in elk fecal samples from Yellowstone National

Park. On the whole, the evidence for occurrence of these parasites is largely negative. Among 32 Roosevelt elk examined, Schwartz (1943) found *Oesophagostomum venulosum* and *Trichurus* sp. in 5 of the animals.

Lungworms *(Dictyocaulus hadweni)* were found in a number of elk in Jackson Hole, Wyo., where they appeared to have contributed to the death of at least one animal, but lungworms do not seem to be widely prevalent in the elk herds and do not constitute a serious menace. John E. Schwartz examined 32 Roosevelt elk and found lungworms in 23, some of those being heavily infested. He expressed the belief that the lungworms made a serious drain on the vitality of the elk, though he had seen no indication of pneumonia. Rush (1932a: 57, 95) and Mills (1936: 251) recorded *D. hadweni* from Yellowstone National Park elk, and the writer found them in the Roosevelt elk of the Olympic Mountains, Wash. There has been no opportunity to examine carcasses of the tule elk of California.

LIVER FLUKES

No liver flukes were found in the Rocky Mountain elk, but Schwartz recorded *Fascioloides magna,* the giant liver fluke, from the elk of the Olympic Mountains, in some localities the fluke infestation reaching 81 per cent of the animals examined.

BOTFLY LARVAE

Larvae of botfly *(Cephenomyia* sp.*)* were found in a limited number of elk in Wyoming and in the Olympic Mountains, Wash. Rush (1932a: 93) suspected that these larvae may have caused heavy losses in the deer and elk of Yellowstone National Park and that they are probably an important factor in the calf-elk losses. In the Jackson Hole herd no positive evidence was found that botfly larvae were implicated in elk mortality; in fact, not a single carcass examined revealed any evidence that these larvae had been involved in the death of the animal.

Inasmuch as botfly larvae appear to be much more prevalent in the mule deer than in the elk, it is possible that the latter may be more heavily infested with these parasites in areas where they associate closely with mule deer in summer, the season when botflies are laying their eggs.

When it is considered that to date no losses from botflies have been reported among the heavily infested northern caribou, practically all of which appear to carry a throat full of the grubs in late winter and early spring, it may be concluded that the comparatively scattered infestation noted in the elk herds cannot cause serious losses, however painful or annoying these parasites may be.

EFFECT OF PARASITES

Summing up the effect of parasites on elk, it may be concluded that some parasitic species are scarce and mainly of academic interest; that others are fairly common but only occasionally serious and that the elk appear to be adjusted to their presence; that a few, such as the various horseflies, are very annoying but have no lasting effect and are taken "philosophically" by the elk; and that at least one, the winter tick, though not in itself usually fatal to the elk, is certainly a drain on vitality and may be a contributing factor in some elk deaths.

There is a possibility that parasites of various kinds tend to multiply and become correspondingly troublesome on overcrowded ranges or areas of concentration. This has not become entirely clear, however, at least so far as internal parasites are concerned, on areas such as the National Elk Refuge, where thousands of elk have been concentrated for many years.

13

DISEASE

ELK ARE SUBJECT to a number of ailments, including specific diseases. Some of these are unimportant; others have a definite bearing on the welfare of the herd.

HERNIA

Hernia is not common among elk and does not seem to be seriously injurious. Several cases of it were examined, some of them in various stages of strangulation. An extreme case was observed on the National Elk Refuge at Jackson Hole, Wyo., before the elk study there was begun. A cow elk, with a great pendent hernia "as big as a water bucket" had appeared year after year. The under surface of the hernia had become calloused and often showed blood, probably where it had scraped as the animal jumped fences.

BLINDNESS

Occasionally blindness is noted in elk, sometimes in only one eye. In the winter of 1932-33 an elk blind in both eyes fed with the rest of the herd on the National Elk Refuge in Jackson Hole, Wyo., finding its feed by its sense of smell. Several blind moose have been found south of Yellowstone National Park, and some years ago a blind caribou was observed in Alaska. Rush (1932a: 95) reported that four totally blind elk in Yellowstone Park might have been affected with infectious keratitis, and Mills (1936: 252) described two cow elk in the same area that appeared to be suffering from the disease called "moon blindness" in horses. In the blind elk examined by the writer, the eyes were clouded over by an opaque milky white film. The cause of the ailment is unknown.

STAPHYLOCOCCUS

A young bull elk that had unusually solid and heavy antlers and died late in the spring was found to be heavily infested with lungworms, numerous sarcocysts in the heart muscles, and *Staphylococcus* bacteria in the lungs and liver. Apparently such bacterial infections are not common, as no other cases were found. Referring to the bacteria, Dr. John S. Buckley, of the Bureau of Animal Industry, made this comment:

"The above organism, with one slight exception, shows the cultural and staining characteristic of the *Staphylococcus epidermidis,* and inasmuch as the organism has proved pathogenic for laboratory animals it is thought quite probable that it may also have been a causative factor in the death of the yearling elk."

SARCOCYSTIS

Sarcocysts were found in the heart muscles of elk in Jackson Hole and in Yellowstone National Park. Mills (1936: 252) reported that 10 cows "examined for *Sarcocystis* in the heart" were all infected, several badly. In the Olympic Mountains Schwartz found sarcocysts in 40 per cent of animals shot in the hunting season; these infestations were usually in the heart, sometimes in the tongue and cheek tissues.

ARTHRITIS

In some winters arthritis seems to be common among elk. It is manifested by an abnormal joint condition that appears among the older animals early in winter, sometimes as soon as December. A knee or hip joint, a foot, or occasionally two joints may swell, forming a prominent lump or lumps. Not all joint swellings represent true arthritis; some of them result from injuries. A number of leg bones from animals that had been variously affected in the joints were sent to the late Dr. Roy L. Moodie, who was interested in prehistoric diseases. One of these showed pronounced arthritis, and some of them revealed injuries. Arthritis is discussed further in the section on necrotic stomatitis (p. 177).

TUMORS

Occasionally miscellaneous tumorlike manifestations filled with a thick, creamy liquid were found in elk. In one instance, at least, the cyst was thought to have been the result of injury. A tumor found within the thoracic cavity of a bull moose was the largest observed, being about the size of a small football.

HEMORRHAGIC SEPTICEMIA

Studies made to date do not indicate that hemorrhagic septicemia is extensive or serious among elk. The organisms causing the disease were demonstrated in only one elk in the Jackson Hole herd. Cultures obtained from this animal by the Bureau of Animal Industry produced acute septicemia and death in guinea pigs but did not have sufficient virulence seriously to affect a horse. Other cases that had all the appearance of the one just cited were noted, but no attempt was made to obtain bacterial cultures. In recent years some veterinarians have advanced the theory that hemorrhagic septicemia is secondary and more or less the result of other abnormal conditions.

INFECTIOUS ABORTION

Early in the investigations of the Jackson Hole, Wyo., elk herd it was learned that abortions were not uncommon among the cows during winter. Accordingly, in the spring of 1930 arrangements were made with Dr. W. E. Cotton, then Superintendent of the Bureau of Animal Industry Experiment Station at Bethesda, Md., for testing blood serum of elk. The agglutination test was applied to samples of blood serum furnished from nine elk at Jackson, Wyo. Of these, three reacted positively for infectious abortion; at least, they reacted to a degree that would be considered positive in bovine samples. One of these samples was from an old bull.

Subsequently Rush (1932a; 1932b: 372) reported this disease in the northern Yellowstone herd; and in the winter of 1935-36 the office of the Wyoming State Veterinarian found evidence of the disease among a number of elk that were slaughtered in the State in a herd-reduction program.

In this connection some field observations are significant. As stated above (p. 142), more than 50 per cent of the cow elk of breeding age that died on the Elk Refuge during the winter and were examined were without fetuses. It is true that many of these cows were old, some very aged, which would account for the barrenness of a certain proportion. However, some in their prime were without calf, and some very old ones were pregnant. It will be recalled, too, that, as stated earlier (p. 142), about 90 per cent of the cow elk examined that had been killed by hunters in the fall and 89.2 per cent of the 344 cow elk killed in January 1936 by the Wyoming State Game Commission were pregnant.

It is apparent from these data that failure to bear calves is not due to failure to breed and that a certain proportion of the calves are lost prematurely in late winter. As a matter of fact, many aborted fetuses have been observed on the winter range; doubtless, too, ravens consume many before their discovery is possible. Sometimes dried fetuses are washed down irrigation ditches, in Jackson Hole, when the water is first turned in during the spring.

Abortion may result from other causes than infection. The cause of death for one cow elk examined was found to be necrotic stomatitis. A few feet away lay an aborted calf. Had the abortion been induced by the stress of this disease, or had the cow been afflicted also with infectious abortion? Frequently in late winter or early spring ranchers found it necessary to drive elk from their pastures. Running the elk at this time of the year was generally deplored as dangerous for the prospective calf crop, but no abortions were observed as a direct result of the running. All facts considered though some barrenness in spring is undoubtedly due to old age and perhaps other similar causes, it appears that infectious abortion must play an important part in the curtailment of the calf crop.

MORTALITY OF NEWBORN CALVES

Some of the calves examined had died from a cause not yet determined. One spring eight such deaths were noted in Jackson Hole, Wyo. Considering how difficult it is to find dead calves in the forest when the herds are scattered in the spring, the fact that this many were actually observed may indicate an appreciable percentage of loss in the elk herd. In other spring seasons in Jackson Hole additional dead elk calves were found incidentally.

Unfortunately, the dead calves are seldom found before decomposition has advanced too far for diagnosis to be made. In these studies there was only one opportunity for diagnosis, and in that calf no disease was apparent.

In his work in Yellowstone National Park, Adolph Murie (1940: 49) recorded a few elk calves that had died shortly after birth from unknown causes. One of these was obviously abnormal in development.

Each spring in the Olympic Mountains Schwartz found a few dead calves, 1 to 3 or more days old.

At present we have no explanation of these spring losses. Recently there has been some evidence that summer losses among the young of mountain sheep and antelope are related to malnutrition of the mother on overstocked winter range, but the question is still in the speculative stage.

NECROTIC STOMATITIS

Necrotic stomatitis appears to be by far the most important elk disease. It was the cause of most of the annual losses of elk in Jackson Hole, Wyo., particularly on the feed grounds. It had been prevalent there for many years prior to these investigations, too, as shown by observations of Jackson Hole residents which match closely the facts later developed.

As a rule, in those winters in which the elk are fed hay on established feed grounds, heavy fatalities begin to occur about the middle of February, although sporadic mortality occurs before that, even as early as December. From the middle of February through March and April the elk deaths continue, and often many occur in May, after feeding of hay has ceased.

Manifestations

Early in winter what appears to be arthritis (p. 174), sometimes becomes evident. To what extent this may be associated with necrotic stomatitis is hard to say, but it has been observed to be at least a small part of the normal "pattern" of the annual affliction. In some winters many of the calves on the feed grounds of the National Elk Refuge appear lame, some apparently with sore feet, suggesting mycotic stomatitis (Mohler, 1923b). This disease has not yet been demonstrated in the elk, however, so that it remains a question whether some of the lameness is not after all connected in some way with necrotic stomatitis. Occasionally, particularly among adult elk, the epiphysis, or terminal cap, of a leg bone may be eroded on its surface. Such an extreme condition is not usual among the calves.

The first definite indication of the disease is a drooping and generally unthrifty appearance of the animal—lowered ears, sometimes drooling—with wasting (Plate 19). Emaciation is more pronounced among the older animals, presumably because the younger ones succumb before they have had time to lose much flesh. Breathing becomes halting, with an audible catch just before exhalation. Finally the animal lies down and is unable to arise. Death usually follows this stage within 24 hours, though old animals sometimes linger for a number of days.

Often an elk is seen with one cheek distended by a large wad of food, usually hay; and rarely, one is seen with both cheeks so stretched. This symptom accompanies necrotic stomatitis in cattle also. In some elk observed the wad of hay appeared to have so irritated the lining of the cheek that necrotic (mortified) lesions had formed. In other cases bad lesions themselves seemed to have caused the lodgment of such a wad.

The most reliable post-mortem indications of necrotic stomatitis are the characteristic lesions in the mouth, grayish-yellow, cheesy patches of varying sizes and shapes, occurring very often in the soft tissues in the back of the mouth, in the angle between the jaws, or along the gums. Quite often they affect the cheek, tongue, or palate. In one victim the entire dental pad was replaced by a necrotic lesion. A break in the tissues had in some way occurred, and presumably the constant abrasion by the lower incisors during the feeding process had helped spread the infection. In adult elk the infection often takes place in the gums, causing ulcers in the tooth sockets with consequent loosening and often loss of teeth.

The lesions often penetrate deeply into the tissues and come in contact with bone, resulting in a bony necrosis that eventually produces an exostosis (abnormal outward bony growth, or tumor) in varying degree. An extreme case was found in which a lower jaw had been completely severed, the two parts being held together secondarily by the huge, spongy exostosis (Plates 20 and

21). At the other extreme are instances in which the lesion had barely reached the bone before death ensued, causing merely a slight roughening of the bone surface to indicate its response to incipient necrosis. These osseous manifestations are useful in detecting the cause of death even when the bleached bones are found on the winter range.

Ulcers—cheesy-yellow patches rimmed thinly with red—were found in quite a few stomachs, and necrotic enlargements were noted in an occasional liver or spleen. What appeared to be necrotic foci were present in some lungs, producing essentially a pneumonic condition. Extensive lung infections were confined chiefly to the older animals, however. Some of these internal infections cause a copious release of exudate in either the abdominal or thoracic cavities, sometimes with extensive adhesions, and some ulcers penetrate the stomach walls.

It should be added that frequently the characteristic lesions were not found, and yet the general condition of the animal and its actions were similar to those of animals carrying the lesions. In other instances no oral lesions were found and only the liver was necrosed. Although such partially negative cases are difficult to determine, they seem to be part of the epizootic. Perhaps in some cases lesions were overlooked, or there may have been some other obscure manifestation. Such cases cannot be dismissed as simple starvation, for they often occurred on the feed grounds toward the end of a hay-feeding period when hay for all the animals was abundant.

Virulence

It has not been possible to learn what percentage of animals recover from necrotic stomatitis. One captive elk got well under treatment. On the other hand, seven deer, two moose, and a number of elk were captured and kept in an enclosure with the intention of shipping them out eventually. They were fed oat hay, and one by one they became sick. As the disease progressed, the elk were released; but all the deer and moose died in spite of the treatment given some of them.

Field observations also indicate that necrotic stomatitis is usually fatal, but that some of the older animals may recover in the wild is indicated by specimens in which one or more molariform teeth have been lost and the gums have healed. In the skull of one animal that had succumbed to a second attack of the disease, an entire upper tooth row was missing but the jaw had healed completely.

Duration

The period of time taken for necrotic stomatitis to run its course varies greatly. It is difficult to establish the time among free-moving and more or less unapproachable elk, but it appears certain that the old elk endure the disease for several weeks, possibly 6 or more. This gives time for producing the more extreme manifestations found in these older animals. Among calves the course of the disease appears to be much shorter, but even with them it is long enough to produce some degree of emaciation and sometimes exostosis of the jaw. Careful study of numerous cases indicates that death may occur in some individuals with very slight lesions, the disease in this acute form evidently producing a powerful toxemia. It is said that among cattle the duration of a fatal course of the disease may be as short as "five to eight days."

Etiology

Necrotic stomatitis involves an infection with a bacterium that appears under the microscope as a rodlike organism. Formerly known as *Bacillus necrophorus*, more recently it has been commonly referred to as *Actinomyces necrophorus*.

Merchant (1942) says:

"It has been customary to place the necrophorus bacillus in the genus *Actinomyces*. This has not been done in the 1939 edition of Bergey's Manual, but the name *Actinomyces necrophorus* has been retained for the organism which is placed in a heterogeneous group of bacteria which need classification. This organism does not conform to many of the characteristics of the other members of the group; for example, it does not form a mycelium, is not branched or does not produce clubbed radiations, but the organism does form long, beaded filaments which disintegrate into coccoid cells which appear to function as conidia. Pathologically, the organism is rather similar to the other *Actinomyces*. *Actinomyces necrophorus* causes abscess formation and marked necrosis of tissue. *Actinomyces bovis* produces the accumulation of mononuclear leucocytes in abscesses and the proliferation of

PLATE 18. Scabby cow elk, Jackson, Wyo. The cow above has lost most of her hair and lies curled up as closely as possible to reduce the area exposed to the cold air. Spring of 1939. The cow below, with one side largely denuded of hair, is dead. April 9, 1928. (U. S. Fish and Wildlife Service photographs.)

PLATE 19. Elk afflicted with necrotic stomatitis, Jackson, Wyo. The elk above is nearly 1 year old, February 3, 1928. When young animals reach the stage where they cannot arise or do so with difficulty, they generally survive about 24 hours. Older animals are often prostrated several days before death ensues. (U. S. Fish and Wildlife Service photograph.)

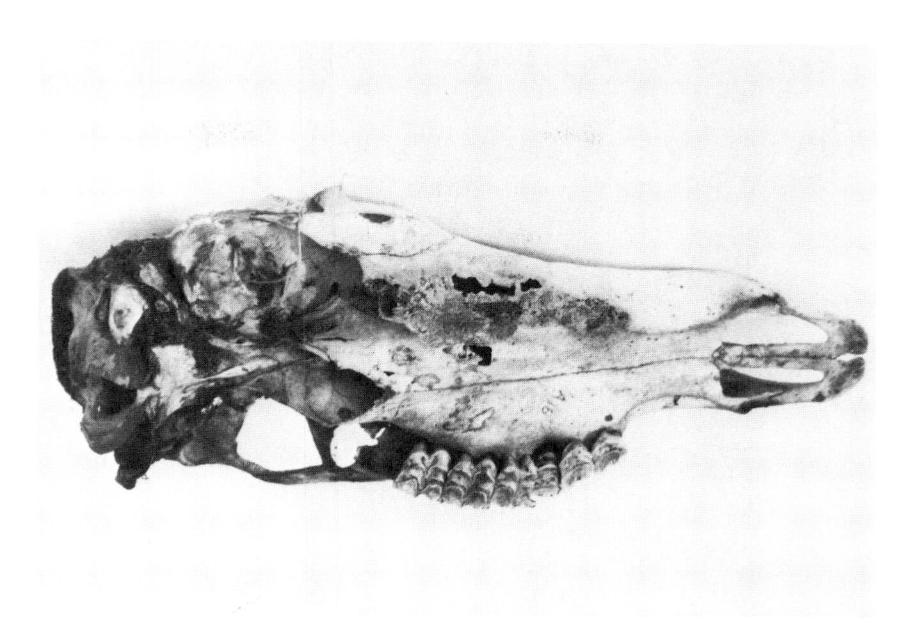

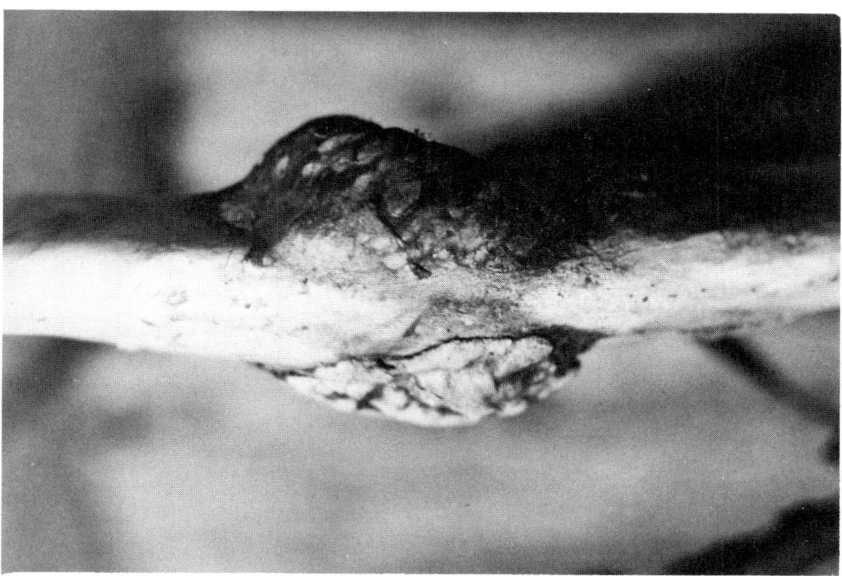

PLATE 20. Some effects of necrotic stomatitis. *Above.* Skull of cow elk that had lost all the upper teeth on one side from necrotic stomatitis, possibly through repeated infections. The gums had healed and the animal survived, only to die of another attack in the lower jaw. *Below.* Characteristic exostosis on the lower jaw of an elk infected by the disease. Such bony manifestations are useful in interpreting the history of a herd on a much-used range. (U. S. Fish and Wildlife Service photographs.)

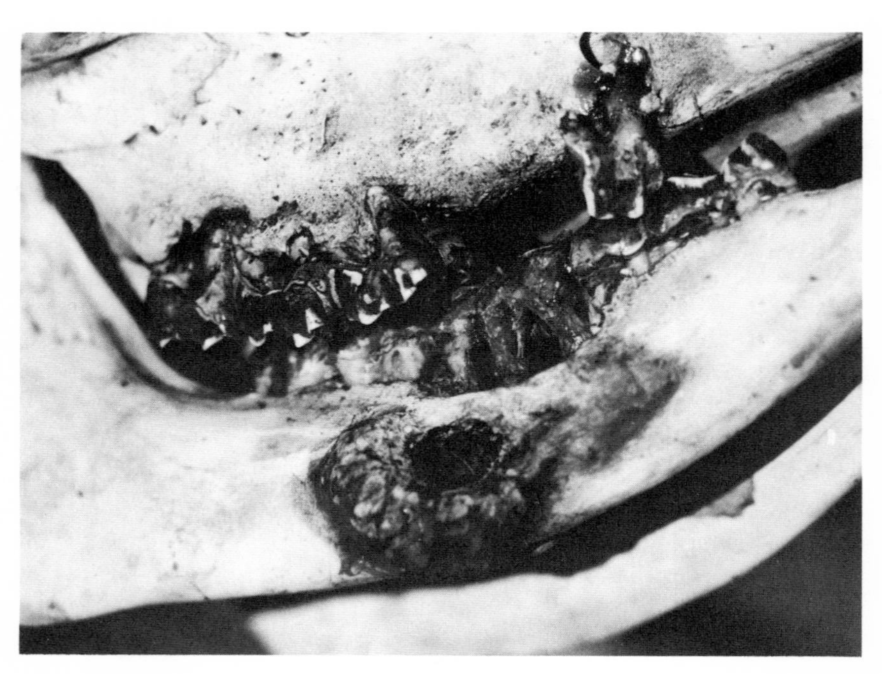

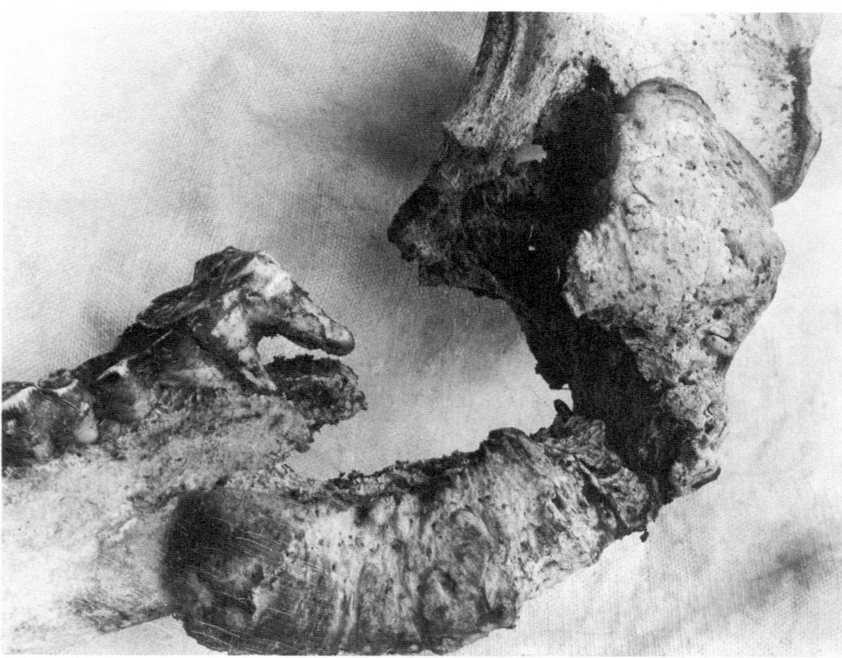

PLATE 21. Some effects of necrotic stomatitis, Jackson, Wyo. *Above.* Lower jaw of cow elk, showing bony excrescence resulting from ulcerated teeth, such infection with the disease having been caused by bobtailed barley in hay. Several teeth are missing as a result of previous infections. November 1928. *Below.* Lower jaw of cow elk infected with necrotic stomatitis. A section of the jaw has been entirely eaten away by necrosis, but the severed parts are united by the extensive exostosis. May 1929. (U. S. Fish and Wildlife Service photographs.)

fibrous connective tissue cells, and it, too, may cause necrosis, particularly of the bone. Until a more appropriate genus is formed it appears reasonable to classify the necrophorus organism as *Actinomyces necrophorus.*"

In line with the above characterization of this organism, it should be stated that it is entirely probable that several other organisms of the *Actinomyces* group—particularly *Actinomyces bovis* (lumpy jaw), and *Actinobacillus lignieresi (Actinobacillosus)*—may be present among elk. Especially lumpy jaw could be expected to occur. In the following discussion references are made interchangeably to necrotic stomatitis and actinomycosis mentioned in literature because they have similar significance, in etiology at least. It should be kept in mind, however, that although etiology is similar in many respects, Actinobacillosus and lumpy jaw are chronic and do not cause many deaths; on the other hand, necrotic stomatitis operates more quickly and is quite deadly. It is obvious from the general pathological picture presented among the game herds that necrotic stomatitis is the principal factor.

On the National Elk Refuge at Jackson, Wyo., the chief agent in creating conditions favorable for the disease has been the seed of squirreltail grass *(Hordeum jubatum* var. *caespitosum),* locally known as foxtail or bobtail barley. This is the seed most commonly found in the lesions and jammed along the teeth, but the seed of downy chess *(Bromus tectorum),* less common in the hay, also plays a part. Certain rushes *(Juncus* sp.) that grow in the margins of the hay meadows in wet places also are involved, as the short, stiff sections of their stems have been found wedged tightly between the gums and the teeth. From numerous observations it appears that not only the awn of the squirreltail grass is dangerous but the sharp seed itself punctures the lining of the mouth and thus facilitates ingress of bacteria. In fact, any sharp seed, any stiff, brittle stems like those of rushes, any very coarse browse, or any type of harsh, coarse forage is likely to cause sufficient abrasion to permit infection.

A feeding experiment was conducted with a small number of elk calves to learn about the action of squirreltail grass. In pen 1, for control, a number of calves were fed good, clean alfalfa, but they escaped before experiment had progressed far; in pen 2, six calves were fed hay with a very light addition of squirreltail; and in pen 3, seven calves were fed hay with a heavy proportion of squirreltail. There were no injuries or losses in pen 2; but in pen 3, four of the seven elk died and all of them had the typical lesions characteristically filled with seeds.

During January and February 1935, to carry out a feeding experiment for the Wyoming state game warden, 560 elk calves were placed in a corral on the National Elk Refuge and were fed

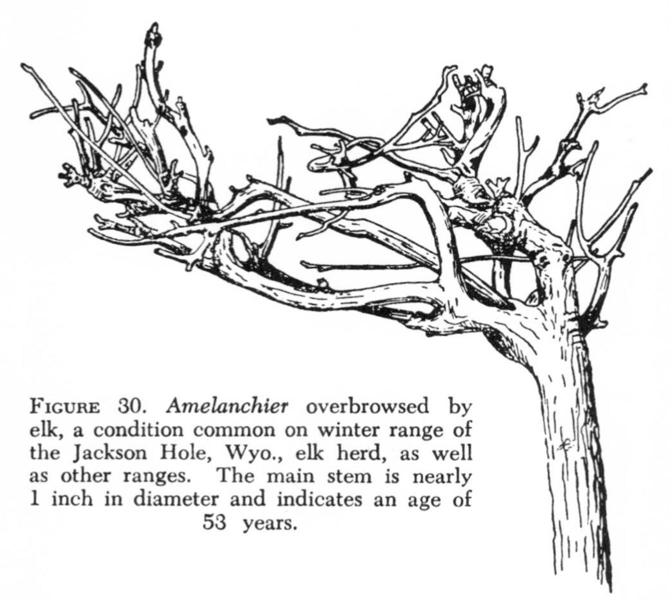

FIGURE 30. *Amelanchier* overbrowsed by elk, a condition common on winter range of the Jackson Hole, Wyo., elk herd, as well as other ranges. The main stem is nearly 1 inch in diameter and indicates an age of 53 years.

second cutting of alfalfa and corn. By February 14, 42 (7.5 per cent) of the calves had died. The carcasses that were not frozen too badly for examination revealed typical lesions of necrotic stomatitis. This loss was puzzling, as no squirreltail or other dangerous grasses were mixed in with the alfalfa hay. Coarse, dry alfalfa stems in the characteristic wad of food lodged in the cheek of one dead animal were found. Thus the alfalfa hay itself was suspected as the cause, and perhaps it was; but a happening in the Zoological Gardens in Frankfort on the Main, Germany (Marx, 1908), suggests that the corn fed to the elk calves were at least partly responsible for the loss. In May 1908 a giant kangaroo *(Macropus giganteus)* died at the zoo. It had a swelling in the jaw that suggested actinomycosis, but the organisms involved were not fungi—as required in that disease—but other bacteria. There were loose molars in the upper jaw with necrosis of the jaw, and suppuration had extended to the cheek and orbit. This is the precise condition found in many necrotic stomatitis cases among elk. The kangaroo had been fed hard corn. When corn was eliminated from the diet of kangaroos in the zoo, there was no further trouble.

Not all elk losses from necrotic stomatitis occur where hay is fed. The disease also prevails on the open range. In some severe winters there were heavy losses from it in the Gros Ventre basin, adjacent to Jackson Hole, Wyo. It is notable that in such winters there was undue grazing on willows and other browse, and twigs of an unusual thickness were eaten because there were too many animals for the supply. Such forage includes sharp splinters that may cause lesions in the mouth or stomach (Figures 30 and 31).

Dangerous browsing conditions are especially notable in the Olympic Mountains, Wash., where in recent years, since some of the watersheds have become overpopulated with elk, winter losses have become greater, especially along the Elwha and Hoh Rivers. When the writer spent the winter of 1916-17, on the Elwha River, he neither found nor received reports of elk losses, but a number of years later a resident guide, Grant Humes, wrote repeatedly of his observations and mentioned the annual winter losses, chiefly among the calves. He estimated that half the elk calves were lost and in one letter (February 18, 1932) remarked: "It is but a repetition of what has taken place each spring for about 7 years in the north part, at least, of the Olympics."

Humes believed that ticks were the responsible agents, but subsequently other informants described lesions found in the mouths of the dead animals and the writer examined a number of specimens, including old skulls found in the woods. In some of these specimens the exostoses characteristic of necrotic stomatitis were present. In short, the winter losses in the Olympics bore all the characteristics of similar occurrences in the Yellowstone National Park-Jackson Hole region.

It is significant that the losses in the Olympics occurred in the

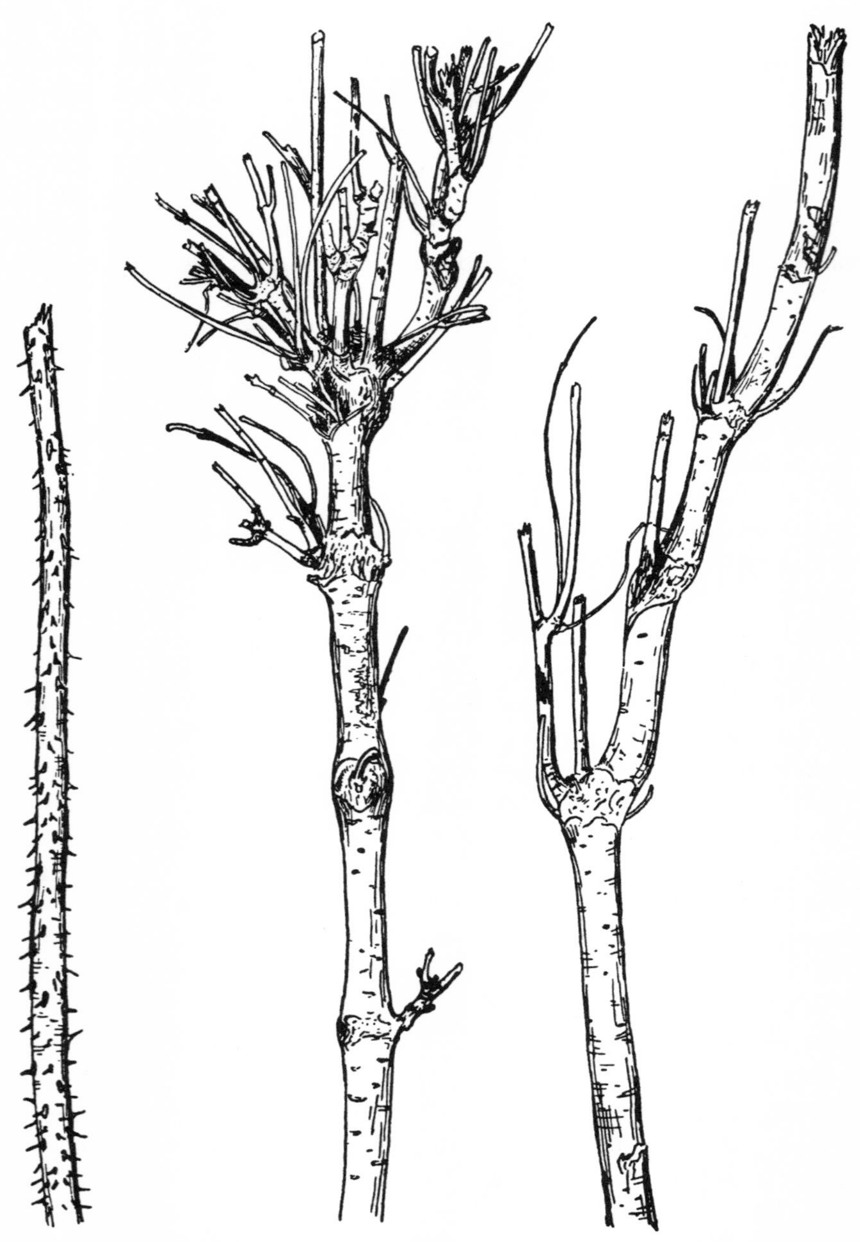

FIGURE 31. *(A)*, Rose stem browsed by elk. Spiny forage does not deter the animals. They also eat gooseberry twigs at times, and on the Pacific coast even the twigs of devilsclub *(B)*, *Cornus stolonifera* overbrowsed by elk. Diameter of main stem near top one half inch. Scars indicate that further browsing was attempted.

areas overstocked with elk and deer. Hay feeding was not involved and neither squirreltail nor other dangerous grasses were present. Browse was heavily utilized, however, and apparently browse of too coarse a type was the primary cause of the trouble.

There are, then, two principal conditions that favor winter losses from necrotic stomatitis. One is an overstocked range, where the animals are forced to eat too coarse forage, thus causing injury to the mouth and opening a channel for infection. The other is artificial feeding. On the National Elk Refuge, Wyo., even with the best of care, the losses have often been greater on the feed grounds than on the range. In 1927-28, when between 7,000 and 8,000 elk were fed, the winter loss on the refuge was 409; in 1928-29, when about 6,000 were fed, but for a short period only, it was 76; in 1929-30, when even fewer were fed, it was about 30; in 1930-31, when no feeding was done, the loss was negligible; and in 1931-32, when more than 11,000 elk were fed throughout the winter, the deaths exceeded 600. Finally, in 1942-43, when 10,000 elk were fed from December 4 to April 3, the losses in the National Elk Refuge area totalled 1,175.

These figures show a distinct correlation between the number of elk lost in winter on the one hand and the number fed and length of time feeding was continued on the other. It may be argued that the character of the winter influenced the death rate, a mild winter in which no feeding was necessary producing little loss, and a severe winter occasioning losses whether or not feeding was done. This is partly true. At times there have been heavy losses in the hills, where feeding could not be done. On the other hand, the evidence indicates that feeding does not remedy the situation, the losses being about the same in each environment but in some instances proportionately greater on the feed grounds.

There are some factors here that are puzzling. Small numbers of elk have been fed in recent years on Black Rock Creek and at the "Elk Ranch" in northern Jackson Hole, yet the losses have been negligible. Even more surprising is the record of elk feeding on upper Green River. In the winter of 1942-43, when 1,175 elk were lost on the National Elk Refuge area, the same kind of hay, obtained from the same source, was fed to elk on upper Green River with no losses, and it is reported that no losses have as yet occurred there during the years that feeding has been practiced. This suggests that elk are more susceptible to infection on feed grounds or overstocked ranges used over a long period of time.

Distribution

Necrotic stomatitis is not confined to elk, nor to one or two localities. It was reported that some deer fawns captured on the Kaibab National Forest, Ariz., for shipment had died of this disease; on the Shasta National Forest, Calif., the writer was informed, occasional deer have succumbed to this malady in the absence of artificial feeding; and on a few visits to wintering ranges of domestic sheep in Wyoming, skulls of adult sheep were found with ulcerated tooth sockets or exostoses on the lower jaws. In Jackson Hole, Wyo., at least one moose skull that exhibited the usual necrotic lesions was found; and among the moose of Isle Royale, Mich., the disease was demonstrated on the basis of skulls collected by Adolph Murie (1934) who studied conditions there. Casually looking over a series of mountain sheep skulls in the Biological Survey's collection in Washington, D. C., the writer noted a number that bore suggestive exostoses. Some of these were from British Columbia, and in this connection the following statement by Sheldon (1932: 198) is of interest.

"Every sheep collected showed traces of a jaw disease that in many cases had so distorted the bone that most of the teeth were gone. In some rams it was actively at work, since I found pus inside the swollen jaws. Mr. A. Bryan Williams, Game Commissioner of British Columbia, writes me that he has seen the same disease among the Cassiar sheep, but all the animals he shot had recovered from it. I doubt very much if it diminishes their numbers; at the most it may shorten their lives a year or two because of loss of teeth. In most cases it seemed to have taken the greatest effect on the lower jaws."

It is possible that in the instance cited here, as apparently many actual fatalities were not observed, the disease may have been *Actinomycosis*.

On September 29, 1946 the writer shot a female pronghorn near South Pass, Wyo., that was afflicted with necrotic stomatitis. As in the case of so many elk observed in Jackson Hole, and commonly occurring among domestic cattle, an accumulated wad of food had greatly distended the right cheek. A large active lesion was present near the gums. A small wad of food had also begun to form in the other cheek. Rush (1932a) records six pronghorns from Yellowstone National Park that had necrotic lesions and Adolph Murie (1940) reports several others from the same locality that had necrosis of the bone about some of the teeth.

From German literature it is learned that "actionomycosis" occurs occasionally among the roe deer and red deer in Europe (Roszbach, 1936; Stichter, 1936; Grumke, 1938; Hölldobler, 1938). Photographs of affected jaws of roe deer show a condition identical with that found in many Wyoming elk.

One or two human deaths have been attributed to this disease.

Finally, there is evidence that necrotic stomatitis has existed among wild animals for countless ages. When a number of affected elk jaws were sent to the late Dr. Roy L. Moodie, who was study-prehistoric diseases, he wrote in reply, on July 29, 1931:

"The diseased elk jaws which you sent to me some time ago have proven unexpectedly of the greatest help in solving pathological conditions in a large Pleistocene mammoth. The jaw has an alveolar abscess large enough to hold three quarts. A ventral fistula drains the abscess and the subperiosteal inflammation has brought about the formation of a huge growth of pathological bone about an inch and a quarter in thickness. The osseous lumps on the side of the elk jaws are exactly the same so that we are able to study the same process in two animals."

Examining fossil material in the extensive collections of Childs Frick of the American Museum of Natural History, the writer found that among the Pleistocene specimens collected in Alaska was an elk ramus bearing typical exostoses, similar to those common today, and a number of bison carrying similar lesions. He also noted several fossil horse skulls that bore these lesions. A lower left mandible from the lower or middle Pliocene had a lesion comparable with that on the mandible in Plate 21, upper figure; a skull of *Hipparion* from the middle Pliocene had indications of necrosis in tooth sockets on both mandibles; a *Pliohippus* had two teeth missing and swollen mandible; and another specimen showed on the jaw, near the teeth, a slight lesion that suggested similar bony infections in some elk specimens in which the lesions in the soft tissues have penetrated to and slightly infected the bone.

The organism, *Actinomyces necrophorus*, is world-wide in distribution, and is said to be present in the oral cavities of animals as well as on the vegetation and on the ground. It is generally assumed that concentration of animals and unsanitary conditions may increase the prevalence of the disease.

Summary

It appears that necrotic stomatitis is very widespread—perhaps world-wide; is pathogenic for a variety of so-called game animals, some domestic stock, and occasionally man; and has come down at least from the Pliocene and Pleistocene. Furthermore, it occurs in epidemic form, producing much of the "winter kill" when a winter range is overstocked. It has not been observed in summer.

STARVATION

Starvation is not always easy to detect, yet it is quite certain that it sometimes occurs. It is true, there is emaciation, but emaciation accompanies some diseases. The animals observed with necrotic stomatitis usually had full stomachs, and those on the National Elk Refuge had been on abundant feed for a long time, yet emaciation was generally present. Starvation and the most serious elk disease would appear to have the same cause—insufficient suitable forage—and this, as a rule, is due to an overstocked range (Plate 22). It is possible, however, that necrotic stomatitis becomes effective in most cases before actual starvation can occur. In northern Minnesota, in the spring of 1935, a number of dead white-tailed deer and several dead moose were examined in which the usual indications of necrotic stomatitis were absent. Possibly these were actual cases of malnutrition or starvation. It is suggestive that in some instances deer have died in some numbers on winter range or on artificial feed without much evidence of necrotic stomatitis but with general symptoms of malnutrition. Elk are generalized in food requirements, while deer on winter range are specialized and hence suffer more directly from malnutrition. The writer examined several deer—clearly victims of malnutrition—that had died on the National Elk Refuge in an area where they had no access to browse. They had died on a diet on which elk were thriving. In carefully conducted experiments with supplemental feeding of mule deer Doman and Rasmussen (1944: 317) found that supplemental feeding of alfalfa or grass hay did not prevent losses of deer from malnutrition.

Elk, on the other hand, thrive fairly well on hay. The danger to elk is injury to the soft tissues by coarse or brittle stems, seeds, or awns, resulting in necrotic stomatitis. It is noteworthy that in the experiment with feeding bobtail barley hay, described on page 181, the losses occurred among the animals that were fed the heavy mixture of bobtail barley, even though the hay itself was of much better quality by ordinary standards than the hay fed to another group of animals in which no losses occurred. The hay fed to the second group contained very little bobtail barley.

On natural range, however, malnutrition is no doubt a factor when the range is overpopulated. There can be no question that the nutritious part of a twig is the slender tip rich in nourishing buds. When the browse forage is so overutilized that the animals are forced to eat an abnormal amount of the coarser part of the twig, with corresponding increase in the proportion of woody tissue, it is logical that the animals are not so well nourished as they should be. Furthermore, it may be that for good health winter diet of naturally cured grass should be varied by at least a small portion of browse.

Generally speaking, we may conclude that the more adaptable elk withstands malnutrition more successfully than deer, but that malnutrition in combination with disease, undoubtedly plays a part in winter losses.

14

VARIED OPINIONS ON THE "ELK PROBLEM"

IDEAS ON THE CAUSES of elk mortality are numerous, and remedies suggested are equally varied. Generally the opinions and advice offered in the course of these studies were based on the handling of cattle, with which the persons concerned had had years of experience.

STARVATION

Starvation probably ranks first in the popular conception of the "elk problem." As explained above, starvation was not demonstrated as an isolated agent in winter losses; yet it is difficult to dissociate it from winter kill as a factor, for the reasons that winter losses (at least on the open range) are caused by diseases brought on by shortage of the best types of feed, and that possibly starvation would eventually take place if these diseases did not kill so quickly. After all, malnutrition and starvation independent of disease may be present on an overstocked range, but they are difficult to detect under field conditions.

ORPHAN CALVES

Some believe that the calves that die in winter are those that lost their mothers in fall or early winter and hence were weaned too early. This theory can hardly be acceptable. The calf elk is grazing regularly as early as July and August, and although some sporadic suckling takes place in autumn and early winter, the small quantity of milk the calf obtains after the rutting season is of no vital consequence. Furthermore, losses begin in Febru ary, after the calves have survived considerable winter hardships, and continue as late as May when the orphans should have become adjusted to the lack of milk. Moreover, some of the largest, strongest-looking calves die and many small, unthrifty ones survive. Thus there is no evidence to show that calves whose mothers are killed in the hunting season die as a result.

SUDDEN CHANGE OF DIET

Some persons contend that the change from natural forage to cured hay will cause the elk to founder, and others that the sudden change to green grass in spring causes the same trouble.

There is evidence that the first statement is sometimes true. One instance of foundering noted in the Gros Ventre basin in the winter of 1937-38 seems to indicate this, and probably there have been other instances of foundering on good hay when the animals were in a weakened condition. This situation does not prevail on a large scale, however, for on the National Elk Refuge, to which this explanation of losses has been specifically directed, elk losses do not begin until long after feeding is under way and often are heaviest toward the end of the feeding period, after the animals have become thoroughly adjusted to the change in diet. Incidentally, on the refuge feeding grounds the quantity of hay fed is increased gradually until the normal ration is reached and in addition the animals graze in the adjacent swamps.

As for the sudden change to green grass in the spring, that does not take place. The elk are not, like cattle, kept on hay until the grass is plentiful and then suddenly turned loose in a luxuriant pasture. They are free to roam and leave the feed grounds to seek the new green growth while it is still scarce. Moreover, the quantity of such food increases slowly, and the proportion of green grass in the stomach contents rises by very gradual stages throughout the spring, as amply shown by numerous field examinations of stomachs.

FEEDING BEGUN TOO LATE

That feeding begun too late is responsible for winter kill is a logical statement that under certain circumstances could have some basis in fact. Observation has shown, though, that very often the elk were in good condition when feeding was begun. In fact, in some instances the elk doubtless would have wintered successfully without feeding, but it was begun anyway, largely to bait the animals away from neighboring ranches where they had become a nuisance. In the areas studied the annual winter losses cannot be attributed to delayed feeding.

TICKS

Ticks have been blamed very commonly for winter kill, but the matter is difficult to demonstrate one way or another. In northwestern Wyoming, at least, elk losses begin before the ticks have become abundant and before the elk show evidence of being greatly annoyed by them. Only in the latter part of the winter-kill period do engorged female ticks become abundant; hence they cannot be held responsible for earlier losses. In view of the findings on the diseases present, the ticks probably can be no more than a contributing cause of death. After all, the effect of ticks on the physiology of elk has not actually been measured experimentally, and the so-called "drain" on the system is merely an assumption—perhaps a logical one—and not yet a proved fact.

SALTING

Salting the elk has been offered not only as a remedy for ticks but also as a general preventive of winter losses. The potency of sulphurated salt as a tick eradicator is still to be demonstrated. Ordinary rock salt was fed experimentally on the National Elk Refuge, and its effect was neutral in every way. For a more comprehensive discussion of salting see p. 309.

CONCLUSIONS

In conclusion it should be pointed out that, after all, necrotic stomatitis has been demonstrated as a potent mortality factor which each hard winter causes a definite winter kill, supplemented as it may be by the other diseases and circumstances described above.

15

FOOD HABITS

METHODS OF STUDY

THE FOODS and food preferences of elk were determined by three methods: (1) examination of stomach contents; (2) direct observation of feeding elk; and (3) examination of unit areas recently grazed by elk.

1. Stomach contents were preserved from a number of elk taken in different seasons. This material was analyzed in the Food Habits Research Section of the Biological Survey by C. C. Sperry, R. H. Imler, F. H. May, W. P. Baldwin, and G. H. Jensen. In addition, rough field analysis of many other elk stomachs were made by the writer from time to time to check on the general character of the vegetation utilized.

2. Numerous observations of grazing elk were made at close range, with binoculars, and by this means many species of plants used as food and the feeding characteristics of individual elk were noted.

3. Frequently a band of elk was found grazing. When the animals were frightened off or moved away of their own accord, the area used by them was determined and within it as large as possible a convenient unit was marked off for study. A record was made of its plant composition by percentage of plant species; the density of the plant growth was noted; and finally a careful examination was made to determine the species eaten and the degree of utilization of each. By this means it was possible to list the

plants eaten by elk and to determine approximately the degree of palatability of each.

The following discussion of forage species rests chiefly on direct field observations and experiments and on analyses of stomach contents; but specific references to Virginia are from Baldwin and Patton (1938), whose study included the examination of the contents of eight elk stomachs collected in November, and Roy K. Wood (unpublished thesis); and references to the Selway Game Preserve of Idaho are from Young and Robinette (1939).

An animal as versatile as the elk may be expected to show little specialization in its food preferences, and such is indeed the case. Little if anything is known about the food of the Merriam elk of the Southwest or of the indigenous elk of the Eastern States. The Rocky Mountain elk, however, has been introduced in many of the localities from which these other subspecies have disappeared and has usually adapted itself to the available forage, and it is possible that any of the other existing forms, if placed in the same environment, would react largely in the same way. Of course, some of the elk—notably the Roosevelt and the Rocky Mountain—today do live in distinctive environments and utilize different forage.

Food habits here are discussed under the separate headings of Rocky Mountain Elk, (including the information available for the elk transplanted to Canada and the Eastern States), Roosevelt Elk, Valley Elk, and Merriam Elk. The indications are that food-habit trends were more or less similar for the elk originally distributed eastward from the Rocky Mountains.

PLANTS EATEN BY THE ROCKY MOUNTAIN ELK

The Rocky Mountain elk *(Cervus canadensis nelsoni)* is the subspecies that has been most intensively studied, and its food habits are used as a standard with which to compare those of other elk. Although many ranges were examined, the most extended work was done on the Jackson Hole-Yellowstone National Park, Wyo., area. For convenience, the food plants are discussed separately under the headings of Grasses and Grasslike Plants, Miscellaneous Herbs, Browse, and Miscellaneous Foods.

Grasses and Grasslike Plants

Under most circumstances the utilization of grasses and sedges is difficult to observe in the field, but the analyses of stomach contents reveal fairly well the degree of use. The percentage of grass, or mixture of grass and sedges, in 78 stomachs examined (39 from Montana, 27 from Wyoming, and 12 from Colorado) ranged from 0 to 100 per cent but was high in most of the stomachs as shown in Table 9. No correlation was found between the state in which the stomachs were collected and the percentage of grass.

TABLE 9. PERCENTAGES OF GRASSES AND SEDGES FOUND IN 78 ELK STOMACHS

Month	Percentages
January	16–94–78 (sedge 16)–92–85–50–3–26–70–72
February	40–0–0–90–81–74–26–98–5–23–91–93
March	0–100–98–24–99–95
April	90–90–80–85–94–49–28–33–56–15–1–22–99–57–85–88–61–25
May	75–85–89–92
June	60
July	95–98–75
August	97–100–90–65–10–97–20
September	90–25–trace
October	5–90
November	70–60–5–60–77–trace–0
December	20–100 (sedge)–96 (sedge 80)–98–95 (sedge 20)

In addition, 12 stomachs received from the Washakie National Forest, Wyo., without exact date but apparently collected during the hunting season in autumn and early winter and 9 undated stomachs from Montana evidently collected in winter, revealed the following percentages of grasses or grasslike plants, which in 2 of the autumn stomachs (100 per cent and 84 per cent) may have been hay:

Autumn: 46–84–68–76–100–100–92–66–100–100–30–58.
Winter: 76–84–27–77–25–11–trace–trace–0

In January 1943, the writer had an opportunity to examine 98 elk stomachs from animals killed in the reduction program on the winter range in the northern part of Yellowstone National Park, with the following results:

Fifty-five stomachs contained 100 per cent grasses or sedges, or a combination of these together with rushes.

Thirty-four stomachs contained 94 to 99 per cent of these same plants.

Nine stomachs contained 65 to 90 per cent of these plants.

The remaining small percentages, or traces, consisted of browse, usually Douglasfir, but sometimes spruce, juniper, or sage, rarely willow or aspen, once a trace of lodgepole pine.

Probably an explanation of the high incidence of grass and grass-like plants in these stomachs is the fact that this winter range is greatly overbrowsed and no longer furnishes significant quantities of such food, with the exception of Douglasfir.

In the autumn of 1943 there was opportunity to examine 32 elk stomachs from animals shot by hunters in the latter part of October or first week in November. Of this number, 17 contained 100 per cent grass, with traces of browse; 13 contained from 90 to 99 per cent grass, also with traces of browse; and 2 stomachs contained 40 and 80 per cent grass. In some of these there were also evidences of sedge. These animals had been in migration down the open valley of Jackson Hole, where grass was readily available.

On the other hand, a cow elk killed October 30, 1939, in the same valley had eaten 100 per cent browse.

Baldwin and Patton (1938) reported that in the 8 stomachs collected in Virginia grasses formed 26 per cent of the diet and led all single food items; and the many field examinations in which only rough approximations could be attained, supported the general trend of the figures given above.

The foregoing indicates that grass is preferred food throughout the year; yet there are modifying circumstances, and in all a complex assemblage of plants is utilized. In the first list, above, note the range in the percentage of grass eaten: from a trace to 90 per cent grass in September; a trace to 77 per cent in November; 0 to 100 per cent in February and March, and 10 to 100 per cent in August. The flora of the locality where the elk feed has an important bearing. Elk that fed where grasses or sedges were not so common utilized browse and other herbaceous species more extensively. All factors considered, however, where grasses and grasslike plants were available, these were the staple food of elk the year round.

Young and Robinette (1939: 39) reported, however, that on the Selway Game Preserve of Idaho the "elk show a decided preference for browse throughout most of the summer, and this fact is not because desirable grasses are not available."

Although palatable willows are common in the Yellowstone-

Jackson Hole region and are utilized by the moose, the elk do not browse them extensively in the summer. It would be worth while if feasible to determine why this difference in feeding in two localities exists and also why on the same range grasses should furnish 100 per cent of the food in one stomach and only a trace or 0 in another.

Grasses

Elk prefer grass ranges, especially in winter, and grass is often the dominant item in the diet at any time of year when it is sufficiently available, although naturally in certain environments, especially when the grass has been grazed down or snow-covered, other types of forage will be the chief food.

In summer elk have decided preferences among the grasses. Perhaps the most palatable are those of the genera *Poa, Bromus, Agropyron, Melica,* and possibly *Festuca,* although many others are taken freely. In winter, although elk may have the same preferences, they are of necessity much less selective, and probably it can safely be said that in that season any grass available furnishes very acceptable forage.

Even in summer, however, when the vegetation present is all available, the elk do not confine themselves to grass by any means; nor do they appear to seek dense grassy meadows for feeding. Often they bed down in a heavy patch of reedgrass *(Calamagrostis)* or sedge *(Carex)* without utilizing it for food to any noticeable extent. Rather, they often seek grasses where these grow more sparingly among herbaceous plants, and generally the diet is a thorough mixture of grasses and so-called "weeds."

In the literature may be found brief references to grass species other than those discussed below, but these were not included here because the degree of palatability or importance was not indicated.

Agropyron. Wheatgrass. Wheatgrass is common on elk ranges in Wyoming and is extensively grazed in both summer and winter. According to some reports, the introduced crested wheatgrass *(Agropyron cristatum)* is somewhat more palatable than slender wheatgrass *(A. pauciflorum),* and field observations indicate that the same is true of bearded bluebunch wheatgrass *(A. spicatum).* In summer, the latter was sometimes utilized 8 to 12 per cent, but at other times showed less use. In winter, of course, wheatgrasses are fully utilized.

Some forms of wheatgrass, if eaten as dry hay, may be dangerous to elk and have been reported to cause sore mouth in domestic sheep. Bluestem wheatgrass *(A. smithi)* has been reported a good elk feed.

Agrostis. Bentgrass. Information on genus *Agrostis* as elk food is meager. It was not identified in the stomachs examined, but use of tickle grass *(A. hiemalis)* by elk was observed, and spike bentgrass *(A. exarata)* is said to be relished by elk. Unidentified species were found utilized slightly on the Whitman National Forest in Oregon. Undoubtedly other species of the genus are used at least occasionally, but none have been so recorded.

Avena. Oats. Oat hay is eagerly consumed by elk. Not only will a field of oats left standing in early winter attract elk from most other types of forage, but the field will be entirely grazed down before the animals leave it.

There are some indications that oat hay, like some wheatgrasses, may cause necrotic stomatitis if fed to the elk in a dry form, but this hay is probably harmless if grazed from a standing field. As related on page 179, a group of captive deer and moose died of this disease, and the only cause to which the disease could be attributed was the oat hay (especially the seeds) fed to them.

Bromus. Bromegrass. Bromegrasses are one of the most palatable groups of the grasses for elk, apparently being surpassed in this respect only by *Poa.* The commonest form on the summer ranges in the Jackson Hole, Wyo., country and the one most often observed is *Bromus marginatus,* which is found in a variety of situations throughout all the open slopes. A characteristic of this brome is its close association with tall larkspur, which is abundant in this region and occurs in large patches, but *B. marginatus* (or related bromes) is of course also found grouped with other plants and in some places is dominant. Locally it may form almost pure stands.

One autumn a dense patch of wheatgrass (chiefly *Agropyron spicatum)* was found untouched, but in it were clumps of *B. marginatus* heavily grazed by elk. The writer's saddle horse and pack mule showed the same preference. Repeatedly the mule nosed around a clump of luxuriant *Agropyron* to seek out the brome stubble nearby, left by elk. Although this seemingly illustrates a preference for brome, it should be stated that this grass was green at the time, whereas the more mature wheatgrass had already acquired the characteristic wine color of autumn, a difference that may have had much to do with relative palatability.

Two other members of the genus may be mentioned. Nodding brome *(B. anomalus)* occurs on elk ranges and is readily eaten, but it is relatively scarce and hence of less importance than mountain brome. Downy chess *(B. tectorum* var. *glabratus)*, which is found in abandoned fields or waste areas in northwest Wyoming, does not normally occur on summer range but is eagerly taken by elk in winter. If grazed naturally it is probably harmless, as most of the seeds are doubtless shattered by the time the elk use it; but when cut for hay it ranks with squirreltail *(Hordeum)* as a distinct menace in producing the fatal sore mouth that may lead to necrotic stomatitis (p. 181). On Whitman National Forest in Oregon *Bromus carinatus* was grazed slightly.

Calamagrostis. Reedgrass. During the present studies little use of bluejoint reedgrass *(Calamagrostis canadensis)* by elk was observed, although this species occurs in dense patches. Its growth is often thick and rank in moist ground, and elk use such places for beds. Examination of the vegetation at such bedding spots showed that grazing on the abundant grass was very slight. Possibly use of this grass is more common in other situations.

Elk generally pass by the pinegrass *(C. rubescens)* in summer in favor of more desirable food, but they eat it in winter when it is not covered by snow. Late one November, when other grasses had dried up, Adolph Murie (field notes) observed that in the Yellowstone National Park some of the pinegrass was still green and was being used by elk. In parts of Washington it was noted that in yellow pine country the pinegrass was less used than many other species but that the few other, more palatable species of herbs and browse were eaten to a damaging degree. DeNio (1938: 424) gives this species a winter palatability rating of 80 per cent in Idaho.

Cinna. Woodreed. This grass has not been found abundant on elk range, but its use by elk has been noted.

Danthonia. Oatgrass. Oatgrass is not abundant, and observations on its use by elk were meager. It has been recorded as being used in early winter.

Deschampsia. Hairgrass. Tufted hairgrass *(Deschampsia caespitosa)* has been seen being grazed by elk in summer, but comparatively little information on its use is available. *D. elongata* has been recorded as used by elk slightly.

Elymus. Wildrye. On the summer range elk were noted utilizing blue wildrye *(Elymus glaucus)* slightly, but observations were limited. Giant wildrye *(E. condensatus),* however, plays a more important role. Although it is a coarse grass and is generally considered poor forage, it produces great volume and stands up well in winter. The elk utilize it rather freely at that season, when their choice of forage is limited. In some parts of Jackson Hole, Wyo., the available supply is grazed gradually until none remains, and in Yellowstone National Park Adolph Murie noted it was eaten down in some places to within 4 inches of the ground as early as November 21. In the series of 98 stomachs collected in January 1943 in Yellowstone National Park, mentioned previously, 10 contained as much as 98 to 100 per cent grass which was chiefly *Elymus,* and 10 others contained similar percentages of grass in which *Elymus* was present in varying proportions. On the whole, this species is a valuable part of the winter forage supply.

Festuca. Fescue. Elk are known to feed on fescues. These grasses, widespread in their various forms, are recognized as valuable forage, yet it so happens that they were identified in only a few stomachs and field observations on their use were few.

Hordeum. Barley. The cultivated barley *(Hordeum vulgare)* is freely utilized by elk in winter when available either in the form of hay or standing in the field. Foxtail barley *(H. jubatum)* is also readily taken. The bobtail barley *(H. j. caespitosum)* is the agent that brings on necrotic stomatitis, the cause of such great losses among the elk in northwestern Wyoming. Any foxtail barley, however, would have the same effect. There is no evidence that foxtail barley without fruiting heads is harmful when grazed from the fields in winter. Meadow barley *(H. nodosum)* occurs sparingly on elk summer range and is utilized, but it is probably not important.

Koeleria cristata. Prairie Junegrass. This was found in at least four elk stomachs, and it was observed used from May to August. It is particularly common, together with *Agropyron* and some other species, on the elk winter ranges on the open foothills of Jackson Hole, Wyo., and is a useful addition to the winter diet.

Melica. Oniongrass. Grasses of this genus, particularly the showy oniongrass *(Melica spectabilis),* are characteristic of the mixed flora of the high, dry mountain slopes of the elk summer range in northwestern Wyoming. They are scattered in growth, and their utilization is hard to observe, particularly as the elk often pull the whole plant up by the roots and eat it entire. The characteristic bulbs were found in every stomach collected in June, July, and August; they are evidently very palatable.

Muhlenbergia. Tanglegrass. Information is meager on the use by elk of any species of this genus. One form appears to have been used once in August.

Oryzopsis. Ricegrass. Ricegrass, especially Indian ricegrass *(Oryzopsis hymenoides),* is plentiful on some of the elk winter ranges. It is generally utilized by elk, and as it produces abundant forage is no doubt an important element in their winter food supply.

Phleum. Timothy. Timothy *(Phleum pratense)* has often been fed to elk as hay and is acceptable in that form, but it is not so palatable as alfalfa and other types of hay. Alpine timothy *(P. alpinum)* is greatly relished by the elk on the high summer range, but it has not been found abundant in any one locality.

Poa. Bluegrass. The bluegrasses appear to be the most palatable of all grasses to elk. Skyline bluegrass *(Poa epilis)* is found in small clumps of sod, sometimes only a few inches in diameter, growing on dry, high slopes. In such situations the blades are fine and short. The elk eagerly seek out these tidbits during the summer, often consuming the plants entire. Nodding bluegrass *(P. reflexa)* and Kentucky bluegrass *(P. pratensis)* are also taken freely, but not so zealously as *P. epilis. P. nervosa* has been noted in elk diet.

Sitanion hystrix. Bottlebrush Squirreltail. This grass was reported used to a slight degree by elk on the Whitman National Forest in Oregon.

Stipa. Needlegrass. On the high mountain slopes the subalpine needlegrass *(Stipa columbiana)* occurs irregularly but commonly, sometimes in pure stands. It was utilized by elk in summer but was evidently less palatable than bluegrass, brome, or wheatgrass. On winter range *Stipa* is undoubtedly very acceptable.

Trisetum spicatum. Spike Trisetum. The limited observations made on the use of this grass by elk show that the animals eat it on summer range. Undoubtedly they relish it in winter also when available.

Triticum. Wheat. Elk have been known to raid wheatfields, thus demonstrating their taste for this cultivated grass.

Zea mays. Indian Corn; Maize. The kernels, cobs, husks, and "silks," of Indian corn were reported readily utilized by elk in parts of Virginia (Baldwin and Patton, 1938: 753). Elk also eat corn that is put out as feed in winter.

Grasslike Plants

Carex. Sedges. Sedges are generally considered inferior to grasses as elk food and possibly rightly so. Yet sedges often play an important part in the elk's food and under certain circumstances surpass the grasses in value. Many *Carex* species are represented in the elk diet, but they are here referred to collectively, with the exception of the elk sedge *(C. geyeri).* During winter, when the choice of food is restricted, it is doubtful whether the elk have much preference as to species.

In early spring the new sedge sprouts are welcomed by the elk, and throughout the summer *Carex* is eaten frequently by them and forms a regular, though small, proportion of their diet. It occurs in a variety of situations on the summer range—on ridges, slopes, and benches, along the borders of pools and watercourses, and in wet, level meadows, where it is the dominant vegetation. Elk generally do not feed in these densely grown wet meadows but prefer to pick sedges occasionally in other situations, from a mixed flora.

The *Carex* species are particularly valuable in that they remain green, at least basally, in late fall after the grasses have dried up. In the extensive swamp near Jackson, Wyo., the elk find these basal green parts very palatable and even dig into muskrat houses to get them. In winter all the sedges, even the brown, dry ones are eaten, and such use of dry, coarse sedge was noted as early as November 13. Sedges were used by elk very extensively in Yellowstone Park in January 1943, and they seemed to thrive on them, for the animals were in prime condition. The sedges at that time were still a little green in the basal parts (Plate 23).

The elk sedge *(C. geyeri),* which forms a clumpy growth over many high slopes of the Rocky Mountains, deserves special mention. Although this species is a rather harsh reed, in late summer and in fall, when other vegetation is largely dry, it remains relatively green, except for dried tips, and the elk then feed on it extensively. In parts of Wyoming this sedge becomes important in helping to hold the elk in the high ranges until heavy snows come, keeping them from going on to the more critical winter range too early. Because of its great abundance and high palatability Young and Robinette (1939: 36, 47) consider it the most important herbaceous species of the Selway, Idaho, region as elk forage. On Whitman National Forest it also proved to be valuable elk forage.

Juncus. Rush. In summer rushes are sometimes eaten by elk with considerable relish, apparently in preference to sedges, but at other times they are definitely ignored and it is clear that grasses are much preferred. On Whitman National Forest rushes were reported in the elk diet, *Juncus parryi* in particular being noted.

Luzula. Woodrush. The use of *Luzula* by elk was observed less often than *Juncus,* principally, no doubt, because it was scarcer in the areas studied.

Scirpus acutus. Tule Bulrush. Where available this bulrush is sometimes utilized slightly by elk in winter, even when other food is present; but on the whole it is very low in palatability.

Triglochin maritima. Arrowgrass. Use of arrowgrass by elk was observed only once, in Yellowstone National Park, Wyo., in June. It had been grazed fairly well and appeared quite acceptable.

Typha. Cattail. Cattails are very low in palatability. The elk appear to pick at them slightly in winter, although much of the cattail growth is merely trampled and shattered by the passing animals.

Miscellaneous Herbs

Although grasses are considered the most important item in the elk diet, certain herbaceous plants also have high value as food and undoubtedly are so eagerly sought at times as to control the movements of the elk. As a matter of fact, Pickford and Reid (1943: 330) report that on the Whitman National Forest in Oregon the elk summer diet consists of 80.9 per cent weeds, 13 per cent grass and grasslike plants, and 6.1 per cent shrubs. They suggest that this may be explained by the fact that the vegetation in the area studied was composed of 76 per cent weeds, 14 per cent grasses, sedges, and rushes, and 10 per cent shrubs. It should be pointed out that in spite of the preponderance of grass in many instances, it is obvious from observation that elk are very fond of a variety of herbs and seek them eagerly. As one woodsman remarked, "They are great weed eaters." A great variety of herbs is consumed, and the following list cannot be considered complete although it probably contains the majority of the most palatable forms.

Acalypha virginica. Virginia Copperleaf. Seeds of this plant were reported by Baldwin and Patton in one elk stomach from Virginia.

Achillea lanulosa. Western Yarrow. Yarrow is fairly common on summer range but usually is eaten only sparingly by elk. It was found in small quantity in several elk stomachs and formed as much as 7 per cent of the contents of one stomach from Montana.

Aconitum. Monkshood. Monkshood is entirely palatable to elk, but on some elk summer ranges it is not particularly abundant. On one study plot 65 per cent of the plants were grazed; and on another, 66 per cent; on each to the average extent of 10 per cent per plant; the total utilization thus being about 6.5 per cent. Utilization was observed only in July and August, but there may have been spring grazing that was overlooked.

Actaea. Baneberry. This herb was found eaten by elk in the Jackson Hole, Wyo., area, only once when several plants were grazed. In the Selway River region in Idaho, *Actaea spicata* was found taken to a considerable extent in late September and early October, as its "foliage remains green and palatable after frost."

Agastache urticifolia. Nettleleaf Gianthyssop. This mint is very palatable. It was found eaten on many occasions in northwest Wyoming, where it grows commonly on many of the open slopes so favorable for elk. On one plot there was 20 per cent total utilization, as all the plants were grazed to an average of 20 per cent each.

Agave. Mescal. The Rocky Mountain elk introduced in the Guadalupe Mountains of New Mexico and Texas are known to feed on Mescal. They eat the green flower stalks as well as the bloom itself and even paw loose the spinetipped fleshy leaves to get the basal portions.

Agoseris. Mountain-dandelion. Probably several species are eaten by elk, but the one observed particularly was the smooth mountain-dandelion *(Agoseris glauca).* The young leaves are eagerly sought from May to August. Often a band of elk will linger on a barren clay slope, apparently devoid of vegetation, feeding persistently for some time. Subsequent examination of the place will reveal that mountain-dandelions, which there find a favorable habitat, were the chief attraction.

Allium. Onion. Wild onions of undetermined species are palatable to elk. Their use has been noted in Wyoming and Montana. *Allium validum* has been reported utilized by elk in Oregon.

Amaranthus. Amaranth. This was reported eaten in winter by elk in Yellowstone National Park, by Skinner (1928: 312) and seeds were found by Baldwin and Patton in one stomach from Virginia.

Ambrosia artemisiifolia. Ragweed. Seeds of ragweed were present in three stomachs of elk in Virginia, according to Baldwin and Patton.

Amsinckia. Fiddleneck. Although not noted in the field as grazed, *Amsinckia* was identified in five elk stomachs.

Angelica. Angelica. Remains of plants of this genus were found in 1 elk stomach in a series of 29. Unfortunately, field observations on the use of this herb were limited. Young and Robinette (1939: 30) stated that "*Angelica lyallii* was seldom taken by elk" on the Selway Game Preserve, Idaho.

Antennaria. Pussytoes. Use of *Antennaria* was not observed in the field, but in a series of 29 stomachs, 3 stomachs (1 of which was collected in November) contained parts of these plants.

Aquilegia. Columbine. Columbine is eaten by elk with evident relish, but it occurs only in scattered situations on the summer range. The species observed particularly was the Colorado columbine *Aquilegia coerulea.*

Arbutus texana. Texas Madrone. This shrub was noted browsed by elk in the Guadalupe Mountains, N. Mex. in 1939. It appeared quite palatable.

Arctium minus. Smaller Burdock. Seeds of this burdock were identified by Baldwin and Patton in one elk stomach from Virginia.

Arnica. Arnica. One species, *Arnica cordifolia,* was often utilized by elk. It is fairly palatable to elk, but they do not seek it with enthusiasm in the forest shade, where it grows most abundantly. Other species also were occasionally eaten, but *Arnicas* are sometimes difficult to determine specifically, especially when cropped. In one elk stomach examined in September *A. arcana* was found, but this species, which often grows in dense patches in high rocky places and thus offers a large volume of forage, is usually passed by untouched.

Artemisia gnaphalodes. Cudweed Sagewort. This herbaceous sage has been reported by the U. S. Forest Service (1937) as occasionally grazed by elk but as not having much forage value for them because of its inferior palatability.

Aster. Aster. In general, herbaceous asters are very palatable to elk, and as a group they are valuable in that they protrude through the snow and are still available in early winter. Thickstem aster *(Aster integrifolius)* and possibly related species that were not identifiable in their early season condition are common on the high summer range and are much utilized by elk. The new growth of basal leaves is particularly sought in spring and summer, and later in the season plants in bloom are utilized, sometimes at least as much as 65 per cent. Two others, *A. engelmanni* and *A. glaucodes,* are also eaten regularly by the elk. *A. adscendens* was also eaten and *A. conspicuus* was used in Montana. *A. perelegans* was found at lower elevations, but its palatability was not observed. On the Selway Game Preserve, Idaho, *A. eatoni* and *A. canescens* were used late in summer and in fall. All eight stomachs collected in autumn in Virginia by Baldwin and Patton contained aster, which in one of the stomachs contributed 1 per cent of the contents.

Astragalus. Loco; Milkvetch; Poisonvetch. Species of *Astragalus* that were common in certain restricted localities were readily taken throughout the season by elk on the Selway Game Preserve in Idaho.

Athyrium filixfemina. Ladyfern. This was eaten freely in late July on the Selway Game Preserve, Idaho.

Balsamorrhiza sagittata. Arrowsmith Balsamroot. This balsamroot is much used by elk. In early spring the first leaves, still curled, are eagerly eaten, and even after the plants mature parts of leaves and some flower heads are grazed rather extensively. On August 21, out of 110 plants in an observation plot 100 had been grazed, averaging 5 per cent use per plant. Balsamroot is acceptable in winter and is available at least as late as early December.

Barbarea verna. Early Wintercress. Seeds were noted in one elk stomach from Virginia examined by Baldwin and Patton.

Bidens bipinnata. Spanish Needles. Seeds were identified by Baldwin and Patton in an elk stomach from Virginia.

Bistorta bistortoides. Bistort. Bistort grows rather inconspicuously in the mountain meadows, and its use is not easy to detect. The elk eat it in summer at least to a slight extent.

Boykinia major. Sierra Boykinia. This is grazed eagerly by elk in the Selway region in Idaho and appears to be one of the more important early summer weeds in their diet.

Calandrinia. Rockpurslane. This plant was found in three elk stomachs, but there were no field observations on its use.

Caltha leptosepala. Elkslip Marshmarigold. When elk were found in wet meadows where this marshmarigold was plentiful, the plant was grazed with evident relish, sometimes to the extent of 50 per cent. It is very local in distribution.

Campanula. Bellflower. The use of *Campanula* was seldom observed, but it was grazed a little at least once in early autumn.

Castilleja. Paintedcup. Several species of *Castilleja* are frequently eaten by elk, but as there is a considerable variety of these brightly colored plants on the summer range, it is uncertain which ones are preferred.

Chenopodium. Goosefoot. Goosefoot is eaten sparingly. It occurred in a single elk stomach from Wyoming, and Baldwin and Patton found seeds in two stomachs from Virginia. The species involved were not determined.

Chrysanthemum leucanthemum. Oxeyedaisy. Seeds of this plant were found in the stomach of an elk from Virginia.

Cirsium. Thistle. Thistles of several species, including *Cirsium hookerianum* and *C. drummondi,* are eaten freely by the elk at any time of year available and appear to be entirely palatable.

Claytonia. Springbeauty. A little *Claytonia* is grazed at times in the high mountain meadows. Young and Robinette (1939: 30) reported that on the Selway Game Preserve in Idaho *C. asarifolia* was taken frequently from late July until late September.

Comandra umbellata. Common Comandra. This plant was eaten by elk in winter, but its widespread use has not been observed.

Conioselinum scopulorum. This species was reported to be moderately utilized on the Selway Game Preserve, Idaho.

Corydalis aurea. Golden Corydalis. This was often eaten during midsummer in the Selway region in Idaho.

Cuscuta. Dodder. Dodder was found in two elk stomachs collected in July, the only time its use was observed.

Daucus carota. Queen-Anne's-lace. This was noted in two elk stomachs from Virginia.

Delphinium. Larkspur. It is possible that several species of larkspur are eaten by elk, but observations were very limited except on the tall Barbey larkspur *(Delphinium barbeyi),* which was given particular attention because of its poisonous properties. This tall larkspur is particularly abundant on the spring and summer ranges in northwest Wyoming; but every spring it was noted that though larkspur was very abundant and easily available and caused losses among range cattle, the elk would not eat it. Day after day scattered bands of elk would graze on aspen slopes where the emerging new green leaves of the tall Barbey larkspur were the dominant vegetation, but in only one or two doubtful instances did they graze it at all during May and June. In fact, they consistently avoided this abundant larkspur and sought the less conspicuous plants.

In the spring of 1932 an experiment was performed in an effort to learn whether elk are immune to larkspur poison. Two young animals were placed in an enclosure relatively free from grass. They were supplied with plenty of water but for food were given only young leaves of tall larkspur, recently picked. They gleaned every vestige of grass and other plants about the borders of the enclosure but did not touch the larkspur. This continued for several days until practically no more grass was available, even by reaching through the bars. Salt was then sprinkled over the fresh larkspur leaves, and a sack of dried larkspur leaves, gathered the previous spring, was added to them. Then the elk ate some of the leaves, both the fresh and dried ones, but seemed to prefer the latter. This continued for several days, but such a small quantity was consumed daily that it was hardly enough to sustain life and the elk was turned back on good feed. A few days later one of the animals died. Diagnosis was unsatisfactory. There had been no particular symptoms except falling off in flesh, and the results were complicated by the possible effects of insufficient feed. The second animal survived and was later released.

There were not sufficient facilities for a precise, carefully controlled experiment, but the results at least illustrate the elks' extreme unwillingness to eat the tall Barbey larkspur in the spring, even though the leaves are then tender and young.

Later in the summer and in fall the story is different. In July, as the plants reach full growth and bloom, an occasional stalk is nipped off by elk and such grazing increases during the month; and during August, when the seed pods are ripening, large quantities of them are consumed. Grazing of larkspur continues through September, after the pods have fully matured. In that month a cow elk was watched systematically nipping off the heads, one after the other, and larkspur over large areas showed extensive use by elk. Usually only the flower head or seed is eaten, sometimes only a part of the head (a single bite being the rule), but occasionally a considerable part of the plant is consumed. Larkspur therefore furnishes much forage in summer and fall, perhaps being most valuable in fall.

Young and Robinette (1939: 30) found that on the Selway area, Idaho, "*Delphinium scopulorum* was taken primarily in June while the young leaves were still growing. After the flower stage, it was taken but sparingly." *D. occidentalis* has been reported utilized by elk in Oregon.

Desmodium. Tickclover. Leaflets were found in one elk stomach from Virginia.

Dianthus armeria. Deptford Pink. Seeds formed a trace of the contents of two elk stomachs from Virginia.

Dodecatheon conjugens. Bird Bill. This species was reported unusually palatable to elk in the Selway region in Idaho from early July until killed by frost.

Drymocallis. (See Potentilla) Several species of *Drymocallis* that were not readily differentiated in the field studies are available on elk summer range and are eaten readily by elk.

Epilobium. Willowweed. Fireweed *(Epilobium angustifolium)* is grazed to some extent in summer, but it is more acceptable in fall when so many herbaceous species have disappeared. Its sturdiness and height permit it to protrude through the snow, and its use by elk has been noted at least as late as December. Red willowweed *(E. latifolium)* was found grazed occasionally, and *E. stramineum* was noted as utilized by elk on one occasion in August.

Erigeron. Fleabane; Wild Daisy. The erigerons include forms that are very palatable to elk and that contribute appreciably to elk summer forage. The animals graze especially the young plants before the flower stalks are well developed. The aster fleabane *(E. salsuginosus)* is widely distributed and is very palatable, as much as 40 per cent utilization of individual plants having been noted; aspen fleabane *(E. macranthus)* also is extensively grazed but possibly somewhat less than the foregoing species; and Bear River fleabane *(E. ursinus)* is palatable and frequently taken. Other species too are undoubtedly used.

Eriogonum. Eriogonum. Some species of *Eriogonum* are grazed by elk. *E. subalpinum* was taken commonly, usually only the flower heads but sometimes the leaves also. Probably in winter eriogonums are even more acceptable. In one clump of 39 fruit stalks of *E. subalpinum* examined one winter, elk had grazed 28. *Eriogonum* were found in 14 of 29 elk stomachs in one series. *E. elatum* was found grazed by elk on the Snoqualmie National Forest in Washington, and *E. heracloides* was reported grazed by elk on the Whitman National Forest in Oregon.

Erodium cicutarium. Alfileria. This plant, introduced from Europe, has been reported eaten by elk in parts of California.

Erythronium. Fawnlily; Dogtooth Violet; Troutlily. Observations on the use of *Erythronium* by elk are meager, but utilization has been noted in the Yellowstone National Park in summer.

PLATE 22. Yellowstone National Park, Wyo. *Above.* A potential "Winter kill." Cow elk in poor condition owing to malnutrition, disease, or both. *Below.* A bull elk eating aspen bark. Such an aspen grove is no longer productive of an appreciable forage supply. (Photographs by Adolph Murie.)

PLATE 23. Yellowstone National Park, Wyo. *Above.* Elk feeding on aquatic vegetation in the manner of the moose. *Below.* These elk have found that snow disappears near the water's edge, exposing sedges and other vegetation on the bank. (Photographs by Adolph Murie.)

Eupatorium. Seeds of this herb were found in three elk stomachs from Virginia.

Euphorbia preslii. Spurge. Capsules and seeds of spurge were found in two elk stomachs from Virginia by Baldwin and Patton.

Fragaria. Strawberry. Strawberry leaves are eaten by elk from April to late June, possibly later, but they appear to be only moderately palatable.

Frasera speciosa. Showy Frasera. In spring, before leaving their winter range in the foothills, the elk in northwest Wyoming obtain some of their first green food by eating the new *Frasera* plants just emerging from the ground. Use of these plants by elk during summer could not be determined; but in fall, when the elk return to winter range, the bulky leaves are acceptable, especially those that have remained green. Even when the plants are dried, as late as November 21, elk have been observed wading through the snow from one plant to another, eating the upper parts of the dried stalks and seed heads.

Galium. Bedstraw. In a series of 29 elk stomachs examined, 3 contained traces of *Galium.* No other data on the use of the plant by elk were obtained.

Gentiana. Gentian. Gentians have been reported as fair forage for elk, but as a group they probably are not very palatable. Small forms were found grazed along the upper Sun River, Mont., and once a gentian, apparently *G. forwoodii,* was found eaten by elk.

Geranium. Geranium. Geraniums are persistently eaten and are an appreciable item in the forage supply. Sticky geranium *(Geranium viscossissimum),* which is abundant on both summer and winter ranges in northwest Wyoming, is eaten with relish by elk in spring and summer. On observed plots from 8 to 59 per cent of the number of plants present in August were grazed, utilization of the individual plants ranging from 7 to 50 per cent.

Gerardia. Gerardia. Capsules containing seeds of *Gerardia* were found in three elk stomachs from Virginia, and Roy K. Wood (unpublished thesis) reports two species of *Gerardia* quite palatable to elk in Virginia.

Gilia. Gilia. Only once was *Gilia* found eaten by elk. It is relatively scarce on the high summer range.

Hedysarum. Sweetvetch. This plant was found grazed by elk

on a few occasions, but pertinent observations were too limited to warrant conclusions as to its palatability.

Helianthella. Helianthella. Only one species, *Helianthella uniflora,* was noted particularly although others occur on summer range. When available, this species—both its basal leaves and part of the flower stalk—is eaten freely by elk. The young leaves that first emerge from the soil are particularly sought.

Helianthus. Sunflower. Sunflower was recorded in the contents of one elk stomach, the only information on hand as to the food value of this plant for elk.

Heracleum lanatum. Cowparsnip. Cowparsnip is eagerly sought by elk, particularly in July and August; and over extensive areas it is completely demolished by their grazing. Utilization varies. In one plot, observed August 25, there were 297 plants, all grazed to an average degree of 80 per cent; whereas on another area, studied on August 15, only 1 of the 45 plants present had been grazed and some other patches either had not been found by elk or had been passed by. Possibly individual plants vary in palatability. It is suggestive that natives in the Aleutian Islands, Alaska, where the peeled leafstems are eaten as a delicacy, have learned to avoid those with reddish color near the base because they have a bitter taste. On the whole, *Heracleum* is very palatable to elk and furnishes an appreciable proportion of their forage supply.

Heuchera. Alumroot. Seeds of alumroot ocurred in one elk stomach from Virginia.

Hieracium. Hawkweed. Hawkweeds are eaten readily by elk, but wild species of this genus are often so scattered in occurrence that they probably do not furnish an appreciable quantity of food. Woollyweed *(Hieracium scouleri)* was particularly noted as used in northwestern Wyoming, and leaves of a hawkweed occurred in three elk stomachs from Virginia. *H. chapacanum* was used to a noticeable degree by elk on Whitman National Forest in Oregon.

Houstonia. Bluets. Capsules containing seeds of this plant were reported found in two elk stomachs from Virginia.

Hydrophyllum capitatum. Ballhead Waterleaf. As this is one of relatively early forms of new plant growth that become available early in the season, it is readily eaten by elk in spring.

Ivesia (Horkelia) gordoni. Gordon Ivesia. This species was found eaten by elk once, in July.

Lappula. Stickseed. Stickseed is abundant on parts of the elk summer range. The elk obviously ignored it on some occasions in summer but at other times grazed a little. Sometimes they eat it in fall after snow has fallen.

Lathyrus. Peavine. Peavine is eaten by elk, but its degree of palatability is still to be determined.

Lespedeza. Lespedeza. Leaflets of bushclovers furnished a trace of the contents of three elk stomachs collected in Virginia.

Ligusticum. Ligusticum. This plant is common on elk ranges in Wyoming and is extensively grazed by elk. One elk had reached back under the branches of a current bush for some, even though other vegetation was available. On August 2, in one observation plot 75 plants out of 175 were grazed. At other times these plants were grazed lightly or passed by. Total utilization in test areas varied from 3 to 43.2 per cent. *L. grayi* was found grazed in Oregon.

Linum. Flax. There is no evidence that flax is very acceptable to elk. It was found in only 2 of a series of 29 elk stomachs taken in September and October, and on at least one occasion in December it was observed to be passed by consistently.

Lithospermum pilosum. Stoneseed. Traces were found in at least two stomachs in early November.

Lobelia. Lobelia. The only information at hand on this plant is that it was identified in a single elk stomach.

Lomatium (Cogswellia). Biscuitroot. Several species of *Cogswellia* are known to be grazed by elk, but these plants do not rank among the important elk forage species. Wyeth biscuitroot *Lomatium ambiguum (Cogswellia ambigua)* is one form noted as utilized by elk in Jackson Hole, Wyo.

Lupinus. Lupine. Lupines of several species are grazed by elk from June to December at least and probably at any time of year available. On Wyoming ranges *Lupinus parviflorus, L. humicola, L. argenteus,* and *L. leucophyllus* were identified among the species eaten by elk. The first two appeared to be the favorites, with some preference for *L. parviflorus.* Many other species were taken, no doubt, but observation was hampered by the difficulty of field identifications. As much as 10 per cent of individual plants was consumed, but the average was less. Lupine was found in 11 of 29 stomachs examined from Wyoming.

Lysimachia. Loosestrife. Leaves of loosestrife occurred in four elk stomachs collected in Virginia.

Medicago sativa. Alfalfa. In common with domestic animals, elk find alfalfa, a universally valuable forage that is usually available in the form of hay, very acceptable. They will graze it eagerly, too, when they find it in winter uncut or in the form of stubble. Undoubtedly other species in the genus would be acceptable if available.

Melilotus. Sweetclover. In fields and along fence-rows elk find sweetclover in winter, both white- and yellow-flowered varieties. Fields of it left uncut on the National Elk Refuge in Wyoming were grazed by elk in winter, being among the first areas invaded by these animals. In some roadside situations, however, the growth became so rank that only the tips were taken at first, although usually as winter progressed all parts of these plants, as well, indeed, as almost any kind of vegetation, were eventually consumed.

Mertensia. Bluebell. Large varieties of bluebells grow abundantly in high mountain meadows, especially in damp depressions, in open gullies, or along borders of high watercourses. Elk often congregate in these places and graze extensively on *Mertensia* usually taking only the fine tips. In some of the patches examined from 4 to 90 per cent of the plants had been grazed and from 5 to 15 per cent of individual plants utilized, the total utilization ranging from 0.2 to 12 per cent.

Mimulus. Monkeyflower. Several species of *Mimulus* are grazed by elk, but *M. lewisii* appears to be the favorite. In one observation plot all the plants had been grazed, each to the extent of about 40 per cent.

Monarda. Beebalm. Seeds of *Monarda,* which has been reported as elk food, were found in one elk stomach from Virginia.

Nolina. Nolina. In the Guadalupe Mountains of New Mexico and Texas, elk were reported to feed on this plant, locally called beargrass. It is said that elk paw the plant loose with their hoofs and eat the unprotected base and the basal part of the leaves.

Nuphar. Spatterdock. Once in a small lake in Jackson Hole, Wyo., a band of elk was seen feeding far out among the lily pads in the same manner as a moose that stood in the same pond. The elk were not positively observed eating the pondlily plants, but they appeared to be doing so. Other aquatic vegetation may have been involved (Plate 23).

Opuntia. Pricklypear. Pricklypear occurs on some winter ranges, and Skinner (1928: 313) reported elk feeding on it in March in Yellowstone National Park.

Osmorhiza. Sweet Cicely; Sweetroot. Sweet cicely was recognized in two elk stomachs and was noted as slightly grazed on the summer range. Apparently it is less palatable than *Ligusticum,* for once *Osmorhiza obtusa* growing conspicuously in a patch of currant bushes had been left untouched by elk that had browsed the currant bushes and had reached far underneath their branches to get some *Ligusticum.* On the Selway Game Preserve in Idaho sweetanise *(O. occidentalis)* was considered of moderate palatability to the elk, and this species has also been reported in the elk diet in Oregon.

Oxalis. Oxalis. Seeds of wood sorrel occurred in two elk stomachs from Virginia.

Pedicularis. Pedicularis. Information is not available on the palatability of many species of *Pedicularis.* In northwestern Wyoming *P. bracteosa* is one of the favorite food plants on elk summer range. In one plot 90 of the 100 plants and in another all 25 plants were grazed by elk. Utilization per plant varied from 40 to 50 per cent, and the total utilization over some areas probably averaged close to 40 per cent. On these same ranges *P. racemosa* is plentiful; but it occurs in greater abundance in the forested areas, which are less frequented by elk. Its utilization was not noted. The species *P. groenlandica* is restricted in distribution, but the few times its use was noted it had been heavily grazed.

Pentstemon. Pentstemon. Several species of *Pentstemon* are grazed by elk, but the plants appear to be of only fair palatability in northwestern Wyoming. Frequently these plants are ignored. They are hard to identify in the field, and preferences the elk may have among the many forms cannot be stated. Young and Robinette (1939: 30, 47) reported, however, that in the Selway region in Idaho, *P. pinetorum* was the most important weed species studied from the standpoint of total forage taken by elk, being especially sought during August and September. Its palatability apparently is not impaired by frost.

Perideridia. Yampa; Squaw-root. A plant of this genus, probably Yampa, Perideridia *(Carum gairdneri),* was utilized to some extent. As its use is difficult to observe, it is not easy to determine its degree of palatability.

Phacelia. Phacelia. Two phacelias, varileaf *(Phacelia heterophylla)* and silverleaf *(P. leucophylla),* are fairly palatable to elk. In one instance these plants were grazed to the extent of 10 per cent.

Phlox. Phlox. Phlox was observed only as winter food. Found in two elk stomachs collected in early spring, it formed a considerable percentage of the contents in one. Phlox species with persistent leaves and stems may be acceptable winter forage on stony slopes in the foothills, but sufficient data on this point have not been obtained.

Phytolacca decandra. Pokeweed; Pokeberry. This species was present in seven elk stomachs from Virginia, in three of which it formed as much as 1 per cent of the contents.

Plantago. Plantain. Remains of plantain were noted in three elk stomachs, and the plant was found grazed on the summer range. The only species identified was *P. tweedyi.*

Polygonum. Knotweed. Some species of *Polygonum* appear to be readily eaten by elk, as plants of this genus were found in 10 of the series of 29 stomachs examined from Wyoming, occurring in considerable quantity in 1 of them. Some species were present in five stomachs from Virginia.

Polygonum phytolaccaefolium was found by Young and Robinette (1939) to be very palatable and eagerly sought by elk on the Selway Game Preserve, Idaho, in late spring and early summer. On the Whitman National Forest in Oregon Pickford and Reid (1943: 330) found that this species was used by elk to the extent of 61.4 per cent of the diet. It is noted that on the ranges studied this plant also constituted 47.4 per cent of the vegetation present.

Potentilla. Cinquefoil. In northwestern Wyoming species of herbaceous cinquefoil furnish an important percentage of elk forage. The animals on the lower ranges feed extensively on the leaves in spring and continue to use the plant, together with a multitude of other herbaceous ones, throughout the summer. All the species utilized by elk have not been determined, but *Potentilla nuttalli* and *P. pulcherrima* were identified in Jackson Hole, Wyo. *P. glandulosa* has been eaten by elk in Oregon.

Prenanthes. Rattlesnakeroot. Seeds of rattlesnakeroot occurred in one elk stomach from Virginia.

Pteridium aquilinum. Bracken. It is reported that this species is very abundant on the Selway Game Preserve in Idaho, but is taken only in June, when the tender fronds are unfolding.

Ranunculus. Buttercup. In the wake of the receding snow in spring numerous buttercups spring up, and in July the wet borders of snowdrifts in the mountains encourage a progressive supply, usually such low-growing forms as *Ranunculus jovis* and *R. adoneus*. These are eaten by elk and are evidently quite acceptable; but they are not grazed heavily and the elk do not especially seek areas where buttercups occur.

Rudbeckia occidentalis. Western Coneflower. Apparently this is not very acceptable to elk. Young leaves are sometimes eaten, and occasionally a flower head is nipped off.

Rumex. Sorrel. An unknown species of *Rumex* was identified in one elk stomach from Wyoming, and seeds and leaves of sheep sorrel *(R. acetosella)* were found in two stomachs from Virginia. The western form is grazed by elk to some extent but is scattered in distribution and probably is not important in their diet.

Salsola kali tenuifolia (S. pestifer). Tumbling Russian-thistle. Grazing by elk on this species of Russian-thistle was noted particularly in Yellowstone National Park, where it was eaten with relish late in November.

Sanguisorba sitchensis. Sitka Burnet. This plant has been reported in the elk diet in Oregon.

Saxifraga. Saxifrage. There is no evidence that the saxifrage group of plants is much sought by elk, but the animals at least eat the plants, as traces of them were noted in three elk stomachs and *Saxifraga arguta* was found grazed by elk.

Senecio. Groundsel. A number of groundsels are eaten by elk and furnish important elements in the summer forage supply, but some species are much more palatable than others. The most palatable one noted during these studies was *Senecio semiamplexicaulis,* which is very common in the mountains of northwestern Wyoming, where it is eagerly grazed. Especially relished are the basal leaves as the plant emerges early in the season, but the plants are eaten persistently throughout the summer, not only the leaves but later also the flower heads and part of the stalk. Sometimes as much as 80 per cent of the plant is taken.

The species S. *triangularis,* which is considered quite palatable for livestock, is used very moderately by elk. In fact, in the Jackson Hole, Wyo., area it is comparatively seldom eaten. It was found used slightly in July and somewhat more commonly in August and September but was never grazed extensively. This form grows in damp spots, whereas *semiamplexicaulis* inhabits drier ground. Often, though, both were found side by side, and almost invariably *triangularis* had been avoided and *semiamplexicaulis* eaten. In the Selway region in Idaho, however, *triangularis* appeared to be used heavily during September and early October, as it is unharmed by early frosts. It was also reported utilized by elk in Oregon.

Other groundsels were found eaten in Wyoming, but field identifications of many were uncertain. Limited observations showed that S. *crocatus* was grazed and that S. *hydrophylus* was eaten in Yellowstone National Park in July and seemed to be decidedly palatable. S. *serra,* with its abundant foliage, is probably acceptable elk food, but in the areas where the basic studies were made it generally inhabits the elevations below the typical summer elk ranges, and its degree of palatability was not determined.

On the Selway Game Preserve in Idaho S. *columbianus* was reported eaten extensively by elk in June.

Sieversia. Sierversia. This was found grazed by elk only once, but it was not common on the high summer range.

Smilacina. Solomonplume (False Solomonseal). One species, *Smilacina stellata,* whose foliage remains green and palatable after frost, was taken to a considerable extent by elk on the Selway Game Preserve of Idaho in late September and early October. Another form, probably S. *racemosa,* was found in two elk stomachs in Virginia.

Solidago. Goldenrod. Traces of goldenrods were found in three stomachs from Wyoming and in six from Virginia. In Wyoming the use of goldenrod was observed on a few occasions.

Sonchus. Sowthistle. Seeds of this herb were found in one elk stomach from Virginia.

Stellaria media. Chickweed. Three elk stomachs from Virginia contained a trace of chickweed.

Taraxacum. Dandelion. Dandelion was palatable to elk in Wyoming, Montana, Washington, and Oregon. It becomes available to them on high mountain ranges, particularly in small meadows that have been overgrazed; and in such situations it is readily eaten.

Thalictrum. Meadowrue. Meadowrue is common on elk summer range, but it was found eaten by elk only once.

Trifolium. Clover. Elk eagerly graze clover whenever available. Alsike *(Trifolium hybridum)* and red *(T. pratense)* clovers have been grown on the National Elk Refuge in Jackson Hole and have been fed to elk in some winters. Ordinarily clovers are available to elk chiefly as hay in hard winters, although small quantities grow on the mountain ranges.

Valeriana. Valerian. Valerians in general are very acceptable to elk. Two species were observed especially, *Valeriana occidentalis* and *V. micrantha.* These were eaten frequently, to the extent of 80 per cent in one small area. Other species would doubtless be equally palatable.

Veratrum. Falsehellebore. This herb was found grazed in summer in Montana, the flower heads being nipped off. Probably the plants are acceptable in winter when available.

Veronica. Speedwell. This plant was found grazed by elk only once.

Vicia. Vetch. Field observations indicate that vetch is undoubtedly palatable to elk, but confusion in field identifications of imperfect specimens leaves some doubt as to the relative merits of the various "wild pea" groups. More study is required.

Viola. Violet. Violets, being small and scattered, usually do not yield a large volume of forage; but they are very acceptable to elk, being present in eight elk stomachs and found consistently grazed from time to time.

Wyethia. Wyethia. It has been reported that elk will graze *Wyethia,* but in the areas covered by these investigations plants of this genus were greatly restricted in distribution and were found below the typical summer range.

Xerophyllum tenax. Beargrass. This plant was found grazed in parts of Montana, both the flower heads and some leaves having been taken. It was eaten also on the Selway area in Idaho, where the flower stalks appeared to be a delicacy. One stomach from Montana, collected as early as November 13, contained beargrass exclusively, but this instance cannot be taken as indicating the usual palatability.

Zigadenus. Deathcamas. An unidentified species of *Zigadenus* was found grazed once in Washington in May.

Browse

One of the most confusing points with reference to elk forage is the relative importance of browse and grass on elk winter range. It must be understood that conditions vary tremendously on different ranges. On summer ranges that are not overstocked, where vegetation is generally available, the proportions of grasses, other herbaceous plants, and browse may be fairly constant. But on winter range they are variable, ranging between wide extremes —from a small admixture of browse to 100 per cent browse—depending on the local availability of grasses. In some localities browse becomes predominant, but browse ranges carry a special danger from overstocking. When too coarse materials are eaten, necrotic stomatitis ensues, and winter losses result. Also, browse is slow to recover from overuse, and range where browse is the main forage may become very unproductive.

Of special interest are the findings of Young and Robinette (1939: 46) to the effect that on the Selway Game Preserve in Idaho elk showed a preference for browse from July 15 to September 15 but preferred grasses and sedges before and after this period.

Many browse species are highly palatable and should not be regarded as emergency feed only, as is sometimes done. Even when provided with good hay, elk will eagerly browse many species. The following paragraphs name the species browsed by the Rocky Mountain elk and discuss their palatability.

Abies. Fir. Fir needles were found in eight stomachs from northwestern Wyoming. Alpine fir is utilized in winter but generally is not so plentiful on winter range as some other conifers are. It probably ranks fairly high in palatability. (See also *Pseudotsuga.*)

Acer glabrum. Rocky Mountain Maple. Rocky Mountain maple is used extensively by elk on winter range, where it often grows abundantly and is frequently found browsed into the characteristic club shape so common on overstocked ranges. Twigs are generally cropped to a thickness of one fourth inch at least. The palatability of this maple for elk is high.

It is likely that other species of *Acer* were extensively used in earlier times by the elk in such states as Wisconsin and Michigan. Avery (1929) wrote that in Michigan ". . . elk winter on rolling hardwood land, where they subsist largely on soft maple browse."

Alnus. Alder. Alder is only moderately used and is low in palatability. Occasionally some of the bark is stripped.

Amelanchier. Serviceberry. Serviceberry is widely distributed on elk winter ranges and is one of the most palatable species. On some ranges occupied by both elk and deer it is closely cropped and eventually is so reduced that it produces little forage. This species suffers in places where the general vegetation is used moderately.

Arctostaphylos uva-ursi. Bearberry. Elk are known to feed on the leaves of bearberry, especially in fall and winter, but observations were not extensive enough to determine the degree of palatability. Deer and caribou eat this plant readily, and probably elk find it equally acceptable.

Artemisia. Sagebrush. It is not known how many species of *Artemisia* are palatable to elk, but at least two are utilized. Fringed sagebrush *(A. frigida)* is eaten readily in winter and perhaps in fall and in early spring also if any remains; and big sagebrush *(A. tridentata)* also furnishes winter food. The former species is widespread but is less abundant than the latter. Although *tridentata* appears to be somewhat less palatable than *frigida,* it is often of greater importance locally because of its abundance and availability. Elk are fond of the matured seed stalks, which project above the snow, and the animals have been seen going from bush to bush, consistently picking them off. Elk also takes the leaves. In some localities sagebrush shrubs are overbrowsed and many bushes killed by game use, including that of elk. This is particularly true at the northern edge of Yellowstone National Park, although there, according to the observations of Adolph Murie (field notes), antelope appear to be the principal offenders.

Although *A. tridentata* may be considered a valuable food resource at a time when other sources fail, it should not be relied on as an indicator of food shortage. The degree of use is important in judging range depletion because elk have been known to eat sagebrush leaves in fall and early winter, at least as early as November 14, when other forage was plentiful. In a series of 14 stomachs taken in winter in Jackson Hole, 6 contained this species of sagebrush. Another series of nine stomachs—four from Washakie National Forest, Wyo., and five from Colorado—contained substantial percentages of sagebrush, as high as 16 per cent by volume in the Wyoming group and 36 per cent in the Colorado group. Of 97 stomachs from Yellowstone National Park in January 1943, 13 contained this sagebrush, varying in amount from a trace to 2 per cent. On the Whitman National Forest in Oregon Pickford and Reid (1943: 330) reported this sage in the elk summer diet to the extent of 3.8 per cent.

Atriplex. Saltbush. Although the use of *Atriplex* by elk was not observed, it is mentioned here as a forage group that in early times undoubtedly was important for elk when they migrated to the so-called sagebrush plains for winter and that would again be valuable should the elk revisit those areas. Two species in particular, *A. canescens* and *A. confertifolia,* are considered valuable forage species for livestock, and elk would no doubt utilize these and others.

Betula. Birch. Bog birch, *Betula glandulosa,* or related forms are quite palatable to elk in winter and in some places have been very heavily browsed. It and the willow often occupy the same or contiguous area, and when the birch occurs in abundance in such situations it is a valuable supplement to the supply of willows. This shrub is utilized freely by the caribou of Alaska and is probably generally useful to the deer family.

Where available, the larger birches too are undoubtedly

browsed by elk. Water birch *(B. fontinalis)* is eaten with great relish. When the elk ranged over parts of Minnesota and Wisconsin, they probably utilized also the larger paper birches that occur there. Leaves of black birch *(B. lenta)* were found in an elk stomach from Virginia.

Carya. Hickory. A trace of hickory leaves was found in an elk stomach from Virginia.

Castanea dentata. American Chestnut. Leaves of this species were identified in a stomach from Virginia, where its use was also observed in the field.

Ceanothus. Ceanothus. Observations on the use by elk of plants of this genus were confined chiefly to snowbrush ceanothus *(Ceanothus velutinus),* which is fairly common on the elk ranges of the Rocky Mountain area. This plant is utilized by both elk and mule deer as early as November 15, at least, and where available on the ranges it is a valuable addition to the winter forage supply. On the Jackson Hole, Wyo., ranges the supply is limited, and only 1 stomach in a series of 14 examined in winter contained any. On the Whitman National Forest in Oregon Pickford and Reid (1943: 330) reported this browse used by elk, apparently in late summer or early fall. Another species known to be on the elk winter diet is *C. sanguineus.* Several other *Ceanothus* species in the Western States are known to be good deer food, being considered, in fact, better than *C. velutinus,* and these undoubtedly would be palatable to elk also. In Virginia the leaves and stems of New Jersey tea ceanothus *(C. americanus)* is eagerly sought by elk and is ranked high in palatability.

Cercocarpus. Mountainmahogany. In the limited areas where observations on its use by elk were possible, mountainmahogany was very palatable. This appraisal refers particularly to *Cercocarpus ledifolius,* which occurs in tree form on some of the rocky bluffs and slopes on the east side of Jackson Hole, Wyo. In this area the trees are trimmed up as high as elk can reach and no reproduction is surviving except a few small plants that failed to grow out of reach in time and that are kept trimmed down in a permanent pincushion form. This species no longer furnishes an appreciable amount of feed in the Jackson Hole area. In the Blue

Mountains of Oregon this browse was one of the key species for elk and deer in winter (Cliff, 1939: 561).

In Arizona a shrubby form, probably *C. intricatus,* is eagerly taken by elk in winter and doubtless *C. arizonicus* also is grazed. The elk now in Arizona, particularly on the Sitgreaves National Forest, are the Rocky Mountain form, but it is very probable that the extinct Merriam elk had similar tastes.

Other species of the mountainmahogany group, such as *C. montanus,* also are utilized by elk and deer, and the genus as a whole furnishes palatable forage.

Chimaphila umbellata. Pipsissewa. DeNio (1938: 424) gives this species a palatability rating of 30 per cent for elk in Idaho.

Chrysothamnus. Rabbitbrush. Rabbitbrush is an important element in elk winter forage. Perhaps the most palatable species are *Chrysothamnus lanceolatus* and related small low-growing forms. These are often abundant on stony ground in mixtures of *Phlox, Artemisia frigida,* and scattered grasses; and such areas are productive winter range for elk. The elk begin to browse the low rabbitbrush in late fall or early winter and indeed have been observed feeding on it eagerly when grass in nearby areas was available. They continue to use it as long as the supply lasts. Mule deer on the same ranges also take it. Tall rabbitbrush *(C. speciosus),* also locally abundant, ranks next to *C. lanceolatus* in palatability. The elk take some of it in early winter but do not browse it closely until late January or February.

On the Jackson Hole winter ranges in northwest Wyoming, rabbitbrush is perhaps one of the most important forage species, probably ranking with grasses in value. It has volume of forage and sturdiness to protrude through moderate snowfall, and it grows on so-called wasteground. It is thoroughly utilized by the end of winter. A range carrying an abundance of such growth may be considered favorable for elk. Tentative experiments have indicated that rabbitbrush, at least the two species here mentioned, withstands repeated winter browsing very well.

Cornus stolonifera. Redosier Dogwood. Observations on this species were limited, but several times it was noted as extensively used by elk in winter, especially in parts of Washington and Montana; and it was reported to be highly palatable in Idaho.

This shrub is eagerly browsed by white-tailed deer in the North Central States, and it seems reasonable to assume that elk formerly utilized it in those areas and would do so again were they reintroduced there.

Corylus. Filbert; Hazel. Information on the use of *Corylus* by elk is very meager. It was reported that hazel was utilized by elk in winter on the Snoqualmie National Forest, Wash. In northern Minnesota hazel is browsed by white-tailed deer, although it is not among the shrubs most palatable to them. Probably it would be taken by elk also in a similar environment.

Cowania stansburiana. Stansbury Cliffrose. This species of cliffrose is one of the favorite winter forage species for elk on the Sitgreaves National Forest in Arizona, the only locality in which there was opportunity to observe its use by elk.

Elaeagnus commutata. Silverberry. This shrub is very palatable to elk and on some winter ranges has been browsed to a degree injurious to the plant. Summer use was not observed, but one elk stomach collected in September contained *Elaeagnus*.

Eurotia lanata. Winterfat. Where available, this is a highly palatable plant on winter ranges. In the Jackson Hole-Yellowstone National Park area in Wyoming winterfat is browsed as early as November.

Galax aphylla. Galax. This evergreen plant is eagerly sought by elk in Virginia, where in palatability it ranks near the top of the list of elk foods. It was found in six Virginia stomachs, in three of which it formed as high as 20, 24, and 25 per cent of the contents.

Gaultheria procumbens. Checkerberry; Wintergreen. Wintergreen is much sought by elk in Virginia, where in one stomach it contributed as much as 30 per cent of the food.

Gaylussacia. Huckleberry. In Virginia this browse is apparently rather consistently though sparingly eaten by elk in autumn, as a small quantity of the twigs, leaves, and berries was found in each of the eight stomachs collected.

Holodiscus (Sericotheca) discolor. Creambush Rockspirea or Oceanspray Rockspirea. Oceanspray is browsed readily by elk, but such use was observed in so few localities that its relative palatability cannot be stated. Young (1938: 132) gives this plant a palatability rating of 5 per cent for certain ranges in northern Idaho.

Juniperus. Juniper. There is great variation in the degree of use of the junipers. In northern Yellowstone National Park, Wyo., Rocky Mountain juniper *(Juniperus scopulorum)* has been browsed as high as the elk can reach; but in Jackson Hole, south of the Park, on dry buttes where thousands of elk congregate each winter, this same species is practically untouched and its dense foliage extends all the way to the ground. In a few places some small trees have been browsed, but this may be the work of deer. On the other hand, in the Snake River bottoms south of Jackson, in the vicinity of a state elk feed ground, junipers are overbrowsed. In the Gros Ventre River drainage, Wyo., the creeping form, *J. horizontalis,* is eaten extensively; and in the Washakie National Forest, Wyo., *J. communis* is taken. On the Sitgreaves National Forest in Arizona, the elk introduced from Yellowstone National Park exhibit a striking choice between two species. They are very fond of the alligator juniper *(J. pachyphloea)* but do not eat the one-seed juniper *(J. monosperma),* although they break down its branches to obtain a mistletoe that is common on it. *J. occidentalis* is a winter elk food in the Blue Mountains of Oregon.

Ledum. Labradortea. A species of Ledum was found browsed once, in Montana; and Young and Robinette concluded that *L. glandulosum* was of negligible palatability on the Selway area, Idaho.

Linnaea borealis. Twinflower. The twinflower was found commonly utilized by elk in Montana (DeNio, 1938).

Lonicera. Honeysuckle. Bearberry honeysuckle *(Lonicera involucrata)* is generally utilized by elk, but there have been no satisfactory opportunities to measure its relative palatability. Utah honeysuckle *(L. utahensis)* is quite palatable, both in summer and in winter. In the Jackson Hole, Wyo., region it was found regularly browsed by elk on parts of the winter range and used by them in June, July, and September, as well.

Mahonia repens. Creeping Mahonia. In Wyoming, creeping mahonia occurred in 11 of the 29 elk stomachs examined in Jackson Hole and in 2 from the Washakie National Forest, forming 20 per cent of the contents in 1 of the 2. It was found also in four stomachs from Colorado. Apparently this evergreen forage is not eaten by elk in summer, but is very acceptable to them in fall, when other vegetation is largely dead, from early September until snowfall becomes too heavy, usually late in November, and again in spring as soon as the snow disappears. It is also eaten in winter when not covered by snow.

Malus pumila. Common Apple. In Virginia it was found that the elk were fond of apple browse, including twigs, leaves, fruit, and seeds, thus indicating what might be expected should the animals visit orchards.

Menziesia ferruginea. Rusty Menziesia. This shrub was found browsed in summer and winter in parts of Washington and Montana. On the Selway Game Preserve, Idaho, Young and Robinette (1939: 37) found it heavily browsed in July, in lesser amounts later, and very little used in areas where many other browse species were present.

Pachistima myrsinites. Myrtle Pachistima. This plant grows abundantly on some winter ranges, particular in wooded portions. Although eaten by elk in summer, it is especially valuable as forage in winter and appears to be very palatable then. It is browsed in fall, winter, and spring. It was present in 8 of a series of 29 stomachs examined in Jackson Hole, Wyo., and in 17 of another series of 35 stomachs from the same locality, early in November.

Philadelphia lewisi. Lewis Mockorange. This shrub was found browsed by elk, but the writer had no opportunity to determine its relative palatability.

Phoradendron juniperinum. Juniper American-mistletoe. This parasitic plant, which grows abundantly on *Juniperus monosperma* in Arizona, is very palatable to elk in winter. The animals break down the juniper limbs to reach the parasite.

Physocarpus malvaceus. Mallow Ninebark. On the elk winter ranges of Jackson Hole, Wyo., this ninebark is very palatable and is heavily browsed. Its use in other areas also has been noted.

Picea. Spruce. Needles of Engelmann spruce *(Picea engelmanni)* and Colorado spruce *(P. pungens)* were found in 16 elk stomachs, to the extent of 20 per cent in one, though generally from a trace to 4 per cent. These were not accidental occurrences, as browsing on spruce by elk has been noted in the field. Indeed, one tree was trimmed to a considerable height. However, spruce must be considered the least palatable of the conifers.

Pinus. Pine. Of 163 elk stomachs examined, 18 contained pine needles. Pine is essentially a winter forage, although a little appears to be taken in summer also. Whitebark pine *(Pinus albicaulis),* limber pine *(P. flexilis),* and lodgepole pine *(P. contorta* var. *latifolia)* are quite palatable to elk. The whitebark has been found stripped of available limbs, and some trees are barked by the elk. The lodgepole is much more available and more extensively used. In some areas most of the young forest growth is "high lined," the foliage stripped off as far up as the elk can reach. In the winters of 1937-38 and 1938-39 lodgepole pine logs had been hauled to the headquarters area on the National Elk Refuge in Jackson Hole, Wyo., for construction purposes. During each winter elk ate the bark from the piled logs and nibbled the strips of bark that had been shaved off by workmen. These animals had been feeding in open fields with a minimum of browse.

In one elk stomach from Colorado yellow pine *(P. ponderosa)* formed 98 per cent of the contents. The use of these species by elk in Idaho has been recorded by DeNio (1938) and in the Blue Mountains of Oregon by Cliff (1939: 562).

Under date of March 30, 1940, Roy Wood informed the writer that in Virginia pitch pine *(P. rigida)* is browsed to some extent by elk.

Populus. Poplar; Aspen; Cottonwood. The golden quaking aspen *(Populus tremuloides* var. *aurea)* is very acceptable to elk and, together with willows, should rank near the top of the browse list in palatability. These aspens are so heavily utilized that on many winter ranges the groves are injured and are dying out in some places. Elk browse the leaves freely in spring, and even as late as July 5 an elk was seen eating them. As a rule, aspen is not used by elk during summer and early fall. One reason may be that it is not common on the higher summer ranges. Late in fall, when most other vegetation is dead, fallen aspen leaves are eaten freely at times. As winter progresses, elk dig through the snow for the old leaves and browse all available twigs, including those of the young plants. Furthermore, old trees are barked so extensively that on most winter ranges the trunks are blackened with numerous old scars (Plate 22). Under such consistent browsing the available parts of the trees are soon consumed; thereafter an aspen grove can furnish little forage. Under heavy browsing, the reproduction is unable to rise above the snow line, and as the older trees have already been trimmed as high as the elk can reach, with the prevailing overstocking of elk it will be increasingly difficult to save the aspen growth.

The larger *Populus* species, generally known as cottonwoods, are also browsed, but usually they are not so common on winter ranges as the aspen. Observations have been confined largely to the narrowleaf cottonwood *(P. angustifolia)* whose green leaves elk have been seen eating in June.

Potentilla fruticosa (Dasiphora f.). Bush Cinquefoil. This shrub, common on many elk ranges, is utilized by elk in winter. Normally it ranks low in palatability but sometimes is used rather extensively if other food is scarce.

Prunus. Chokecherry; Cherry. Chokecherry is very palatable for elk and is eaten extensively in winter. In the spring, when new growth is beginning, the elk seek out the fresh, partly unfolded leaves. Bitter cherry *(Prunus emarginata)* was found browsed only moderately on the Snoqualmie National Forest in Washington. Apparently it is more palatable on the Selway area in Idaho. Black chokecherry *(Prunus virginiana melanocarpa)* was identified in the stomach contents of one elk from Colorado.

Pseudotsuga taxifolia. Douglasfir. With the possible exception of certain junipers in some areas, Douglasfir is probably the most palatable of the conifers (Plate 24). Sixty-three elk stomachs contained needles and twigs of this species, some of them to the extent of 20, 35, and even 80 per cent of the contents. Striking evidence of the elk's liking for this forage is the frequent occurrence of heavily browsed young trees that have been trimmed up as high as the elk can reach. Elk also eat the bark of young trees. Locally there are areas where the young second growth has all been neatly trimmed, although the tops of many Douglasfirs have managed to grow beyond reach of the elk.

Purshia tridentata. Antelope Bitterbrush. This shrub, known to be sought eagerly by mule deer, was found to be an equally attractive winter forage for elk wherever it was available. In areas where both deer and elk winter, this shrub is often overbrowsed.

Pyrularia pubera. Alleghany Oilnut. Roy K. Wood (unpublished thesis) found this to be a preferred browse for elk in Virginia.

Quercus. Oak. Acorns, leaves, and twigs of many species of oak are known to be eaten by elk, chiefly in fall and winter. Acorns are particularly acceptable. Gambel oak *(Quercus gambeli)* is browsed in parts of northern Arizona, and oak (unidentified) is browsed rather commonly in Colorado, where seven stomachs collected on the Gunnison National Forest showed the following percentages of oak products: December, trace; January, 2 and 84 per cent; February, 6, 59, and 80 per cent; and April, trace. In Virginia the acorns or leaves or both of *Q. alba, Q. borealis* var. *maxima, Q. coccinea, Q. ilicifolia,* and *Q. muhlenbergi* (?) were each found in one or two stomachs; and acorns, twigs, and leaves of other oaks not specifically determined were noted in six.

Rhamnus purshiana. Cascara Buckthorn. DeNio (1938: 424) gave this a palatability rating of 15 per cent for elk in Idaho.

Rhododendron. Azalea. An elk stomach from Virginia showed a trace of azalea. The degree of use of this plant was not determined.

Rhus. Sumac. The use of sumac by elk has been recorded in Colorado. Roy Wood found that elk in Virginia ate the terminal twigs of *Rhus* and stripped the bark from the larger stems.

Ribes. Currant. Currant bushes, especially some species that have been observed, are very acceptable to elk. Although not so palatable as many other browse species, *Ribes petiolare, R. cereum, R. viscosissimum,* and *R. lacustre* are eaten fairly consistently in winter, and the last two are taken frequently in summer. Once about 6 per cent of the leaves of one large bush of *R. lacustre* was browsed in a place where other palatable vegetation was easily available.

Ribes (Grossularia). Gooseberry. Gooseberry bushes are browsed by elk in winter and apparently are entirely acceptable, but they are not eaten so readily as currant *(Ribes)* browse.

Robinia pseudoacacia. Black Locust. Twigs and leaflets of this browse formed traces of the contents of two elk stomachs from Virginia.

Rosa. Rose. Wild roses of a variety of species are distributed generally on elk ranges. They are universally taken rather freely as winter forage, and in view of their abundance in some places the plants undoubtedly furnish an important quantity of food. Often only the smaller tips are nipped off by the elk, but in some cases as much as one fourth of the growth is consumed (Figure 31).

Rubus parviflorus. Western Thimbleberry. The leaves of thimbleberry are eaten by elk in summer, but this browse has not been found abundant on the ranges of the Rocky Mountain elk; probably it is more important on the Pacific coast.

Salix. Willow. Willows are utilized by elk at all times of the year but chiefly in winter. Because of varying circumstances it is difficult to give a comparative palatability rating for any plant, but willows should undoubtedly rank near the top of elk browses —if not actually highest—in palatability. On the Sitgreaves National Forest in Arizona willows were so eagerly browsed that they were fenced in for protection from the elk. In the vicinity of feed grounds where elk are given hay, willows cannot stand the browsing pressure and eventually die out; and even on the open winter range the willow growth is severely injured wherever elk become numerous. On the winter range in Yellowstone National Park the willows are being destroyed by elk browsing.

In spring and early summer elk stand in a willow patch and persistently feed on the new leaves and occasional young shoots. Throughout the rest of the summer, in fall, and even in early winter, while engaged in seeking the grasses and various herbs, the elk browse the willows only moderately as a rule; but as the snow deepens, and especially when the grasses have been consumed, the elk concentrate on the available willow patches.

There is definite choice among species at times: on some bushes twigs up to a thickness of one half inch were browsed, whereas on other species nearby slender twigs were left untouched. This point has not been worked out for the numerous willow species, however, and on overcrowded range all species are taken. The following willows have been positively identified in the diet of the Rocky Mountain elk: *S. barclayi* (Wyoming, in August; Washington, in May); *S. hebbiana perrostrata* (Wyoming, in winter); *S. geyeriana* (in Wyoming, fall and winter); *S. planifolia* (Wyoming, June); *S. pseudocordata* (Wyoming, fall and winter); and *S. scouleriana* (Wyoming, fall and winter).

Sambucus. Elder. Utilization of bunchberry elder *(Sambucus microbotrys)* is somewhat irregular, but the plant is quite palatable and is extensively browsed on the summer range. Total utilization on observed plots varied from 3.5 to 75 per cent. Sometimes one bush was eaten down closely and stripped of all leaves while nearby bushes were untouched. Utilization was observed chiefly in July and August but was noted also in fall and late in winter. On the Selway Game Preserve of Idaho, *S. melanocarpa* was found to be extremely palatable but not abundant enough in any given area to be considered an important forage plant.

Sassafras albidum molle (S. variifolium). Sassafras. This browse has proved to be quite palatable to the elk in Virginia.

Shepherdia canadensis. Russet Buffaloberry. Elk eat the leaves of buffaloberry in summer and the twigs in winter, and as the plant is fairly plentiful on some winter ranges, it tends to add considerably to the winter forage supply. On some ranges in Wyoming, however, this browse is not eaten with enthusiasm, and often most of it remains unused at the end of the winter.

Smilax. Greenbrier. Berries and leaves of *Smilax* were found in the stomachs of four elk from Virginia. The species may have contributed rather materially to the forage of eastern elk in primitive times.

Sorbus. Mountain-ash. Mountain-ash is irregularly distributed on elk ranges. It was found only moderately browsed by elk in summer in the Jackson Hole, Wyo., area, though the stomach of one elk shot by a hunter on winter range in the fall of 1940 contained at least 80 per cent mountain-ash leaves and berries. *Sorbus scopulina* is reported very palatable to elk on the Selway Game Preserve in Idaho.

Spiraea. Spiraea. This small bush is eaten by elk in both winter and summer, but it was not found used very extensively. Apparently its palatability is fairly low, though the plant may be important locally where it is abundant.

Symphoricarpos. Snowberry. Snowberry must be rated very low in palatability for elk. On many winter ranges it is almost untouched, and even on overbrowsed and crowded ranges it shows the least use and generally is not damaged. Occasionally, however, the elk appear to eat snowberry freely; in some places on the east slope of the Cascades, in Washington, it was overbrowsed, some bushes having been eaten 100 per cent, but this was on a greatly overstocked winter range. Moderate use of the leaves in summer was noted.

Tetradymia canescens inermis. Spineless Gray Horsebrush. In the Jackson Hole area, Wyo., this plant, though not plentiful, was found readily browsed by elk in winter. Its use was not observed elsewhere.

Vaccinium. Blueberry. The big whortleberry *(Vaccinium membranaceum)*, generally referred to as huckleberry, was quite palatable to elk in summer and winter, and elk were once seen eating it in September after the leaves had turned yellow. On the Selway area in Idaho, it was reported browsed freely during June but more rarely thereafter. The grouse whortleberry, or "low huckleberry" *(V. scoparium)*, is not used extensively, although it is often dominant ground cover in lodgepole pine forests. Elk do feed on it, however, in winter and early spring. Its evergreen terminal twigs are probably especially welcome in early spring, the time of year when green forage is at a premium. Probably the species would be even more extensively used in winter if it were not so generally covered with deep snow. Twigs, leaves, seeds, and berries of a species of Vaccinium were found eaten in autumn by six Virginia elk.

Viburnum acerifolium. Maple-leaf Viburnum. Seeds of this species were found in one elk stomach from Virginia.

Vitis. Grape. Seeds and leaves of grapes were found in two elk stomachs from Virginia.

Miscellaneous Foods

Lichens. Reports indicate that Rocky Mountain elk, like the Roosevelt elk on the Pacific coast, eat lichens (probably *Usnea*) from trees in winter. Lichens were found in one stomach collected in February.

When at one time there arose the question of the possible competition with caribou if elk were introduced in the north, several sacks of lichens mixed with mosses were obtained from Alaska for experimental use on the National Elk Refuge in Jackson Hole, Wyo. The lichens were principally species of *Cladonia,* the type of forage so commonly sought by caribou. Unfortunately, the material was stored for 2 years before there was opportunity to use it and undoubtedly lost something in palatability.

In the experiment, several animals were placed in an enclosure and a good supply of lichens and hay was made available. For some time the elk did not touch the lichens, but once when the hay supply was low they consumed about half the lot. The results were not conclusive in several respects. The thoroughly dried-out lichens were far from being fresh and were, moreover, mixed with mosses. Furthermore, the choice of food was not a fair trial, as well-cured hay is naturally more attractive than dried grasses under snow on the open range. The elk did eat some of the lichens, and in view of the use of lichens by free-ranging elk it is safe to conclude that elk do eat the so-called "reindeer moss" in winter when other food is relatively scarce or inferior. Where fresh, live lichens are available, elk might consume a considerable quantity.

Equisetum. Horsetail. In one elk stomach from Colorado, this plant constituted 20 per cent of the contents. It was found grazed once. Horsetail was not plentiful on the range most often occupied by elk, however, so it may be more palatable than these few observations indicate.

Lycopodium. Clubmoss. This was found eaten once by elk in Virginia.

Mushrooms. Mushrooms were in an elk stomach obtained in August and were found generally utilized by elk and moose. In Virginia this product of the rainy spells was found in three elk stomachs, totaling 1 per cent of the contents.

Manure. Adolph Murie (field notes) reported that at one place in Yellowstone National Park, Wyo., the elk consumed virtually all the manure from two horses kept at a barn there in the winter. This manure-eating habit is often observed among cattle.

Meat. Donnelly (Yo et al, 1889: 431), who kept a few tame elk in Wyoming, reported that one of his calves became sick in September and was finally killed. Three pieces of bone had lodged in the throat, and bits of bone and meat were found in the stomach. It was thought that the animal had fed on refuse elk meat lying about the place. Elk appear to share the carnivorous propensities of some of the other deer, notably the caribou, which eats lemmings and fishes.

Elk Licks

Like other members of the deer family, elk are attracted to natural sources of minerals. The summer range of the Jackson Hole, Wyo., elk herd, comprising parts of Yellowstone National Park and the Teton National Forest is unusually well supplied with mineral licks, large and small, and from each well-used lick radiate elk trails, some of them running into other trail systems centering in other licks (Plate 15). The Upper Yellowstone area is poorly supplied with licks, but the elk appear to thrive as well.

The contents of natural licks are by no means exclusively common salt ($NaCl$); in fact, sometimes this mineral is practically absent. Analyses of lick contents have been made from time to time, but the possible significance or value of the contents is not yet understood. A number of mineral salts are present. In one group of lick samples from Jackson Hole were found calcium, magnesium, sulfate, phosphate, and only minute traces of chlorides. In the most extensively used lick the minerals occurring in greatest quantities are sulfate (SO_4) and calcium (Ca). For discussion of mineral requirements and tastes, see the section on salting (p. 309).

The Year-round on Elk Range

To illustrate the year-long feeding activities of elk, the Jackson Hole, Wyo., area is chosen, as it is the location of the most intensive studies. On ranges that differ from this area in climate, terrain, and vegetation, there will be some striking differences in food habits, as the elk must subsist on the available forage, whatever it is.

April.—Sometimes as early as March but more commonly in April, the elk begin to find more acceptable food than that predominant during winter. What snow is left thaws rapidly, and new forage becomes available. Dried grasses are still the chief food and, together with some browse, form the bulk of the diet; but there is an increasing admixture of green tidbits in the food especially toward the end of April, as by this time the elk can get at the ground to pick the evergreen leaves of myrtle pachistima, creeping mahonia and even a little green grass, as well as new shoots of balsamroot. If the chokecherries begin to sprout leaves, these are nibbled off. If the elk have been on artificial feed, it is generally in April that they voluntarily desert the feed grounds and scatter over the nearby slopes, evidently attracted by traces of green vegetation. It is then, too, that they seek the bare knoll tops and perhaps injure the new growth there by trampling. In normal years, therefore, April gives the elk definite release from winter hardship.

May.—Progressive elimination of snow and steady transition from severe winter conditions to those of summer bring forth abundant green vegetation in May. The elk begin to migrate, following the receding snow, and often even reach the high summer range during the latter part of the month. Early in May dried grasses may still be an important part of the diet, but as the month advances green food becomes available in increasing quantity and eventually predominates.

Of course there are variations in the conduct and travels of the elk because of special features in different localities. In areas such as those about Two Ocean Lake and Emma Matilda Lake, cinquefoil *(Potentilla nuttalli, P. pulcherrima* and related forms) dominates the diet for a while; many elk linger there a long time, while others move on to much higher elevations at or beyond the immediate snow line. In this month the emerging still folded leaves of balsamroot are eagerly consumed and a number of other herbaceous plants, as well as grass, are eaten. Although the elk have largely abandoned winter browse, they spend considerable time in the stream bottoms, eating the new leaves of aspen or of the larger cottonwoods. On higher slopes, where vegetation is more backward, they spend much time feeding on the abundant elk sedge *Carex geyeri,* which is comparatively palatable as the basal parts of the blades are still fairly green.

Although many elk leave behind the luxuriant vegetation of the valley bottom and persistently follow the snow line back into the mountains, May marks the advent of food abundance for them; and if they choose, they may live in an entirely green landscape with a bounteous selection of food at hand.

June.—This month is marked by luxuriant vegetation that furnishes a wealth of choice food for the elk. Grasses, herbaceous cinquefoil, leaves of balsamroot (many of which are still tender), and a variety of other plant species are eaten. Occasionally the animals still eat willow leaves and—on the higher slopes—the elk sedge *(Carex geyeri).*

July.—June and July are the months of green food in greatest variety, and July marks the culmination of plant growth, both in size and in number of species. In July the complex assortment of plants favored by elk has reached high development, and the effect is shown in the sleekness and fine condition of the animals. Elk frequent the high, open and rather bare slopes, where a large variety of herbaceous species are available and where a generous growth of grasses furnishes the staple food supply. There they find the basal leaves of certain late-maturing forms of *Senecio, Helianthella, Aster, Erigeron* and *Agoseris.* They will stop to browse a currant bush, wallow in luxuriant patches of tall *Mertensia,* nibble at *Ligusticum,* and begin to pick at the maturing flower heads of tall larkspurs *(Delphinium).* Cow-parsnip *(Heracleum lanatum)* is eagerly devoured, and "Indian warrior," bracted pedicularis, is grazed wherever found. During rainy spells mushrooms are added to the diet. This is the season of abundance of the staple grasses, and the elk obtain all they wish of their favorites—bromegrass, wheatgrass, and, particularly, certain fine bladed bluegrasses. There is also a satisfying variety of favored weeds and tidbits.

As early as late March or in April, when the snow is rapidly disappearing, elk are prone to lie on snowdrifts to rest on pleasant days, and they seek the highlands early. As warm weather develops and reaches its height in July, accompanied by the fly torment, there are fewer and fewer snowdrifts and the elk seek either the shade of the characteristic isolated clumps of firs, which generally have a hollow space within the tight thicket formation, or wet, sedgy places in the open. Often they bed down in a draw, almost concealed by a heavy growth of *Mertensia.* At other times they simply lie in an open, dry meadow, or a few may forage through the heavier woodlands. In the mornings and evenings, particularly, the elk may seek the nearest lick. They usually spend the middle of the day lying down, but at any hour a number of animals of a band may be seen on their feet, grazing. About 4 p.m., when the shadows begin to lengthen, there is a general stir and soon all the elk are active, feeding. The calves have been growing rapidly, and they join their elders in all these activities.

Thus, in the height of the good season, the elk take their comfort and flourish.

August.—The elk are very fat in August, probably at their best. This month also is characterized by plenty of food, although even early in August severe frosts may occur. Some plants have matured and are already beginning to dry, and some are in fruit. There is a distinct change in the vegetation, but many of the same general plant groups are utilized by the elk. Besides the usual grasses, many of the herbaceous plants are still taken—geranium leaves (even after they have autumn colors), senecios, asters, erigerons, mint, the ripening pods of tall larkspur (which are even more attractive than in July), and many others. However, although the same general plant groups are used, their ripening and the effects of frost have caused the elk to seek them in areas where they are still relatively unaffected, so that the elk frequent the woods more generally than before. They often disappear from areas where they were plentiful in July, and one wonders where they have gone.

September.—In September grasses more generally ripen, other plants are in seed, and the vegetation is drying up. There may be early snow, which later disappears. There is little change in diet from that of August, but the elk lose no opportunity to pick out any vegetation that still remains green. Food is still plentiful and of good quality, but more selection is necessary to find the best. From now on palatability is on a comparative basis. Plants to which the elk may have been indifferent early in summer may now step up in palatability because they have remained greener than others. The elk seek *Carex geyeri* more persistently. Plants of one species may be passed by in one place because they have dried up and be eaten in other places where they have longer, continued, vigorous growth.

September is the time of the rut, when the bulls begin the exhausting quest and rivalry of the breeding season. They spend less time foraging and as time goes on lose flesh. The cows and younger animals continue grazing in normal manner.

October.—This month is similar to September in many ways, but vegetation is drier and snow occasionally falls. After the middle of the month the rut is waning and there is even a little advance migration, although the main movement does not occur until later. The elk sedge *(Carex geyeri)* is eaten extensively on slopes where it is available, and green grass blades are yet found in many places. Most of the favorite herbaceous plants of summer are still present and, though dried, are utilized.

November.—The first touch of winter occurs in this month. Ordinarily snowfall becomes sufficiently heavy to induce mass migration to lower elevations and to more favorable locations to the south. Grasses become more important. Always the choice of the elk, they are now a necessity, for a great many herbaceous plants have withered away and those still available—such as fireweed, asters, and other composites that protrude from the snow—are not present in sufficient quantity for an exclusive diet, though they are still readily eaten when available. Some browse is utilized. When the elk reach winter range, they eat the dead, yellow leaves of cottonwood and aspen, leaves of the tall *Frasera*, and rabbitbrush (some *Chrysothamnus speciosus*, but particularly *C. lanceolatus*). They feed a little on willow, myrtle, pachistima, creeping mahonia, ceanothus and even a little pine and Douglasfir.

The rut is over, and by the end of November the elk are fairly well established on winter range. A few, mostly old bulls, have remained high in the mountains, on wind-swept areas, such as The Trident on the Upper Yellowstone; some have stopped in the area about lower Buffalo Fork; many more have sought the Gros Ventre River Basin; the greatest herd has migrated to the valley in the vicinity of Jackson; and still others have gone on to familiar ranges south of Jackson, as far as the Hoback River and beyond. Each locality has different features that affect the movements and conduct of the elk.

December.—Although this month marks the advent of real winter, usually with deepening snow and storms, the elk do not suffer; food is still plentiful. The winter diet of dry grass and browse is now established, but any fragments of green vegetation are doubly welcome. Some elk glean the green or partly green basal parts of the sedges in the swamp or the leaves of such plants as creeping mahonia, myrtle pachistima *(Pachistima myrsinites)* or *Ceanothus* in the woods. The great dependence, however, is on grass; and the elk are widely scattered over the grasslands, the open foothills, and the wind-blown hills and slopes of the Gros Ventre Basin, seeking the cream of the winter forage. In addition to grass, the animals seek the short species of rabbitbrush *(Chrysothamnus)*, the fringed sagebrush *(Artemisia frigida)*, and other browse that is easily available.

January.—Generally the elk do fairly well in this month, but in many winters they have by this time gone over the grass ranges once and will be found more extensively in the woods, pawing for the grass and other vegetation under the snow. There is still good forage available, but the elk have to work harder to get it. Consequently, they rely more and more on browse, taking first the choicest parts, as they did with the grass.

If the winter is normal and the range not too badly overstocked, there is only the minimum, unavoidable winter loss of elk. Each year, in December and January, some aged cows succumb and there are the usual deaths from scabies, normally confined chiefly to adult bulls. Scabies usually takes some toll before this, in October and November.

February.—Although January may sometimes be severe, February is the critical month, as the best of the forage has been consumed. The elk go through the swamp sedges again and take what is left. Once more they cover the grass ranges. Rabbitbrush *(Chrysothamnus)* and the coarse wildrye *(Elymus)* are now eaten down close. Even some tules and cattails are nibbled. In the Gros Ventre Basin the elk invade the willows and when they do so in numbers, demolish large patches. Douglasfir and lodgepole pine, heretofore merely tasted, are now eaten more extensively. The herds go up on the hillsides to forage grasses they may have overlooked, and they penetrate wooded draws to dig for pinegrass and find there more rose and chokecherry.

About the middle of February elk losses become more noticeable. Ticks appear, diseases take effect, and calf mortality begins.

This description applies to an overstocked range. Where the animals are less numerous, they have a much easier time. They do not suffer if there is forage under the snow, for they can paw down to a depth of 30 inches at least (Plate 25). Also, in an open winter, when the elk herds are scattered on higher terrain, they fare much better. On the other hand, on those rare occasions when winter rains or alternate thawing and freezing cause a hard crust to form as in 1937-38, the animals suffer severely.

The account here does not fit conditions on all winter ranges. Different vegetation and other factors modify the situation. On some winter ranges grass may continue to be the dominant item of diet throughout the winter; on others, the food must be at times 100 per cent browse. In fact, even in one area, as in Jackson Hole, Wyo., some individual elk may subsist entirely on sedges while in neighboring brushy canyons others may eat practically nothing but browse.

March.—Winter conditions in general are ameliorated in March. The snow begins to thaw, making available many new areas of forage hitherto snow-covered. The diet continues about the same, but there may be a little shifting of the herds. Antlers are shed by many bulls, the pelage is rough, and the elk begin to look ragged although their spirit is obviously renewed. From this time on foraging conditions improve until the new growth actually begins to appear in April and May, as described above.

PLANTS EATEN BY THE ROOSEVELT ELK

The Roosevelt elk *(Cervus canadensis roosevelti)*, which originally inhabited much of the rain forest of the Pacific coast, depends much more upon browse than do the other elk. Possibly this is due to necessity, especially in the winter months, as extensive grass ranges are not available. It has been found, for instance, that even in winter this elk feeds rather extensively on herbaceous plants when they are not covered by snow. The Rocky Mountain elk sometimes depends largely on browse for winter food in certain localities. Therefore, it is unwise to state categorically that one race of elk is more partial to browse than another.

Less time was spent in the study of the diet of the Roosevelt elk than in that of the Rocky Mountain elk, and the following data are based chiefly on investigations on the Olympic Peninsula, Wash., supplemented by more limited observations in the redwood forests of northwestern California and on Afognak Island, Alaska, where this elk has been introduced. When no other locality is mentioned, statements refer to the Olympic Peninsula. Unfortunately, there has been no opportunity to visit Vancouver Island to study this race of elk there also.

The writer has had access to the field reports of John E. Schwartz, who spent considerable time on the elk problem in the Olympics and whose findings supported the writer's conclusions and furnished added information.

Browse

Abies grandis. Grand Fir. This species was identified in one elk stomach collected by Schwartz.

Acer. Maple. Less than half a dozen browse species constitute the bulk of the winter forage for the Olympic herds. Among these vine maple *(Acer circinatum)* ranks high and is possibly the most important, for although it is perhaps slightly less palatable than the bigleaf maple *(A. macrophyllum)*, it is more available and persistent. *A. circinatum* is plentiful on the winter range, and although it is heavily utilized, the upper limbs are out of reach and the tree is not easily killed. At times, too, the high limbs are bent down by heavy snows, giving the elk added forage. However, this species, in common with other woody plants, may become dangerous forage if browsed too closely, so that splinters from sizable branches injure the mouth of the elk and thus give rise to necrotic stomatitis. Such mishaps may be a factor in some of the winter losses that have occurred on overstocked watersheds of the Olympics in recent years. The elk not only eat the twigs of this species in winter but in spring and summer feed on the leaves. One stomach, collected in March, contained 99 per cent vine maple mixed with some bigleaf maple.

Characteristic of the lowlands and river bottoms, the bigleaf maple, draped with moss and often carrying the fern *Polypodium vulgare*, is apparently more palatable to elk than vine maple and is probably near the head of the palatability list. Twigs are eaten

PLATE 24. *Above.* A bull elk in Yellowstone National Park is beating the limbs of a Douglasfir with his antlers, causing the broken twigs to fall on the snow. *Below.* Is feeding on the twigs he has knocked down. These limbs are out of reach of the antlerless female nearby. (Photographs by Adolph Murie.)

PLATE 25. Elk winter range among the aspens of the foothills, showing the pits dug in the snow to reach the grass underneath. Jackson Hole, Wyo., January 10, 1928. (U. S. Fish and Wildlife Service photograph.)

whenever they are in reach. When storms throw down limbs, the twigs are trimmed off at once and the bark stripped from the remainder. The fallen blossoms are eaten in the spring and the fallen leaves in autumn. The young growth is eaten so greedily that along the Hoh River, where elk are overabundant, no maple reproduction could be found.

Alnus. Alder. Red alder, *A. rubra*, is only moderately palatable to elk. They utilize it chiefly in winter, when they take the twigs and some bark, but also in spring, when they eat the leaves. Schwartz and Adolph Murie reported *A. sinuata* more palatable and utilized rather extensively in summer and early fall.

Amelanchier. Serviceberry. Adolph Murie noted a little of this shrub heavily utilized. Wherever available, it is doubtless as palatable to this elk as to the Rocky Mountain elk.

Baccharis pilularis. Kidneywort Baccharis. This shrub was found browsed by elk in northern California, but the degree of palatability is not known.

Ceanothus. Ceanothus. Schwartz found both *Ceanothus sanguineus* and *C. velutinus* to be quite acceptable forage in the Olympic Mountains. The former appeared to be the more palatable. Both are used in winter, and *C. velutinus* also in the spring. In the redwood forests of California *Ceanothus thyrsiflorus*, blue blossom, is an abundant shrub, and the leaves and twigs are a favorite food of the elk.

Cornus. Dogwood. Some flowering dogwood *(C. nuttalli)* was found browsed in May, but elk were not abundant on the lowlands where this tree was most common. Schwartz found bunchberry dogwood *(C. canadensis)* in three stomachs.

Gaultheria shallon. Salal. Salal, irregularly distributed but very abundant on some watersheds, is readily eaten by the elk; often shrubs are stripped of their leaves. It is probably less palatable than many other species but nevertheless is a useful source of forage. It contributed 3 per cent of the contents in each of two elk stomachs; 6 per cent, in another; and 60 per cent, in yet another.

Mahonia. Mahonia. Cascades mahonia *(Mahonia nervosus)* and Oregongrape *(M. aquifolium)* are both eaten by elk in fall, winter, and spring, just as *Mahonia* is eaten in the Rocky Mountain region.

In one stomach one of these forms supplied 5 per cent of the contents.

Malus pumila. Apple. Just as with the Rocky Mountain elk, apples and the twigs of these trees are very palatable to Roosevelt elk.

Menziesia ferruginea. Rusty Menziesia. This plant is eaten by elk, but its degree of palatability has not been determined.

Myrica. Waxmyrtle. Sweetgale *(Myrica gale)* was found browsed in only one locality, on Afognak Island, Alaska, and its status in elk diet is not otherwise known. Pacific waxmyrtle *(Myrica californica)* similarly was found browsed by elk in the redwood forests of northern California.

Oplopanax horridus (Echinopanax h.). American Devilsclub. Although this spiny plant must present a forbidding aspect to browsing animals, elk eat it with some relish, nipping off the tips of twigs or stripping off the bark. The leaves are eaten in spring and early summer.

Picea sitchensis. Sitka Spruce. This tree is browsed much less than some other conifers, although the twigs are eaten rather frequently and appear to be quite palatable. In one stomach Sitka spruce furnished 20 per cent of the contents. Twigs up to one-fourth-inch diameter at least were used.

Populus trichocarpa. Black Cottonwood. Schwartz found this tree quite palatable to the Roosevelt elk, though not extensively available. Fallen limbs were utilized by the elk.

Pseudotsuga taxifolia. Douglasfir. Douglasfir is very palatable to the Roosevelt elk, as it is to the Rocky Mountain elk, and is browsed when available. On the winter ranges studied in the Olympic Mountains comparatively little was within reach. The elk browsed the twigs of the limbs that fell from the trees. In one elk stomach Douglasfir contributed as much as 30 per cent of the contents and in another 12 per cent.

Rhamnus purshiana. Cascara Buckthorn. Adolph Murie found this browsed in the upper Hoh River valley.

Ribes. Currant. Currant bushes were browsed, but only limited observations were made. Schwartz found that *Ribes bracteosum* is the form most frequently utilized, often heavily browsed on summer range.

Ribes (Grossularia). Gooseberry. Observations on the use of gooseberry were very limited, but elk had browsed it on at least a few occasions.

Rosa. Rose. Not much rose was found on winter range in the Olympic Mountains of Washington, but it was eaten on several occasions. Rose was generally quite palatable. In the redwood district of California several varieties of domestic rose bushes had been extensively cropped on some abandoned home sites.

Rubus. Blackberry; Thimbleberry; Salmonberry. The leaves and fine twigs of trailing blackberry *(Rubus macropetalus)* are eaten freely, but on the Olympic Peninsula these plants are more common in the open logged-off, burned areas, where elk are not now plentiful. Wild blackberry vines are also eaten in the redwood district of California, where it is reported that the "evergreen blackberry" and the introduced "Siberian blackberry" are also browsed freely by elk.

Western thimbleberry *(R. parviflorus)* is eaten by elk but apparently is less attractive than salmonberry. Both the leaves and twigs of salmonberry *(R. spectabilis)* are eaten, and this plant ranks among the most palatable species. It is easily destroyed, especially by trampling, when elk become numerous and may be almost eliminated on an overstocked range. For this reason it cannot be rated a dependable resource for elk if the herds become large.

Salix. Willow. Willows are eagerly browsed and rank high among the forage plants. They were comparatively scarce on the Olympic elk winter ranges but where they occurred—on river bars—were utilized by elk. *Salix sitchensis* and *S. scouleriana*, particularly, were browsed by elk, the latter severely. Large willows were barked. Willows were also browsed by the elk on Afognak Island, Alaska, where leaves were eaten as early as September. Schwartz found the subalpine *S. commutata* browsed to some extent in summer in the Olympic Mountains.

Sambucus. Elder. In the Olympics, as elsewhere, elder is highly palatable to elk. Large fallen stems of *Sambucus callicarpa* were completely barked. A species of elder native to Afognak Island, Alaska, was also eaten extensively. In some areas in the Olympics almost the only elders that survived use by elk were those growing high up on stumps or on the high ends of logs.

Sorbus. Mountain Ash. Schwartz found two forms, *Sorbus dumosa* and *S. occidentalis*, very palatable. Adolph Murie found it so heavily browsed in places that it seemed likely to be eliminated from the flora.

Spirea. Spiraea. Adolph Murie found at least two species with moderate to high palatability in various places. It appears to be quite acceptable to elk.

Symphoricarpos. Snowberry. Schwartz found this shrub utilized but not very important.

Taxus brevifolia. Pacific Yew. Yew is reported to have been eaten by elk, but data on its utilization are limited. Adolph Murie found some of it heavily browsed on the Quinault River, suggesting high palatability where it is available.

Thuja plicata. Giant Arborvitae. This plant, locally called western red cedar, is very palatable. Although in a mature forest little of it is normally available to the elk by direct browsing, many limbs fall to the ground and the elk miss no opportunities to trim off the twigs.

Tsuga heterophylla. Western Hemlock. Hemlock ranks high in palatability. Elk browse the young trees, which are plentiful, and trim up many fallen limbs. Hemlock supplied 65 per cent of the contents in one stomach; 20 per cent, in two; 12 per cent, in one; and a trace in another. It seems likely that hemlock ranks below cedar and Douglasfir in palatability, but as hemlock is much more available it is probably the most important coniferous forage.

Vaccinium. Blueberry. Huckleberry, or more properly blueberry, is abundant in the coastal ranges of the Roosevelt elk and is one of the more important forage plants for elk. At least four species were found in this forage class: *Vaccinium parvifolium*, *V. ovalifolium*, *V. membranaceum*, and *V. ovatum*. The first two were the most abundant on the winter ranges studied in the Olympic Mountains and were most utilized by elk. On the Hoh River drainage these shrubs were so heavily browsed that some plants were dying. In one stomach *Vaccinium* made up 20 per cent of the contents. It is probable that in the Olympic Mountains, at least, these two species of blueberry, vine maple, and salmonberry are the most palatable browse species that are common enough to be a staple source of forage, though salmonberry tends to disappear rather readily under heavy use.

On some elk ranges studied in the redwood forests in Humboldt County, Calif., *V. ovatum* was more abundant than other species, but it is much less palatable to elk and is eaten only moderately. The elk on Afognak Island, Alaska, also browse the local *Vaccinium* in winter.

Viburnum pauciflorum. Mooseberry Viburnum. This plant is browsed by elk in winter on Afognak Island, Alaska.

Miscellaneous Herbs

The following list must be very incomplete because studies were concerned largely with winter conditions. There would, of course, be many species used in summer that are not mentioned here.

Arnica betonicaefolia. Betony Arnica. Fairly palatable in summer.

Bistorta (see *Polygonum*)

Caltha biflora. Twinflower Marigold. Quite palatable, though probably not sought in any great quantity.

Claytonia sibirica. Springbeauty. Springbeauty is palatable and is extensively grazed by elk in spring and early summer.

Dicentra formosa. Pacific Bleedingheart. Found much used by elk on one occasion.

Dryopteris dilatata. Mountain Woodfern. Elk ate several species of ferns, and specimens provisionally referred to this species were among them.

Epilobium angustifolium (Chamaenerion a.). Fireweed. Fireweed is extremely abundant on Afognak Island, Alaska, where elk were observed eating it and where it should furnish a large supply of forage.

Equisetum. Horsetail. Undetermined species of Equisetum were eaten by elk in the Olympic Mountains and on Afognak Island, Alaska.

Eriogonum. Eriogonum. A few plants were found eaten by elk.

Heracleum lanatum. Cowparsnip. Cowparsnip was relished by elk on Afognak Island, Alaska, as well as in the Olympic Mountains.

Leptarrhena amplexifolia. Leptarrhena. This plant has been eaten by elk, but there is no evidence of high palatability.

Leptaxis menziesii. Youth-on-age. Common in the Olympic flora, found well represented in one elk stomach. Schwartz found this commonly eaten by elk.

Ligusticum scoticum. Scotch Ligusticum. This plant was grazed by elk on Afognak Island, Alaska.

Linnaea borealis, var. *longiflora.* Longtube Twinflower. A few plants of this species were eaten by elk in the Olympic Mountains.

Lupinus. Lupine. Lupines were grazed by elk on Afognak Island, Alaska.

Lysichitum (Lysichiton) americanum. American Yellow Skunk-cabbage. This species was reported to be grazed extensively by elk in the redwood forests of northwest California.

Maianthemum dilatatum. Beadruby. This was identified in one elk stomach.

Mitella ovalis. Bishopscap. A small quantity of this appeared in one elk stomach.

Oxalis oregana. Oregon Oxalis. This plant is abundant and is said to be a favorite food of elk. It was found eaten to some extent, but in spots where elk were feeding much of it was left untouched.

Pedicularis. Pedicularis. Adolph Murie found a *Pedicularis,* apparently *racemosa* or closely related form, eaten with great relish, though it was not abundant.

Polygonum bistortoides. American Bistort. This is eaten by the Roosevelt elk, but the degree of palatability is uncertain; probably fair.

Polystichum munitum. Western Swordfern. This abundant swordfern is eaten by the elk, but it is not nearly so palatable as the elk fern *(Struthiopteris spicant).* It was sometimes found eaten down only about half length where elk fern had been almost entirely consumed.

Pteridium aquilinum var. *lanuginosum.* Bracken. This bracken, so abundant in the lowlands, occurs more sparingly farther up the stream valleys. As the tender, curled shoots emerge from the ground in early spring they are grazed eagerly by elk. The mature plants are grazed moderately in the fall.

Ranunculus. Buttercup. Elk are known to eat these plants in summer and fall, but the species taken and the degree of palatability were not determined.

Stellaria. Starwort. Identified in two stomachs.

Struthiopteris spicant. Elk Fern. This evergreen plant is important in the winter diet and is the elk's favorite fern. It has been grazed so extensively in some places that the stand seems to have been reduced.

Tiarella trifoliata. Trefoil Foamflower. This species is quite palatable and is eaten at various times of year when available.

Naleriana sitchensis. Sitka Nalerian. This is very palatable in the Olympic Mountains, though not abundant.

Veratrum. False Hellebore. This was found eaten by elk on Afognak Island, Alaska, as well as in the Olympic Mountains, where it was evidently quite palatable.

Xerophyllum tenax. Beargrass. The Roosevelt elk, as well as the Rocky Mountain elk, eat this plant in moderate degree.

Grasses and Grasslike Plants

Very little was learned about particular grasses used by the Roosevelt elk. In winter, when the snow is deep, the animals do not obtain much of this forage; but when the snow disappears, elk graze on green grasses and sedges under the vine maples, as well as on the mountain meadows. In spring this type of forage becomes important. Only the following species can be listed at this time.

Anthoxanthum odoratum. Sweet Vernalgrass. This is especially sought by the elk in low, wet places in the vine maple thickets or openings in the forest.

Calamagrostis. Reedgrass. An undetermined species was grazed by elk on Afognak Island, Alaska.

Carex. Sedge. Sedges, some of them being large coarse species, are commonly eaten by elk in the Olympic Mountains.

Juncus. Rush. Rushes were eaten by elk, but they appeared much less palatable than sedges or grasses.

Miscellaneous Foods

Usnea. Tree Lichens. This hanging lichen is common throughout the Olympic Mountains and is very palatable. Elk wintering on high ridges sometimes find quantities of it on limbs that are blown down by storms. In 1916, on the Elwha River, these lichens had been utilized so persistently by elk that a "high line"—below which the lichens had been nearly cleaned off the tree trunks—could easily be discerned in the forest.

TABLE 10. CHEMICAL COMPOSITION OF ELK FOOD FROM THE OLYMPIC MOUNTAINS, WASH.

By Otto C. McCreary, University of Wyoming

Plant	Moisture	Ash	CaO	P_2O_5	Crude fat ether extract	Crude fiber	Crude protein	Nitrogen free extract
	Per cent	Per cent	Per cent	Per cent	Per cent	Per cent	Per cent	Per cent
Polystichum munitum (western swordfern):								
Old leaves, April 7, 1934	9.37	4.26	0.64	0.27	1.03	39.75	9.41	36.18
New leaves, May 18, 1934	9.25	5.95	.34	.63	1.76	35.26	12.62	35.19
Gaultheria shallon								
Salal	8.84	4.89	1.59	.23	2.86	30.39	6.80	46.22
Pteridium aquilinum var.								
lanuginosum (bracken)	11.58	8.96	.15	1.56	2.17	20.28	29.84	27.17
Struthiopteris spicant								
Elk fern	10.70	12.48	.72	.33	.85	25.28	8.34	42.35
Acer macrophyllum								
(bigleaf maple)	9.18	5.32	1.40	.46	2.70	23.39	12.07	47.34
Circinatum (vine maple)	8.14	4.38	1.06	.71	3.04	17.97	17.55	48.92
Vaccinium parvifolium	6.84	4.98	1.39	.40	2.72	33.40	11.67	40.39
Alnus rubra (red alder)	8.57	4.65	1.13	.46	4.23	13.28	17.38	51.89
Tsuga heterophylla								
(western hemlock)	6.90	3.10	.81	.44	5.29	24.84	6.19	53.68
Old rotten log (western hemlock)	8.04	1.86	.35	.03	1.18	....	2.80	....
Rotten log, slimy and soft	6.69	.96	.30	.04	.59	....	1.58	....

Mushrooms. On Afognak Island, Alaska, elk were found eating unidentified mushrooms on September 2, 1936. Undoubtedly the Roosevelt elk is similar to other members of the deer family in its fondness for such vegetation.

Rotten wood. In the Olympic Mountains there appeared to be a scarcity of true mineral licks, but the elk had the habit of licking and eating the rotten wood of two types of logs. One type was firm; examples observed were hemlock. The elk preferred the under side, and once in an effort to get under a huge log, they dug small pits in the ground and got down on their knees to reach under the log and nibble the soggy wood. Sometimes they chose the rotten wood of an old stump. The other type, either maple, cottonwood, hemlock, or spruce, was smaller and so mushy that one could poke a finger into it. The wood was slimy and appeared impregnated with fungus mycelia. In this type, which was common, the elk bit out and swallowed small pieces. Some of these small pieces were identified in one stomach.

Why the elk eat rotten wood or what makes it tasty, is not known. In the second type, possibly the mycelia furnish an attractive taste. Table 10, showing the chemical composition of several food plants, as well as of samples of these rotten logs, does not indicate any notable quantities of mineral salts, for the ash contents are low.

PLANTS EATEN BY THE TULE ELK

Field studies of the tule elk *(Cervus nannodes)* were so limited that only the briefest mention of a few food plants can be given. One herd of these elk (the Buttonwillow herd) is now under fence at Tupman, Calif., and the other is in the Owens Valley. A short visit to the fenced herd produced very little information, for the elk were kept continuously on artificial feed. There was no semblance to the original range (now unoccupied by these animals), so that what the food habits of the race might have been in primitive times was not learned.

Cynodon dactylon. Bermudagrass. Although Bermudagrass was abundant in the elk pasture, the animals were not eating it, and the keeper stated that he had not seen them graze it. An unidentified grass that occurred in smaller quantity was being grazed.

Erodium cicutarium. Alfileria. The tule elk eagerly feed on alfileria when it is available.

Frankenia grandifolia var. *campestris.* This shrub was browsed by elk on the refuge.

Prosopis glandulosa. Honey Mesquite. It is generally understood that mesquite was one of the forage species for tule elk when they roamed their natural habitat.

Salix. Willow. There were willow trees *(Salix gooddingi vallicola)* as well as young shoots on the refuge, but there was no satisfactory evidence that the tule elk were eating them. Perhaps the artificial feeding kept them from doing so.

Urtica. Nettle. The keeper at the refuge reported that he had seen the elk eating "stinging nettles," which is not surprising as these plants are palatable to other animals.

PLANTS EATEN BY THE MERRIAM ELK

Any discussion of the food habits of the extinct Merriam elk *(Cervus canadensis merriami)* must, of course, be largely hypothetical. However, there are certain considerations that are indicative. Examination of areas formerly occupied by this race shows that it lived in a habitat not very different from that of the Rocky Mountain elk. In the southern part of the Guadalupe Mountains, in Texas, for instance, are areas that are suggestive of elk habitats farther north, with Douglasfir, yellow pine, oak, and even a slight trace of aspen.

The Rocky Mountain elk that were introduced into the Guadalupe of New Mexico and Texas have adapted themselves to the native vegetation, including mescal *(Agave)* and Texas madrone *(Arbutus texana)* among others. It is reasonable to believe that these animals select about the same plants as the extinct form did. The indigenous mule deer browses these same plants. In parts of Arizona, too, it seems very likely that the Merriam elk enjoyed certain cedars, cliffrose, rabbitbrush, and other species that are so eagerly sought today by the introduced Rocky Mountain elk.

The North American elk, as a whole, have very generalized food habits. Considering the habitat of the Merriam elk, one must conclude that it very likely had the same food habits as the present introduced Rocky Mountain elk.

FOOD COMPETITION ON THE ELK RANGE

There are two kinds of food competition on an elk range—competition with other game animals and competition with domestic stock.

The food habits of our big game are not sufficiently specialized to prevent competition among the species. Perhaps the elk, being the most omnivorous big-game animal, is the most likely to bring on harmful rivalry. On the winter ranges at the northern border of Yellowstone National Park, Wyo., there is serious competition of elk with mountain sheep on one hand and with antelope on the other, and the multiple use of those ranges has caused a serious shortage of winter food. This situation has been described in detail by Adolph Murie (1940), who also pointed out that years ago, in a well-meant attempt to provide more winter forage for elk, lands in Yellowstone Valley were seeded to grass. The elk utilized the grass, as expected, and were to that extent helped, but in going to and from these grassy fields they crossed the sage lands and fed to some extent on the sage. Sagebrush is the principal forage for antelope, and this particular area has been so depleted that the sage has nearly died out in some parts. Thus, help for elk was an inadvertent impact against already hard-pressed antelope. In the Jackson Hole, Wyo., region the elk compete with mule deer and with mountain sheep on their respective ranges, but there is not much direct competition between elk and moose, for, although the moose is a heavy browser and has a somewhat specialized diet, and the elk, too, is a browser and inordinately fond of willows (the mainstay of the moose), the moose uses as its winter range certain brushy areas in deep snow that the elk have not yet invaded to any great extent. In the Gros Ventre Basin, however, moose are coming in and are meeting the competition of elk.

Comparison was made (p. 188) between the specialized winter food habits of the deer and the generalized food habits of the elk, showing that deer will die on a winter diet on which elk may thrive. This fact comes into play in some situations where both animals are present in numbers on the same winter range. The elk can raise the "browse line" above the reach of deer and at the same time can paw for grass and thrive on it. The deer do not normally feed on dead grass, do not thrive on it, and are dependent on browse. Where sage or other low browse is present in quantity, the deer will hold their own, but some situations arise where this is not possible.

A striking instance of deer-elk competition has been recorded by Edward P. Cliff (1939), showing that in the Blue Mountains of Oregon, in an area where these species were together on winter range, both increased until the deer, having reached a high of 19,500, suddenly dropped to 8,600 in an adverse winter. The deer began to increase again. Meantime the elk population had continued on a steady upswing, without any setback.

In allowing for range room for big-game animals, it must be assumed that all the animals have nearly the same food habits and that the population of a mixed stand should not exceed the number that would be permissible for a single species. On no other basis can overstocking be avoided. For instance, if a given range is well stocked with antelope bitterbrush *(Purshia tridentata)*, a favorite deer food, it cannot be assumed that deer will eat that while the elk will eat grasses; elk, too, are fond of *Purshia*. Beavers cannot be planted in the vicinity of an aspen grove and then be expected to leave any of these trees for elk or deer. In a willow swamp there is direct competition between moose and beaver; and if elk also have access to the swamp, the competition is three-sided. The pressure of elk on a given range is all the more severe because the elk tend to congregate in large numbers.

For a number of years there has been a proposal to introduce elk into the interior of Alaska and the question of possible competition with the native Alaskan animals has been given much thought. Assuming that the introduced elk would thrive and become numerous, the consequences of having two species of herd animals (elk and caribou, both of which tend to become numerous) occupying the same range would have to be faced. Whether the elk would graze extensively on the lichens that are indispensable to the caribou, is still unknown. Elk will eat lichens, and as the elk in the Olympic Mountains in Washington are known to have stripped *Usnea* from trees over extensive areas, it is more than likely that the introduced elk would encroach on the supply of lichens on which the Alaskan caribou depend. The elk would certainly compete with the moose for willows; and although Alaska is well supplied with these shrubs, large herds of elk would undoubtedly browse them heavily. As the elk multiplied, many would be almost sure to invade the mountain sheep ranges, for elk naturally seek high country in summer and tend to utilize windblown slopes. Mountain sheep are in need of all the winter forage that their environment can produce and should not be put under the burden of competition with elk.

Then, too, elk harbor certain diseases and parasites, the introduction of which into interior Alaska might be disastrous to the native game. For example, it is not known whether the Alaskan game herds are afflicted with infectious abortion, but elk are. Moreover, the mere increase of animal numbers in a given area may bring on a disease hazard.

The result of the introduction of an exotic species cannot be foreseen, but the above are some of the factors that should be given much consideration before such an introduction is made.

As the elk and domestic stock consume practically the same plants, the competition between them is direct. There appears to be somewhat more similarity in the diets of elk and sheep than in those of elk and cattle, but in practice the food taken by one of these groups removes just that much forage suitable for the others.

Occasionally elk and cattle may occupy different parts of a range. For example, cattle are likely to prefer grassy bottomlands, whereas elk generally prefer hillsides. Again, some highlands not at present occupied by elk are grazed by cattle in summer; in winter these high ranges may be deeply covered with snow and thus be unavailable to elk, which move to lower elevations. However, such segregation is generally not clear-cut. Cattle do at times cover hillsides as well as bottomlands, thus utilizing parts of the elk winter range, and elk do use bottomlands also in winter if forage is present. It is sometimes thought that elk will not use timbered north slopes in winter, preferring the open south ones, and that, therefore cattle may as well take the forage on the north slopes in summer. This is only partly true; elk do prefer the open wind-blown slopes, but, having taken off the "cream" of the forage there in the early part of winter, they later enter the woods where the snow is deep and loose and paw for pine grass or any other available vegetation.

All things considered, it must be concluded that domestic stock and elk exert the same type of pressure on the range, which should therefore be stocked accordingly.

The question sometimes arises as to whether there is actual antipathy between a game species and domestic stock. This is very difficult to determine. On winter range, elk do not appear to mind the presence of horses or cattle, but in winter elk are under greater stress than at other seasons and become accustomed even to the presence of man. It may be a different matter on summer range, where forage is more available and there is greater freedom of movement. It is likely, however, that the presence of horses and cattle does not affect the elk materially unless the cattle are rather numerous. The reaction of elk to domestic sheep was not determined.

16

ELK HABITS

GENERAL HABITS

THE ELK has the same fundamental habits and temperament, senses, and reactions as other members of the deer family. In general, under natural conditions it is very alert but not particularly difficult to stalk. There is one reaction in which the elk is similar to other deer and perhaps to most ungulates. When it has had a glimpse of what may prove to be a foe, it turns its head quickly so as to face the intruder squarely and often stares for some time before going off in a panic. If the intruder remains motionless, the elk will sometimes stand a long time, puzzled, and will perhaps bark repeatedly. The females are generally more alert and suspicious than the old males, although at times the latter too are fully awake to their surroundings.

Elk display curiosity over an unknown object and sometimes express their puzzlement by frequent barking. In the writer's experience, however, they have not shown so great a curiosity as that of the pronghorn, or of the loon, which permit themselves to be decoyed within close range by strange antics.

Some members of the deer family have a trait that could be expected to be their undoing under certain circumstances. During a study of the Alaskan caribou the writer sometimes found individuals sleeping soundly. One mild September day a drowsy bull was quietly resting, his head drooping lower and lower until an antler rested on the ground, and he was approached quite closely. Once a sleeping female caribou was actually captured and tied with a rope.

Cameron (1923: 67) reports several instances when the red deer of Scotland was found fast asleep.

Though this habit could be expected in the related North American wapiti, the writer has never observed it. Many times a band of these animals has been found resting on a high meadow, lying about in comfortable attitudes, and occasionally one stretched out its head and neck along the ground. Assuredly elk take their ease, but never has one been seen fast asleep, nor has any published record of it been found.

There have been many accounts of the stubbornness of pronghorn in keeping to an adopted course, sometimes crossing close in front of a horse or vehicle if the two lines of travel converge, and the Alaskan caribou sometimes crosses in front of a traveling dog team. This trait is well developed in elk, also, though perhaps it has been less often observed in them. Once the writer was traveling by saddle horse along a lane between two buck fences. A band of elk that had been feeding in a field nearby took alarm and fled in a direction almost parallel with that of the horse but designed to cross the fence some distance ahead. With a view to observing their actions the writer speeded up the horse. As it had been a cow pony at one time, it thoroughly enjoyed this and entered into the spirit of the chase. The elk kept stubbornly on their course, running almost parallel with the horse. Eventually some of them were overtaken and changed direction, but a number of them kept right on and actually jumped the fence and crossed the road only a short distance in front of the horse. Repeatedly, too, when the writer has been traveling on snowshoes or skis over the winter range and a band of elk has sensed that he was about to pass dangerously near them, the animals have started for the nearest hill on a course that would cross that of the intruder. Generally they are not greatly alarmed and travel at a moderate pace, so that often it is possible to approach quite close before the entire band has crossed. The nearer one approaches, the faster becomes their flight, though, and frequently the last few elk lose courage and turn back.

THE HERDING HABIT

The elk, like the caribou, is essentially a herd animal, and this fact conditions many of its responses, but the extent of herding varies greatly. In the spring the animals are scattered and the cows are likely to be alone or in groups of two or three when the calves are born; but as soon as the calves are able to travel readily, the animals more or less band together. In fact, as calving extends throughout a month or more, small bands may be found at any time during that period. As summer advances, the elk remain banded and groups of 20, 30, or more are common. Occasionally a herd numbers 300 or 400 elk. These are generally cows, calves, and some of the younger bulls (Plate 13). At this time the older bulls are generally by themselves, singly or in smaller bands, three to six being found together frequently. During the rutting season, of course, the bulls and cows and calves are together, but the larger bands are broken up because in the face of constant competition one bull can hold together only a moderate number of animals. At this time, too, small bands of spike bulls are occasionally together, these having been driven off by the old bulls. Often these outcast spikes linger in the vicinity of their herd, and as the season advances they are again tolerated in the herd. During migration the elk are again grouped in larger herds and may remain so during the winter, depending on the character of the winter range. At any time, though, summer or winter, there may be single animals or small groups, as well as larger bands on the range.

Like other herd species, the elk are subject to panic and wild stampedes. Such species take their cues from neighboring animals in the band and run off at a sign of danger that has been perceived by only one or a few. The stampeding habit is particularly noticeable on the National Elk Refuge in Wyoming, where the elk often gather in large bands. When a few animals take flight at an approaching person and run off, neighboring bands perceive the action, and group after group joins in until a whole field full of elk are in wild flight, giving the impression that they will run out of sight. Suddenly a few of them stop, and the rest progressively come to a standstill. Rarely do they flee for any great distance.

PLAY

Elk are often playful. This is especially noticeable among the young calves in spring, but it occurs also among older animals. Sometimes in March or April, when the snow is melting and people speak of the "feel of spring" in the air, elk have been observed running and kicking up their heels, not only some that are just under a year in age but also some that are a little older.

Again, in summer, when a band of cows and calves and young bulls find a shallow pond, most of the animals go into the water and obviously enjoy it (Plate 26). The calves, and some of the older animals as well, run through the shallows near shore, kicking and splashing. This bathing is distinct from the habit of wallowing, which takes place in sloppy mud and appears to be practiced only by rutting bulls for the cooling effect or the quieting of the rutting fever. Bathing may be for the purpose of cooling off as an escape from warm weather, but it is done in the spirit of play.

On June 23, 1944, the writer observed characteristic behavior in a herd of 182 cow and calf elk near Pelican Creek in Yellowstone National Park.

The following is quoted from the field notes:

"The sun was low, and the elk obviously felt the exhilaration in the air that comes with a lowering sun in spring. They milled around, ran off in sudden stampedes, all pouring across the meadow into the woods, then suddenly reversing and all coming back again. Some of the cows would run in crazy fashion, shaking their heads and leaping in a zig-zag manner, just feeling good. Occasionally two cows would leap around each other or chase each other in mock hostility. Some of them came on a pool of water. They jumped in the water, pranced around, or pawed the water vigorously with a front hoof."

Seton (1929: 42) described a circular, dancelike performance of elk that would appear to be a more organized type of play, but it was not observed in the present study.

REACTION TO CLIMATIC CONDITIONS

Elk are influenced to some degree by weather. Naturally, when heavy snow covers the summer range the animals migrate to lower levels, but aside from this obvious necessity they respond to other weather conditions. Extreme cold apparently does not greatly affect them unless they are sick. Healthy animals do not appear to mind the cold and do not seek shelter from it, but cold weather probably is a potent factor in the death of scabby individuals. A scabby cow that had lost more than half of her protective pelage was observed shivering miserably in a cold wind. She lay down and curled up as tightly as possible for warmth but did not survive (Plate 18).

Storms may drive elk to shelter. Although many of them feed in the open during storms, a hard wind is disagreeable to elk and often causes them to seek the shelter of trees to avoid it.

Elk are quite susceptible to heat. Even early in spring, late in March or in April, when the snow is disappearing they often bed down on snowdrifts, and they seek the highlands early, traveling over snow, as described on page 62, to reach high ranges, even though the valley below is green. There are exceptions to this, as a few bands will linger on the winter range. In summer, even at high altitudes, the elk often seek the shade of trees (but not universally) or lie in wet meadows, where they chew their cuds and fight flies.

Mention has been made of the playfulness of elk in early spring, apparently in response to the subtle stimulus of the mild humidity of the season.

An even more subtle stimulus has been noted during the rut in the fall. Although bugling may be heard at any time of day or night, it is more frequent in early morning and in the evening and is often particularly noticeable on bright moonlight nights. Man himself is subject to these influences that do not necessarily involve his intelligence but perhaps his more elemental responses, and when he speaks of the "freshness" of early morning, the "pleasantness" of evening, or the "particular appeal" of moonlight, he is probably moved primarily by atmospheric stimuli, the same as other animals. At any rate, the elk are stimulated by climatic conditions and establish their daily rhythm of activity accordingly.

In summer and fall elk feed in the morning, gradually ceasing toward 10 or 11 o'clock; spend the middle of the day in resting; and become active again about 4 o'clock when the shadows lengthen, feeding from then until dark. They are relatively inactive during the middle of the night. Of course, there are exceptions, and in any given band of elk a few may be seen stirring about at any time of the day.

In winter, if food is scarce and hard to obtain, the daily foraging continues over a longer period, but under normal conditions the same general rhythm of activity prevails.

REACTION TO MAN

In the presence of man elk are timid. As a rule, they do not show the truculence that sometimes is encountered in the moose. Even when a young elk calf is handled and squeals with fright, the mother often comes running up fairly close but does not attack.

A few isolated instances of a bull elk attacking a man have been reported. Theodore Roosevelt (1902: 143) recorded two such cases, one being a wounded animal. There have been reports also that male elk and deer have attacked women particularly. Domesticated elk are more prone to attack than wild ones. One bull that was being raised on the National Elk Refuge at Jackson Wyo., would threaten with his antlers when a tidbit was withheld from him, and when at 4½ years of age he showed signs of becoming dangerous, he was shot.

Seton (1929: 47) reports a remarkable instance of elk pugnacity. It appears that John Legg, a Wyoming rancher, was riding behind a pack horse along a narrow mountain trail. In a particularly dangerous spot a bull elk charged the pack horse and hurled it over the cliff to its death below. The elk then charged the rider, who managed to kill it with his revolver.

Caton (1877) reports that two men were treed by elk and that another was killed by a truculent park elk. Elk confined in an enclosure are likely to become pugnacious. Even freshly caught elk, of either sex, when being handled in a corral after prolonged driving and urging, are likely to become very truculent, stubborn, and dangerous. This happens, however, only when they are confined in a limited space, such as a small holding pen at the entrance to a chute. In a larger enclosure freshly caught elk will dash wildly about in great panic and are likely to injure themselves in blind leaps against the fence.

When an elk is so sick that it is unable to run away, it first attempts to hide by lying with head and neck extended on the ground, exactly in the attitude of a newborn calf. Even an old elk cow will do this. When it knows it is discovered, however, and is approached closely, the animal may attack if it is able to get to its feet. Sick yearling elk, still able to move a little, have often been seen charging an intruder. In any event, when closely approached, an elk thus found at a disadvantage will grind its teeth in a display of hostility.

Grinding of the teeth seems to be an offensive or warning gesture made in an attempt to intimidate the enemy. The sound has sometimes been attributed to the action of the canine teeth, but obviously this is impossible, for the canines have no opposing teeth and could rub against the tongue or inner surface of the lip. The lower jaw is moved from side to side, the lower lip is slid up a little first on one side then on the other, while the grinding sound occurs. Apparently the cheek teeth on one side are rubbed together, then those on the other side. This is repeated over and over while the intruder is present, but if no overt motions are made the grinding stops after a few moments.

↑ 17 ↑

THE ELK POPULATION

IN PLANT ECOLOGY an association of plants is often treated not only as individuals but also as a "population" unit having characteristics of its own. Similarly, it is advantageous for some purposes to consider an animal population as a unit having group behavior and characteristics. Especially it is important to treat certain statistics from the standpoint of the group unit. The elk lend themselves admirably to such treatment because they are a herd species.

THE HERD HABIT

An important consideration in the study of the elk population is the fact of herding. Some animal species are solitary; others are gregarious. In some cases, as among the social insects, the gregarious habit permits cooperation and a division of labor among a large number of individuals; but among herbivores, such as elk, it may be permissible to reason that the gregarious instinct has developed simply as a result of the particular ecological niche occupied by the species.

The musk ox benefits directly by herd organization in that the group can effectively protect its young from carnivores and has specialized habits for that purpose. Among the elk such mutual protection is not so obvious, although as soon as the young calf is able to travel with its elders it has the protection of their presence. A herd of animals benefits also by the combined watchfulness of many individuals, and more information of a fundamental sort would possibly disclose that this feature is of great importance and may have contributed toward the organization of the herd.

There are other circumstances that have a bearing and may be important. Examination of the food habits of various species of ungulates reveals a tendency toward gregariousness among the grass eaters, but it is necessary to examine some of the particulars.

Horses depend chiefly on grass, and they tend to run in herds. The zebra and numerous other African ungulates that run in large herds are generally assumed to depend on grass ranges also. Mountain sheep, where abundant, assemble in bands, sometimes several hundred together, but this animal has chosen a specialized habitat that tends to restrict its general distribution. The bison, another grass eater, is highly gregarious. The elk, too, depends largely upon grass when it is available.

On the other hand, the moose is a solitary animal, and although in summer it feeds on grass and other herbaceous vegetation, as well as on browse, in winter, the critical period, it depends on browse. It is highly specialized in its food habits and has become specialized physically as an adaptation. If the moose were to live in large herds and still retain its present food habits, it would rapidly ruin its forage supply to the injury of its population. In the critical winter period, this animal depends on willows and other "high" browse species and so far as known does not paw for its food. Once overbrowsed, such food plants become unproductive, and the supply is not replenished rapidly. Although willows are very abundant locally and withstand heavy use better than many other browse species, too heavy and prolonged use will eventually eliminate them. Other species used by moose, such as aspen, birch and conifers, replenish the forage supply comparatively slowly. This type of forage persists better when utilized by a scattered animal population.

The elk, too, are browsers as well as grazers, but they have the least specialized food habits of all the deer and are extremely adaptable. They can adjust themselves to a 100 per cent browse diet in winter when necessary, at least for a time; yet they also do well on grass ranges and can roam over the plains (in the absence of man). Although large herds may consume all the forage on a winter range, the grasses quickly recover. It is true that lasting injury may be done to browse on winter range in the mountains, but in primitive times Nature promptly reduced elk populations in such areas to carrying capacity, the survivors being those that used the more favorable, or grassy, areas. Thus the gregarious tendency was permitted to develop.

The antelope and deer are in a somewhat intermediate position. The antelope has the herd instinct, yet recent information indicates that at least in winter it does not use grass extensively in some parts of its range but feeds largely on herbaceous plants and low browse species. The food plants taken, however, are of a kind that will repeatedly furnish forage and still reproduce rather quickly; they are not to be compared with trees and tall shrubs in snow country, which so readily become unproductive. In migration and in winter the mule deer, and to some extent the white-tailed deer, gather in moderate herds; but at other times these animals are essentially solitary. The herd organization of the elk is the only one that is operative the year round. Although the deer feed on green grass at certain seasons, they are more dependent on browse than are the elk, especially in the critical winter period. Normally the mule deer will not eat dried grass in winter but subsists on browse. It thus approaches the moose in tastes but has the advantage over the moose in that it migrates to areas where it may find low browse species, such as the small rabbit-brushes, bitterbrush, and sagebrush.

It would appear that the gregarious habit of the elk is the result of the evolutionary trend taken by the race in its various "habits," keeping it generalized and highly adaptable in food habits and permitting it to exist in large numbers.

SIGNIFICANCE OF THE BIRTH RATE

The birth rates of animal species are adjusted to the hazards in the life of each. The physiological mechanism of such adjustment is of no concern here, but it is obvious that certain species that are preyed upon by numerous birds and mammals produce large litters several times a year. Thus rabbits and field mice are key species in the lives of certain raptors and carnivores. On the other hand, certain animals that are not regularly preyed upon by others have a low birth rate. The elephant, for instance, has an extremely low rate of increase.

Deer frequently have twins, and under favorable circumstance deer populations build up very rapidly. On the other hand, deer are the favorite prey of certain carnivores, among them the mountain lion. The elk, being larger than the deer, resists some types of predation more successfully and apparently finds it necessary to produce only one young a year for continuance of race, as twin elk are exceedingly rare.

It might be objected that the moose is a still larger animal than the elk yet frequently produces twins. What makes the moose more vulnerable than the elk, thus necessitating a higher rate of increase? Several factors suggest themselves.

For one thing, the moose is comparatively solitary, whereas the elk is social in habit; so that per unit of occupied territory, losses among moose are proportionately more serious than among the more numerous elk. Moose have one advantage in being more solitary: they are less apt to overutilize their range, or at least to do so as readily as elk, and so have fewer losses from that cause. Isle Royale, Mich., furnishes a striking example of overutilization by moose, but the animals there were on an island with no chance for dispersal and with no effective carnivores present. Overbrowsing has also occurred on Kenai Peninsula, Alaska.

Moose have a disadvantage in the greater hazards of their deep-snow ranges, with occasional crusting, which consequently increases their vulnerability to predatory species, including man. In the far north, when snow depth and consistency are favorable. Indians readily run down these animals. An Eskimo companion of the writer in Alaska once readily approached a moose at close range and killed it with a .22 special rifle, and he remarked that his people commonly used such guns for moose. Deep snows, too, often cover the willow growth in valleys, such as the Upper Yellowstone in Wyoming, and in such winters the moose move onto the mountain slopes and feed on conifers. In many other areas part of the low browse is snow-covered, thus temporarily reducing the forage supply. The elk have their difficulties on winter range, of course, but in the writer's opinion elk would normally seek areas where the climatic hazards would be somewhat less frequent and less severe; elk can do this as they are most tolerant of diversified habitats.

The elk is a robust animal; although it is subject to numerous diseases and parasitic infestations, there have been some indications that elk resistance to them is somewhat greater than that of the moose. No instances have been found in which it could be determined that elk succumbed directly to the effect of ticks, but there have been a few instances in which the ticks apparently caused the death of moose.

Another possible factor is that the young elk calf is apparently more precocious than that of the moose, although precise data on calf predation are lacking.

In considering this question, only the circumstances that prevailed through the many centuries prior to the coming of the white man can be given weight, for it was during that time that the reproduction rates were established. As information on the ecology of the animals mentioned here is very imperfect, this discussion is necessarily somewhat theoretical; yet the data available seem to show that the average of numerous ecological influences accounts for the birth rates of the deer, moose, and elk.

The rate of increase of the elk of Jackson Hole, Wyo., has been discussed (pp. 138 to 143). The rate for other elk herds is not known, but there is a general tendency to overpopulation among elk throughout the country.

SEX RATIOS

Numerous observations had indicated pretty definitely that the male and female offspring of elk are about equal in numbers, and the examination of numerous fetuses at random over a period of years and the sexing of elk calves that died during their first winter, further indicated that this was true.

In the winter of 1935-36 the Wyoming game authorities slaughtered 541 elk in a herd-reduction program in Jackson Hole. At least 312 were cows of breeding age. Upon autopsy all but 8 of these were found to be pregnant, and 1 carried a dead, disintegrating fetus; of 283 fetuses of which the sex was recorded, 144 were male and 139 female.

During a similar reduction program in Yellowstone National Park in January 1943, the writer noted 141 elk fetuses. Of these, the sex of 11 was not recorded, 68 were males, and 62 were females.

On the other hand, Schwartz reported that of 94 calves examined in the Olympic Mountains, from the fetal stage to about 10 months of age, 61 per cent were females and 39 per cent males.

Rosenhaupt (1939: 163) stated that among humans male births exceed female in the ratio of 106 to 100 and that apparently after a war this male preponderance is even greater. He discussed a number of causative factors. He reported, though, that the mortality among boy infants far exceeds that among girl infants and that stillbirths among the males exceed the females by 25 to 30 per cent. Such "weakening" influences as sexual exhaustion of the male, exhaustion through age, and underfeeding are thought to be conducive to male births. He quoted (p. 164) Ploss as remarking: "The male surplus rises with the prices of food."

Rosenhaupt (p. 164) quoted from published accounts that with the advancing age of female horses and sheep "there is a corresponding increase in the surplus of male offspring" and that a similar result is observed in cattle "when the parent bulls have become exhausted by excessive service" whereas "in herds containing a large number of bulls, which are consequently less overworked, there is said to be a majority of cow calves"; also that "mothers of ewe lambs show on an average a heavier weight than those which give birth to male lambs," and that "Landois, experimenting with many thousands of larvae of the *Vanessa urticae* (small tortoise-shell butterfly) succeeded in breeding males or females at will, according as he fed them well or not."

Do such factors influence big-game species, including elk, on their natural range? In the first place, it should be noted that statistics on elk cannot be so complete as those for the human race. Where there are only a few hundred notations on elk fetuses, for mankind there are many thousands.

Thus for man it is possible to state a sex ratio as precisely as "106 to 100," whereas for elk it can be given merely as "about 50-50." The statistics for elk given above actually show a small preponderance of males, but the sampling is too small to rely on.

It should be noted, moreover, that the factors mentioned, such as exhaustion and undernourishment, apparently are not absolute or all-powerful but are merely influences that shade the results one way or another; perhaps chance and some unknown causes are even more important.

From what is known about elk it would not appear that these causes are usually operative in producing surplus males. During the summer the elk are well fed. This is probably generally true, for if the summer range is not sufficient, the winter range would ordinarily be worse and surplus animals would be eliminated. In August the animals are in good health, fat, and in good condition to enter the rut. Thus at the breeding period the animals are not lacking in nourishment. The poor condition during the winter, which appears to be the natural thing, does not, of course, affect the sex of the embryos then being carried. Although the bulls become run down during the rut, this is probably not from excessive breeding but rather from great activity and neglect of eating. Whether this would be a "weakening" effect on the sex ratio is hard to say, but it would seem to be balanced by the calm attitude and well-fed condition of the female. Probably some very old cows breed, and according to the factors discussed above, these should produce male offspring. In the group of 312 autopsied cows mentioned above, 9 were aged: of these, 3 contained male and 4 female fetuses, 1 was barren, and 1 had a dead fetus. This does not indicate a male preponderance, but of course the sampling is very small.

Possibly the factors described by Rosenhaupt do operate among elk, but there has been no opportunity yet to measure their effect. It appears, at any rate, that all factors must strike a balance under natural conditions, for the sex ratio at birth among elk does appear to be about 50-50.

Differential sex survival appears to be another matter. Many have remarked that elk bulls are much less abundant than cows, and both classified counts and general observations seem to bear out this opinion

In Jackson Hole, Wyo., counts of elk have been made over a period of years, by airplane back in the hills and by team and sleigh on the feed grounds. On the feed grounds the count has often been partially by sex and age, although under the conditions of such a count it is impracticable to distinguish the sex of animals less than a year old. In tallies of winter losses, however, as well as in some other random samplings, the sexes have proved to be present in about equal numbers up to the age of 1 year. Animals in their second winter are not abundant, as there are usually heavy losses during the first winter. After that elimination these coming 2-year-olds seem to be more resistant to disease and other destructive agencies, so that among the "winter kills" no large series of animals of this age class has been available for examination. The limited figures available indicate no significant difference in numbers between the males and females.

The classified counts on the feed grounds of the National Elk Refuge at Jackson, Wyo., shown in Table 11 are instructive in regard to the sex ratio. These counts show that the ratio for 7 years is 35,388 cows to 8,767 bulls, excluding calves in their first winter, or a ratio of almost 4 to 1.

TABLE 11. SEX DISTRIBUTION OF ADULT ELK ON FEED GROUNDS, IN WYOMING

Sex	1928	1931	1932	1933	1935	1936	1938	1941	Total
National Elk Refuge, Jackson Hole									
Females	4,835	2,425	4,305	6,975	6,626	....	4,346	5,876	35,388
Males ..	1,161	172	1,354	2,149	1,566	....	653	1,712	8,767
On State Feed Grounds, South Park									
Females	348	....	211	1,595	484	411	125	....	3,174
Males ..	322	....	375	1,253	605	225	216	....	1,996
On State Feed Grounds, Horse Creek									
Females	....	....	10	85	63	245	81	....	484
Males ..	....	....	122	502	332	174	183	....	1,313

Is this sex ratio universal among adult elk? It is hard to say. Great variations are known locally. Most of the elk included in the above computations were on the Elk Refuge, where the difference between numbers of males and females is always great. Yet on some smaller feed grounds maintained by the State of Wyoming in the southern part of Jackson Hole the relative numbers of males are often greater, as shown in Table 11. On Horse Creek the males are invariably in the majority, but the total numbers are small and do not materially affect a general great preponderance of females. If the figures on all feed grounds are included, the result is still over 3 to 1 preponderance of cows. To offset this is the fact that in the winter of 1943-44, among 354 elk being fed at Black Rock Ranger Station, on Teton National Forest, there were 279 cows and 17 bulls.

The female of the third winter is not satisfactorily distinguished in the field, but the males of this age are recognizable. Although there are no precise means, therefore, of determining the sex ratio, bulls with their second antlers are relatively numerous in a given herd and create the impression that there has not been a disproportionate loss of males up to this stage. Nevertheless, there must be a greater mortality of males at some time to account for the relative scarcity of mature bulls. As there is no evidence that females predominate among elk up to the age of 2 years, or perhaps a little older, the unequal sex proportion therefore characterizes the breeding age class.

It has been observed that elk wintering on high, wind-swept areas are chiefly males. Here again, however, only a relatively small proportion of the herd is represented and it is doubtful that inclusion of these isolated groups would greatly change the general female preponderance of about 4 to 1 in the Jackson Hole area.

Statistics show that the hunter kill in modern times is nonselective, and that there is no preference for bulls on a large scale to account for a decrease in that sex class. The law permits the shooting of either sex or any age. In the 1932 hunting season in Jackson Hole, Wyo., the following animals were killed: adults, 777 males and 770 females; calves, 59 males and 62 females. The statistics for 1943, the latest available, show that of 4,742 elk that were classified, 1,395 were males, 2,384 were females, and 9.63 were calves. The calves were not classified as to sex.

The Protection Division of Yellowstone National Park compiled information on classified hunter kill in the area just north of the Park that shows, for six seasons, the following classification summary of a total of 18,836 elk; bulls, 6,913 (32.9 per cent); cows, 9,561 (50.8 per cent); and calves, 3,082 (16.3 per cent).

All hunting data available are at least neutral and by no means indicate a preponderant kill of bulls.

In the Olympic Mountains, Wash., also, females predominated among the elk actually seen. The figures are not entirely satisfactory, however, because in this area males are more scattered than females, occur more often at higher elevations, and are generally more difficult to find except at the time of the rut—facts which undoubtedly affect impressions as to the ratio.

On the whole, it appears to be fairly well established that among mature elk females greatly exceed males in numbers, but the cause is hard to find. There are some pertinent circumstances connected with the annual winter kill that may well be reviewed. Among the animals dying of necrotic stomatitis females are sometimes more common than males, but as they are more abundant, nearly 4 to 1, it is reasonable to suppose that they would preponderate in the losses from any nonselective agent.

There is another annual drain on the elk population, however, that is selective with reference to sex. Scabies, which is very common among some elk herds, is more prevalent among adult males than among adult females. In many winters the ailment is almost confined to the older bulls and afflicts very few cows. As suggested on page 167, it may be that the older bulls that have been weakened by the stress of the rutting season fall prey to scabies more readily than any other class. Some of these bulls even succumb in the autumn in advance of winter hardships. Scabies, then, is a selective factor in male reduction that does something toward explaining the unequal sex ratio among mature elk.

It is understood in the human race that longevity is greater, on the average, in the female than in the male. Although the writer does not have an accurate compilation on the question, in the handling of several thousand dead elk that succumbed during a number of hard winters, he gained the general impression that individuals in extreme old age, with teeth worn to the gums, were more numerous among the females than among the males. Possibly (and at present it can be given only as a possibility) length of life may be enough greater in the female to produce a proportionately greater number surviving in the old-age class.

PLATE 26. Typical elk high summer range north of Jackson Hole, Wyo. *Above.* Elk trails appear in the foreground. *Below.* Elk "swimming hole," the type of pond in which elk run and splash in obvious spirit of frolic. (U. S. Fish and Wildlife Service photographs.)

and this, with the possibility of greater resistance to various lethal factors, may give the elk cow an important "edge" over the male in survival.

The preponderance of females of breeding age is normal for a polygamous species and the relative scarcity of males does not affect elk reproduction. There are still enough bulls present to cause considerable competition, and their ranks are recruited each autumn from the group reaching the required degree of sexual maturity—a group that has not previously suffered heavy losses. The breeding rate in the autumn is effective in keeping up the elk population, and the disproportionate sex ratio seems in no way detrimental.

FLUCTUATIONS

Fluctuations, often more or less periodic, are well known among the smaller species that have a rapid rate of reproduction. That they may take place among the larger species also, notably the deer family and mountain sheep, is very probable, and there is some evidence to indicate it; but they probably occur at longer intervals and may be irregular in character.

Population fluctuations in such comparatively long-lived animals as elk are very confusing and not readily subject to quantitative analysis. Care must be taken to distinguish depletion in numbers through unusual slaughter by man and to consider only variations created by environmental or other natural causes. The general lack of precision in historical records is also a handicap.

Striking changes in population levels, such as occurred among the Kaibab deer, are well known; but that particular instance was abnormal in that effective natural controls were removed and no killing by man was done until the deer population went out of bounds. Yet even in this and similar cases there are factors that may be significant also in natural fluctuations.

In the springs of 1934 and 1935 the writer traveled through interior British Columbia and learned that at that time deer, and especially moose, were unusually abundant. Information was obtained specifically in the Chilcotin district, Williams Lake, and Quesnel Lake areas, and north as far as Fort St. James, near Stuart Lake. In some sections, according to certain informants, there had been a great influx of moose since about 1910 or 1912. Indians had testified to the absence of moose previous to that and they were not very familiar with the animals when they first came into these areas. At the time of the writer's visit, however, the number of moose had increased enormously and it was reported that 40 had been seen at one time in a meadow near Alexis Creek and that the wife of a trader at One Eye Lake had seen "about 75" in one day's travel along the lake shore. The southern limit of this moose invasion was not determined. There were reports of some moose dying in winter (See also McCabe, 1928a and 1928b; and Brooks, 1926 and 1928).

In Alaska, some years ago, Indians told the writer moose and caribou had alternately occupied certain areas, especially about the upper Kuskokwim River. As one Indian put it: "Caribou come, moose go!"

For many years white-tailed deer were absent from the ranges of northern Minnesota, particularly in the area now established as the Superior National Forest. Now they are abundant there.

In the latter part of the nineteenth century moose were scarce in the Jackson Hole country in Wyoming, but they began to increase, even in the face of human occupation and have extended their range into adjacent parts of Idaho and even south in Utah (a few records). Mule deer were not plentiful in Jackson Hole in 1927 when the writer began studies on the elk, but except for a slight setback by disease in 1936-37, they have increased steadily and are still abundant.

The elk of North America have been on a numerical decline so persistently and so long, owing to hunting and progressive occupation of their ranges, that there has hardly been a chance to observe whether any natural fluctuation has occurred. The trend has been in one direction, toward extermination. There has been, of course, a great increase in elk locally in recent years, but that has been caused by deliberate reintroduction by man and has nothing to do with the principle of natural fluctuations. There can be no doubt, though, that elk in nature would exhibit the same population phenomena as other deer populations, and there is, indeed, good evidence that such is the case.

What are the causes of population changes? Several suggest

themselves—predators, variations in food supply, and diseases. From time to time some or all of these have been considered of importance in relation to local elk populations.

How potent, for example, are predators? There is no universal answer. On the Kaibab National Forest, Ariz., intensive reduction of the number of predators, especially mountain lions, has been considered an important factor in the phenomenal increase of the mule deer. In the Olypmic Mountains, Wash., where the cougar had been hunted with increasing intensity for many years, and where the wolf had been eliminated, the Roosevelt elk and the deer increased in such numbers that overpopulations resulted.

In recent years there has been an almost universal reduction in numbers of mountain sheep throughout the Western States, and it has been pretty well established in those areas where intensive studies were made that some factors other than predators were the principal cause.

On the other hand, the mountain sheep of the Tarryall Mountains of Colorado were down to a total of seven individuals some years ago, according to Forest Service records, but there has been remarkable increase, and now the herd has apparently reached full range capacity and is one of the most flourishing mountain sheep herds in the country.

Again, Adolph Murie (1944: 62) describes in detail a drastic reduction of the mountain sheep herds of the Mt. McKinley region of Alaska, caused by abnormally heavy snows in two hard winters and probably aggravated by wolf predation, as the wolf population had reached a high level at that time.

On Isle Royale, Mich., where there were no effective predators and no opportunity for the moose to spread into other areas, the moose so increased in numbers that by 1930 much of the forage supply was injured. Adolph Murie (1934: 41) predicted a "die-off" of these animals, and this occurred in the next few years.

Freuchen (1935: 128) recorded that in 1901 in the Thule District in Greenland: ". . . nearly all the caribou in the area died of hunger on account of the hardness of the snow. For 2 years no foxes were trapped there, as they had enough to do in consuming the carcasses, which lay about everywhere." Such cases have been observed among elk also. In the winter of 1937-38 abnormal weather in the Yellowstone National Park-Jackson Hole, Wyo., region produced heavy crust and ice, resulting in great elk losses. Again the record snowfall of the winter of 1942-43 caused another heavy elk reduction in this same region.

A great many similar instances could be enumerated, and examination of the circumstances seems to reveal familiar operative factors, including predation, disease, and malnutrition. It is difficult to assign to each factor its relative importance. It has been noted in some cases that reduction of effective predators has tended to increase the game herds unduly, but the degree of predator control that will produce this increase is not known. Cowan (1947: 172) concluded that "under certain circumstances, predators are powerless to prevent game irruptions."

In the Olympic Mountains, Wash., the cougar population has been greatly reduced only in recent years; yet elk losses from overpopulation began to occur on the Elwha River as early as 1925—at a time when cougars had not yet been unduly reduced in numbers—and became a regular, annual occurrence thereafter. Bears and bobcats, often stated to be a drain on the elk calf crop, are still numerous, yet the overabundance of elk has increased.

Evidence indicates that carnivores themselves fluctuate in numbers, and perhaps only a slight diminution of their population is sufficient to permit increase of the prey species.

Disease assuredly is a factor in fluctuations, but possibly it is only a secondary factor that takes effect when forage is insufficient or when the game population has become too crowded. At any rate, range condition—using that term in a comprehensive sense—appears to be the factor that is most potent in affecting the size of populations, including those of elk. This circumstance has been observed so often that it can hardly be doubted. The writer has seen winter ranges, once productive, depleted to such a degree that they will no longer carry even a normal population. There is good evidence that under natural conditions a depleted range, with the help of an unusually unfavorable winter and the secondary effects of disease, will eventually trim down the elk population to a remnant small enough to permit range recovery. In primitive times, when effective predators were more common, they may have had an important influence at such critical periods.

It appears logical that the processes discussed have operated to create the winter distribution of elk one often finds; that is, a few hardy animals remaining for the winter in favorable spots at high altitudes, the numbers increasing at lower elevations, and the main herds seeking abundant forage on the plains. Thus the facilities of each ecological niche, combined with the limiting forces and the strongly implanted migratory habit, tend to limit the winter elk population everywhere to carrying capacity.

NATURAL SELECTION AND BALANCE OF NATURE

These concepts have caused much discussion and disagreement and are discarded from practical consideration by many (McAtee, 1936 and 1937). One difficulty is the fact that actual proof of such subtle influences is hard to obtain; their operation, granting that they exist, must of necessity be very obscure; and they are confused with apparent contradictions.

Another difficulty is a misconception as to the nature of a so-called "balance of nature." The phrase calls to mind a fixed beam balance, with all forces in a state of equilibrium. Such a conception of a static balance cannot exist in a world of living things containing the elements of growth and decline, and evolution, which itself implies change.

The only reasonable concept would take into account the ebb and flow of animal life, together with similar changes in the plant world, all correlated with environmental influences and an intricate variety of changing relationships. Such a concept, taking into account fluctuations, becomes tenable, and indeed there is much evidence to support it. After all, population uncontrolled would soon fill the earth.

As related above, in recent decades observations have repeatedly been made of cervine populations temporarily released that have exceeded normal proportions and have thereby become more vulnerable to limiting influences and have promptly declined. On the other hand, it is reasonable to deduce that when a herd declines numerically below normal, the surviving individuals fare unusually well after recovering from their share of the misfortune. Perhaps it is harder for predators to find such survivors and their food would again become abundant. It has been observed repeatedly that in open winters the elk on a heavily used range become widely scattered and suffer extremely light losses. This is because they readily find good forage on areas that in the usual winters are protected by deeper snow.

The balance of nature, then, should be conceived of as a teetering one having fluctuations (or temporary setbacks) that sometimes on the surface may appear to be contradictions of the general theory. It is sometimes thought that only one main force, disease, for instance, is effective in regulating the population of a given species, particularly among mammals; yet it is the writer's conviction that very often other agents contribute important influences. This thought has been well expressed by McAtee (1928), who stated (p. 80): "Birds are one of the natural forces regulating the numbers of insects, and it takes all of them combined to do the job." His discussion pertained specifically to bird-insect relationships, but the thought may be applied to elk. Even though certain carnivores, if as abundant as in primitive times, might reduce elk populations locally, recent research indicates that such control would probably be temporary and would exist only in combination with other mortality factors.

Granted the fact of a natural balance, it would seem that the forces involved would affect natural selection. Again it is not necessary to consider here the mechanism of operation from the genetic standpoint; but it is important to know whether or not natural selection exists. In regard to elk, proof is difficult to find. One naturally thinks of the carnivore-prey relationships, and there are seldom opportunities for significant observations of these. There is the general fact of development of speed, which is always useful. It is generally thought that speed has been developed by deer, horses, and other animals as a survival factor. The elk has also developed or inherited a size that furnishes efficient protection against certain carnivores, such as coyotes and bobcats. In common with other deer, the elk can use its front hoofs for defense, has good hearing and sense of smell, and is possessed of remarkable alertness in the wild state.

Its newborn young instinctively lie in hiding. It is reasonable to believe that all such abilities have been acquired in response to need. It has often been remarked that the predatory animals put the deer family up on its toes.

How does selection operate? It seems impossible to see its operation. Our concept of natural selection probably has been confused by our insistence that it proceed smoothly and steadily in the desired direction without contradictory events. Extensive observation in the field does not bear out such a concept. There are plenty of incidents that appear to be detrimental to the species, as when a cougar strikes down a fine animal, robust, and in good condition. One cow elk may keep very efficient watch over her calf, but no coyote or bear may chance to come near, so that her watchfulness may be wasted. A "poor" mother may be able to raise her young by sheer chance. Furthermore, the offspring of the "good" mother may meet death later in some other way, so that its mother's watchfulness is transmitted no further in the genetic line.

On the other hand, it would be logical that predatory animals will usually take the individuals that are the most easily captured; that for the one "poor" mother that raised her calf, there would be several others that lost theirs; that for the one calf raised by the watchful mother but lost through other causes later, there would be others that survived through the mother's efficiency and later used those qualities themselves in a similar way. The average of the complex circumstances and operative ecological factors could be expected to influence the development of the species over a sufficient length of time.

This process is not comparable to artificial selection in some respects. Man arbitrarily sets a goal and culls the breeding organisms to that end, which may not be and often is not, to the interest of the species itself.

Whether or not winter losses among elk involve selection is an open question. Such losses largely involve elk in their first year. Where classified counts have been made, 66 to 70 per cent of winter losses have been the calves, and another portion the aged animals, with a relatively small percentage of the fully mature ones in the prime of life. If the death is due to starvation or malnutrition, the weaker ones might be expected to succumb first. In the case of disease caused by infection, such as necrotic stomatitis, the matter is not so clear.

Webster (1939) reported that experiments with mice indicated that resistance to disease varies greatly among individual animals; that the degree of resistance is inherited; and that survivors of an epidemic are not the "lucky" ones but almost entirely those that had an inherent resistance to the disease at the outset of the epidemic. He found that infections artificially applied to resistant populations did not result in disease or epidemics. He further remarked (p. 72) that it appears that resistance is not "so much manifest at the portal of entry or bloodstream as in the specific tissues to which the agent has a special predilection," citing encephalitis in mice.

Among elk many have been observed that apparently had been infected by necrotic stomatitis but had not succumbed to it. One elk cow had lost all the cheek teeth in one upper jaw, and the jaw had healed perfectly. The disease appears less selective among the calves that die. Some large robust calves die and small ones often survive. The accident of infection here plays a part, and the whole situation is too complex for any casual study.

In the present state of our knowledge it is extremely difficult to be dogmatic about natural selection, either for or against. We need much more knowledge about what actually takes place in the life of a species, as its environment impinges upon it. Meantime, it is extremely important that we keep an open mind on such an important, far-reaching subject.

18

ELK HUNTING

IN COMMON with other herbivores, the elk have always been pursued by man for food. Explorers of the new continent spoke in glowing terms of the hordes of game. Typical of the buoyant spirit of early travelers in new lands is the following from Ingersoll (1883: 139):

"It was a constant delight to ride across the green ridges at the foot of the mountains (in western Wyoming). They were dotted with gaudy flowers, growing in such profusion as to throw great patches of color upon the hillside, and a faint sweet smell of tender grasses and myriads of wild blossoms was wafted to us upon each breeze. Then we were entertained by the unceasing company of big game. Antelopes were always in sight, and we might easily have shot a score. Every now and then we would start small bands of elk, or an elk cow and her calf, out of the bushes of some little valley as we came over the hill, and these would make no haste to get out of harm's way, seeming not to understand that we were to be feared. Many others were seen at a distance, for the gulches leading up into the mountains were full of them, as also of deer; but we did not fire a shot, not needing the flesh!"

Later in the book (p. 199) Ingersoll stated that there was never a day during the trip that they did not have at least two varieties of venison in camp.

The pioneers shot for food, as a necessity, and thereby enjoyed a type of hunting that has practically vanished. Elk and other game were easy to approach and easy to obtain in parts of the country where they were still common. The enjoyment consisted not so much in outwitting an animal, though that too entered in, as in providing for camp and thus insuring the welfare of the company and fruition of their plans.

Killing urge of quite another sort ran riot in the latter half of the nineteenth century, when hide hunters took toll of the buffalo and elk to make money; at the close of the century, after the elk herds had dwindled, tusk hunters invaded the last strongholds of these animals, killing the bulls for their canine teeth alone, as there was a profitable market for them among members of the Elk's Lodge. For a time this traffic was persistent. In the Jackson Hole, Wyo., region some of these poachers devised a contrivance equipped with elk hoofs to fasten on the soles of their shoes, in order to deceive the game wardens. Eventually, official action of the Elk's Lodge to discourage the marketing of elk teeth and increased protection by wardens eliminated commercial tusk hunting. Even today, though, the elk canine is looked upon as a trophy of a sort and is extracted from any dead elk found in winter by the chance passer-by.

In the meantime sport shooting developed, and that kind of hunting, by many considered the highest type, is pursued today. The trophy hunter is generally one who organizes a pack train, with guide and helpers, and penetrates the wilderness parts of the elk range to establish a permanent camp. Generally a comparatively long period is spent in the mountain, and the sportsman is not in a hurry. He is in search of the finest head obtainable, and the killing done for this reason is a negligible factor in relation to elk conservation.

With the development of wilderness roads and the great increase in automobile travel, another type of elk hunting has grown and now exceeds sport hunting. It is often spoken of as "meat hunting." In this, although the hunters may spend some time obtaining their elk, the rule is to get their elk as quickly as possible and return home. Often they are not people of great means. Many of them spend a single day in the field and are satisfied with bull, cow, or calf, the first one that can be had. Indeed, for meat, a cow or calf is preferable.

The elk is usually not difficult to hunt. It is not nearly so elusive as the mule deer that often occupies the same territory. There are some places, however, where elk hunting is harder, not, perhaps, because the animals are much more wary but because they are scarcer and thus harder to find. The writer has on occasion found both caribou and elk elusive and unusually alert.

Early in the season, the time when sport hunters are generally in the field (speaking of Jackson Hole, Wyo.), the bull elk are bugling and there is little difficulty in finding an animal. The old bulls are not wary at this time and are rather easily called by imitating the bugling. Some of the Indians, it is said, formerly constructed a simple elk call for use at this season. On the other hand, the females are still wary; a hidden elk cow, whose presence is unsuspected, frequently gives the alarm, causing the little band, including the bull, to flee.

Some hunters and guides have the ability to whistle with a reedy, vibrant quality that makes a good imitation of the elk bugle when heard in the distance. On one occasion the accidental neighing of a saddle horse brought a rutting bull on the run, close up to the hunter.

Later in the fall, when the weather is cold and the elk are coming down from the high country and hence are more available, the so-called meat hunters take the field. Many of these, it should be remarked also desire a head for a trophy. At this time it is largely a matter of luck whether the hunters encounter a traveling band of elk. If they happen to, success then depends chiefly on marksmanship.

Many experienced woodsmen have remarked that "shooting elk is like shooting cattle," the same remark that the writer has heard applied to another herd animal, the caribou in Alaska. One experienced sportsman from the Olympic Mountains region of Washington expressed little enthusiasm for elk hunting but became eloquent in his accounts of deer hunting, which he considered much greater sport.

Why, then, do earlier accounts sometimes speak so highly of the sport of elk hunting? It must be remembered that in the days of Theodore Roosevelt and other well-known sportsmen, the hunting fields were not crowded, game and other animals were in many places still plentiful, and the hunter enjoyed the flavor of the wilderness. This has much to do with one's estimate of such matters.

Even today there are places still comparatively inaccessible, hence not crowded, where the hunter may in some measure hark back to the frontier atmosphere.

19

ELK MANAGEMENT

OBJECTIVES

"WILDLIFE MANAGEMENT" concerns the maintenance of the wildlife supply. There are many complexities, and the wildlife population impinges on many land uses and viewpoints. Therefore it is desirable to clearly define management objectives.

One obvious objective is to produce huntable game, such as that on national forests and public domain. This should be tempered by other wildlife "uses," such as observation and study by nonshooters, and considerations arising out of special circumstances. In national parks a game species is but a member of a fauna, though encouragement is given to hunting the surplus where such species might migrate outside the boundaries to winter range. Planning for range use by elk must have consideration for soil conservation, preservation of plant life, not only as forage but also as interesting flora and landscape feature in recreation areas. Greater stress is being placed on balancing a given game species with nongame species. Perhaps it is not even proper to speak of "elk management" by itself. It needs to be integrated with management of other species, for the elk is only one member of the fauna. Often mule deer, coast blacktails, white-tailed deer, mountain sheep, even pronghorns, not to mention smaller creatures, are fellow members.

Esthetics, exceedingly hard to manage under modern conditions, is recognized as an increasingly important part of management.

Not only does the sportsman crave some element of outdoor enjoyment associated with the killing of his elk, but he, as well as nonshooters, wants to see, photograph, or study the animal under authentic, natural conditions. Dude ranches and guiding groups particularly appreciate this demand for esthetics.

In view of these social and biological complexities, the only basis on which to formulate reasonable objectives for elk management is reliance on the principles of ecology and multiple use.

ELK CENSUS

As a first step in management the administrator desires to know how many animals are on the range. Yet an exact annual census is not necessary. Fluctuations in numbers are bound to result from the climatic, physiological, or other influences operative on a game herd. Experience has shown that it is not practicable to hold a game herd at a precise figure. The administrator, and if possible the public, should not rely too implicitly on an exact number. Range inspection, accumulation of data over a period of years, and experience, provide a basis for estimating the proper hunter take each year. Slight overshooting or undershooting need not cause concern. It may be corrected the following season. The average, and the trend, are the important considerations since the natural status of a wildlife population is a fluctuating one.

An occasional census is desirable, however, and some techniques for these have been worked out. Locating elk in winter is relatively easy since they are in herds. An elk census should be taken in late winter when distribution is more or less stabilized. If no other facilities are at hand, a number of observers on skis or snowshoes may count the elk at the same time over the range. An area with well-known, natural boundaries is assigned to each observer. This is laborious procedure and requires capable, active participants, experienced in locating game.

In the Rocky Mountain region the airplane has proved to be most effective in counting elk. A pilot experienced in game census work is an asset, and the same applies to the observers. Usually the winter distribution of elk is pretty well known to rangers or wardens, and the territory can be partitioned into convenient flying strips according to ridges and drainages. One, or preferably two, observers keep on the lookout for elk and keep a tally of those observed. Over a fairly large herd the pilot will circle until the count is completed. Trails in the snow are conspicuous and guide the eye to the location of the animals. It is an advantage to make the count at a time when elk are in willow bottoms or otherwise out in the open. They are difficult to find in timber. Veteran counters claim 90 per cent efficiency with this method. Caution is needed in flying over mountainous terrain. The terrain is in better view from a moderately high elevation.

On a hay feeding ground, such as at the National Elk Refuge in Wyoming, a very accurate census is possible. Starting from the same point, two loads of hay are distributed in a large circle, until the two teams meet again. This causes the herd to be lined out in a circle, feeding on the hay. Each haysled, with several observers on either side, starts back through the feeding elk in opposite directions until they again meet. The elk move out on either side and close in behind, as the haysleds pass. One observer on either side, with tally register in hand, counts every animal that passes his side. The animals pass rather rapidly at times, so usually it is found convenient to record every fifth animal, multiplying the total so obtained by five.

Another pair of observers counts the cows, another the calves, and still another the bulls. In this way a classified count is possible, which is helpful in determining calf crop, sex ratio, yearling survival, and so on.

In dense forests, such as in the Pacific Coast region, airplane census is probably out of the question. In such areas reliance must be placed on estimates by rangers and other fieldmen who spend considerable time in the elk habitat.

A word should be said about the classified summer count. On summer range a total census is not practicable, but certain ratios may be arrived at. Cow-calf ratios would be hard to obtain, due to the difficulty of distinguishing in the field the animals too young to bear calves. However, a fair estimate of the ratio of calves to the total herd may be obtained.

Such observations should begin after the calving season, preferably after July first. At every opportunity, when the observer is certain that all the individuals of a band of elk are in view, he should count all animals and then pick out the calves. Such counts, repeated often enough throughout the season to insure adequate sampling, produce an approximate ratio of calves to the herd. There will be some duplications of samples, but if the total number is large enough to be significant the ratio obtained can be very useful.

CARRYING CAPACITY

Elk on summer range have so much country available, move so freely, and are so widely scattered that it has proved next to impossible to determine area needed per head to sustain them.

It has become a truism that the winter range is the measure of the possible elk herd. Also, if sufficient winter range is available one can safely assume there will be adequate summer range.

The two types of range, of course, have diverse characteristics. Vegetation on summer range is growing and the impact of grazing is more severe at that season. On the other hand, green forage has more nutrient value for the elk.

In winter there is less direct injury to plant growth by grazing, though it is still necessary to have a percentage of the dead vegetation return to the soil.

In winter, too, there are climatic fluctuations, such as depth of snow and crusting, that may affect the availability of forage in any given year. The forage crop for any given winter may have been affected during its previous growing season by drought or excessive moisture. Elk may remain longer on winter range one season than in another, depending on how soon in the fall the heavy snows come, and on when the spring breakup occurs. Add to this the fact that in early spring elk tend to congregate on the first bare spots and there seek the new growth at a time when it can least withstand heavy use. These are some of the perplexities that enter into a carrying capacity formula.

An approach to this problem was made in two ways: By experimental feeding to determine food requirements; and by direct examination of a known area of winter range on which a known number of elk had wintered.

Upon inquiry, the average daily ration per head fed to elk in several zoological parks was found to be as follows:

National Zoological Park, Washington, D. C.—10 pounds mixed hay, 4 quarts oats and corn, 1 pint bran.

New York Zoological Park—7 pounds clover hay, 6 quarts plus, crushed oats, some vegetables (more grain in winter).

Philadelphia Zoological Garden—10-15 pounds mixed hay, 3 quarts cracked corn, crushed oats, bran.

These rations were for varying proportions of young and old, males and females, and are inapplicable to the wild because they include concentrated food that elk do not obtain on the range.

On the National Elk Refuge in Wyoming, where hay feeding has been conducted for many years, estimates of the average number of elk fed over a known period of time and the gross tonnage of hay supplied for each period indicate a daily hay consumption of 7 to 10 pounds per animal. In the winter of 1938-39, for example, when baled hay was fed, the ration appeared to be about 8 pounds per animal. This figure is not precise, for the number of elk was not constant throughout the period and the animals did some grazing; but it is a figure arrived at, for a mixed herd of both sexes and all ages, under actual operating conditions.

In the winter of 1940-41 an experiment was made on the National Elk Refuge, under controlled conditions, to determine the food requirements per day, per individual and per hundredweight, over a period of 43 days. In one pen were placed 29 elk calves (about 9 months of age) and in another pen 25 adults (7 bulls and 18 cows). The animals were weighed at the beginning of the experiment on February 5 and again when they were released on March 9. The hay given to them was weighed each day. At the same time four head of Hereford stock (three cows and one steer) were being fed hay from the same hay shed, by a neighboring rancher who cooperated by weighing his animals and feeding the weighed hay.

Omitting the details, the following results were obtained, expressed as average amounts consumed per day:

Elk calves consumed 7.8 pounds each, or 3.11 pounds per cwt.

Adult elk consumed 12.5 pounds each, or 2.5 pounds per cwt.

Elk, all ages, consumed 9.8 pounds each, or 2.7 pounds per cwt.

Adult cattle consumed 31 pounds each, or 3.6 pounds per cwt.

The figures for the adults in each case suggest animal unit

values for elk and cattle in the ratio of 12.5 to 31, or roughly 2½ elk to one head of domestic cattle. Comparison of elk of all ages, herd run, with adult cattle gives a ratio of practically 3 to 1.

During the experiment it was noted that although all animals were given all the hay they would eat, the cattle gained in weight, while the elk not only lost weight but progressively ate less. There had been a progressive rise in temperature, and those who fed the hay to the elk commented that elk eat less in warmer weather. Apparently warmer weather did not influence the cattle used in this experiment.

In 1937 Assistant Chief Ranger Maynard Barrows, of Yellowstone National Park, had weighed 139 elk paunches. Based on the dried contents, he obtained the following averages:

Bulls —average dry contents—13.753 lbs.
Cows —average dry contents— 9.498 lbs.
Calves —average dry contents— 7.512 lbs.

From these data it was concluded that for the period represented by the stomach contents, adult elk require about 11½ pounds and calves 8 or 9 months old about 7½ pounds, per day.

The data arrived at by these two totally different methods show remarkable agreement. It may be safe to say that for purposes of range stocking, three elk, herd run, are equivalent to one adult domestic cow.

An attempt was also made to appraise the elk carrying capacity of the National Elk Refuge and adjacent foothills. The area includes roughly 25,000 acres on the Elk Refuge proper and 3,840 acres in adjacent foothills. This total of about 28,840 acres was classified as follows:

3,000 acres under irrigation.
1,550 acres marsh land
240 acres willows and scattered timber
24,050 acres aspen groves, stony ground, miscellaneous

28,840 acres, total.

Calculations based on the 11,000 elk present, for a 6-month period, and the food requirements determined experimentally, revealed an impossibly low figure—2.6 acres per head.

A more practical approach was made by considering the fact

that 11,000 elk were wintering on the area and that midway in the winter feeding period the forage was gone and it was judged that feeding of hay had become necessary. This indicated twice as many elk as the forage produced by the area could support and that the safe carrying capacity would be 5,500 elk. If the hay harvested on this area the previous summer had been left standing it would probably have supported another 400 head.

We might roughly estimate that the 28,840 acres of miscellaneous lands here described could probably support about 6,000 elk for a 6-month winter period. It should be noted particularly that 3,000 acres of this are under irrigation and highly productive.

On ranges where irrigated meadows are not included carrying capacity would be much less, probably nearer 12 acres per head for the winter. On arid sagebrush ranges, where elk once wintered, the unit acreage would be much greater.

Whatever the degree of stocking that might be in practice, periodic checking of ranges is necessary, in order to modify practice to attain full range productivity.

Special problems arise. Even under fairly moderate range stocking with a herd animal such as the elk, certain highly palatable browse species may be injured or removed entirely. This applies especially to aspen groves, cottonwoods, and in some situations, to willows. Where willow patches are scarce, as in some mountain habitats in the Southwest, they may be overbrowsed even in summer.

It is especially difficult to apply a remedy in places were excessive numbers of elk in previous years have already reduced the browse stand to a point where reproduction is not succeeding. Recovery of browse in such instances would mean herd reduction to an impossibly low figure, or elimination of the elk for a period of years. It would appear that a hard decision is presented where the browse in question involves prominent features of the landscape—such as the picturesque cottonwoods along the upper Lamar River in Yellowstone Park or the scenic aspen groves of southern Jackson Hole. Reproduction is not taking place in these instances and with the death of the mature trees they will not be represented in those particular scenes.

In some instances it may be possible to induce migration to a new area for a time. This will be discussed further.

Finally, it is most important to consider the carrying capacity of a winter range not only from the standpoint of elk alone, but also all the other big-game species occupying that range, particularly deer and moose. Here enters the principle of multiple use, which has been discussed under the heading of Management Objectives. And for proposed reintroduction of elk into new territory, it is vital that winter range facilities be fully considered, not omitting possible conflict with agricultural needs.

COMPETITION FOR FOOD ON ELK RANGE

Two kinds of competition are to be considered on elk range: competition with livestock, and with other wildlife.

Competition with Livestock

Elk and domestic stock compete directly where both occupy the same range. It has been noted how nonselective elk are in their feeding habits. Grasses, other herbaceous plants, and browse are all greatly relished, and any of the three plant groups may receive particular attention where one of them is dominant in the vegetation. There is perhaps least competition with horses, since they are predominantly grass eaters.

Competition with cattle is somewhat greater than with horses, since cattle are more catholic in their tastes. In the case of cattle, there is a tendency to feed extensively on stream bottoms, in lush meadows, and to a less extent on higher and rougher slopes. Elk, on the other hand, tend to scatter more on the slopes, and reach the rougher country. This should not be taken to mean that cattle and elk are exclusive of each other on a range. But there is the tendency that produces a favorable distribution on a jointly occupied range and to a slight extent minimizes competition.

It is sometimes reasoned that in winter elk do not utilize the wooded, north slopes, but confine their foraging to the more open country, and hence in summer cattle could just as well take the forage in these forested areas.

This is only partially true. Elk do prefer the open, wind-blown slopes in winter. But, having taken the cream of the forage there in early winter, they later enter the woods where the snow is deep and loose and paw for pine-grass and any other available vegetation.

In the case of domestic sheep we do not find this favorable shade of selectivity. Sheep are as "omnivorous" in their selection of plant foods as elk and they get over all kinds of terrain. Here we must recognize direct competition, from every standpoint.

There is a seasonal consideration that may be important in some localities. A high mountain range utilized by livestock in summer may not be suitable for winter range for elk because of depth of snow, and it may not be of primary importance as elk summer range. In such cases competition is at a minimum.

There are other ranges on national forests which are not occupied to an important extent by elk in summer, but which are vital to them in winter.

The upper Gros Ventre River area in northwest Wyoming is an illustration of both of these circumstances. The high country is good cattle range, and competition there is slight, since winter snow depth precludes a favorable elk winter range. But the somewhat lower slopes and stream bottoms of the Gros Ventre basin itself are a strategic wintering ground for elk. This provides a problem, since it is necessary to drive the cattle across this elk winter range to and from the high cattle range beyond. This provides an administration problem that requires understanding and cooperation of interested groups of people.

It should be stressed that the problem of range competition becomes greater on more arid lands, where total carrying capacity is so much less. We have tended to overstock such lands.

In our approach to this problem it is better for all concerned to frankly acknowledge that elk and livestock compete for food, to stock our ranges on the basis of total carrying capacity, to apportion this among the types of animals using the range, taking advantage of selective habits where these apply and zoning land use where public interest calls for it.

Competition with Other Wildlife

Too often we concentrate our attention on carrying capacity for a single game species, without allowing for other species occupying the same range. This is particularly true in our concern for elk.

Food habits of big game are not sufficiently specialized to prevent competition. The elk, being the most omnivorous of them all, is the most likely to bring on harmful rivalry. Elk compete with mountain sheep, antelope, deer, and moose, particularly on winter range.

Elk and mountain sheep. Because of their habitat choice mountain sheep are more restricted in winter than most other big-game species. Elk wander widely, and often overrun the mountain sheep winter range in bands, early in the winter. This has often occurred on the Flat Creek mountain sheep range in Jackson Hole, Wyo. On the talus slopes and ridge tops near the cliffs where the sheep assemble in winter, elk often come in the fall and early winter and skim over the forage, taking what they like best, leaving what remains for the mountain sheep to winter on.

A similar situation has existed in the northern Yellowstone region. Adolph Murie (1940: 115) describes it thus: "The competition for food each winter is severe. On Mount Everts the bighorn, during the last half of the winter, subsist on a range so heavily utilized that the elk for the most part avoid it, after taking the cream' of the forage. The bighorn in winter pick at discarded seed stems lying on the ground and at the already closely grazed grass."

Even on the rims above the Yellowstone River farther south in Yellowstone Park, and on some of the adjacent slopes, elk share the range of mountain sheep.

It has been suspected that one of the factors in the decline of mountain sheep in recent years might well be the increased competition in winter by other big-game herds.

Elk and antelope. Again, in the northern part of Yellowstone National Park, we find elk competing for food with antelope, where they occupy the same range to some extent. The competition in this instance concerns browse, such as sagebrush and *Atriplex*. Such browse species are vital to the antelope, and are also eaten by elk to some extent.

Probably few situations in the country offer a situation such as this where elk and antelope are in serious range competition.

Elk and deer. Mule deer in winter are primarily browsers. They eat a minimum of dry grass. The elk eats both and where the supply of browse becomes critical, the elk can reach higher and secure such food when it is out of reach of deer.

A striking instance of deer-elk competition has been recorded by Edward P. Cliff (1939) showing that in the Blue Mountains of Oregon, in an area where these species were together on winter range, both increased until the deer, having reached a high of 19,500, suddenly dropped to 8,600 in an adverse winter. The deer began to increase again. Meantime the elk population had continued on a steady upswing, without any setback.

In the south end of Jackson Hole, Wyo., the deer herds have reached a high point in numbers several times with resultant large winter die-offs. Elk occupy much of the same winter range. The deer eat little or no grass at that time of year, but both eat browse and the range has lost much of its productiveness for either species.

In planning for deer and elk it is necessary to keep in mind that while elk will utilize practically any type of vegetation present on winter range, deer are confined to browse. Deer are the first to suffer, in serious competition, and the range suffers to the detriment of both species by a too drastic combined onslaught on its forage resource.

Elk and moose. In many observed instances the moose are able to winter back in deep snow country where elk do not thrive, and normally these two animals do not compete seriously. Certain circumstances, however, may bring them together. In the Jackson Hole-Yellowstone country the moose population has been increasing for many years, with a widespread dispersal into new territory. Normally they have wintered away from the elk herds but in recent years they have been increasing in numbers in the upper Gros Ventre valley in Jackson Hole, which is a critical wintering ground for elk. Moose are persistent browsers and will make heavy inroads on the browse of this area, which in this instance is vital to the elk.

There is another situation in Jackson Hole which could easily become a problem. The north end of the valley is not, normally, winter range for elk because of heavy snow. But on certain willow bottoms adjacent to the Snake River and the Buffalo Fork the moose find ideal wintering places and congregate there in unusual numbers. It has been the practice for some time to feed hay to a band of elk in that part of the valley. Undoubtedly at one time a very limited number of elk wintered there, but artificial feeding invariably raises the number of elk far above the normal carrying capacity. In time there will be serious competition with moose. The area here described is primarily moose range and it would be good management to confine it to that purpose.

The balanced fauna. Since man has become all-pervasive in wildlife habitat, we have reached the point where our wildlife does not have the opportunity to swing in the direction of equilibrium. For instance, at one time beaver could strip an area, then go on to another. In areas where natural conditions are restricted in extent, this is not so feasible as formerly. Where elk have already removed all aspen reproduction, it is not helpful to encourage beaver colonies there to cut down the remaining mature aspens. However, it must be admitted that another viewpoint is permissible if those groves are doomed anyway, because of our inability to protect them from browsing.

On the National Elk Refuge in Wyoming there were, at one time, numbers of sharp-tailed grouse. Great concentrations of elk have practically removed willow growth. This could not be avoided, but it seems more than mere coincidence that the grouse disappeared with the willows.

It is difficult to anticipate what food competition may arise among wildlife species.

A specific question that came up years ago may be given here. It concerned a proposal to introduce elk into the caribou range of interior Alaska.

Assuming that elk would succeed there, we would face the consequences of having two species of herd animals, both of which tend to become numerous, occupying the same range. Both eat grass and browse in winter. In addition, lichens are important for the welfare of the caribou, and because of their slow growth, present a range problem. Elk are known to eat certain lichens, as observed particularly in the Olympic Mountains, and preliminary experiments with this food showed they would probably take to that diet readily. There would be competition with moose, and elk would no doubt in time invade the mountain sheep range.

The probability is that such an introduction of elk, if successful, would eventually play havoc with the native wildlife.

It should be remarked that elk have been successfully introduced on Afognak Island, where no member of the deer family was originally present, and no food competition is involved.

PRESERVING THE RANGE

There are many objectives in range preservation. An obvious one is to keep the range productive, so that it may continue to support game animals under optimum conditions. Manifestly, the normal carrying capacity of an overpopulated range is less than that of one that has not been injured, and to promote recovery an abnormally low population must be maintained.

There are other considerations. A given elk range produces not only forage, but a cover of vegetation that tends to retard erosion, a supply of timber, a forest growth that has distinct landscape value in recreation areas, and cover and habitat for birds and other animals, as well as a flora that in itself has recreational and scientific interest.

A vexing problem that confronts the manager of an area is the preservation of the browse. It is now well known that, in spite of the fact that in general the elk depend largely on grass in winter, the browse species tend to disappear from winter range where animals are present in any numbers. This is particularly true of the areas occupied by the Rocky Mountain elk. On one range in Arizona certain willow patches were fenced off. In Jackson Hole much of the willow growth has disappeared on winter range and available aspen growths are doomed. In the northern part of Yellowstone Park browse species are being reduced to a minimum.

Browse here referred to is of the taller kinds, including tree growth. The smaller shrubs, such as rabbitbrush, sage, *Atriplex*, and similar types, reproduce more rapidly and survive more successfully.

The question is often asked, "How small must the herd be to let the browse recover?" So far there is no answer. It must be sought on a trial basis. One thing should be remembered. When a class of forage has been reduced virtually to the point of extermination, even any elk at all, so to speak, will tend to keep it at that point and prevent recovery; whereas if the type of forage in question could be brought back to normal, a considerable elk population could be accommodated provided it were kept within bounds and concentration due to artificial feeding were avoided.

With the present range conditions and demands of the sportsmen groups, the wildlife manager is in a dilemma. Should he sacrifice the browse species and let the winter range revert to grass? With a large elk herd the forest edge would in time recede to an elevation where snow depth would make foraging impossible. Or should efforts be made to save the browse species? The question is involved. If deer are present, one must keep in mind that deer *must* have browse in winter, and reversion to grass would eliminate the deer. Then, too, we do not know to what extent the elk themselves need a certain percentage of browse in the winter diet. The writer has been verbally informed of some studies in Colorado indicating that domestic livestock thrive better on a mixture of grass, weeds, and browse than on pure grass (see also *The Record Stockman*, Feb. 19, 1942).

This question has become acute on Teton National Forest in Jackson Hole and on the winter range in Yellowstone National Park. We do not yet have the answer, but it should be forthcoming within the lifetime of the present adult trees on the ranges where there is no forest reproduction. The section on artificial fluctuations suggests a principle that applies here. Clearly, the forester is the one who has the facts necessary for appraisal of the range. He should have the services of the forest biologist for the necessary ecological appraisal. On information obtained by such experts, against a background of general recreational and other needs of a community, must proper management rest.

ARTIFICIAL FLUCTUATIONS

Heretofore most efforts have been directed toward maintaining a stable game population. The attempt has been made to keep a constant level, directed at a uniform, maximum annual bag of game.

The present studies, as well as other available modern information, emphasize a different concept, one in which population levels are not constant but normally fluctuate (pp. 277-281), the fluctuation being the normal process under natural conditions caused by the unequal surges of ecological elements. It is obvious to the field observer that certain types of vegetation recede in the face of growing game populations; hence it should also be obvious that there must be a recession in game numbers for certain plant types to recover lost ground.

Under present conditions, where man has so greatly interfered with this natural tug of war, there is danger that the vegetation will not be given the opportunity to recover. On many ranges that have been depleted the game population has been prevented from spreading to other ranges, and at the same time natural reduction factors have been removed and artificial feeding and other devices have been practiced in order to maintain the same population level. There are local ranges now that are in need of rest from game pressure. Many aspen groves have entirely disappeared; others are on the way out, merely awaiting the natural removal of remaining mature trees through old age; and no aspen reproduction is succeeding (Plates 22, 24, 25, and 29). In some areas only a few aged mountain-mahogany trees remain, and as there is no reproduction, with their going will perish their race. In the Olympic Mountains in Washington some areas have been swept bare of salmonberry and even the prolific bigleaf maple is not reproducing. Elsewhere, locally, the forest edge appears to be receding.

It is not intended to convey the impression that irreparable damage has already been done on a large scale; but it is logical to believe that in primitive times there would have been a reduction or shift in game population before range depletion had reached the point indicated. Such natural cures for overpopulation are not now permitted to operate in places where artificial feeding is used to maintain the abnormal numbers, and continued pressure from these herds may eliminate certain plant forms from local floras to the permanent detriment of ranges. Grasses and other herbaceous plants that recover relatively soon, even low shrubs are not meant in this discussion, but rather the tall browse species and forest growth, which usually recover only through a long period of time.

A procedure that should rectify such range abuse would be to simulate the natural process and permit, or even induce, fluctuations in herd numbers. In spite of the efforts to the contrary, there are actually changes in elk population today. The Jackson Hole, Wyo., elk herd has ranged from a high of about 22,000 to about 17,000 in recent years, and carrying capacity is gradually declining. Fluctuations in the game supply are not new, and it is suggested here that a new concept be entertained, namely, that of regulating hunting and other factors in such a way that there be maintained a herd of fluctuating size appropriate to permit range recovery from time to time. This would mean that at times the game bag would be lowered and at times increased, for management would be directed primarily not toward producing the maximum number of animals year after year but rather toward preserving a range in perpetuity and at the same time providing for the sportsman a reasonable number of animals, which would vary periodically. Whenever tall shrub species have been permitted to send their reproduction up far enough to insure survival, the range could again support a somewhat larger animal population without permanent injury to forest growth.

Experience in the extermination of wildlife species has been so bitter that naturally any decline in numbers of any animal causes apprehension, and rightly so. Many species are in danger today. Elk, though, have been re-established, and a moderate herd reduction—even a drastic reduction in some places—need arouse no fear. There should be special concern, however, over reduction of habitat area or depletion of range, as territory adequate in size and quality is now known to be vital for the continued existence of a species.

In other words, the suggestion here given is to transfer the primary concern to perpetuation of the range and habitat as the fundamental basis of management of the elk herds. Some steps in this direction have already been taken. By agreement of the public agencies concerned, the desired size of the northern Yellowstone elk herd has been set at 7,000 on a tentative basis. In the Jackson Hole area similar efforts are being made, and at the time of writing (1946) this herd has been tentatively reduced to 15,000.

ARTIFICIAL FEEDING

Artificial feeding of game animals is now rightly looked upon as the last resort in wildlife management, a practice to be adopted only when all efforts to provide suitable winter range have failed.

The chapter on elk diseases (p. 177) presents a sordid picture of a game animal on an overstocked range and on a diet of hay. Direct losses, rising to about 10 per cent of the herd on the feed lots that have been studied, seem directly traceable to artificial feeding. It is difficult to eliminate squirreltail barley or other injurious elements from the hay. There are some indications that even coarse stems may cause lesions and resulting disease. Such losses from hay seem more marked on long-established feed grounds, where large numbers of elk are being fed.

Another effect of feeding is to keep the elk inactive. They soon learn to come for their feed, then go off a short distance to lie down. Hay fed animals appear less virile and thrifty than wild ranging animals rustling for themselves on adequate range, a condition that has been noted frequently by game wardens and other experienced observers.

One of the most critical situations the wildlife administrator has to face is a public demand to feed elk on an overstocked range. The temptation is strong to take the line of least resistance, haul in a supply of hay and cottonseed cake, and establish a feed ground. The result is a loss of any favorable animal distribution there may have been. The elk become concentrated in an area of feeding. No matter how heavily hay is fed, the browse in that vicinity will be destroyed. Once feeding is begun, popular pressure will be such that the supply of hay, rather than carrying capacity of range, will become the measure of the herd.

There are circumstances, however, under which it becomes virtually necessary to establish a feeding program. The classical example is the National Elk Refuge in Jackson Hole, Wyo. With a large herd of elk, numbering over 20,000, denied access to the desert ranges and long ago weaned from them, this important herd was left without adequate winter range. As an emergency measure a federal refuge was established to raise hay for elk.

In this instance we can hardily avoid the feeding program. But every effort is being made to lessen its evils. In recent years hay meadows have been left uncut and emergency supplies of hay have been obtained elsewhere and hauled to the refuge for use when needed. Thus the elk are scattered, are permittd to graze naturally, with much lessened danger of infection with *Necrotic stomatitis,* and with the promise of retaining more of their natural alertness and vigor. At least the hay feeding period is appreciably lessened, and if this experimental procedure succeeds, hay feeding may be unnecessary in some winters.

When feeding becomes unavoidable, certain procedures have been proved effective. The hay is hauled out on sleighs and pitched off on the snow in long parallel lines, with an effort to scatter the animals as much as possible. With a large herd, hay is distributed morning and afternoon. It has been found that elk (all ages and both sexes combined) will eat daily from 8 to 11 pounds of hay (see section on Carrying Capacity, p. 292).

Concentrations of animals year after year in the same place is generally assumed to be conducive to greater hazard from disease and parasites. As a precaution it is wise to change the location of the feed lot each year, if possible.

It is highly important to provide hay that is free from squirreltail barley or cheat grass, coarse brittle stems, or any awned or sharp-seeded hay types that may injure the mouth tissues.

It must be remembered that once feeding is begun it must be continued until spring, for the animals become dependent on it and will not leave. Therefore an adequate supply of hay should be assured before feeding is started.

ELK DOMESTICATION

Elk are readily domesticated. Numerous zoological gardens and city parks maintain some of the animals, and on private preserves herds have been kept for market and for pleasure. Kellogg (1887: 126) wrote enthusiastically: "Probably no animal in existence is naturally fitted to take so kindly to domestication as this noble creature so rapidly disappearing from the face of the earth." (See also "Anonymous" 1905.)

Caton, who had extensive experience with the deer family in domestication, wrote (Brown and Caton, 1880: 396):

"My elk *(C. canadensis)* continue to do well and are so prolific that I have had repeatedly to reduce their number, and would be glad now to dispose of at least 30. I have on an average about one old buck a year killed in battle, and sometimes another by some casualty, but all are healthy. Mine grow very large, and of all the Cervidae they seem best adapted to domestication."

Lantz (1910) summarized the experience of many owners of private elk preserves. Most of the animals reported upon were kept in enclosures; others were thoroughly tamed and were given their freedom, and the latter did not stray off with wandering bands of wild elk. Tame elk raised on the National Elk Refuge in Wyoming showed the same disinclination to leave, even though in winter several thousands of wild elk were present

Ignar Olson, of Quinault, Wash., informed the writer that in early days his family domesticated the Roosevelt elk at their ranch in the Olympic Mountains. They found the young calves by the use of hounds and raised them by hand. These tame elk scattered out on the range with the cattle in summer but always returned to the ranch when the cattle did.

In *Outdoor Life* for October 1941, is shown a photograph of elk one of them a spike bull, all with halters and carrying packs. There is the following caption: "Court Du Rand, Martinsdale, Mont., dude rancher, not only tamed these elks; he's turned them into a fine pack train for deer hunting."

HANDLING ELK

Elk are not particularly hard to handle. Even wild ones may be driven. In Jackson Hole, Wyo., where bands of elk have repeatedly raided haystacks and interfered generally on ranches, the local game wardens have made it a practice to drive the marauders to the National Elk Refuge where hay feeding tends to hold them. A group of mounted men get behind the band of elk and quietly urge them on. The men, scattered in a line, remain at a considerable distance from the animals and slowly guide them in the desired direction. One or two horsemen can do much in this way under favorable circumstances.

Occasionally it is desired to capture the younger elk or to segregate them from the older ones. For such purposes the Biological Survey designed a series of panels with which to build a holding corral. Some of the panels have openings, or "creeps," 11 inches wide, large enough to admit only the elk calves (in winter). If the openings are wider, yearlings attempt to pass through, with the result that they get stuck and injure themselves.

In earlier years special feeding corrals provided with "creeps" that permitted the entrance of calves only were put up each winter on the National Elk Refuge in Jackson Hole, as it was thought that winter losses might have been caused by the competition of the older animals, preventing the calves from getting enough hay. After the true nature of the winter losses was learned, these corrals were abandoned.

When elk are held in a corral or chute, they become greatly excited at the approach of people. They dash wildly about and often leap blindly against the walls of the enclosure, at times becoming so panic-stricken that they injure themselves. Moose, in similar circumstances, are comparatively docile and are easier to handle. The remedy is to provide an enclosure from which the elk cannot see out; for the moment the elk enter such a place, they become quiet. This was amply demonstrated when single animals were maneuvered into a small, tight shed built over scales, for the purpose of weighing them. The moment the door was slammed behind an elk, it stood perfectly still and was leisurely weighed while it was examined through a peep hole. Elk may be readily transported in a covered truck, without individual crating.

Elk fences are in use in some places. Whatever their purposes, for a holding corral or drift fence, they should be at least 7 feet high to hold elk safely. An elk corral should be built of poles or similar substantial material, for in close quarters elk will charge into woven wire and injure themselves (Plate 27). However, a long drift fence may be built of heavy woven wire. Barbed wire should be strictly avoided in any fencing for elk, for it results in many casualties. Both elk and moose have cut their throats trying to get through barbed wire, and frequently a foot will become twisted in the strands, holding the animal until it dies.

An ordinary buck and pole fence is no obstacle, as the elk easily jump over it. Such drift fences built on national forests to control the distribution of cattle do not prevent the migration of elk.

PLATE 27. Elk trap built by the U. S. Forest Service at Elwha, Wash., to capture surplus Roosevelt elk for transplanting. A tightly built fence of this kind, limiting the vision, tends to discourage efforts of the elk to break through. (U. S. Fish and Wildlife Service photograph.)

PLATE 28. View on the National Elk Refuge, Jackson, Wyo., where elk are fed hay in severe winters. (U. S. Fish and Wildlife Service photograph.)

PLATE 29. National Elk Refuge, Jackson, Wyo. *Above.* Aspen "high line" produced by elk browsing. In spite of extensive hay feeding the elk forage on the available browse, which is now badly depleted. (Photograph by Adolph Murie.) *Below.* Open foothills that furnish natural winter range. (U. S. Fish and Wildlife Service photograph.)

SALTING

Appraisal of the values in salting must be based on its purpose. Two objectives that have been sought by those who have put out salt for game are to insure the health of the animal and to distribute the animals on the range.

The Need for Salt

That wild game species need salt is open to serious question. It may be argued that the chlorine in salt is required for the formation of hydrochloric acid in the stomach and that the animals naturally seek salt licks. Furthermore, stockmen agree that cattle, especially dairy breeds, should be given salt. The undesirability of an animal consuming the great quantity of earth that it must ingest at a natural lick in order to obtain the amount of salt supposedly required for its well being, has been cited as a reason for giving the animal pure salt.

In the light of what is at present known concerning the metabolism of wild animals and the factors underlying their apparent adaptation to their natural food, it is necessary to approach the salting question with caution.

Even for cattle, apparently, salt is not always a physiological necessity. Lommasson (1930) indicated that commercial salt is not required for keeping cattle on alkali ranges in southeastern Montana in good condition. Moreover, domestic stock has been specialized for particular purposes and through selective breeding has had certain physiological functions, such as milk production, highly developed. The animal's metabolism has undoubtedly changed to meet these new requirements. Therefore, with respect to salting, livestock is not strictly comparable to wild game.

It might be mentioned here that humans in certain occupations or in excessively hot situations take salt purposely to compensate for the large quantity of it lost through perspiration. This is, however, an abnormal situation.

It is true that elk and other animals frequent mineral licks, but as previously stated (p. 237) these licks rarely consist essentially of sodium chloride, and in fact sometimes contain very little of it. Many other mineral salts are present as well. Moreover, it has been repeatedly noted that when salt has been made available to elk or moose, the animals have not been greatly attracted to it at first, although later they acquire a taste for it. An experiment conducted one winter on the National Elk Refuge in Jackson Hole, Wyo., brought out this same reaction; and at the end of the winter it was found that the elk, several thousands of which were present, had consumed a relatively small amount of salt compared with what would have been taken by cattle.

Experimenting with the elk of the Oympic Mountains in Washington, John E. Schwartz put out six salt stations on January 20, 1936. On several subsequent visits he found no indications that the salt had been used. In a field report he wrote:

"The last inspection of the salt licks was made on April 18. Fresh tracks were found in the vicinity of three stations, one set passing directly past a box; but there was still no evidence that any elk or deer has so much as noticed the salt."

"These boxes will be left in their present location during the summer months to determine if the elk will have greater tendency to take salt during that period when most of the feed is of the fresh, green type."

It is true, of course, that wild animals that have become accustomed to eating salt continue to seek it eagerly and eat it with obvious relish. It may be well to tell here of a tame elk which had perfect freedom to range where it pleased, that was inordinately fond of tobacco and would eat cigarettes as long as they were given to him. Does it therefore follow that tobacco should have been provided for this animal as necessary ration?

Van Loon (1937), writing the story of salt in popular form, presented some interesting facts and viewpoints that indicated that salt should be considered a condiment rather than an aliment. In discussing the human need for salt, he stated (p. 86) as follows:

"But as medical science began to probe more deeply into the mysteries of animal metabolism, its exponents grew increasingly skeptical of this alleged minimum physiological need [an absolute and more or less constant minimum supply of sodium chloride]. The time came when it was irrefutably demonstrated that common salt was only one of several equally important minerals necessary to the human body, and that the basic requirement was so small and indirect assimilation so easily incurred due to the profuse distribution of salt in nature, that it was seriously doubted whether salt itself was an essential aliment at all."

Van Loon quoted at length from Dr. Vilhjalmur Stefansson, who found that his men in the Arctic got along well without salt and that to break them of the salt habit was just as difficult as to break them of the tobacco habit. In the writer's experience with Eskimos it was found that salt and tobacco were two of the desirable articles of trade or for use as gifts. Tastes for both were clearly acquired.

Van Loon (1937: 87) stated further as follows:

"Biologists have reported instances where the growth of chickens and small animals was greatly aided by the addition of salt to their diet, and in fact have demonstrated that dogs entirely deprived of salt soon die. However, this was an artificial case. The dogs' food was deliberately treated to remove all sodium chloride, whereas in the natural state there is no record of any absolutely salt-free diet having ever been found. As against this evidence, Stefansson submits the case of certain members of the deer family both in Montana and Maine. The Montana deer seem to desire salt because they seek saltlicks. But, so far as Stefansson knows, there is no evidence of similar licks nor of any search for them by deer in Maine's interior. Yet both groups seem equally prosperous . . .

"Stefansson has come to believe that the predilection for salt, like that for tobacco, is almost entirely a conditioned response, since salt is sufficiently abundant throughout Nature to satisfy the chemical needs of the body."

The following free translation of a statement by Dungern-Oberau (1935: 233) is pertinent, as it applies to deer of Germany, where artificial salt licks are in general use:

"It is a fact that the placing of salt licks for ungulates as a necessity for their care in preserves has been developed in the last 10 years. Now one can scarcely enter a hoofed game preserve without coming upon an old or a new salt lick. Has the wild game's habituating itself to salt worked to its own benefits? We must make it clear that an artificial habit is involved. Then where, least of all in north German districts, does one find natural salt licks? It is a fact that in the remote as well as the more recent past a strong and healthy game population was to be found in Germany. This game was certainly better venison and apparently had better antlers than at present. This game had no mineral salts at its disposal. Yet all the ruminants in northern parts of Germany were thriving. Therefore we are not dealing with natural conditions when we give salt to our hoofed game."

"Without doubt it is demonstrated that the game eagerly accepts the salt and that where salt is furnished the game assembles and is readily held there. It may now be asked, does this salt contribute to the health of the game, has it any wholesome effect whatsoever, or is it only a bait or luxury? On the other hand it could easily be possible that salt has a harmful effect on the constitution of our ruminant game."

"If one put out a sugar lick in a given town children would assuredly gather, though one could not hope that this would be wholesome for them. It would be the same with springs of beer, schnaps, or wine, or offering of coffee or nicotine. That the game is attracted to the salt lick is by no means proof that salt is good for it."

The normal food supply of elk undoubtedly contains the various mineral salts in quantity sufficient for the animals' physiological needs. Numerous chemical analyses of food plants have shown the presence of minerals, and these probably are present in adequate amounts. There may be areas in which the vegetation is somewhat deficient in some minerals, but this would have to be determined in each case, as well as the apparent effect on elk in the locality. The animals may be relied upon to choose an adequate diet if the condition of the range permits.

So far the evidence points to a sufficiency of mineral salts in elk forage. Well-meaning efforts to improve the condition of the elk by feeding salt may be simply introducing a new appetite that further study may show is useless, if not harmful, in the long run.

Salting for Distribution

It is the practice among stockmen to place salt on the range to effect a desired distribution of cattle. A similar practice has been adopted in a few cases among elk herds.

In a great many instances, however, the natural elk distribution is satisfactory and salting should not be introduced. There may be special local circumstances, where elk are being kept in competition with other animals, for instance, or where other adverse conditions are present, under which it is advisable to use salt as a bait; but in the light of present knowledge on the subject it is not justifiable to adopt the practice as a general management feature or in a casual manner, with the rationalization that "at any rate, it will be good for the elk."

DAMAGE BY ELK

Elk will invade vegetable gardens, pastures, grainfields, cornfields, and orchards. They will browse available orchard and shade trees. They have been known to damage dahlias and root crops in California, Colorado, and Washington. In Wyoming and several other states they have raided haystacks in winter and persistently returned to feed lots established for cattle. In spring, bands of elk large enough to be destructive have grazed persistently in pastures intended for cattle. Damage in ranching country sometimes reaches serious proportions.

Remedies are hard to apply. Under certain circumstances a drift fence has proved to be effective. This must be fully 7 feet high and strongly built. The drift fence built on two sides of the National Elk Refuge consists of heavy woven wire, 90 inches high. Further experiments with the electric fence may furnish suitable and convenient means of keeping out elk, but at present much remains to be done on it and at the present state of development it is not recommended.

The ordinary pole corral designed for excluding cattle from haystacks and built to satisfy legal requirements, will not exclude elk, nor will the ordinary pasture fence; but haystacks can be fenced effectively by use of high, well-built panels of 2-inch material, a method well demonstrated at the National Elk Refuge. Although a panel hay corral costs more to construct than the usual pole corral, it lasts much longer and saves the annual expense of rebuilding.

Several repellents in the form of a spray, furnished by the Control Methods Research Laboratory in Denver, were tried on the elk at the National Elk Refuge but no spray was found that could be recommended.

There are other types of damage by elk. The bulls thresh the bushes and saplings with their antlers. Many small trees are killed in this way but in extensive forested areas the effect is negligible.

Another kind of injury may be more serious, an injury identified with a type of range depletion, but which may be included here. On an overcrowded elk range aspen groves, maples, second growth Douglasfir, and other forest trees may be destroyed by the elk keeping down reproduction until the older trees are gone and there is no young growth to replace them. Actual instances of disappearance of such groves have been observed. This type of damage is more than range depletion as generally understood, for it is actual, eventual removal of forest cover and may affect the aspect of the landscape in recreation areas. The only remedy appears to be to limit the elk population, or effect a different distribution, for although on a small scale areas may be fenced until young growth grows out of reach, fencing on a large scale is hardly practicable.

Other tree species also may suffer from overuse, among them the bigleaf maple *(Acer macrophyllum)* on the West Coast, although the mature maples may survive longer than aspen.

PREDATORY ANIMAL CONTROL

Predatory animal control as a management measure should be based on the need as shown by adequate information. The golden eagle has been seen attacking a young elk in winter, yet it could not be considered a limiting factor on the elk population.

Coyotes are known to take young elk calves in the absence of the mother. In the Yellowstone-Jackson Hole region such sporadic predation by coyotes does not prevent production of surplus elk in numbers that are sometimes hard to harvest, and there is no difficulty in maintaining as large an elk herd as the ranges will carry. Under such circumstances the coyote is no problem. However, it is conceivable that where a small plant of elk has been made in new territory, and they do not appear to be increasing satisfactorily, the coyote may be a factor. If so, control would be advisable until the herd is established.

The mountain lion, or cougar, is an effective predator on elk. On ranges where this animal occurs it is desirable to consider the entire big-game population, whether deer and other prey species are also present, and whether there has been difficulty in keeping game animals down to range capacity. Since the mountain lion is itself a game animal, a balanced situation should be the rule in normal instances. Only if sport hunting of the mountain lion fails to keep it in normal numbers should professional control be adopted.

Wolves are no longer a problem in most areas. However, Cowan (1947) has reported on the wolves in Canadian national parks, where they range with elk, deer, moose, and other ungulates. He found that the presence of wolves in that particular game population is not detrimental, but adds that if reduction of the present overpopulation of elk is accomplished, the present favorable balance may be upset.

Grizzlies and black bears will prey on elk calves, and grizzlies occasionally kill adults. However, in an adequate elk population these animals are not an effective hazard.

In short, predation on elk under normal circumstances is not serious. Some of the predators are themselves game species or interesting members of the fauna. In an elk population struggling for survival, predation may be a factor. If an elk population is flourishing and affords a reasonable amount of hunting, predation can generally be ignored.

In a well-balanced management program it is advisable to utilize public agencies for any needed predator control. These agencies are experienced and equipped, and can be directed to a specific local situation.

There is no dearth of situations in which an elk herd is taxing the ingenuity of everyone concerned to produce a satisfactory management program. Invariably it is a range problem—a land-use problem. We are required to maintain satisfactory elk numbers in the face of growing hunter demand and restricted land areas.

HUNTING REGULATIONS

Although open seasons, length of season, and open and closed territories are handled by the various state game authorities in the light of local conditions and local sources of information, certain suggestions of general application may be worth while.

It is important that regulations be flexible and that the game commission have the power to make necessary changes in emergencies. In the case of a migratory animal like the elk, weather conditions may delay migration so that the animals do not reach hunting areas in sufficient numbers during the regular open season; or the animals may arrive too early, resulting in excessive kill.

In some localities it may be desirable to issue a limited number of permits and to designate in what parts of a given range the animals should be taken. Sometimes when a limited area is opened to hunting for the removal of a limited number of animals, the temptation is to try to accomplish the objectives by permitting a very short hunting season, sometimes only a few days. The result is an abnormal concentration of hunters, tending to create a human hazard and to destroy the esthetics of the hunt and possibly resulting in a greater kill than desired. Issuance of a limited number of licenses over a relatively long season would be a better procedure.

The traditional method of conserving a big-game species has been the "buck law." Recently it has been found advisable to abandon this in some places. In the Jackson Hole, Wyo., area no restrictions have been placed on the hunter with reference to sex or age of the elk that may be taken so that each season a large number of calves and cows, as well as bulls, are shot. After the hunting season what appears to be a normal sex ratio remains and the elk herd as a whole is well able to maintain itself. In fact, there is still a surplus beyond carrying capacity, resulting in periodic winter losses.

Indiscriminate shooting of all classes by the hunter has the added advantage that the finest animals are not constantly being shot—an important consideration.

Postwar developments in aviation have introduced a disturbing element in the field of big-game hunting. Some game and fish commissions have already taken steps to limit the use of aircraft in taking game. Such use may have far-reaching effects in changing the whole aspect of big-game hunting, including elk hunting, unless the role of aircraft in sport is carefully studied and defined.

The same may be said of the jeep. Steps have been taken to discourage shooting from an automobile, or from a highway, as not being conducive to sportsmanship. The jeep is in the same category, but is capable of penetrating areas hitherto inaccessible by car.

ELK REFUGES

In the earlier era of nation-wide destruction of game herds the first concern was, naturally, to check the steady decline of wildlife populations. Game laws were developed, poaching was discouraged by all possible means, and refuges were set aside wherever it was possible. As related before, the elk had suffered an alarming decrease in number and had been exterminated over most of their range. There were a few strongholds in the West where they were making their last stand. One of the most important of these was the Jackson Hole country in Wyoming. Tusk hunters operated here for a time, but these were eventually discouraged.

Then a new phase in conservation began. It became obvious in a few instances even many years ago that some of our game species lacked winter range. This was brought forcibly to our attention in the case of the Jackson Hole elk. The northern Yellowstone elk herd had considerable range within Yellowstone Park. The Jackson Hole herd was definitely dependent on private lands in winter.

In an Agricultural Bill passed by Congress on August 10, 1912, and March 4, 1913, $50,000 was appropriated. From this fund 1,760 acres of privately owned land near the town of Jackson were purchased, and together with 1,000 acres of adjoining public lands, these permitted the creation of the National Elk Refuge, placed under the administration of the U. S. Biological Survey, now known as the Fish and Wildlife Service. In 1925 the Izaak Walton League raised a fund of $36,000 by nation-wide public subscription and purchased 1,760 additional acres. In 1927 Congress authorized acceptance of this gift, which has been known as the Izaak Walton League Addition to the National Elk Refuge. Again, in 1935, from an appropriation available at that time, more private lands were purchased, bringing the acreage up to the present total of about 23,000 (Plates 28 and 29).The recent bill enlarging Grand Teton National Park, contributed additional land along the Gros Ventre River, so that now the elk have the use of the area extending from the willow bottoms of the Gros Ventre River south to the town of Jackson. For a number of years the State of

Wyoming has also maintained some feed grounds south of Jackson.

This, very briefly, gives the history of one of our earlier attempts to supply range exclusively for wildlife. Since then the need for that type of refuge has become more generally recognized, and a nation-wide system of refuges to care for all kinds of wildlife is being developed.

In recent years, however, inviolate sanctuaries have been under fire. Sometimes this has been justified, as in instances where a game herd has increased far beyond the capacity of the range it inhabits. Nevertheless, the refuge system, properly administered, has an important place in wildlife management. The type of refuge that permanently supplies a habitat exclusively for wildlife is becoming more and more necessary if we are to maintain wildlife. Examples are given elsewhere to show that if strategically placed, another type of refuge, the area simply closed to hunting, may serve to effect better game distribution.

In the northern end of Jackson Hole, Wyo., on Teton National Forest, is an area comprising chiefly the watershed of Pacific Creek. This, together with some additional territory, had been designated the Teton Game Preserve, and for a great many years had been kept by the State of Wyoming as an inviolate sanctuary for elk, an action that is still recognized by many as commendable. The territory lies in the heart of the elk summer range. Here, after spring migration, a large number of the calves are born, here the herds spend the summer, and here the rut takes place. Some years ago this Game Preserve was opened to hunting. There has been some concern over the effect of this action. One result is that the elk herds that formerly were seen by recreation parties in summer are less in evidence. Probably those that summered on the southern portion were eliminated by hunting, or there has been a northward shift in the summer elk population. Much of this abandoned summer range carries elk sedge, a good late summer and fall forage. Mass hunting that prevails today may discourage use of this range.

One difficulty is that in order to take the annual increase, a comparatively large number of hunters must be accommodated, and in hunting territory that does not invite mass killing of elk.

Experience in this particular instance should be instructive and the ultimate effect will be watched with much interest.

The sanctuary will unquestionably remain in wildlife management. It should be applied for a definite purpose and as a result of close study of any given situation.

Another type of refuge is the territory that is temporarily closed. From year to year certain sections are closed to hunting, such areas more or less alternating with areas that are open to hunting. This is designed to prevent a slaughter of elk by closing places that are too accessible to hunters and where hunters might, otherwise, concentrate in too great numbers.

Similar closed areas were found to be beneficial in the vicinity of Gardiner, Mont., where, by closing the Decker Flat to hunting, the elk coming out of Yellowstone National Park are given an opportunity to become somewhat scattered before they encounter the concentration of hunters assembled there. It is likely that by setting aside other limited areas in that general vicinity to serve as escape areas, some of the elk could be induced to migrate farther north and thus eventually to establish a longer migration and wider dispersal on winter range.

REINTRODUCING ELK

When the American elk had reached the point where the species was endangered and virtually its last stronghold remained in limited portions of the Rocky Mountains and a few places on the Pacific coast, the first concern was to save it from extinction. First efforts were directed toward legal protection and the next toward restocking former ranges.

A recent report compiled by the protection department of Yellowstone National Park, National Park Service, presents some interesting facts. From 1892 to 1939 a total of 5,210 elk from Yellowstone Park were shipped to 36 states, District of Columbia, Canada, Argentine. From a herd at Gardiner, the State of Montana shipped 5 carloads in 1912 and 90 individuals in 1916. Shipments have been made also from the Jackson Hole, Wyo., herd and from the Olympic Mountains, in Washington. A number of these various consignments went to city parks and zoological gardens, but a great many were used for restocking areas where elk had been exterminated (Figure 32).

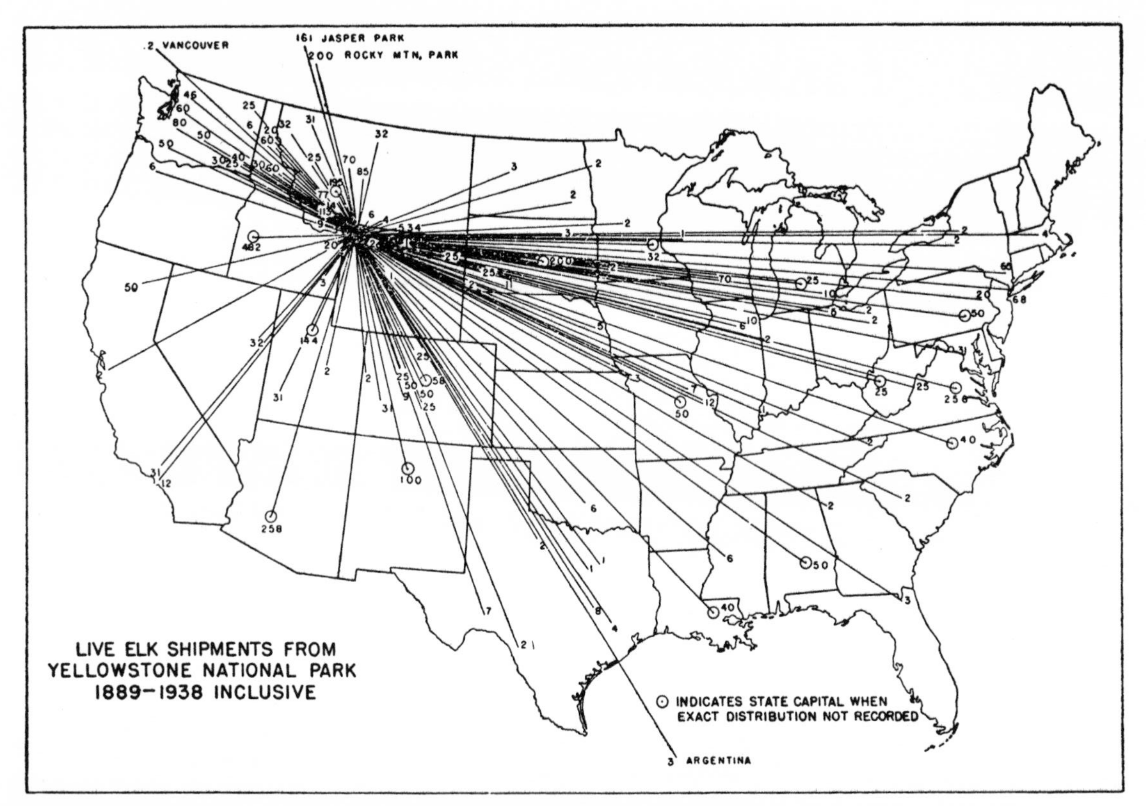

FIGURE 32. In addition to shipments indicated on this map a considerable number of live elk have also been shipped from Yellowstone National Park, Wyo.

As a result of the restocking program over a long period of time elk again have a nation-wide, if scattered, distribution. In some areas where elk had been exterminated, sportsmen now enjoy an open season. The Merriam elk, for instance, disappeared from Arizona, but the Yellowstone elk that replaced them have adapted themselves to the environment and recently have been numerous enough to warrant an open season. Thus, in a measure, it has been possible to undo the damage done in earlier years.

Jackson (1944: 21) has summarized the recovery of elk numbers. By 1910, there were some 60,000 in 7 states. Today elk occur in 25 states, and in 1942 they were considered huntable game in 11 states. Jackson summarizes increases as follows: "52,000 in 1920 (Palmer, 1922: 24); 165,764 in 1937; and 236,787 in 1941."

Restocking with elk has not been universally successful. In some instances the introduced nucleus has not increased according to expectation for reasons often obscure. Gerstell (1936) discussed the experiment of restocking Pennsylvania with elk. At first, from 1913 to 1923 at least, the herd maintained or exceeded the expected normal increase, but in the following years the rate of increase declined and finally the species was again in danger of disappearance. Gerstell believed that the building of roads, which made the wilder parts of the mountains more accessible, and so increased human interference, was the cause of the decline. Could it be also that the changing forage supply brought about by the large deer population in Pennsylvania and the changing forest cover seriously affected the welfare of the elk? The writer has not had opportunity to visit those ranges.

In contemplating new elk plantings, several important factors should be considered. First, is there adequate winter range for the future elk population? It should be determined how far the public will go in providing adequate winter range. Elk will eventually work down into adjacent lowlands for winter, as that is their tendency, and if the lowlands are an important farming country, the elk will interfere with economic interests.

Perhaps next in importance is the factor of competition. If a given range, even on a national forest or public domain, is already heavily utilized by a combination of other game species and domestic stock, the introduction of elk will only aggravate range competition, especially if such animals as mountain sheep, that have difficulty in holding their own, or, in some situations, moose, are involved. In any event, competition with other game species should always be taken into account, and any questions arising from it should be thoroughly considered by the groups that are contemplating stocking an area with elk.

In each transplantation, too, effort should be made to anticipate the type of available range that the elk is likely to occupy and to plan accordingly, for elk will not always remain in the district in which they are released. Sooner or later they may drift into preferred territory. They are likely to invade any reasonably close farming section and to go high into any available mountains in summer.

Consideration should be given to the strongly ingrained migratory habit, as it is possible that the time of year and the age of animals may affect the success of establishing a new planting. If placed on the desired summer range in spring or summer, the animals could be expected to adjust themselves to adjacent winter range and the suitable migration would be developed. Some introduced animals have disappeared, and it is possible that if their original home is not too distant, they may have found their way back to it. Statistics on this question are needed.

SOME ELK MANAGEMENT PROBLEMS

Jackson Hole Herd

This is a herd which has presented a problem ever since it became nationally known. As a local resident recently said: "Thirty years ago we were saving the elk of Jackson Hole. Today we are still saving the elk. What is the answer?"

Perhaps no one can give the full answer at this time. One difficulty is differences of opinions on what are the adverse factors and what should be our objectives.

In previous sections the history of this herd has been presented, such as the shrinking of its winter range, the crowding of animals with resultant further range deterioration and lessened carrying capacity. The drawbacks to artificial feeding have been explained. The following suggestions are offered to alleviate some of the difficulties and to serve as examples of a helpful approach to management.

1. Give a thorough trial of the forage management that has just recently been put into effect. This means to irrigate the proper hay meadows on the Elk Refuge, but leave them uncut for the elk to graze naturally the following winter. Meantime, fill the hay sheds with baled hay from the hay ranch at the north end of Jackson Hole, on the Grand Teton National Park. This will tend to lessen the amount of hay feeding, will cause the elk to become more widely distributed over the Refuge, actually provide a larger volume of forage, and tend toward a somewhat more healthful and natural situation.

2. It would be unwise to continue feeding elk at the north end of Jackson Hole. This is not good elk winter range, but is normal winter range for moose and would best be used for that purpose. Establishing an elk feed ground in that area would inevitably build a herd that would destroy browse resources there sooner or later, and thus compete with the moose.

3. In the Gros Ventre River basin the elk winter range is in precarious condition. Erosion has become accelerated and both grass and browse forage has had destructive use. Reseeding experiments are now under way. The soil is seleniferous and if present studies so indicate, certain selenium bearing plants which have appeared on the range should be eradicated before it is too late. Furthermore, moose have spread to this range and have become plentiful. Under the peculiar circumstances prevailing in the Jackson Hole region, elk probably should be given preference in the Gros Ventre basin and the moose be greatly reduced in numbers. There is direct competition for browse here, where this type of forage is already in bad condition. It would be a mistake to establish a feed ground in the Gros Ventre basin. But acquisition of certain privately owned meadows, for natural grazing, would be a help.

4. The annual hunting of elk has become a problem. There is much opposition to hunting in the open valley, where hunters assemble with automobiles and create what has become known as the firing line. Hunting should by all means be encouraged to take place on the surrounding forest. However, there is a small group of elk that summers in the northern Tetons and that migrates across the open valley. It has been feared that without

hunting this group may become too large. Congressional action enlarging Grand Teton National Park makes provision for herd reduction in that area. Such reduction probably will not be necessary except at long intervals. Experience should point that way.

5. In recent years elk have shown a slight tendency to migrate out of Jackson Hole into the Green River basin. It is entirely probable that if encouraged, this tendency could very well result in the re-establishment of the ancestral migration to the plains. However, ranches and use by livestock in the Green River basin are complicating factors and one cannot advocate active encouragement of this migration until much further study has been made.

Northern Yellowstone Herd

This herd has been a problem for many years, as have the other major elk herds. At one time the elk undoubtedly migrated out of the park, down the Yellowstone drainage, much more extensively than they do now. But in the earlier years, when hay feeding was popular as a game-management measure, feed grounds were established within the park boundaries. Also, as in other localities, the elk migrating to lower ground were hunted persistently. As a result, the herd became more or less sedentary and the high winter ranges became overused. The browse here too suffered severely. As a matter of fact, some of the features of the landscape will change. The aspen groves are disappearing and the picturesque cottonwoods along the Lamar River will go. There is no reproduction.

1. Artificial feeding of elk within the Yellowstone National Park was discontinued many years ago. Therefore, the first step toward dispersal of the elk in the winter problem area has already been taken. Furthermore, in cooperation with the Montana Fish and Game Commission for disposal of meat, some official herd reduction was undertaken inside the park boundaries. This was done in 1943, fortunately in a winter when there was a big migration out of the park. Thus the removal of the least migratory individuals was assured, an important consideration. Further reduction took place in 1950.

2. The solution for the management of the northern Yellowstone elk herd must lie in the encouraging of migration; in making of this a migratory instead of a sedentary herd. Artificial feeding

and persistent hunting of the more migratory individuals that left the park had produced the wrong trend in early years, concentrating animals on the higher restricted winter range.

It is suggested that early migrants from the park be given safe passage to winter range farther north; that each season for a period of years the migratory individuals be allowed to increase in numbers on winter range outside the park. This could be done by judicious postponement of the hunting season and establishment of closed areas in strategic places to insure the perpetuation of the early migrants. It would entail restraint on the part of hunters for a time, and some faith in biological management; but a good trial of this principle could possibly produce much better hunting eventually and a much more favorable elk distribution in winter.

Such an effort would rest on the well known strength of habit in animals such as elk. It is true, there are many exceptions and animals that do not migrate one year may migrate at another time. But we must rely on *average* behavior when dealing with animal masses. We know that many of the animals will remain in high country, but that when winter forces become severe, they will migrate down. On the other hand, we know that a certain proportion go out *every* winter, and some of these go early, before conditions are severe. These early moving elk hold out the best promise for building a migratory herd. It is with these we should work.

Wyoming-Idaho Elk

In primitive times we know that some of the elk that summered in the southwestern Yellowstone area found their way over into what is now Idaho, in the Henry's Lake district and neighboring areas on the west side of the Teton Mountains. Such elk movements have now largely ceased, probably because of hunting of those particular migrants over a long period of time. There would be much gained by encouraging such a movement once more.

It may be worthwhile for Idaho to survey the possibilities for some winter range in that particular section. It may be too late in the developmental stage of that part of the State to make room for elk. But if it is not too late, and if a small elk population there is desired, such a project could be encouraged. The first step—to determine the presence of some winter range. If none is available, there is no need to plan further. If there is range available determine whether a few elk are still tending to come that way. If that is the case, give absolute protection over a period of years. There are indications that poaching and lack of interest in such a project have kept the elk out of this part of Idaho.

The thought behind this suggestion is this. We have winter room for only a limited elk herd in Jackson Hole. We have found it necessary to reduce the numbers so drastically that travelers on the summer range do not have the opportunity to observe the animals as readily as they formerly did. Additional winter range on either side of the state boundary would help by that much to maintain a satisfactory summer herd. We have already found it difficult to harvest the surplus in Wyoming without a firing line. There can be plenty of elk for all—but we must be sure there is room for them when winter comes.

The Wind River Herd

The Wind River Mountains of Wyoming contain one of our finest primitive areas. At one time these mountains supported many elk, in the years when the famous artist, Carl Rungius, traveled and painted there. But as in the case of other elk populations, they had been denied lowland winter range.

In recent years attention has again been directed toward the elk of the Wind River Mountains. The Wyoming Game and Fish Commission, and a number of interested local people, have become encouraged to revive this elk herd. One difficulty has been elk depredations on ranchers' haystacks, a matter that cannot be ignored. The State of Wyoming has acquired a strategic hay producing ranch as one step in the program. In this case feeding of hay may be a first necessary procedure to establish suitable migration for winter. If experimental procedure now in progress on the National Elk Refuge should prove promising, raising of hay for natural winter grazing may be worth trying with the Wind River herd.

Management of this herd must be exploratory, due to conflict with ranching interests. But certainly the purchase of a key ranch was a wise step in the right direction. However, free ranging, wild animals should be the objective and advantage should be taken of every possibility that arises toward that end.

The California Redwoods Elk

At one time the Roosevelt elk was plentiful down the northern California coast. There is a nucleus still present within the Prairie Creek Redwoods State Park. Much interest has been shown in preserving this herd by the recent establishment of an elk refuge there.

This is a worthy objective, but it may be worth while to point out the difficulties to be overcome. Through the enthusiastic work of Chief Ranger C. L. Milne, with the support of his associates, a sizable herd has developed along the meadows at Prairie Creek itself, an attraction for the many visitors, and a good start toward the recovery of the elk population of the redwoods country. However, these meadows are now fully stocked, even overstocked, with elk. And the adjacent redwoods forest does not supply sufficient forage for many animals. Elk occur only singly or in small numbers throughout the shaded stands of redwoods. This would seem to present a discouraging prospect, but certain other circumstances hold out promise.

There has been some logging in parts of the area. Such land will undoubtedly grow up into heavy stands of brush—with much greater carrying capacity for elk. And out along the coast are brushy areas along the bluffs and many grassy spots. These, too, are good elk range.

We may note, then, that certain logging operations have already taken place, and what may be regrettable from the standpoint of preservation of the redwoods, may prove a boon to the elk. The plan of management here is to have public control of sufficient area to include the elk forage to be found in the cut-over blocks in the redwoods forest and the valuable forage spots along the coastal bluffs. Dedication of such optimum feed producing units to the elk is a necessary part of any program for their preservation in redwoods country.

This redwoods state park has now been dedicated to provide habitat for these elk.

The examples of elk problems outlined above necessarily have been treated very briefly, and are not offered as detailed programs or complete procedures. There are many difficulties in applying any management program, and often the chief difficulty is to convince the local public. But these examples may serve to suggest an approach by fully utilizing known biological principles.

MAINTAINING STANDARDS

In any type of mass production, quality is likely to suffer, and today, with nation-wide enthusiasm for wildlife production and general resort to "management," there is danger that quality will again be sacrificed. Quality in this field has at least two applications: quality of the game animal itself and quality of man's enjoyment of the species.

Too often only the production of numbers is stressed. Pressed by an army of hunters who want something to shoot and by the limited hunting areas, the game administrator naturally becomes very conscious of the need for "more game." He turns to various short cuts to produce animals—artificial feeding or any other means available—short cuts which often affect the "gameness" of big-game species and probably result in an inferior animal.

If biological principles are to be utilized in game management, biological laws must be allowed to function. One of these laws is selection. It is known from experience that with domestic stock attention to breeding is essential to maintenance of standards and that neglect of selective breeding produces mongrel, substandard animals. For wild animals there are no artificial standards set by an industry, but the typical characters of a species are accepted as standard. Here, too, there is no great concern over a set of "points" or specifications adopted by an interested group, but there is, and should be, concern over the preservation of the animal's virility and gameness and over the fitness of the species.

It is possible, through an understanding of the life history of the elk and its environmental needs, so to adjust human impact on the species that it does not destroy the animal's normal habits, natural breeding, migrations, and adaptability to environment. Generally speaking, sportsmen tend to kill the best of a game population, although in recent years of mass hunting this tendency is not so pronounced. Some hunters still seek a trophy, the finest they can find, but the majority are more easily satisfied and a great many of them kill cows and calves as well as bulls. This is a favorable trend, as it lessens the unfavorable selection, and is an important argument against the "buck law" as applied to elk.

There are other modern practices, though, that may be adverse to the welfare of the species. Feeding of hay is known to be detrimental when continued over a period of years, as it tends to produce less desirable animals. The natural human desire to take care of animals as pets and to remove all hazards does not promote the greatest strength and adaptability of the animals. It is the responsibility of game management to adopt a policy that looks far into the future, over periods of time through which biological factors can be effective.

A number of authorities have been concerned over the future of the human race itself for the reason that the modern tendency to avoid exertion has introduced a dangerous "softness," as Carrel (1935: 111-112, 115-116) put it:

". . . Civilized men . . . need a way of life involving constant struggle, mental and muscular effort, physiological and moral discipline, and some privations. Such conditions inure the body to fatigue and to sorrows. They protect it against disease, and especially against nervous disease. They irresistibly drive humanity to the conquest of the external world.

". . . Although modern hygiene has made human existence far safer, longer, and more pleasant, diseases have not been mastered. They have simply changed in nature.

"This change comes undoubtedly from the elimination of infections. But it may be due also to modifications in the constitution of tissues under the influence of the new modes of life. The organism seems to have become more susceptible to degenerative diseases. It is continually subjected to nervous and mental shocks, to toxic substances manufactured by disturbed organs, to those contained in food and air. It is also affected by the deficiencies of the essential physiological and mental functions. The staple foods may not contain the same nutritive substances as in former times. Mass production has modified the composition of wheat, eggs, milk, fruit, and butter, although these articles have retained their familiar appearance. Chemical fertilizers, by increasing the abundance of the crops without replacing all the exhausted elements of the soil, have indirectly contributed to change the nutritive value of cereal grains and of vegetables . . . Hygienists have not paid sufficient attention to the genesis of diseases. Their studies of conditions of life and diet, and of their effects on the physiological and mental state of modern man, are superficial, incomplete, and of too short duration. They have, thus, contributed to the weakening of our body and our soul."

In the case of domestic animals the concern is not with their racial future. One animal is developed solely for meat, another perhaps primarily for wool, another for eggs, and so on. All else is subordinated to production of that one item of human use.

Wildlife, however, falls into a different category. Man's enjoyment of it is bound up with its success as a species, and management of it should be guided by that viewpoint, keeping in mind that the biological influences that affect man, such as those suggested by Dr. Carrel as cited above, are equally operative on other animals. A local woodsman, noting the demoralizing effects of large-scale feeding of elk, expressed the same sentiments in a more homely way: "They'd be better off if they gave 'em range and then just let 'em rustle."

There is another phase of quality that involves human use of wildlife resources. Sportsmen themselves often deplore the degeneration of the sport of hunting. It is true that the enjoyment of wildlife in the manner of the old time "mountain man" is now a thing of the past, but many people desire that every effort be made to preserve the "next best" type of enjoyment. Usually this means wild game in the true sense, as well as other indigenous species that add interest; as nearly as possible a wilderness flavor; and room enough so that hunters are not concentrated. The "firing line" that has sometimes been a feature in the hunting of elk does not comport with this ideal, and in other areas elk hunters have become so concentrated that they have been a menace to each other.

Aldo Leopold, who gave the greatest impetus to the modern enthusiasm for wildlife management in America, fully realized the danger of too much management. In his book *Game Management* he formulated several theorems, among which are these two (1933: 394):

"2. The recreational value of a head of game is inverse to the artificiality of its origin, and hence in a broad way to the intensiveness of the system of game management which produced it.

"3. A proper game policy seeks a happy medium between the intensity of management necessary to maintain a game supply and that which would deteriorate its quality or recreational value."

In writing of the esthetics of conservation, the same author said (Leopold 1938: 104): "However, when carried too far, this stepping-up of yields is subject to a law of diminishing returns. Very intensive game or fish-management lowers the unit value of the trophy by artificializing it." Also, in the same article, discussing certain types of fish planting, he stated (1938: 104): "Artificialized management has, in effect, bought fishing at the expense of another and perhaps higher recreation; it has paid dividends to one citizen out of capital stock belonging to all." Again (1938: 107): "The only true development in American recreational resources is the development of the perceptive faculty in Americans."

These and other opinions expressed in recent years are indicative of a trend toward natural wildlife management. In accordance with the growth of such public desires and of a multiplicity of interests in wildlife and its environment, the management of an elk herd and its habitat must proceed with a minimum destruction of the many intangible values. To a large class of individuals, part of the value of the elk is its ecological significance, although they may not express it in those words. Elk trails leading to natural licks have more significance and much more interest than a group of elk at a block of salt put out in a meadow. The adaptations of elk to cope with winter on ancestral range is biologically significant. The differences between the habits of deer and elk, as well as those of the moose, each finding its own niche in a common environment, are part of a complex story in natural history that has great educational value and interest. Nor are such interests confined only to the nonshooting public. The outing of the sportsman himself is enhanced by the same factors.

Fortunately, many elk ranges are still in fairly primitive state and values indicated above have not been destroyed. Looking to the future, in view of the needs of elk and the exacting requirements of recreation based on multiple use, the safest course is to model elk management along natural lines, not only to preserve the elk as a living animal, but also, so far as is reasonably possible, to preserve its distinctive habits as well as its habitat.

BIBLIOGRAPHY

Anonymous

1905. Raising elk for profit. Country Life in Amer. 8: 554, 556, 558.

1922a. Elk moved. Calif. Fish and Game 8: 229-230.

1922b. Elk herds and winter forage. Forest and Stream 92: 495.

1930. What is the fastest animal? Amer. Forests and Forest Life 36: 178.

1931a. Unusual winter data on elk. Calif. Fish and Game 17: 325.

1931b. Grasses kill mammals. Calif. Fish and Game 17: 326.

1935. Playgrounds for all. Nat. Mag. 25: 241-256.

1936. Few elk remain. Penna. Game News 7 (4): 13.

Ainsworth, A. R.

1932. The tule elk. Calif. Fish and Game 18: 81-83, illus.

Allen, Glover Morrill

1938-40. The mammals of China and Mongolia. Amer. Mus. Nat. Hist. 2 vols., illus.

Allen, Joel Asaph

1871a. Notes on the mammals of Iowa. Boston Soc. Nat. Hist. Proc. 13 (1869-1871): 178-194.

1871b. The fauna of the prairies. Amer. Nat. 5: 4-9.

1875. Notes on the mammals of portions of Kansas, Colorado, Wyoming and Utah. Essex Inst. Bull. 6 (1874): 43-66.

1876. Description of some remains of an extinct species of wolf and an extinct species of deer from the lead region of the upper Mississippi. Amer. Jour. Sci. and Arts (ser. 3) 11: 47-51.

1902. North American ruminants. Guide Leaflet 5, Sup. Amer. Mus. Jour. 2 (3), 29 pp., illus.

Altsheler, Brent

1908. A pair of massive elk antlers. Outdoor Life 21: 171-173, illus.

Anderson, Rudolph Martin
1934. The distribution, abundance, and economic importance of the game and fur-bearing mammals of western North America. 5th Pacific Sci. Cong., Canada (1933) Proc. vol. 5: 4055-4075, illus.
1937. Faunas of Canada. In Canada Yearbook 1937, ch. 1, pt. 5: 29-52, illus.

Anthony, Harold Elmer
1928. Field book of North American mammals . . . 625 pp., illus. New York and London.
1929. Horns and antlers. Their occurrence, development and function in the Mammalia. Part 2. N. Y. Zool. Soc. Bull. 32: 2-33, illus.

-----, and E. Roland Harriman
1939. The wapiti. *In* North American big game; a book of the Boone and Crockett Club: 225-237, illus. New York and London.

Archer [Stockwell, G. A.*]
1877. Fauna of Michigan. Forest and Stream 8: 177, 192, 224, 241, 261, 281, 300, 361, 380; 9: 5.

Ashe, Thomas
1808. Travels in America. Performed in 1806 for the purpose of exploring the Rivers Alleghany, Monongahela, Ohio, and Mississippi, and ascertaining the produce and condition of their banks and vicinity. 366 pp., London.

Askins, Charles
1933. The elk. Outdoor Life 72 (4): 12, 51-53, illus.

Audubon, John James, and John Bachman
1846-54. The quadrupeds of North America. 3 vols., illus. New York.

Audubon, Maria Rebecca
1897. Audubon and his journals, with zoological and other notes by Elliott Coues. 2 vols., illus. New York.

Avery, Carlos [edited by]
1929. Elk in Michigan. Field and Stream 34 (Oct.): 11.

Bailey, Vernon
1888. Report on some of the results of a trip through parts of Minnesota and Dakota. U. S. Commr. Agr. Rpt. 1887: 426-454.
1896. List of mammals of the District of Columbia. Biol. Soc. Wash. Proc. 10: 93-101.

------

* Interpretation of pseudonym as given in Bailey, H. B.: "Forest and Stream" bird notes. An index and summary of all the ornithological matter contained in "Forest and Stream" vols. 1-12. New York. 1881.

1905. Biological survey of Texas. North Amer. Fauna 25, 222 pp., illus.

1926. A biological survey of North Dakota. I. Physiography and life zones. II. The mammals. North Amer. Fauna 49, 226 pp., illus.

1929. Mammals of Bath and Highland Counties, Virginia. Virginia Game and Fish Conservationist May-June. 9 (1): 4-6.

1930. Animal life of Yellowstone National Park. 241 pp., illus. Springfield, Ill., Baltimore, Md.

1931. Mammals of New Mexico. North Amer. Fauna 53, 412 pp., illus.

1935. A new name [*Cervus canadensis nelsoni*] for the Rocky Mountain elk. Biol. Soc. Wash. Proc. 48: 187-189.

1936. The mammals and life zones of Oregon. North Amer. Fauna 55, 416 pp. illus.

1937. A typical specimen of the eastern elk from Pennsylvania. Jour. Mammal. 18 (1): 104.

Baillie-Grohman, William Adolph

1900. Fifteen years' sport and life in the hunting grounds of western America and British Columbia. 403 pp., illus. London.

Baird, Spencer Fullerton

1857. Mammals. *In* Reports of explorations and surveys, to ascertain . . . route for a railroad from the Mississippi River to the Pacific Ocean . . . 1853-6 . . . Vol. 8, pt. 1, 757 pp., illus. Washington, D. C.

Baldwin, W. P., and C. P. Patton

1938. A preliminary study of the food habits of elk in Virginia. Third North Amer. Wildlife Conf. Trans. (1938): 747-755.

Banfield, A. W. F.

1949. An irruption of elk in Riding Mountain National Park, Manitoba. Jour. Wildlife Mgt. 13 (1): 127-134.

Bannon, Henry Towne

1928. The Yellowstone elk herd. Outdoor Life—Outdoor Recreation 61 (6): 31, 104-105, illus.

Barclay, Edgar N.

1935. The red deer of the Caucasus. Zool. Soc. London Proc. 1934: 789-798, illus.

Barlow, J. W., and D. P. Heap

1872. Report of a reconnaissance in Wyoming and Montana Territories, 1871 . . . 43 pp., illus. Govt. Print. Off., Washington.

Barnes, Claude T.
1927. Utah mammals. Utah Univ. Bull. vol. 17 (12), rev. ed., 183 pp., illus.

[Bartram, William]
1928. The travels of William Bartram. Edited by Mark Van Doren. 414 pp. [New York.]

Beaman, David C.
1888. On the White River Plateau of Colorado. Amer. Field 30: 155-156, illus.

Bergtold, W. H.
1929. Bison in Colorado. Jour. Mammal. 10: 170.

Beverly, Robert
1722. The history of Virginia, in four parts . . . 284 pp., illus. Ed. 2, rev. and enlarged. London.

Billings, E.
1856. On the wapiti, or Canadian stag, *(Elephus canadensis)*. Canad. Nat. and Geol. 1: 81-87, illus.

Bird, Ralph Durham
1933. A three-horned wapiti *(Cervus canadensis canadensis)*. Jour. Mammal. 14: 164-166.

Birdseye, Clarence (See also Henshaw and Birdseye.)
1912. Some common mammals of western Montana in relation to agriculture and spotted fever. U. S. Dept. Agr. Farmers' Bull. 484, 46 pp., illus.

Bishopp, Fred Corry, and H. P. Wood
1913. The biology of some North American ticks of the genus *Dermacentor*. Parasitology 6: 153-187, illus.

Blair, William Reid
1907. Actinomycosis in the black mountain sheep. N. Y. Zool. Soc. Ann. Rpt. 1906: 136-141, illus.

Boas, Johan Erik Vesti
1923. Dansk forstzoologi. Ed. 2. 761 pp., illus. Kobenhavn, Kristiania, London, and Berlin.

Brackett, A. G.
1883. The elk *(Cervus canadensis)*. Amer. Field 20: 296-297.

Bradley, Frank H.
1873. Report of Frank H. Bradley, geologist of the Snake River Division. *In* Hayden, Ferdinand Vandiveer, Sixth annual report of the United States Geological Survey of the Territories, embracing portions of Montana, Idaho, Wyoming, and Utah; . . . for the year 1872. Pt. 1: 189-271, illus. Govt. Print. Off., Washington.

Brayton, A. M.
1882. Report on the Mammals of Ohio. Ohio Geol. Survey Rpt. vol. 4, pt. 1: 1-185.

Bridges, William
1935. Patriarchs in the Zoological Park; a census of some of the older inhabitants. N. Y. Zool. Soc. Bull. 38: 89-98, illus.
Brooks, Allan
1910. The "wapiti" of the Crees. Forest and Stream 74: 92.
1926. Past and present big-game conditions in British Columbia and the predatory mammal question. Jour. Mammal. 7: 37-40.
1928. The invasion of moose in British Columbia. Murrelet 9: 43-44.
Brooks, Alonzo Beecher
1932. The mammals of West Virginia. W. Va. Wild Life 10 (5): 1; (6): 2-4.
Brooks, Fred E.
1911. The Mammals of West Virginia; notes on the distribution and habits of all our known native species. W. Va. State Bd. Agr. Rpt. for quarter ending Dec. 30, 1910. Forestry: 9-30.
Brown, Arthur E., and John Dean Caton
1880. The domestication of certain ruminants and aquatic birds. Amer. Nat. 14: 393-398.
Brown, C. Emerson
1925. Longevity of mammals in the Philadelphia Zoological Garden. Jour. Mammal. 6: 264-267.
1936. Rearing wild animals in captivity and gestation periods. Jour. Mammal. 17: 10-13.
Burket, H. F.
1896. Editor Recreation. [Note from Ten Sleep, Wyo., re hunting.] Recreation 5: 265.
Burtch, Lewis A.
1934. The Kern County Elk Refuge. Calif. Fish and Game 20: 140-147, illus.
Butler, Amos William
1895. The mammals of Indiana. Ind. Acad. Sci. Proc. 1894: 81-86.
1934. Wild and domesticated elk in the early days of Franklin County, Indiana. Jour. Mammal. 15: 246-248.
Cahalane, Victor Harrison
1932. Age variation in the teeth and skull of the whitetail deer. Cranbrook Inst. Sci., Sci. Pub. 2, Nov. 14 pp., illus.
1938. The annual northern Yellowstone elk herd count. Third North Amer. Wildlife Conf. Trans. (1938): 388-389.
Cameron, Allan Gordon
1923. The wild red deer of Scotland; notes from an island forest on deer, deer stalking, and deer forest in the Scottish highlands. 248 pp., illus. Edinburgh and London.

Carpenter, W. L.
1876. Field notes on the natural history of the Big Horn Mountains. Forest and Stream 7: 196.

Carrel, Alexis
1935. Man, the unknown. 346 pp., New York and London.

Cary, Merritt
1911. A biological survey of Colorado. North Amer. Fauna 33, 256 pp., illus.
1917. Life zone investigations in Wyoming. North Amer. Fauna 42, 95 pp., illus.

Case, George W.
1938. The use of salt in controlling the distribution of game. Jour. Wildlife Mgt. 2: 79-81, illus.

Cass, Jules S.
1947. Buccal food impactions in whitetailed deer and Actinomyces necrophorus in big game. Jour. Wildlife Mgt. 11 (1): 91-94.

Caton, John Dean (See also Brown and Caton.)
1877. The antelope and deer of America . . . Ed. 2, 426 pp., illus., New York.

Chapman, E. J.
1861. Some notes on the drift deposits of western Canada, and on the ancient extension of the lake area of that region. Canad. Jour. Indus., Sci., and Art (n.s.) 6: 221-229.

Chittenden, Hiram Martin
1940. Yellowstone National Park; historical and descriptive. Rev. by Eleanor Chittenden Cross and Isabelle F. Story . . . 1933 ed.,with corrections and additions to 1940. 286 pp., illus. Stanford University, Calif., and London.

Clarke, Frank I.
1910. Game animals, birds and fishes of British Columbia, Canada. Bur. Prov. Inform. Bull. 17 (ed. 6), 67 pp., illus.

Cliff, Edward P.
1939. Relationship between elk and mule deer in the Blue Mountains of Oregon, Trans. 4th North Amer. Wildlife Conf.: 560-569.

Colman, Jeremiah
1926. Notes and queries on sheep and deer fraternizing. Field (London) 148: 690.

Cooper, E.
1926. Sheep and deer fraternizing. Field (London) 148: 690.

Cooper, J. G.
1860. Report (upon the mammals collected on the survey) of J. G. Cooper. *In* Reports of Explorations and surveys, to ascertain . . . route for a railroad from the Missis-

sippi River to the Pacific Ocean. 1853-5, . . . Vol. 12, book 2, illus. Washington, D. C.

Cope, E. D.
1873a. The slaughter of the buffalo. Amer. Nat. 7: 113-114.
1873b. The spike-horned mule deer. Amer. Nat. 7: 169-170.
1887. The classification and phylogeny of the Artiodactyla. Amer. Phil. Soc. Proc. 24: 377-400.

Corson-White, Ellen Pawling
1927. Spinal cord diseases. [Diseases of elk in Philadelphia Zoological Garden.] Zool. Soc. Phila., Lab. and Mus. Compar. Path. Rpt. (1926-27): 28-31.

Cory, Charles B.
1912. The Mammals of Illinois and Wisconsin. Field Mus. Nat. Hist. Pub. 153, Zool. Ser. vol. 11, 505 pp., illus.

Couch, Leo King
1935. Chronological data on elk introduction into Oregon and Washington. Murrelet 16: 2-6, illus.

Cowan, Ian McTaggart
1947. The timber wolf in the Rocky Mountain national parks of Canada. Canadian Jour. Research, D, 25: 139-174. National Research Council of Canada, illus.
1948. The occurrence of the granular tape-worm *Echinococcus granulosus* in wild game in North America. Jour. Wildlife Mgt. 12 (1): 105-106.

-----, and V. C. Brink
1949. Natural game licks in the Rocky Mountain national parks of Canada. Jour. Mammal. 30 (4): 379-387.

Cox, W. T.
1916. The elk or wapiti returns [to Minnesota]. Fins, Feathers and Fur, No. 5 (Mar.): 4, illus.

Crampton, L. W.
1886. [Letter from Fort Bridger] Forest and Stream 26: 85.

Cross, E. C., and John Richardson Dymond
1929. The mammals of Ontario. Roy. Ontario Mus. Zool. Handb. 1, 55 pp., illus.

Cruikshank, James A.
1914. In the Hudson Bay country . . . Forest and Stream 82: 819-820, 853-854, illus.

Dayton, William Adams
1931. Important western browse plants. U. S. Dept. Agr. Misc. Pub. 101, 214 pp., illus.

De Kay, James Ellsworth
1842. Zoology of New York, or the New-York fauna; . . . Pt. 1. Mammalia. 146 pp., illus. *In* Natural history of New York. New York, Boston, Albany.

DeNio, R. M.
1938. Elk and deer foods and feeding habits. Third North Amer. Wildlife Conf. Trans. (1938): 421-427.
DeWeese, Dall
1927. The Jackson's Hole refuge. Outdoor Amer., March: 18-19, 73-75, 80-83, 87-88, illus.
Dice, Lee Raymond (See also Wood and Dice.)
1919. The mammals of southeastern Washington. Jour. Mammal. 1: 10-22, illus.
Dilg, Will H.
1925. The Jackson Hole, a national recreation area. Outdoor Amer. May: 20-23, 91-92, illus.
Doane, Gustavus C.
1871. The report of Lieutenant Gustavus C. Doane upon the so-called Yellowstone Expedition of 1870. 40 pp. Washington, D. C.
Dodge, Richard Irving
1877. The plains of the great West and their inhabitants; being a description of the plains, game, Indians, etc., of the great North American desert. 448 pp., illus. New York.
Doman, Everett R., and D. I. Rasmussen
1944. Supplemental winter feeding of mule deer in Northern Utah; Jour. Wildlife Mgt. 8 (4): 317-338.
Donne, T. E.
1924. The game animals of New Zealand; an account of their introduction, acclimatization, and development. 322 pp., illus. London.
Dow, G. W.
1934. More tule elk planted in Owens Valley. Calif. Fish and Game 20: 288-290, illus.
Dowell, Overton, Jr.
1916. Value of the game refuge. Oreg. Sportsman 4: 254-255.
Dugmore, Arthur Radclyffe
1930. In the heart of the northern forests. 242 pp., illus. London.
Dungern-Oberau, Otto Frhr. von
1935 (?). Salzlecken (?), Deut. Jagd 3: 233-234, illus.
Dunraven, Windham Thomas Wyndham-Quin, 4th earl of
1925. Hunting in the Yellowstone; on the trail of the wapiti with Texas Jack in the land of geysers. 333 pp., New York.
Dymond, John Richardson (See also Cross and Dymond.)
1926. The wapiti in Ontario. Canad. Field-Nat. 40: 140.
East, Ben
1937. Quarry of the camera. N. Y. Zool. Soc. Bull. 40: 89-96, illus.

Edmunds, Ralph
1917-18. Thru the heart of the Bitter Roots. Outdoor Life 40: 374-375, 459-462, 546-549; 41: 13-16, 97-99, illus.

Elliot, Daniel Giraud
1899. Catalogue of mammals from the Olympic Mountains, Washington, with descriptions of new species. Field Columbian Mus. Pub. 32, Zool. ser. vol. 1: 239-276, illus.

Ellsworth, Rodney S.
1930. Hunting elk for the market in the forties. Calif. Fish and Game 16: 367.
1931. Elk in Suisun marshes in late fifties. Calif. Fish and Game 17: 224-225, illus.

Endlich, F. M.
1879. Report on the geology of the Sweetwater District. *In* Hayden, Ferdinand Vandiveer, Eleventh annual report of United States Geological and Geographical Survey of the Territories, embracing Idaho and Wyoming, . . . for the year 1877. Pt. 1: 3-158, illus. Govt Print. Off., Washington.

Evarts, Hal George
1924. The final rally. Outdoor Amer., December: 8-12, 41-43, illus.
1925. Allies now—the $ and sentiment. Outdoor Amer., February: 5-6, 59, illus.

Evermann, Barton Warren
1915. An attempt to save the California elk. Calif. Fish and Game 1: 86-96, illus.
1916. The California valley elk. Calif. Fish and Game 2: 70-77, illus.

Featherstonhaugh, G. W.
1835. Geological report of an examination made in 1834, of the elevated country between the Missouri and Red Rivers. 97 pp., illus. Washington, D. C.

Finley, William Lovell, and Irene Finley
1925. The return of the wapiti. Nature Mag. 5: 69-73, illus.

Fleming, C. E., and N. F. Peterson
1919. Don't feed fox-tail hay to lambing ewes! Nev. Agr. Expt. Sta. Bull. 97, 18 pp., illus.

Fowler, Jacob
1898. Journal: narrating an adventure from Arkansas through the Indian Territory, Oklahoma, Kansas, Colorado, and New Mexico to the sources of the Rio Grande del Norte, 1821-22. Edited with notes by Elliott Coues. 183 pp. New York.

Fremont, John Charles
1846. Narrative of the exploring expedition to the Rocky Mountains, in the year 1842, and to Oregon and north California, in the years 1843-44. 324 pp., illus. London.
Freuchen, Peter
1935. Part II. Field notes and biological observations, pp. 68-278. In Report of the Fifth Thule Expedition 1921-24. The Danish Expedition to Arctic North America in charge of Knud Rasmussen, Ph. D. 2 (4-5): 1-278.
Frick, Childs
1937. Horned ruminants of North America. Amer. Mus. Nat. Hist. Bull. 69, 669 pp., illus.
Fryxell, Fritiof Melvin
1926. A new high altitude limit for the American bison. Jour. Mammal. 7: 102-109.
1928. The former range of the bison in the Rocky Mountains. Jour. Mammal. 9: 129-139.
G 1887. Hunting the elk. Forest and Stream 29: 265.
Gaffney, William S.
1941. The effects of winter elk browsing, South Fork of the Flathead River, Montana. Jour. Wildlife Mgt. 5: 427-453, illus.
Gerstell, Richard
1936. The elk in Pennsylvania, its extermination and reintroduction. Penn. Game News 7 (7): 6-7, 26, illus.
Gilman, S. C.
1896. The Olympic country. Nat. Geog. Mag. 7 (4): 133-140.
Godman, John Davidson
1826. American natural history. 2 vols. Philadelphia.
Goldman, Edward Alphonso
1927. What to do with the Yellowstone elk? Amer. Forests and Forest Life 33: 279-282, illus.
Goodwin, George G.
1936. Big game animals in the northeastern United States Jour. Mammal. 17 (1): 48-50.
1940. Eastern elk antlers from Pennsylvania. Jour. Mammal. 21 (1): 95.
Goodwin, H. A. C.
1926. Sheep fraternizing with deer. Field (London) 148: 543.
Grant, Madison
1904. The origin and relationships of the large mammals of North America. N. Y. Zool. Soc. 8th Ann. Rpt. 1903: 182-207.

Graves, Henry Solon, and Edward William Nelson
1919. Our national elk herds; a program for conserving the elk on national forests about the Yellowstone National Park. U. S. Dept. Agr. Circ. 51, 34 pp., illus.
Green, H. U.
1933. The wapiti of Riding Mountain, Manitoba. Canadian Field Naturalist. Vol. 47 (6): 105-111, (7): 122-132, (8): 150-157, (9): 172-174.
1949. Occurrence of *Ecchinococcus granulosis* in elk *(Cervus canadensis nelsoni)*, Banff National Park. The Canadian Field-Naturalist 63 (5): 204-205.
Grimm, Rudolph L.
1939. Northern Yellowstone winter range studies. Jour. Wildlife Management 3 (4): 295-306.
Grinnell, George Bird (See also Yo, et al.)
1875. Zoological report. *In* Ludlow, William, Report of a reconnaissance of the Black Hills of Dakota, made in the summer of 1874: 79-102. Govt. Print. Off., Washington.
1876. Zoological report. *In* Ludlow, William, Report of a reconnaissance from Carroll, Montana Territory, on the upper Missouri, to the Yellowstone National Park, and return, made in the summer of 1875: 59-92. Govt. Print. Off., Washington.
1925. Old-time range of Virginia deer, moose, and elk. Nat. Hist. 25: 136-142.
Grinnell, Joseph
1913. A distributional list of the mammals of California. Calif. Acad. Sci. Proc. (ser. 4) 3: 265-390, illus.
1933. Review of the recent mammal fauna of California. Calif. Univ. Pub. Zool. 40: 71-234.
Grubb, E. H.
1882. Shooting in the Elk Mountains. Amer. Field 17: 380-381, 399.
Grumke, Dr.
1938. Zungenaktinomykose bei einem Rehbok. Deut. Jagd 8: 804, illus.
Hahn, Walter Louis
1909. The mammals of Indiana. A descriptive catalogue of the mammals occurring in Indiana in recent times. *In* Ind. Dept. Geol. and Nat. Resources 33d Ann. Rpt. (1908): 417-663, illus.
Hale, John Peter
1886. Trans. Allegheny Pioneers Cincinnati. 350 pp.
Hall, E. Raymond
1946. Mammals of Nevada. i-xi, 1-710. Illus. Univ. of Cal. Press.

Hallock, Charles
1877. Notes on the breeding habits of game. Forest and Stream 8: 36.
1892. Marooning in high altitudes. Forest and Stream 39: 401-402, 487-488, illus.

Hallowell, C. R.
1925. I. W. L. A. now buying hay land in Jackson's Hole. Outdoor Amer., June: 32-34, illus.

Hansen, Frank M.
1922. The elk problem. Outdoor Life 50: 418, illus.

Hanson, Joseph Mills
1909. Conquest of the Missouri, being the story of the life and exploits of Captain Grant Marsh. 458 pp., illus.

Hare, L. R.
1878. Report of Lieut. L. R. Hare, acting engineer officer, of march of the seventh cavalry during summer and fall of 1877. *In* Report of the Secretary of War; . . . Vol. 2, pt. 3: 1672-1680. Govt. Print. Off., Washington.

Harkin, James Bernard
1928. Wildlife. *In* Rpt. Commr. Nat. Parks, Canada, for year ending Mar. 31, 1927: 10-13, illus.

Hatch, Edward
1886. [Letter to editor from Fort McKinney, Wyo.] Forest and Stream 26: 85.

Hatt, R. T.
1949. Wapiti in Delaware. Jour. Mammal. 30 (2): 201.

Hayden, Ferdinand Vandiveer
1871. Preliminary Report of the United States Geological Survey of Wyoming and Portions of Contiguous Territories. Govt. Print. Off., Washington. pp. 83-188.

Hays, W. J.
1871. Notes on the range of some of the animals in America at the time of the arrival of the white men. Amer. Nat. 5: 387-392.

Heck, Lutz
1934. Vom Abwerfen des Edelhirsches. Deut. Jagd. 1: 388-390, illus.

Heinroth, O.
1908. Trächtigkeits- und Brutdauern. Zool. Beob. 49: 14-25.

Heller, Edmund
1925. The big game animals of Yellowstone National Park. Roosevelt Wild Life Bull. 2: 405-467, illus.

Henry, Alexander, and David Thompson

1897. New light on the early history of the greater northwest. The manuscript journals of Alexander Henry . . . and David Thompson . . . 1799-1814 . . . Edited with copious critical commentary by Elliott Coues. 3 vols. New York.

Henshaw, Henry Wetherbee, and Clarence Birdseye

1911. The mammals of Bitterroot Valley, Montana, in their relation to spotted fever. U. S. Bur. Biol. Survey Circ. 82, 24 pp., illus.

Herrick, Clarence Luther

1892. The mammals of Minnesota; a scientific and popular account of their features and habits. Minn. Geol. and Nat. Hist. Survey Bull. 7, 300 pp., illus.

Hetrick, R. R.

1907. Bison in Nebraska. Forest and Stream 68: 652.

Hewitt, Charles Gordon

1921. The conservation of the wild life of Canada. 344 pp., illus. New York.

Hibbard, Claude W.

1944. A checklist of Kansas Mammals. Trans. Kansas Academy of Science, 47: 61-88.

Hofer, Elwood

1887. Winter in wonderland. Through the Yellowstone Park on snowshoes. Forest and Stream 28: 222-223, 246-247, 270-271, 294-295, 318-319, illus.

1889. The Yellowstone National Park. Forest and Stream 31: 478. (Author given as "H," but article shows that "H" is Hofer.)

1892. Catching wild animals. Forest and Stream 38: 173, 195, 224, 247-248, 271, illus.

1903. Yellowstone Park game. Forest and Stream 60: 104.

Hoffmeister, Donald F.

1947. Early observations on the elk in Kansas. Trans. Kansas Acad. Sci. 50 (1): 75-76.

Holland, R. P. [edited by]

1920. Elk swim the Columbia. Outer's Recreation 63: 421.

Hölldobler, Karl

1938. Ueber Strahlenpilzerkrankung (Aktinomykose) beim Rehwild. Deut. Jagd 8: 908, illus.

Holmes, W. H.

1883. Report on the geology of the Yellowstone National Park. *In* Hayden, Ferdinand Vandiveer, Twelfth annual report of the United States Geological and Geographical

Survey of the Territories: . . . in Wyoming and Idaho for the year 1878. Pt. 2: 1-62. illus. Govt. Print. off., Wash.

Hooton, Earnest Albert
1937. Apes, men, and morons. 307 pp., New York.

Hornaday, William Temple
1904. The American natural history; a foundation of useful knowledge of the higher animals of North America. 449 pp., illus. New York.

Hough, Donald
1925. As I found the elk in Jackson Hole. Outdoor Amer., February: 3: 24-28.
1929. When the elk move in. Outdoor Life-Outdoor Recreation 64 (3): 32-33, 58-59, illus.

Hough, Emerson
1894. Elk perishing. Plenty of elk. Forest and Stream 42: 401.
1900. Wyoming big game. Forest and Stream 55: 450.

Hoy, P. R.
1882. The larger wild animals that have become extinct in Wisconsin. Wis. Acad. Sci., Arts, and Letters Trans. vol. 5 (1877-81): 255-257.

Huntington, Dwight Williams
1904. Our big game; a book for sportsmen and nature lovers. 347 pp., illus. New York and London.

Huxley, Julian Sorell
1926. The annual increment of the antlers of the red deer *(Cervus elaphus)*. Zool. Soc. London Proc. 1926: 1021-1035, illus.
1931. The relative size of antlers in deer. Zool. Soc. London Proc. 1931: 819-864, illus.

Ingebrigtsen, Olaf
1924. Das norwegische Rotwild *(Cervus elaphus* L.): eine kraniometrische Untersuchung. Bergens Mus. Aarbok 1922-23. Naturvidensk. raekke nr. 7, 262, pp., illus.

Ingersoll, Ernest
1883. Knocking round the Rockies. 220 pp., illus. New York.

Irving, Washington
1836. Astoria; or anecdotes of an enterprise beyond the Rocky Mountains. 2 vols. Philadelphia.
1843. The adventures of Captain Bonneville. pp. I-X; 389 pp. Handy volume edition. New York and London.
1865. A tour on the prairies. *In* Irving, Washington, The Crayon Miscellany (rev. ed.): 11-239. New York.

Jackson, Hartley Harrad Thompson
1908. A preliminary list of Wisconsin mammals. Wis. Nat. Hist. Soc. Bull. (n. s.) 6: 13-34, illus.

1941. Transplanted elk wanders. Jour. Mammal. 22 (4); 448.
1944. Big-game resources of the United States 1937-1942. Research Rept. 8, Fish and Wildlife Service. 56 pp.

James, Preston Everett
1936. Regional planning in the Jackson Hole country. Geog. Rev. 26: 439-453, illus.

Jaques, E. Parker
1907. Buffalo and grouse. Forest and Stream 68: 854.

Jefferies, Richard
1884. Red deer. 207 pp. London and New York.

Johnson, Charles E.
1916. A brief descriptive list of Minnesota mammals. Fins, Feathers and Fur, No. 8 (Dec.): 1-8, illus.

Johnson, Walter Adams
1905. Elk—the last of the big-game herds. Country Life in Amer. 8: 506-511, illus.

Jones, J. R. (See also Riley and Jones.)
1925. The elk of Jackson's Hole. Outdoor Amer., March: 6-10, illus.

Jones, William A.
1875. Report upon the reconnaissance of northwestern Wyoming, including Yellowstone National Park, made in the summer of 1873. 331 pp., illus. Govt. Print. Off., Washington.

K, A. A.
1885. Hunting for meat. Forest and Stream 25: 264-265.

Kavanagh, Edward N.
1930. The Roosevelt elk, *Cervus roosevelti* Merriam. Jour. Forestry 28: 659-663.

Kellar, Andrew J.
1894. A trip to the Big Horn Mountains. Forest and Stream. 43: 288.

Kellogg, Remington
1939. Annotated list of Tennessee mammals. U. S. Nat. Mus. Proc. 86: 245-303.

Kellogg, T. D.
1887. Domestication of the elk. Amer. Field 28: 126-127.

Kennicott, Robert
1855. Catalogue of animals observed in Cook County, Illinois. Ill. State Agr. Soc. Trans. 1 (1853-54): 577-595.

Kirtland, Jared P.
1838. A catalogue of the Mammalia, birds, reptiles, fishes, Testacea, and Crustacea in Ohio. Ohio Geol. Survey 2d Ann. Rpt.: 157-200.

Koch, Elers
1941. Big-game in Montana from early historical records. Jour. Wildlife Management 5 (4): 357-370.
Landon, Monroe
1931. Elk remains in Norfolk County. Canad. Field-Nat. 45: 40.
Lantz, David Ernest
1905a. Kansas mammals in their relation to agriculture. Kans. Agr. Coll. Expt. Sta. Bull. 129: 331-404, illus.
1905b. A list of Kansas mammals. Kans. Acad. Sci. Trans. 19: 171-178.
1910. Raising deer and other large game animals in the United States. U. S. Bur. Biol. Survey Bull. 36, 62 pp., illus.
Larom, I. H.
1921. Elk season—1919 vs. 1920. Outer's Recreation 64(2): 84-85.
Larsen, Al G.
1936. A mountain tragedy. Outdoor Life 77(5): 14, illus.
Lederer, John
1672. The discoveries of John Lederer in Three Feveral Marches from Virginia to the West of Carolina and Other Parts of the Continent. Gray-Inne-Gate in Holborn. 30 pp.
Leek, Stephen Nelson
1909. The starving elk of Wyoming. Outdoor Life 24: 121-133, illus.
1911a. The problem of the elk. Collier's 46(21): 22.
1911b. The starving elk of Wyoming. Outdoor Life 27: 441-452, illus.
1918. The life of an elk. Outdoor Life 42: 357-360, illus.
Leffler, Ross L.
1940. Conservation mistakes. Penn. Game News 11(2): 3.
Leopold, Aldo
1933. Game management. 481 pp., illus. New York and London.
1938. Conservation esthetic. Bird-Lore. 40: 101-109, illus.
Le Raye, Charles
1812. The journal of Mr. Charles Le Raye, while a captive with the Sioux Nations, on the waters of the Missouri River. *In* Cutler, Jervase, A topographical description of Ohio, Indian Territory and Louisiana . . .: 158-204. Boston.
Lett, William Pittman
1884. The deer of the Ottawa valley. Ottawa Field-Nat. Club Trans. vol. 2, 101-117.
Lewis, Meriwether, and William Clark
1893. History of the expedition under the command of Lewis and Clark . . . 1804-5-6. New ed. by Elliott Coues. 4 vols., illus. New York.

Lexden
1914. Across the continent in "the sixties." Forest and Stream 83: 269-272, 295, 301-303, 333-335, 366-368, 396-397, illus.

Lockington, W. N.
1879. Notes on Pacific coast mammals. Amer. Nat. 13: 708.

Lommasson, Thomas
1927. Elk forage in Montana. The Forestry Kaimin 1927: 19-22, illus.
1930. The value of salt on alkali ranges in southeastern Montana. Northwest Sci. 4: 74-76.

Love, Charles R.
1930. Elk thriving in Shasta County. Calif. Fish and Game 16: 82.

Ludlow, Fitz Hugh
1870. The heart of the continent; a record of travel across the plains and in Oregon, with an examination of the Mormon principle. 568 pp., illus.

Lydekker, Richard
1898. The deer of all lands; a history of the family Cervidae living and extinct. 329 pp., illus. London.
1900. The great and small game of India, Burma, and Tibet. 416 pp., illus. London.
1901. The great and small game of Europe, western and northern Asia, and America. 445 pp., illus. London.
1910. On a wapiti and a muntjac. Zool. Soc. London Proc. 1910: 987-991, illus.

M 1880. Nebraska game. Forest and Stream 15: 311.

McAtee, Waldo Lee
1918. A sketch of the natural history of the District of Columbia. Biol. Soc. of Washington, Bull. 1: 52.
1928. Birds and other checks upon insects. Sci. Monthly 27: 77-80.
1936. The postulated resemblance of natural to artificial selection. Ohio Jour. Sci. 36: 242-252.
1937. Survival of the ordinary. Quart. Rev. Biol. 12: 47-64, illus.

McCabe, Thomas T., and Elinor Bolles McCabe
1928a. The Bowron Lake moose: their history and status. Murrelet 9: 1-9.
1928b. The British Columbia moose again. Murrelet 9: 60-63.

McChesney, Charles E.
1879. Report on the mammals and birds of the general region of the Big Horn River and Mountains of Montana Territory, . . . 1878. *In* Report of the Secretary of War; . . . Vol. 2, pt. 3: 2371-2395. Govt. Print. Off., Washington.

McCowan, Dan

1926. Unusual phases of animal behavior. Forest and Stream 96: 10-11, 59, illus.

1927. American big game trophies. Record heads and trophies from American big game fields and inland waters. Forest and Stream 97: 81, 114, illus.

McCreary, Otto C.

1931. Wyoming forage plants and their chemical composition. Studies No. 9. Wyo. Univ. Agr. Expt. Sta. Bull. 184, 24 pp.

McGaffey, Ernest

1933. Why not restore the elk? American Field. Jan. 27.

MacDonald, Henry

1888. Early days on the Missouri. I—Game and hunters. Forest and Stream 31: 4-5.

MacEwen, William

1920. The growth and shedding of the antlers of the deer; the histological phenomena and their relation to the growth of bone. 108 pp., illus. Glasgow.

MacPherson, Hugh Alexander: Cameron of Lochiel: Viscount Ebrington: and Alexander Innes Shand

1896. Red deer. 320 pp., illus. Fur and Feather Series. London, New York, and Bombay.

Maguire, Edward

1879. Annual report of Lieutenant Edward Maguire, corps of engineers, for the fiscal year ending June 30, 1879. *In* Report of the Secretary of War; . . . Vol. 2, pt. 3: 2359-2362. Govt. Print. Off., Washington.

Marx, E.

1908. Über Zahn- und Kiefererkrankung eines Riesenkänguruhs des Zoologischen Gartens zu Frankfort a. M. Zool. Beob. 49: 193-196.

Mason, Michael Henry

1924. The Arctic forests. 320 pp., illus. London.

Matthew, William D.

1908. Osteology of *Blastomeryx* and phylogeny of the American Cervidae. Amer. Mus. Nat. Hist. Bull. 24: 535-562, illus.

1934. A phylogenetic chart of the Artiodactyla. Jour. Mammal. 15: 207-209, illus.

Mead, J. R.

1899. Some natural-history notes of 1859. Kans. Acad. Sci. Trans. (1897-98) 16: 280-281.

Mearns, Edgar Alexander
1907. Mammals of the Mexican boundary of the United States. A descriptive catalogue of the species of mammals occurring in the region . . . Part 1: Families Didelphiidae to Muridae. U. S. Nat. Mus. Bull. 56, 530 pp., illus.
Merchant, Ival Arthur
1942. Veterinary bacteriology, I-X. Iowa State Coll. Press. 640 pp.
Merriam, Clinton Hart
1884. The mammals of the Adirondack region, northeastern New York . . . 316 pp. New York.
1891. Mammals of Idaho. *In* Merriam, Clinton Hart, and Stejneger, Leonhard, Results of a biological reconnaissance of south-central Idaho. North Amer. Fauna 5: 31-87, illus.
1899. Results of a biological survey of Mount Shasta, California. North Amer. Fauna 16, 179 pp., illus.
1921. A California elk drive. Sci. Monthly 13: 465-475, illus.
Merrill, Fred S.
1908. The American elk *(Cervus canadensis)*. Outdoor Life 21: 221-227, illus.
Michaux, André
1904. Journal of André Michaux, 1793-1796 [travels into Kentucky]. *In* Thwaites, Reuben Gold [editor], Early western travels, 1748-1846; a series of annotated reprints of some of the best and rarest contemporary volumes of travel . . . Vol. 3: 25-104. Cleveland.
Millais, John Guille
1915. *In* Carruthers, D., et al. The gun at home and abroad; the big game of Asia and North America. 433 pp. London.
Miller, Gerrit Smith, Jr.
1900. Key to the land mammals of northeastern North America. N. Y. State Mus. Bull. 8: 59-160.
1912. List of North American land mammals in the United States National Museum, 1911. U. S. Nat. Mus. Bull. 79, 455 pp.
1924. List of North American recent mammals, 1923. U. S. Nat. Mus. Bull. 128, 673 pp.
1931. The primate basis of human sexual behavior. Quart. Rev. Biol. 6: 379-410.
Mills, Harlow B.
1936. Observations on Yellowstone elk. Jour. Mammal. 17: 250-253.
Moffitt, James
1934. History of the Yosemite elk herd. Calif. Fish and Game 20: 37-51, illus.

Mohler, John Robbins

1923a. Infectious diseases of cattle. *In* Atkinson et al., Special report on diseases of cattle, rev. ed. 1923: 358-501, illus. Govt. Print. Off., Washington.

1923b. Mycotic stomatitis. *In* Atkinson et al., Special report on diseases of cattle, rev. ed. 1923: 532-537. Govt. Print. Off., Washington.

Moody, Charles Stuart

1910. Where rolls the Kooskia. Forest and Stream 75: 528-531, 568-571, 608-610, 648-650, 688-690, illus.

Moore, William H.

1931. Notes on antler growth of Cervidae. Jour. Mammal. 12: 169-170.

Mountaineer

1894. Lynx and wolverine. Forest and Stream 42: 380.

Murie, Adolph

1934. The moose of Isle Royale. Mich. Univ. Mus. Zool. Misc. Pub. 25, 44 pp., illus.

1940. Ecology of the coyote in the Yellowstone. Fauna of the Nat. Parks of U. S. No. 4, 206 pp., illus.

1944. The wolves of Mount McKinley. Fauna of the National Parks of the United States, Fauna Series No. 5, 238 pp.

Murie, Olaus Johan

1929. Summering with the elk. Amer. Forests and Forest Life 35: 694-697, illus.

1930. An epizootic disease of elk. Jour. Mammal. 11: 214-222, illus.

1932a. Big game trails in research. Outdoor Amer., January: 8-9, 40, illus.

1932b. Elk calls. Jour. Mammal. 13: 331-336.

1934. Studies in elk management. Twentieth Amer. Game Conf. Trans.: 355-359.

1935a. The elk of Jackson Hole. Nat. Hist. 35: 237-247, illus.

1935b. Food habits of the coyote in Jackson Hole, Wyo. U. S. Dept. Agr. Circ. 362, 24 pp.

1937. Natural elk management. Nature Magazine, 30(5): 293-295, illus.

1944. Our big game in winter. Trans. 9th North Amer. Wildlife Conf.: 173-176.

1945. Jackson Hole National Monument and the elk. Nat. Parks Magazine No. 83: 13-17, illus.

1947. The firing line. Am. For. 53(9): 392-394, 422.

1948. What do we have in Jackson Hole? Pacific Discovery, 1(1): 11-14, illus.

Nelson, Aven
1894. Squirrel-tail grass. (Fox-tail.) One of the stock pests of Wyoming. Wyo. Agr. Expt. Sta. Bull. 19: 73-79, illus.

Nelson, Edward William
1902. New species of elk from Arizona. Amer. Mus. Nat. Hist. Bull. 16: 7-11, illus.
1916. The larger North American mammals; with illustrations from paintings by Louis Agassiz Fuertes. Nat. Geog. Mag. 30: 385-472, illus.

Noback, Charles V.
1929. The internal structure and seasonal growth-changes of deer antlers. N. Y. Zool. Soc. Bull. 32: 34-40, illus.

——————, and Walter Modell
1930. Direct bone formation in the antler tines of two of the American Cervidae, Virginia deer *(Odocoileus virginianus)* and wapiti *(Cervus canadensis)*, with an introduction on the gross structure of antlers. Zoologica 11: 19-60, illus.

Norris, Philetus W.
1881. Annual report of the superintendent of the Yellowstone National Park to the Secretary of the Interior for the year 1880. 64 pp., illus.

Oberholser, Harry Church
1905. The mammals and summer birds of western North Carolina. 24 pp., pub. by Biltmore Forest School, Biltmore, N. C.

Old Timer
1898. On the sunk lands. Recreation 8: 209.

Olson, A. L.
1938. The firing line in the management of the northern elk herd. Univ. of Idaho Bull. 33 (22): 36-42.

Olson [Olsen], Orange A.
1936. Elk management. Utah Juniper 7: 10-15, illus.

Orr, Robert T.
1937. Notes on the life history of the Roosevelt elk in California. Jour. Mammal. 18 (1): 62-66.

Packard, Fred Mallery
1947. A study of the deer and elk herds of Rocky Mountain National Park, Colorado. Jour. Mammal. 28 (1): 4-12.

Parkman, Francis
1910. The Oregon Trail. Boston. 433 pp.

Parsell, Jack
1938. The elk problem in the Selway. Univ. of Idaho Bull. 33 (22): 23-25.

Patterson, T. J.
1894. Rambling in Wyoming. Forest and Stream 42: 354, 378-379, illus.

Pattie, James Ohio
1905. The personal narrative of James O. Pattie, of Kentucky, during an expedition from St. Louis through the vast regions between that place and the Pacific Ocean, and thence back through the City of Mexico to Vera Cruz, during journeyings of six years . . . A reprint of the original edition of 1831. *In* Thwaites, Reuben Gold [editor], Early Western travels, 1748-1846; a series of annotated reprints of some of the best and rarest contemporary volumes of travel . . . Vol. 18: 1-324, illus. Cleveland.

Perry, W. A.
1890. Habits of elk. Forest and Stream 35: 147.

Pickford, G. D., and Albert H. Reid
1943. Competition of elk and domestic livestock for summer range forage. Jour. Wildlife Mgt. 7 (3): 328-332.

Pike, Zebulon Montgomery
1895. The expeditions of Zebulon Montgomery Pike, to headwaters of the Mississippi River, through Louisiana Territory, and in New Spain, during the years 1805-6-7. New ed., by Elliott Coues. 3 vols., illus. New York.

Pocock, R. I.
1912. On antler-growth in the Cervidae, with special reference to *Elaphurus* and *Odocoileus (Dorcelaphus)*. Zool. Soc. London Proc. 1912: 773-783, illus.

Preble, Edward Alexander
1908. A biological investigation of the Athabaska-Mackenzie region. North Amer. Fauna 27. 574 pp., illus.
1911. Report on condition of elk in Jackson Hole, Wyoming, in 1911. U. S. Biol. Survey Bull. 40, 23 pp., illus.

Presnall, C. C.
1938. Mammals of Zion-Bryce and Cedar Breaks. Nat. Park Service, Zion-Bryce Museum, Bulletin No. 2, 20 pp.

Rheinfels
1937. Merkwürdige Kämpfe unter dem Wilde. Deut. Jagd 7: 397-399, illus.

Rhoads, Samuel Nicholson
1895. Notes on the mammals of Monroe and Pike Counties, Pennsylvania. Acad. Nat. Sci. Phila. Proc. 1894: 387-396.
1897. Contributions to the zoology of Tennessee. No. 3, Mammals. Acad. Nat. Sci. Phila. Proc. 1896: 175-205.

1898. A contribution to the mammalogy of central Pennsylvania. Acad. Nat. Sci. Phila. Proc. 1897: 204-226.

Rhoda, Franklin

1877. Topographical report on the southeastern district. *In* Hayden, Ferdinand Vandiveer, Ninth annual report of the United States Geological and Geographical Survey of the Territories, embracing Colorado and parts of adjacent Territories . . . for the year 1875. Pt. 2: 302-333. illus. Govt. Print. Off., Washington.

Richardson, John

1836. Zoological remarks. *In* Back, George, Narrative of the Arctic land expedition to the mouth of the great Fish River, and along the shores of the Arctic Ocean, in the years 1833, 1834, and 1835: 475-522, illus. London.

Ridley, Margaret A.

1909. A woman on the trap-trail. Incidents of outdoor winter life in the high Sierras of Idaho. Forest and Stream 72: 248-250, 288-290, 328-330, illus.

Riley, Smith, and J. R. Jones

1918. The Wyoming elk situation. Outdoor Life 42: 173-176, illus.

Roberts, Paul H.

1930. The Sitgreaves elk herd. Jour. Forestry 28: 655-658.

Robinson, Beverley William

1924. Elk hunting in the Rockies. Forest and Stream 94: 528-530, 570-571, illus.

Roosevelt, Theodore

1885. Hunting trips of a ranchman; sketches of sport on the northern cattle plains. 318 pp., illus. New York and London.

1893. The wilderness hunter; an account of the big game of the United States and its chase with horse, hound, and rifle. 472 pp., illus. New York and London.

1902. The deer and antelope of North America. *In* Roosevelt, Theodore, et al., The deer family, 131 pp., illus. New York and London.

Rosenhaupt, Heinrich

1939. The male birth surplus. Sci. Monthly 48: 163-169.

Ross, Alexander

1855. The fur hunters of the far West; a narrative of adventures in the Oregon and Rocky mountains. 2 vols. London.

Roszbach

1936. Aktinomykose beim Rehwild. Deut. Jagd 5: 123.

Rowan, William

1923. An unusual wapiti head. Jour. Mammal. 4: 112, illus.

Rowley, John
1902. The mammals of Westchester County, New York. Linn. Soc. N. Y., Abst. Proc. (1900-1902), Nos. 13-14: 31-60.
Rush, William Marshall
1927a. Harvesting the annual elk crop. Forest and Stream 97: 724-726, illus.
1927b. Notes on diseases in wild game animals. Jour. Mammal. 8: 163-165.
1929. What is to become of our northern elk herd? Amer. Forests and Forest Life 35: 93-95, 125-126, illus.
1930. Montana center of U. S. elk study. Mont. Wild Life 2(8): 4-5, illus.
1931. The northern Yellowstone elk herd. Mont. Wild Life 3(8): 21-23, illus.
[1932a.] Northern Yellowstone elk study. Mont. Fish and Game Comm. 131 pp., illus.
1932b. Bang's disease in the Yellowstone National Park buffalo and elk herds. Jour. Mammal. 13: 372.
Russell, Osborne
1921. Journal of a trapper, 1834-43. Boise, Idaho. 149 pp.
S., C. L.
1889. A month in the Rocky Mountains. Forest and Stream 32: 62-63, 82-83, 106, 130, 150.
Sabine, Joseph
1823. Zoological appendix. *In* Franklin, R. N. Narrative of a journey to the shores of the Polar Sea in the years 1819, 20, 21, and 22: 647-703. London.
Sampson, Arthur William
1917. Important range plants: their life history and forage value. U. S. Dept. Agr. Bull. 545, 63 pp., illus.
1919. Effect of grazing upon aspen reproduction. U. S. Dept. Agr. Bull. 741, 29 pp., illus.
Sanborn, Elwin R. [photographer]
1929. The growth of a wapiti antler. [Series of photographs illustrating progressive stages in development of the wapiti antler.] N. Y. Zool. Soc. Bull. 32: 25-33, illus.
Saunders, William Edwin
1932. Notes on the mammals of Ontario. Roy. Canad. Inst. Trans. 18: 271-309.
Schafer, Joseph
1918. A history of the Pacific northwest. Rev. ed., 323 pp., illus. New York.
Schwartz, John E.
1943. Range conditions and management of the Roosevelt elk on the Olympic Peninsula. U. S. Dept. of Agr., 65 pp.

1904. The wapiti. Forest and Stream 63: 279-280, illus.
Wapiti
1896. [Note on elk in the Olympics.] Recreation 5: 265.
Ward, Rowland
1910. Records of big game; with their distribution, characteristics, dimensions, weights, and horn and tusk measurements. Ed. 6, 531 pp., illus. London.
Warren, Edward Royal
1910. The mammals of Colorado . . . 300 pp., illus. New York and London.
1927. Altitude limit of bison. Jour. Mammal. 8: 60-61.
Webster, Edward B.
1920. The king of the Olympics; the Roosevelt elk and other mammals of the Olympic Mountains. 227 pp., illus. Port Angeles, Wash.
1921. Elk as affected by the great storm in the Olympic Peninsula. Murrelet 2 (2): 6.
Webster, Leslie Tillotson
1939. Inborn resistance to infectious disease. Sci. Monthly 48: 69-72, illus.
West, I. E.
1881. Dakota game. Forest and Stream 17: 271.
Whiteaves, Joseph Frederick
1889. Palaeontology and zoology. *In* Report A . . . Summary reports of the operations of the Biological Survey for the years 1887 and 1888, by the director [Alfred R. C. Selwyn]. Canada, Geol. and Nat. Hist. Survey Ann. Rpt. (n.s.) 3, pt. 1; Reports A, B, C, E, F (1887-1888): 47A-53A; 105A-113A. Montreal.
Whitney, Caspar
1911. The sportsman's view-point. Collier's 46 (25): 22, 24.
Wied [Neuwied], Maximilian [Alexander Philipp], Prinz Zu
1839-41. Reise in das Innere Nord-Amerika in den Jahren 1832 bis 1834. 2 vols., illus. Coblenz.
Wilcox, Alvin H.
1907. A pioneer history of Becker County. St. Paul, Minn. 757 pp.
Winchell, N. H.
1875. Geological report. *In* Ludlow, William, Report of a reconnaissance of the Black Hills of Dakota, made in the summer of 1874: 21-66, illus. Govt. Print. Off., Washington.
Wingate, George Wood
1886. Through the Yellowstone Park on horseback. 250 pp., illus. New York.

Wintemberg, William John
1926. Archaeological evidence of the presence of the wapiti in southwestern Ontario. Canad. Field-Nat. 40: 58.

Wong-Quincey, J.
Chinese hunter. Illus. John Day, New York. No date given. (Printed in Great Britain by Western Printing Service, Ltd., Bristol.) 383 pp.

Wood, Frank Elmer
1910. A study of the mammals of Champaign County, Illinois. Ill. State Lab. Nat. Hist. Bull. vol. 8 (1908-1910): 501-613, illus.

Wood, Norman Asa
1911. Mammals. *In* A biological survey of the sand dune region on the south shore of Saginaw Bay, Michigan, prepared under the direction of Alexander G. Ruthven, Mich. Geol. and Biol. Survey Pub. 4, Biol. Ser. 2: 309-312.

———, and Lee Raymond Dice
1924. Records of the distribution of Michigan mammals. Mich. Acad. Sci., Arts, and Letters, Papers vol. 3: 425-469.

Wyeth, Nathaniel Jarvis
1899. The Correspondence and Journals of Capt. Nathaniel J. Wyeth, 1831-6. Eugene, Ore., 262 pp.

Wyman, Jeffries
1868. An account of some Kjoekkenmoeddings, or shell-heaps, in Maine and Massachusetts. Amer. Nat. 1: 561-584, illus.

Yo [George Bird Grinnell], et al.
1889. Days with the elk. Forest and Stream 33: 430-432, illus.

Young, Vernon Alphus
1938. The carrying capacity of big game range. Jour. Wildlife Mgt. 2: 131-134, illus.

———, and W. Leslie Robinette
1939. A study of the range habits of elk on the Selway Game Preserve. Univ. Idaho Bull. 34 (16): 48 pp., illus.

INDEX